Lycopersicon esculentum

Indeterminate tomato. The terminal bud does not set fruit. The vine can grow indefinitely if not killed by frost. Most of the varieties trained on stakes or in wire cages are in this group.

Determinate tomato. (Commonly called Bush tomato.) Terminal bud sets fruit, stops stem growth. The plant is self-topping. Seldom needs staking.

Determinate plant has less than three leaves between the flowering cluster. When tomato is grown from seed in place, a strong tap root develops. Transplants develop strong spreading root systems.

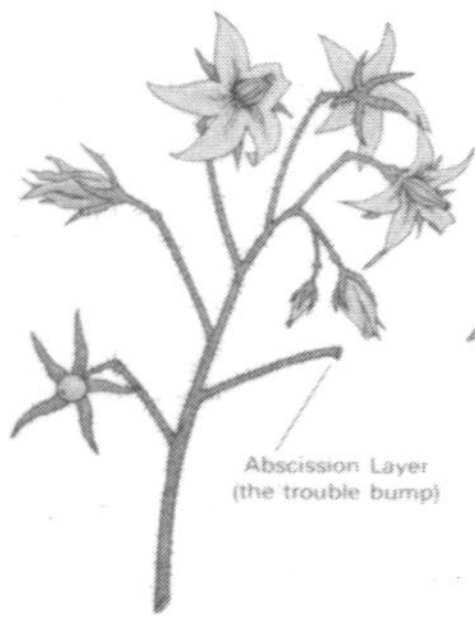

The trouble bump. When you talk about blossom drop or failure to set fruit this is where it happens. Temperatures too low or too high, too much or too little water may be the cause.

When growing tomatoes on stakes, the side shoots or suckers in the leaf axils of the stem are broken off. If pinched out above the first leaf, the plant will have better foliage cover.

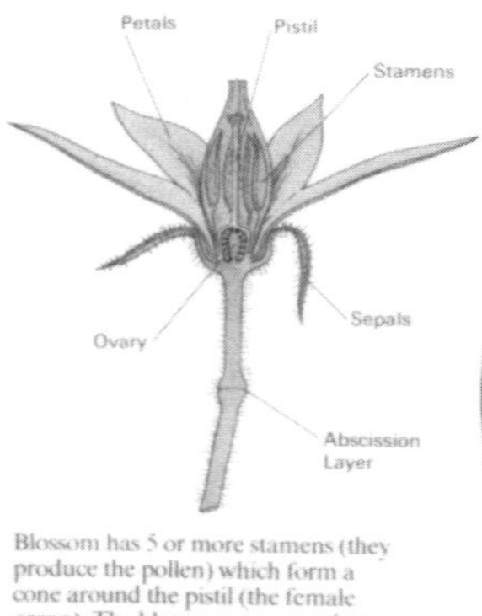

Blossom has 5 or more stamens (they produce the pollen) which form a cone around the pistil (the female organ). The blossoms are complete flowers and mostly self-fertilizing.

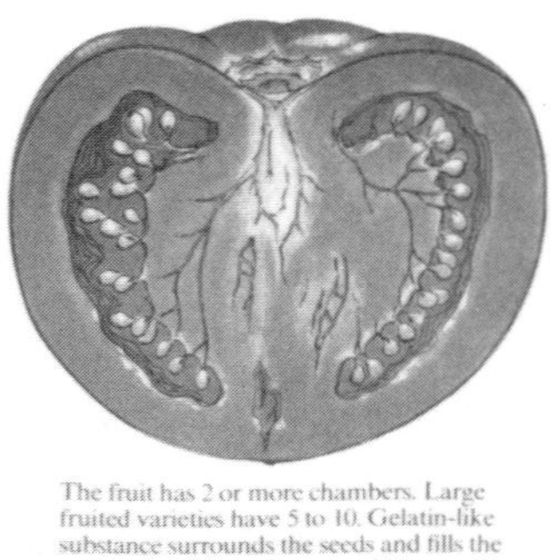

The fruit has 2 or more chambers. Large fruited varieties have 5 to 10. Gelatin-like substance surrounds the seeds and fills the seed cavity. Jelly content should be small in comparison to the amount of flesh.

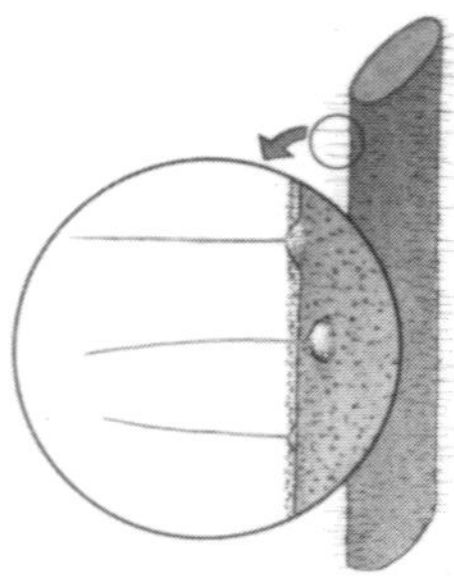

Glandular hairs are found on stems and leaves. When broken, they give off the oil that has the characteristic tomato odor. Roots form all along stem when it's in contact with soil.

TOMATO PRODUCTION, PROCESSING & TECHNOLOGY
Third Edition

by
Wilbur A. Gould, Ph.D.
Food Industries Consultant
Emitrus Professor of Food Processing & Technology,
Department of Horticulture,
Ohio State University,
Ohio Agricultural Research & Development Center,
Former Director, Food Industries Center,
The Ohio State University and
Executive Director Mid-America Food Processors Association,
Worthington, Ohio

CTI Publications, Inc.
2 Oakway Road, Timonium, Maryland 21093-4247 USA
Voice 410-308-2080 FAX 410-308-2079

CTI Publications, Inc.

2 Oakway Road, Timonium, Maryland 21093-4247 USA

Voice 410-308-2080 FAX 410-308-2079

Printed in The United States Of America

by Bookcrafters, Fredericksburg, VA

ISBN Numbers are as follows:

0-930027-18-3

Library of Congress Catalog – in – Publication Data

Gould, Wilbur A., 1920–

Tomato production, processing & technology/by Wilbur A. Gould.

p. cm.

Rev. ed. of: Tomato production, processing, and quality evaluation.

Includes bibliographical references and index.

ISBN 0-930027-18-3

1. Tomatoes. 2. Tomato products. I. Gould, Wilbur A., 1920- Tomato production, processing, and quality evaluation. II. Title. III. Title: Tomato production, processing, and technology.

SB349.G68 1991 91-43484

664'.805642--dc20 CIP

Cover Photo Courtesy
Mike Brown/Ferry Morse Seed Company
Variety: Hybrid 960N

Frontispiece
Anatomy of the Tomato
Taken from Ortho Chemical Co.

CTI Publications, Inc.
2 Oakway Road, Timonium, Maryland 21093-4247 USA
Voice 410-308-2080 FAX 410-308-2079

Other Title's From CTI Publications

FOOD PRODUCTION/MANAGEMENT editorially serves those in the Canning, Glasspacking, Freezing and Aseptic Packaged Food Industries.

Editorial topics cover the range of Basic Management Policies, from the growing of the Raw Products through Processing, Production and Distribution for the following products: fruits; vegetables; dried and dehydrated fruit (including vegetables and soup mixes); juices, preserves; pickles and pickled products; sauces and salad dressings; catsup and tomato products; soups; cured fish and seafood, baby foods; seasonings and other specialty items. (Monthly Magazine). ISSN: 0191-6181

A COMPLETE COURSE IN CANNING, 12th edition, are technical reference and textbooks for Students of Food Technology; Food Plant Managers; Products Research and Development Specialists; Food Equipment Manufacturers and Salesmen; Brokers; and Food Industry Suppliers. The three books total 1,300 pages.
ISBN: 0-930027-00-0.

TOTAL QUALITY ASSURANCE FOR THE FOOD INDUSTRIES is 400 pages of A to Z technology and practical application of the latest methods and detailed procedure in developing total quality assurance in all food plants, including sanitary standards, as well as bacteriological procedures. This is the complete instruction book, easily followed, yet technically complete for the advanced Food Technologist. ISBN: 0-930027-14-0.

CURRENT GOOD MANUFACTURING PRACTICES/FOOD PLANT SANITATION covers all Current Food Manufacturing practices as prescribed by the United States Department of Agriculture, Food and Drug Administration, as it applies to food processing and manufacturing. A total of 21 chapters, covering all phases of sanitation.
ISBN: 0-930027-15-9

GLOSSARY FOR THE FOOD INDUSTRIES is a definitive list of food abbreviations, terms, terminologies and acronyms. Also included are 20 handy reference tables and charts for the food industry. ISBN: 0-930027-16-7.

RESEARCH & DEVELOPMENT GUIDELINES FOR THE FOOD INDUSTRIES is a compilation of all Research and Development principles and objectives. Easily understood by the student or the professional this text is a practical "How To Do It and Why To Do It" reference.
ISBN: 0-930027-17-5.

CTI Publications, Inc.

2 Oakway Road, Timonium, Maryland 21093-4247 USA
Voice 410-308-2080 FAX 410-308-2079

This copy of
Tomato Production, Processing & Technology
belongs to:

Contents

PART 1–PRODUCTION 1

Chapter 1. – INTRODUCTION & HISTORY OF THE TOMATO INDUSTRY 3
- Organization for a Tomato Processing Plant 4
- Cultivars of Tomatoes 1868–1937 7
- Consumption of Tomato & Tomato Products 9
- Acreage, Yield 10
- World Production 14
- Statistical Production Summary In U.S. 15

Chapter 2. — TOMATO CULTURE & PRODUCTION FOR PROCESSING 19
- Field Selection 19
- Climate, Geography, Soil Selection 21
- Land Preparation 22
- Soil Nutrients 24
- Soil Testing 27
- Liming 27
- Fertilizers 28
- Starter Solutions 30
- Cultivars 32
- Planting 34
- Cultivation 39
- Weed Control 40
- Irrigation 41
- Sun-Gard 41
- Diseases 41
  - Symptoms of Early Blight (photos) 42
  - Early Blight (description) 44
  - Late Blight (description) 45
  - Symptoms of Late Blight (photos) 46
  - Symptoms of Septoria (photos) 48
  - Septoria Leaf Spot (description) 50
  - Bacterial Speck (description) 51
  - Symptoms of Bacterial Speck (photos) 52
  - Early/Advanced Symptoms of Bacterial Spot (photos) 54
  - Bacterial Spot (description) 56
  - Bacterial Canker(description) 57
  - Symptoms of Tomato Bacterial Canker (photo) 58
  - Symptoms of Bacterial Wilt (photo) 60
  - Bacterial Wilt/Southern Bacterial Wilt (description) 62
  - Southern Blight or Sclerotium Rot (description) 62
  - Anthracnose (description) 63
  - Black Mold (Alternaria) (description) 63

Contents

Chapter 2. — Continued

Symptoms of Major Midwest Fruit Rots (photos)	64
Soil Rot of Rhizoctonia (photo)	64
Buckeye Rot (photo)	65
Pythium Rot (photo)	65
Gray Mold or Botrytis (description)	65
Soil Rot or Rhizoctonia (description)	66
Buckeye Rot (description)	66
Pythium Rot (description)	67
Gray Mold or Botrytis (description)	67
Photos covering various aspects of the Tomato industry	68-71
Insect Control	76
Preparing for Harvest	78

Chapter 3. – GENETICS IN BREEDING OF PROCESSING TOMATOES	**83**
Classification & Crossing Relationships of Tomato	84
Tomato Genetics Cooperative	85
Methods of Tomato Breeding	85
Breeding Objectives	88
Breeding Improvements	88
Future Challenges	94
Regulation of Plant Breeding	96

Chapter 4. – TOMATO HARVESTING, SYSTEMS AND METHODS	**103**
The Harvester	104
Operation of Harvester	105
When to Harvest	106
Importance of Sorting	106
Mechanical Harvesting Problems	107
Cost of Mechanical Harvesting	109

Chapter 5. – TOMATO HANDLING	**117**
Hampers	117
Field Boxes	117
Plastic Boxes	118
Bulk Containers	118
Water Tanks	119
Bulk Trailers	122

Contents

Chapter 6. – TOMATO GRADING 125
History and Development of Grades 125
Sampling 125
Inspectors and Inspections 128
Grading Platforms 131
Grade Standards 132
Extraneous Material 138
Definitions 138
Grade Determination By Color 140
Agtron Color Measurement 141
Hunter Color Measurement 148
Firmness 151

Chapter 7. – PREPARATION OF TOMATOES FOR PROCESSING 153
Dry Sort 153
Size Grading 153
Washing 154
Final Sorting and Trimming 158
Coring 161
Peeling 164
Steam Peeling 164
Lye Peeling 165
Infrared Peeling 169
Other Peeling Methods 173
Inspection 175

PART II – PROCESSING 179

Chapter 8. – CANNING TOMATOES 181
Filling 181
Salting and Firming 183
Exhausting 189
Process Time & Temperature 190
Cooling 192
Acidification 192
Other Tomato Products 196

Chapter 9. – TOMATO JUICE MANUFACTURE 201
Preparation for Processing 202
Crushing or Chopping 202
Extraction 205
Deaeration 207
Acidification 208

Contents

Chapter 9 — Continued

Salting and Filling 208
Containers 209
Homogenization 210
Thermal Processing 210
Tomato Juice from Concentrate 214
New Products 215

Chapter 10. – TOMATO PULP AND PASTE MANUFACTURE 219

Definition 219
Manufacture of Tomato Pulp 220
Determination of Total Solids 223
Tomato Paste 224
Filling 227
Bulk Storage 227

Chapter 11. – TOMATO CATSUP AND CHILI SAUCE MANUFACTURE 233

Tomato Catsup 233
Manufacturing Tomato Catsup 234
Pulping 235
Constituents of Catsup 235
Formula 236
Cooking 237
Milling 238
Filling and Sterilization 239
Cooling 240
Quality Control of Catsup 240
Chili Sauce 241

Chapter 12. – TOMATO SOUP 243

Formulation 244
Procedure 245
Cooking 247

Chapter 13. – TOMATO WASTES 249

Part III – TECHNOLOGY 251

Chapter 14. – QUALITY ASSURANCE 253

Definition of Quality 254
Standards for Quality 254

Contents

Chapter 14 — Continued

Legal Standards 254
Company or Voluntary Label Standards 256
Grade/Industrial/Consumer Standards 256
Methods for Determining Quality 256
Purposes of QA Program 257
Bases of QA Program 258
Standards & Specifications 261
The Laboratory 264
Reports 271
Interpretation 271
Definitions of Terms Used in Statistacl QC 278

Chapter 15. – QUALITY CONTROL 285

Problem Solving Techniques 287
Brainstorming Principles 287
Pareto Principles 288
Cause & Effect Diagram 289

Chapter 16. – QUALITY EVALUATION OF PROCESSED TOMATOES AND TOMATO PRODUCTS 293

Determination of the Standard of Fill of Container 294
Procedure A: General Method for Water Capacity of Containeers 295
Procedure B: General Method for Fill of Containers 295
Procedure C: Percentage of The Total Capacity of the Can 296

Chapter 17. – COLOR AND COLOR MEASUREMENT 297

Factors Contributing to Tomato Color 298
Color Perception 299
Light & Lighting 300
Systems of Color Measurement 302
- Ridgeway Charts 302
- Maerz and Paul Color Dictionary 302
- Munsell Color Systems and Charts 302
- CIE or ICI System 303
- Macbeth Munsell Disc Colorimeter 303
- Hunter Lab Color and Color Difference Meter 307

Chapter 18. – TOMATO SOLIDS 313

Composition of the Tomato 313
Total Solids 313
Degree Brix/Soluble Solids 314
Water Soluble Solids 317

Contents

Chapter 18 — Continued

Alcohol Insoluble Solids	317
Blotter Test	318
Precipitate Weight Ratio	319
Serum Separation	319
Specific Gravity	320
Refractive Index	321

Chapter 19. – CONSISTENCY (VISCOSITY) OF TOMATO PRODUCTS 323

Classification	323
Measurements	325
Tomato Juice	325
Modified Efflux-Tube Viscometer and GOSUC Consistometer	325
USDA Viscometer	326
Capillary Viscometer	326
Stormer Viscometer	327
Catsup	328
Bostwick Consistometer	329
Brookfield Viscometer	331
Adams Consistometer	334
Blotter Test	334
Continuous Measurement of Catsup	336
Tomato Paste	336
Tomato Pulp	339
Tomato Soup	340
Brookfield Viscometer	341
FMC Consistometer	341
Fisher Electro Viscometer	341
Gardner Mobilmeter	342
Factors Effecting Consistency in Tomato Products	343

Chapter 20. – TOTAL ACIDITY AND pH 345

pH Determination	347

Chapter 21. – DEFECTS AND MATERIAL OTHER THAN TOMATOES 353

MOT and other Material	353
Sand and Inorganic Residues	354
Dark Specks, Seeds, Pieces of Seeds; Peel, Hard Core Material	354
Defects in Catsup	356

Contents

Chapter 22. – FLAVOR AND FLAVOR EVALUATION	**359**
Judging	360
For Each Judge	363
For Each Treatment	363
All Treatments/All Judges	365
Chapter 23. DROSOPHILA AND INSECT CONTROL	**369**
Life Cycle Habits and Other Functions	369
Drosophila Control Before and During Harvesting	372
Drosophila Control at the Plant and During Processing	373
Methods of Detection	375
GOSUL Method	376
AOAC Method	378
Staining Method	380
Determination of Insect Fragments in Tomato Products	380
Summary	384
Chapter 24. – MOLD–COUNTING METHODS AND PRINCIPLES	**387**
The Microscope	387
Construction of the Microscope	388
Proper use of the Microscope	389
Care of the Microscope	391
Histology of the Tomato	392
Parts of the Tomato	394
Types of Mold	394
Characteristics of Mold Hyphae	395
Filaments Often Confused with Mold	395
Howard Mold Count Method of Tomato Products	396
Characteristics of Mold	401
Genera of Molds Frequently Encountered	403
Alternaria	404
Aspergillus	404
Colletotrichum	405
Fusarium	406
Mucor and Rhizopus	407
Oospora (Oidium)	407
Penicillium	408
Phytophthora	408
AOAC Mold Count Procedure	408
Preparation of Sample	408
Equipment Materials and Reagents	409
Procedure	411
Modification and Slide Preparation	411

Contents

Chapter 24 — Continued

Counting Procedures	412
Application	413
Lost Acceptance Criteria	413
Regulatory Action Guidance	416

Chapter 25. – SPOILAGE OF CANNED TOMATOES AND TOMATO PRODUCTS **419**

Flat Sour Spoilage	420
Characterics of Flat-Sour Spoilage of Tomato Juice	420
Heat Resistance of Spores	421
Causes of Flat-Sour Spoilage	421
Controlling Flat-Sour Spoilage	422
Water Activity	424
Spoilage of Canned Tomatoes	425
Spoilage of Catsup	426

Chapter 26. – COMPOSITION OF TOMATOES **433**

Solids	433
Carbohydrates	433
Proteins and Amino Acids	434
Acids	434
Minerals	436
Pectin in Tomato	436
Nutrient Composition of Tomatos and Tomato Products	439
Factors Affecting the Nutrient Composition of Fresh Tomatoes	440
Factors Affecting Retention of Nutrients in Tomato Products	442
Retention of Vitamins During Storage	443
Tomato Flavor	447

APPENDIX A – U.S. STANDARDS IF IDENTITY & GRADES, FILL OF CONTAINER, FACTORS OF QUALITY, DEFINITIONS, INSPECTION & SCORE SHEETS **453**

Tomato Catsup	453
Chili Sauce	459
Tomato Sauce	463
Canned Tomatoes & Okra	468
Canned Tomato Juice	475
Tomato Paste	482
Tomato Puree	488
Canned Tomatoes	494
Stewed Tomatoes	501

Contents

APPENDIX B – FOOD & DRUG ADMINISTRATION PART 53 – TOMATO PRODUCTS 507
Tomato Juice 507
Yellow Tomato Juice 507
Catsup 508
Tomato Puree 509
Tomato Paste 510
Canned Tomatoes 512

APPENDIX C – QUALITY CONTROL AND EVALUATION FORMS 517
Raw Product Recieving Report 518
Daily Preparation Quality Control Report 519
Daily Canned Tomato Quality Control Report 520
Daily Tomato Juice Quality Control Report 521
Daily Double Seam Quality Control Report 522
Daily Process Record Quality Control Report 523
Canned Tomato Score Sheet 524
Canned Tomato Juice Score Sheet 525
Tomato Catsup Score Sheet 526
Tomato Puree Score Sheet 527
Daily Sanitation Report 528
Tomato Variety Evaluation Report 529

INDEX 531

Preface to the 1st Edition

This book is written to summarize basic information on the main factors involved in the production, processing and quality control and evaluation of tomatoes and tomato products. The purpose of this book is to bring together the many interrelationships between production and processing for the manufacture of high quality products. It is hoped that the information contained in this book will also help to emphasize the areas needing more research and define and characterize the scope of the problems confronting the industry as they presently exist.

The main objectives of this book include: (1) furnishing management with the basic and underlying principles for the preparation and preservation of tomatoes and tomato products, (2) summarizing the methods for quality evaluation, control and technology in a concise format, and (3) providing students and technologists with the interrelationships of production and processing for quality packs.

The book is organized into three parts with the first part covering the key areas involved in producing tomatoes. The second part covers the major unit operations included in processing tomatoes and the manufacture of tomato juice and products. The third part deals with the technology and quality control and quality assurance area. This part is more concerned with the scientific aspects of the tomato industry.

My interest in writing the book stemmed from three sources: (1) early work in saving tomato seeds for a large seed company in the early 1940s and observing all the juice going to waste, (2) 30 years of working with researchers in the tomato industry concerning aspects from breeding, unit operations and quality evaluation and control methods, and (3) the encouragement by past and present students to pull together and publish the vast amount of material in my files and other literature.

The author is deeply indebted to the many Food Processors, specialists and supply firms in the food industry who have willingly provided me with literature, photographs, technical information and illustrative material used throughout this book.

Sincere acknowledgment is expressed to my many colleagues, former students and my friends in the tomato industry for their advice, and encouragement. I am particularly indebted to Dr. Winston D. Bash, Dr. Stanley A. Berry, Dr. David C. Crean, Dr. J.R. Geisman, Dr. William George, Ms. Rebecca Gould, Dr. R.W. Hepler, Mr. M. Mahmoud, Mr. E.C. Wittmeyer, and Mr. Jerry Wright for their many contributions and assistance. Special thanks are gratefully accorded Ms. Jacquelyn Gould for the illustrations, charts and stenographic help in the preparation of this book.

I would also like to thank Dr. Donald K. Tressler, Mr. John J. O'Neil, Christine A. Lapke and Deborah J. O'Neill of the AVI Publishing Company for their assistance and cooperation.

WILBUR A. GOULD

Preface to the 2nd Edition

The tomato processing industry is still changing and updating procedures and processes. These innovations have kept the industry out front as a leader in processing technology. In preparing this edition, I have attempted to make appropriate changes where possible and update all the information. Further, I have expanded some of the material to keep the book current with the newer technologies in the production and processing of tomatoes. In some cases, I have modified a whole section to more adequately cover the subject.

Many helpful suggestions have been received and I sincerely thank all concerned. I am particularly appreciative of the efforts of Mr. Traver Smith, Magnuson Engineers, San Jose, CA; Mr. Yukio Ishiguro, Kagome Co., Ltd., Tokyo, Japan; Dr. Richard Basel, Consulting Food Technologist, Columbus, OH; and Mrs. Robert Updegraff for outstanding secretarial work. Further, I am deeply indebted to Mr. Wilfred W. Tressler and Dr. James R. Ice of the AVI Publishing Company and Dr. Norman W. Desrosier for their help and encouragement.

WILBUR A. GOULD

Preface to the 3rd Edition

Many changes are still taking place in the tomato industry with increased emphasis on improvements in tomato production practices. New varieties/cultivars are being developed with hybrids coming to the forefront with much improved yields. Tonnage today is being reported in excess of 50 tons per acre in some growing areas. Major improvements in mechanical harvesting systems with color and dirt sorters eliminating much of the harvesting labor are already in extensive use.

Today the majority of the tomatoes are crushed and pulped under rigid temperature controlled conditions to assure consistency and product yield. The crushed tomatoes may be directly manufactured into consumer products, but more likely than not it is concentrated 2 to 7 fold and either held in tank farms or packed aseptically into 55 gallon drums, 500 pound plastic lined pallet boxes, or aseptically filled into rail cars and tankers for shipment to secondary processors. These secondary processors manufacture many styles and types of sauces, ketchups, soups, dressings, and tomato juice from concentrate. This new trend has allowed the industry to greatly increase output to better than 500,000 tons per year during the prior 10 years on a World wide basis. South and Central America (Argentina, Chile, and Mexico) are the major contributors to the fast growth outside of the United States. USA production (predominantly California, although Ohio, Indiana and Michigan produce some 12-15% annually) is increasing at some 30% every ten years with production in excess of 10 million tons today.

Along with these improvements and changes, quality is becoming somewhat improved and more uniform. The use of better methods of evaluation of tomato cultivars including the insoluble fraction and its relationship to consistency, color segregation and control, and the practical elimination of insect and mold problems allows the consumer a product that better meets their expectations.

With all of these changes, this book has been revised and brought up to date including photos, text, and data. My emphasis has been to keep it a practical book, but as technical as necessary to understand the tomato, the tomato industry, and the many tomato products being manufactured.

My sincere appreciation to many firms in this industry for their help, particularly, Terra-Vegetable Crops Division, Heinz USA, FMC Corporation, and many others as indicated in the text. My special thanks to all those firms that have allowed me access to their fields, their factories and their laboratories and to my colleagues at The Ohio State University for all their help. I wish to sincerely acknowledge the support and encouragement of Art Judge II and Randy Gerstmyer of CTI Publications.

WILBUR A. GOULD

Part I — Production

CHAPTER 1

Introduction & History of the Tomato Industry

Tomatoes rank second to potatoes in dollar value among all vegetables produced in the United States and in other parts of the World where they are grown. In terms of per capita consumption tomatoes are the leading processed vegetables. The average American now consumes over 25 lbs of processed tomatoes exclusive of catsup and sauces per year compared with a total of 60 lbs. for all commercially processed vegetables. In addition probably an equal amount is consumed from home production and processing.

The tomato belongs to one of the nine species of the genus *Lyscopersicum.* As customarily used, the tomato is a vegetable. Botanically speaking, however, it is a fruit based on its plant parts. Technically it is a berry, being pulpy and containing one or more seeds.

Consumption of tomatoes is limited to *Lycopersicum esculentum,* fruits of the wild species *L. cersiform* and *L. pimpellifolium.* According to Rick most of the other species are quite distasteful, however, the other wild species have genes that are resistant to many diseases, useful for color improvement, and have desirable quality attributes. Many of these wild species are useful in breeding programs for the constant improvement of existing cultivars and the development of new cultivars.

The following history of the tomato as a garden vegetable is quoted from Morrison (1938):

The early history of the tomato is not known with certainty. It appears to have originated in tropical America, probably in Mexico or in Peru. Some look upon the cherry tomato as the original type from which our cultivated forms have sprung. However, Tracy has called attention to a much larger fruited form which likewise is found growing wild in South America and which our cultivated sorts have been developed [sic]. Tracy says the name "tomato" is of South American origin and is derived from the Aztec word "xitomate" or "zitotomate." Bancroft states that the fruit was eaten by the wild tribes of Mexico who called it "tomati." According to Humboldt it was called "tomati" by Mexicans who sowed it among maize. The tomato appears to have been taken to Europe from Mexico or Peru during the early 16th century. The earliest mention of the plant by European botanists is in the Herbal of Matthiolus (1554), who says it had

ORGANIZATION PLAN FOR A TOMATO PROCESSING PLANT
SHOWING DEPARTMENTS AND ACTIVITIES

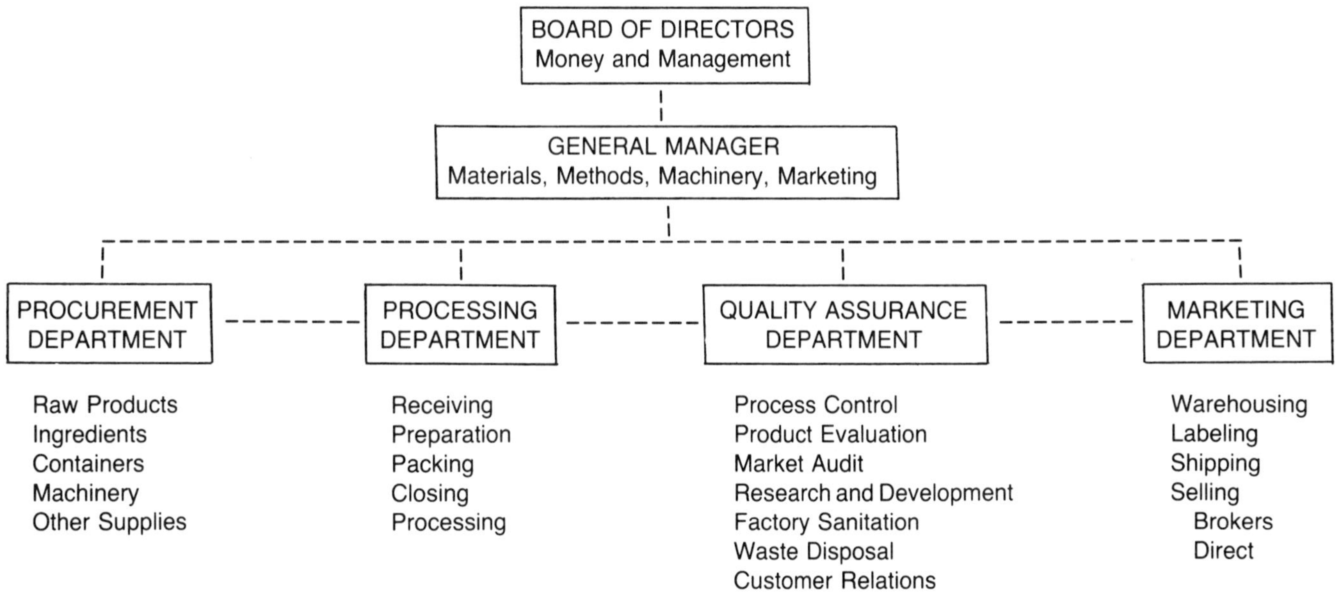

recently appeared in Italy where it was known as *pomi d'oro* (golden apple). Subsequently it became popular in France as *pomme d'amour* (love apple). The preferred name in France is now "tomate."

The tomato was grown extensively in Italy long before it had become a curiosity in England and America. English authors speak of it as an ornamental plant as early as 1578. In 1853, the fruit was eaten in Europe, dressed with pepper, salt, and oil. Gerard, who had tomatoes in his garden in England in 1596, said: "These 'love apples' are eaten abroad." His comments on their nutritive value are very uncomplimentary and in contrast with the high esteem with which we now have reason to regard this vegetable. As early as 1623, four sorts were known: the yellow, golden, red, and white. Tournefort in 1700 mentions seven types including one large smooth red type. In 1752, Miller recorded the use of tomatoes in England for flavoring soups. The cultivation of the tomato for market dates from about 1800 in Europe, but its true value was not realized until 1822 when Sabine wrote about it and gave details for its cultivation. At that time four red and two yellow varieties were in use: These were the Large, Small, Large Yellow, Pear Shaped, Cherry, and Yellow Cherry Love Apples.

First mention of tomato cultivation in the United States was made by Thomas Jefferson in 1781. Tracy comments of the unsuccessful efforts of the enthusiastic growers of that early period to get people to use the fruit. It was brought to Philadelphia in 1798 by a French refugee from Santo Domingo but was not sold in the market until 1829. In 1802, it was introduced at Salem, Massachusetts, by an Italian painter, but he found it difficult to persuade people even to taste the fruit. Gardiner and Hepburn give instruction in "The American Gardener" regarding the out-of-door culture of "love apples" and say "the fruit is used for soups and pickles." M'Mahon lists tomatoes or love apples in *The American Gardener's Calendar* and speaks of them as being highly esteemed for culinary purposes. In 1812, tomatoes were in use as a food in New Orleans.

However, it appears the tomato was still very little known as an edible vegetable in this country until 1830 to 1840. It was during this period that the tomato was acquiring that popularity which makes it almost indispensable today. In 1835 tomatoes were sold by the dozen in Quincy Market, Boston. In 1837 Thomas Bridgeman listed Large Squash Shape and Cherry Shape, but in 1847 he had added Large Yellow and Pear Shape to his list. Buist in 1858, speaking of the tomato, says: "In taking retrospect of the past eighteen years, there is no vegetable on the catalogue that has obtained such popularity in so short a period as the one now under consideration. In 1828–29, it was almost detested; in ten years most every variety of pill and panacea was extract of tomato. It now occupies as great a surface of ground as cabbage, and is cultivated the length and breadth of the country." Buist at that time listed the varieties Large Smooth Red, Large Red, Pear Shaped, Cherry Shaped, and several other fancy sorts "for those who want variety."

Peter Henderson, in *Gardening for Profit* published in 1867, states: "There are always some one or more varieties, said to be earlier than others, sent out every spring, but it must be confessed that the varieties that we cultivated twenty years ago are not in earliness a day behind those issued as vastly superior in 1866." He described nine of the many varieties grown at that time. They were Early Smooth Red, The Cook's Favorite, Tilden, Powell's Early, Fejee, Large Red, Large Yellow, Red and Yellow Plum, and Tree Tomato. His preference was the Early Smooth Red, which he considered as "a very old variety."

The increasing popularity of the tomato for table use encouraged the production of new varieties. Burr listed 23 varieties in 1863. It is said that Trophy was the first of the large, fairly early, smooth, apple-shaped varieties and that when it was introduced in 1870, the seed was sold at five dollars per packet of 20 seeds. A. W. Livingston, a practical gardener and seedsman, observed the need of constructive breeding. He realized that tomatoes could be most readily fixed in type by using desirable specimen plants rather than specimen fruits as the basis of selection. It was his aim to grow tomatoes smoother in contour, more uniform in size, and better in flavor. By adhering to the principle of single plant selection to meet clearly defined demands arising in the tomato trade, Livingston developed and introduced 13 varieties between 1870 and 1893.

The interest in tomatoes was such that within a few decades the number of varieties available to growers increased to several hundred. This increase was due largely to the (1) introduction of European varieties, many of which were subsequently renamed or designated by their English equivalents; (2) development of new American varieties; (3) tendency of seedsmen to list as distinct varieties stocks that differed little or none from already named varieties; and (4) reluctance of seedsmen to shorten their lists because of the insistent demand of conservative customers that they continue to be furnished with seed of the old varieties on which they continue to rely.

The varieties of tomato became so numerous and their names and descriptions so perplexing to gardeners that in 1886 and 1887 Bailey of the Michigan Agricultural College took steps to clarify the situation. Bailey grew 76 varieties in 1886. One hundred seventy sorts offered by American seedsmen as well as these offered by a leading seedsman of England, one of France, and one of Germany were included in his 1887 trials. These variety tests were continued by L. R. Taft, who grew 200 varieties in 1888, 128 varieties in 1889, and 100 in 1890. The 170 samples grown by Bailey represented 110 so-called varieties, not counting those French and German names that were simply equivalents of English names. The samples were observed critically, classified, and described with regard to type of plant and foliage as well as form, size, and color of fruits. Bailey's report indicates that much of the confusion of varieties was due to indiscriminate renaming. It was determined that 170 samples represented only 61 varieties and that many of these were similar to one another.

Some measure of the progress in tomato improvement is afforded by the data presented in Table 1.1 showing the length of time a number of varieties remained in popular demand, as measured by their being listed in the catalogs of a

TABLE 1.1. CULTIVARS OF TOMATO POPULAR DURING THE PERIOD 1868–1937

Cultivar	Period Listed	No. Years	Cultivar	Period Listed	No. Years
Large Red	1868–1885	18	Mikado[c]	1889–1902	14
Ferry's Improved Early Large Smooth Red	1868–1888	21	Red Apple	1889	1
Large Yellow	1868–1881	14	Atlantic Prize	1891–1907	17
Tilden's	1868–1878	11	Ignotum	1891–1898	8
Cherry Red	1868–1936	69[a]	Peach	1891–1930	40
Large White China Sugar	1868	1	Dwarf Champion	1892–1936	45[a]
Large Red Fegee	1868–1883	16	Royal Red	1893–1907	15
Keyes' Early Prolific	1969–1874	6	Stone	1893–1936	44[a]
General Grant	1871–1883	13	Buckeye State	1895–1915	21
Hubbard's Curled Leaf	1872–1886	15	Dwarf Aristocrat	1893–1903	8
Dwarf Orangefield	1872–1874	3	Imperial	1896–1898	3
Red Pear Shaped	1872–1936	65[a]	Honor Bright	1898–1909	12
Cedar Hill	1872–1873	2	Magnus	1901–1914	14
Trophy	1872–1926	55	Matchless	1901–1922	22
Charter Oak	1872–1874	3	Nolte's Earliest	1902–1907	6
Canada Victor	1874–1892	19	Yellow Pear Shaped	1902–1936	35[a]
Arlington	1874–1878	5	Earliana	1904–1936	33[a]
Hathaway's Excelsior	1876–1886	11	Chalk's Early Jewel	1905–1936	32[a]
Early Conqueror	1876–1893	18	Quarter Century	1905–1908	4
Little Gem	1879–1883	5	Dwarf Stone	1905–1936	32[a]
Green Gage	1879–1883	5	Purple Dwarf	1905–1908	4
Triumph	1879–1880	2	Globe	1906–1936	31[a]
Acme	1879–1930	52	Ponderosa	1906–1936	31[a]
Paragon	1880–1892	13	June Pink	1907–1936	30[a]
Essex Early Hybrid	1881–1912	32	Early Detroit	1909–1936	28[a]
Golden Trophy	1879–1882	4	Coreless	1911–1921	11
Turk's Turban	1880–1882	3	Bonny Best	1916–1936	21[a]
Early Trophy	1881	1	Avon Early	1921–1936	16[a]
Hundred Day	1881–1890	10	Gulf State Market	1921–1936	16[a]
Perfection	1882–1922	41	Greater Baltimore	1925–1936	12[a]
Alpha	1882–1883	2	Cooper's Special	1926–1936	11[a]
Favorite	1883–1907	25	Marglobe	1927–1936	10[a]
Queen	1883–1890	8	Morse's Special 498	1931–1936	6[a]
Optimus	1885–1911	27	Break O'Day	1932–1936	5[a]
Golden Queen	1886–1936	51[a]	Ox Heart	1932–1936	5[a]
Beauty	1887–1929	43	Pritchard (Scarlet Topper)	1932–1936	5[a]
Cincinnati Purple	1887–1896	10	Supreme Gulf State Market	1935–1936	2[a]
Cardinal	1887–1888	2	Supreme Marglobe	1935–1936	2[a]
Yellow Plum	1887–1936	50[a]	Michigan State Forcing	1936	1[a]
White Apple	1887–1930	44	Grothen's Globe	1936	1[a]
Yellow Cherry	1887–1930	44	Norton	1937	—[a]
Early Michigan[b]	1889–1930	42	Rutgers	1937	—[a]

[a] Important present-day cultivars.
[b] Known as Early Red Apple prior to 1892.
[c] Also known as Turner's Hybrid and so listed prior to 1891.

firm engaged in the nationwide distribution of seeds since the early days of the industry.

The history of the tomato processing industry dates back to the year 1847. Harrison Woodhull Crosby, Assistant Steward and Chief Gardener of Lafayette College, Easton, Pennsylvania, turned the refectory of the college into a laboratory, soldered tin lids onto small tin pails, stuffed some "love-apples," or tomatoes, through holes in the lids, soldered tin plates over these holes, and immersed the sealed cans in boiling water until their contents were sterilized. According to E. J. Cameron, Assistant Director of the Research Laboratories, National Canners Association, he emerged from his impoverished laboratory as the first practical tomato canner in authenticated history (Judge 1914).

Bitting (1912) describes as follows the method then in use for canning tomatoes.

Tomatoes are now used in enormous quantities in the fresh state and head the list of all vegetables as a canned product. Thousands of bushels are also used in the manufacture of ketchups, chili sauce, and soups. The tomato is produced over a larger part of the United States than any other vegetable. It may be handled with few and simple appliances, and may therefore be canned in the home and in small factories where little capital is required, as well as in the large factories.

The development of a tomato suitable for canning purposes has been a specialty in itself. For this purpose the fruit should be moderately large, smooth, so that it will peel readily, ripened evenly to the stem, of a clear, red color, and have a large proportion of solid meat of good flavor. Varieties which ripen unevenly or are irregular in outline are difficult to peel and the percentage of waste is too high. Tomatoes which are yellow or purple do not have an attractive appearance on opening, and those with excessive seed cells or which are soft and watery will give the can the appearance of being slack-filled or packed with water. A good pack is therefore dependent upon having a variety possessing the right qualities. The canner cannot accept tomatoes of a half dozen or more varieties and furnish the plants for his growers. The production of plants in hotbeds and cold frames to supply several hundred acres is of itself a very large task. The plants are grown in the field, the same as other crops, and a single large cannery will use the product of 1,000 acres. One ketchup manufacturer takes the entire product from more than 5,000 acres. A fair yield is 5 tons of fruit for an acre, but good cultivation and fertilization sometimes brings this up to 20 tons or more. Thirty-three bushels weigh about one ton.

At harvest time the fruit must be picked every day, or every other day, in order to insure collecting it when it is in its prime—just ripe, without green butts, and not overripe. It is preferable that the tomatoes be put in crates, which are wide and flat rather than deep, and which will hold not more than a bushel. They can be delivered to the factory in better condition in the flat crates than in the deep ones or in baskets, as the fruit will crush if piled in too many layers. The arrival in good condition lessens the time required for peeling as well as the loss in parts cut away. The tomatoes should be delivered to the factory promptly, as deterioration begins soon upon standing.

When the tomatoes are delivered at the factory they are weighed, and inspection should be made of each load. One crate is taken out at random and dumped into a tank of water. All defective fruit can be detected at once, picked out, weighed separately, and the load docked accordingly. Rotten fruit cannot be

FIGURE 1.1 — CONSUMPTION OF TOMATO AND TOMATO PRODUCTS

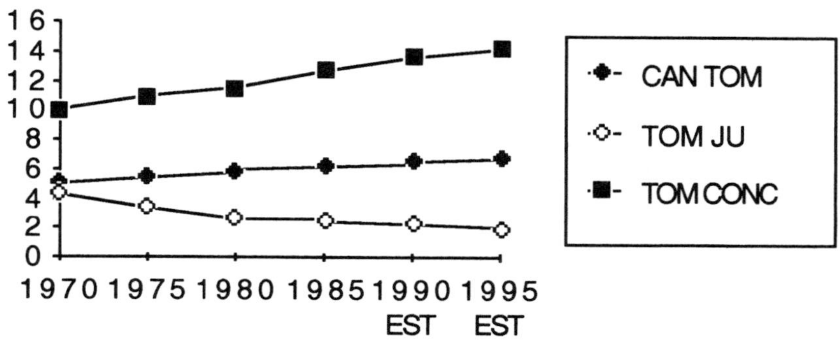

used and green fruit must be held to ripen. The separation at the factory entails extra expense in the inspection and sorting. The rotten fruit should not have been picked and the green should have been left in the field; the only way to reduce this waste to a minimum is by means of a system of dockage.

The first step in manufacture should be proper sorting. This can be done better by a few persons than by the many peelers. Tomatoes which are green should be taken out and held in crates for one or two days, as may be necessary, but small green spots can be cut out by the peelers. The tomatoes with rot should be discarded. Tomatoes which are small, rough, misshapen, and sound, but which will not peel well, can be set aside for pulp. Such a separation will lessen the work and waste in the factory and in the end be economical. The sorting is best done upon a conveyer table, the tomatoes which are passed being fed directly into the washer.

The washing should be thorough and done without bruising or crushing the fruit. It is preferable that the fruit be dropped into a tank of water and rolled over and over gently, either by actually turning the tomato or by strongly agitating the water, and then spraying under a strong pressure as they emerge from the water. This latter operation is of greater importance than is generally supposed. As before stated, a comparatively large volume of water without force behind it is far less efficacious than a much smaller volume having force. The latter cuts the dirt and organisms off, the former only wets the skin and makes it look bright. Allowing tomatoes to dry in the sun after washing by each method will clearly demonstrate the difference. The water in the tank should be changed continuously by the addition of the water used in the spray, an overflow being provided for the tank. The majority of tomato washing machines are inefficient.

The tomatoes are scalded, while passing slowly through a tank or steam chamber, by the continuous action of hot water or steam. The scalding is only sufficient to loosen the skin and not to heat or soften the tomato. As the tomato emerges from the scalder it is sprayed with cold water, which causes the skin to split and arrests the heating of the fruit.

FIGURE 1.2—TOMATOES: 5-YEAR MOVING AVERAGE ACREAGE OF TOMATOES IN 000.

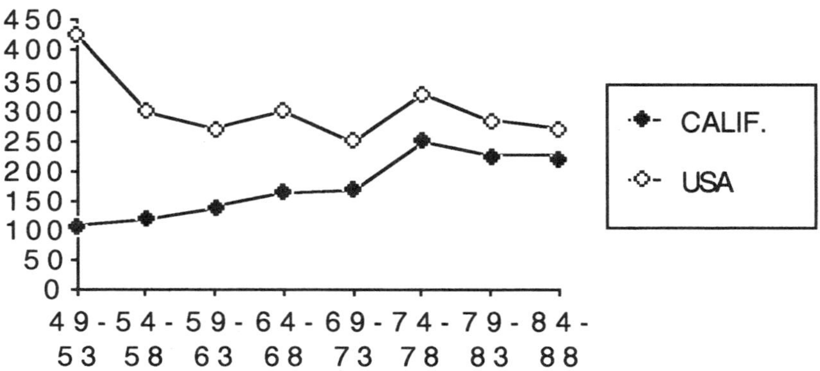

The clean-scalded tomatoes are delivered to the peelers in various ways, in pails and pans by carriers or belts, or by moving table tops, or they are delivered to the tables directly upon belts. Various devices have been used to carry the tomatoes to and from the peelers and to care for the waste, the object being to secure cleanliness and careful handling of the fruit. The bucket system is an old one and is in general use at small factories. The bucket is filled with scalded tomatoes and the peeler works from one bucket into another, dropping the refuse into a third bucket or into a trough under the table. The objection to the bucket is that the fruit on the bottom is mashed more or less before being reached by the peeler, and the same is true of the peeled fruit. Wide, shallow pans have an advantage over the bucket in this respect. In peeling from the special tables, the tendency is to heap the bowls too full, which produces the same disadvantages found in using the bucket. Some paint the buckets different colors to indicate whether they are to be used for scalded tomatoes, peeled tomatoes, or refuse. All buckets or pans should be washed each time they are used, no matter how many times a day that may be. All tables and conveyers should be washed each time the plant stops, and oftener when needed.

The peelers hold the tomatoes with the stem toward the palm of the hand, pull the skin back from the blossom end, and close the operation by removing the core with the point of the knife, keeping it well directed toward the center so as not to open the seed cells. This is not only the quickest way to peel the tomato, but keeps it whole. Green and undesirable spots are cut out.

The cans are filled either by hand or by machine. The sanitary or open-top cans are filled by hand, as it gives a better appearance to the finished product. In this class the cans are weighed to insure the desired fill. If filled too full, which may easily happen, 'springers' or 'flippers,' have the appearance of a swell, but are not due to fermentation. Solder-topped cans seldom bulge in this way for the

FIGURE 1.3—TOMATOES: 5 YEAR MOVING AVERAGE ACREAGE OF TOMATOES (000 OMITTED)

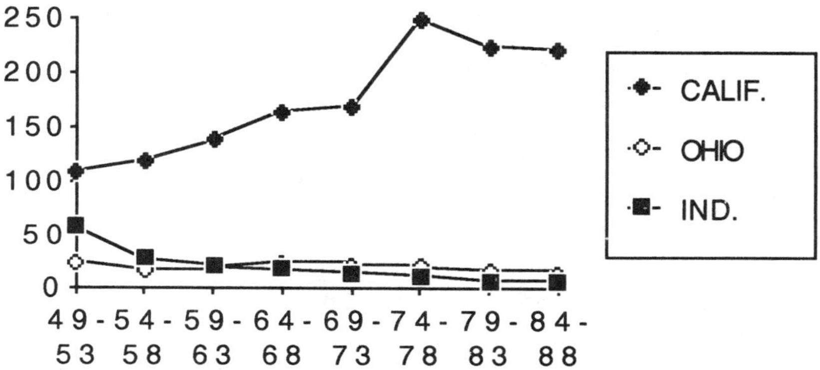

reason that they cannot be sealed when too full, and, as a rule, they weigh from 3 to 4 ounces less than the hand-filled cans. Overfilling also necessitates a longer process, breaking up the fruit and detracting from the appearance of the product. In order to bring out the flavor some canners add one teaspoonful of a mixture of equal parts of salt and sugar, or of one part of salt to two parts of sugar to each can. This is rarely done except upon high-grade goods and must be done by hand in order to insure uniformity.

There are several types of filling machines for solder-topped cans, which consist usually of a cylinder holding the quantity of tomatoes necessary to fill a can and a piston to force them in. The result is more or less badly broken fruit,

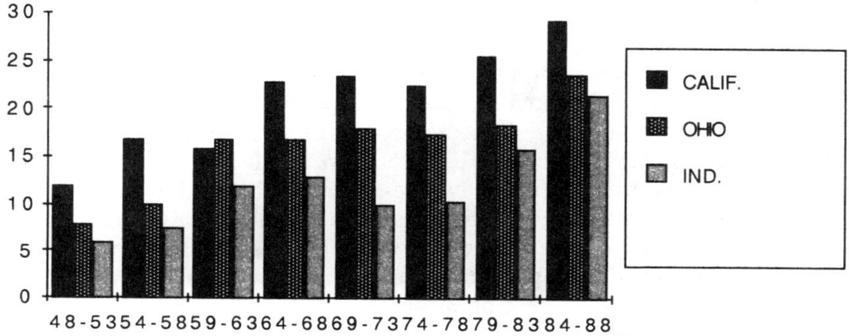

FIGURE 1.4 — TOMATOES: 5 YEAR MOVING AVERAGE YIELD PER ACRE

though the contents are just as good as in the hand-packed. Some of the newer machines fill the cans on the principle of a collapsible tube, and the result is a decidedly better appearance. In all machine-filling the measure is by volume rather than by weight. Cans which are filled full of whole tomatoes by hand are known as 'hand-packed' or 'solid-packed' in distinction from those filled by machine, or filled part full of whole tomatoes and juice added. The adding of juice is done for two purposes, one in high-grade stock to preserve the tomato whole or nearly whole, and in standard grade to complete the machine fill or to utilize the entire product. In the first case the juice is taken from whole tomatoes and usually condensed slightly by boiling. In the latter case it is made from the trimmings and often of inferior quality. The use of water in canning tomatoes is unnecessary and is an adulteration.

Somewhat too much stress is being placed upon the quality of solid meat which will be present after draining on a quarter-inch screen. A very high percentage of solid meat may mean the use of a variety which is hard and

FIGURE 1.5—5 YEAR MOVING AVERAGE OF TOMATO SUPPLY IN TONS (000 OMITTED) FOR U.S. 1949 THRU 1993

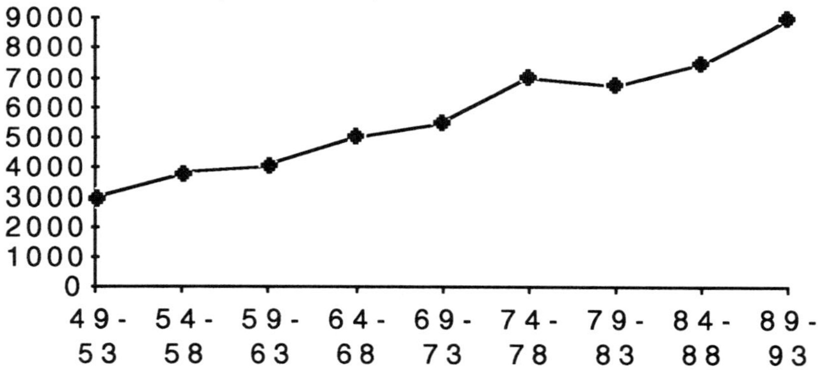

inferior, or fruit which is slightly green, in which even the flavor is deficient. The full rich flavor of the tomato is not developed until it is thoroughly ripe, so ripe that the processing will cause a portion of the tissue to break down, and after long shipments they may be badly broken. While it is desirable to have a considerable proportion of the fruit whole or nearly whole, a broken condition is not of itself evidence of improper methods or poor quality. The cans are next run through an exhaust box, where they are subjected to steam heat for from 2 to 3 min, after which they are capped in the usual way. Tomatoes are given a process in boiling water for from 35 to 55 minutes.

Tomatoes are packed in No. 3 cans as a general rule, though they are also packed in all sizes from special cans for individual service on dining cars and cafes to the No. 10, or so-called gallon cans for hotel trade. Some of the latter are put up unpeeled. The No. 3 comes in the regular size and in what is known as extra tall. The tomato is also put up as condensed tomato soup, paste, and purée.

FIGURE 1.6 — CONSUMPTION OF TOMATO AND TOMATO PRODUCTS

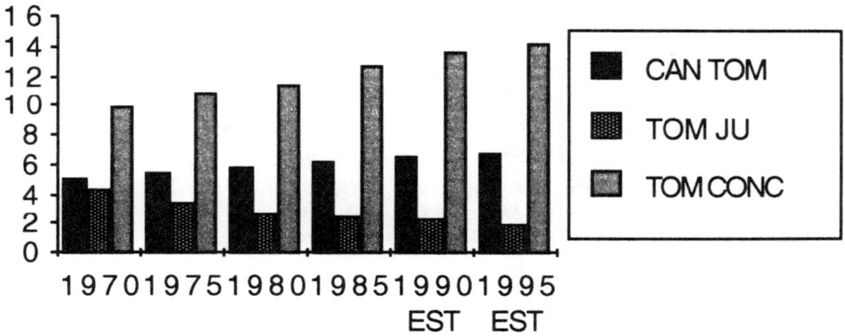

To produce these, the tomato is run through a 'cyclone' to remove the hard portions and seeds, and then concentrated to different degrees. The use of condensed tomato or purée prepared from sound material has many advantages for some purposes over the regular canned article, and its use should be cultivated, especially for soups, etc. At the price paid for the standard grade of tomatoes a better article can be obtained as a purée or paste. Some purée is made from peel and waste from the canning. If the material is clean and sound there is no objection to its use, but too often this is not the case, as is made evident by the presence of microorganisms, broken tissue, and products of decomposition. A paste which is made from the whole tomato and from trimmings by a system of spontaneous fermentation and salting is used largely by foreigners. This article is no longer permissible in interstate trade. Another grade of paste is made by evaporating the pulp until it becomes very stiff and heavy. The straining of the juice or pulp from the seeds and hard portions can be done better and with less waste by special machinery than in the kitchen.

Tomatoes are sold under various trade grades, as extra choice, extra select, choice, select, extra standard, standard, and seconds. It is unfortunate that there are so many ways of designating the contents of a can, particularly when the prefix is meaningless. What one packer calls his 'extra choice' or 'extra select' may be no better than an extra standard or a standard of another packer. The real grade at present is dependent upon the packer's name, not upon what he claims. There should be but two grades—selected or first grade, and standard or field run for the second. A can of first grade tomatoes should be from selected, prime, ripe fruit, having a fleshy body, well-developed flavor, and uniform color. The can when opened should be full and most of the tomatoes whole or in large pieces, free from all peel, core, or defects. The net weight should not be less than 32 ounces in a No. 3 can.

A can of standard tomatoes should be from sound, ripe fruit, having a fair body and good flavor. The can when opened should be full, and part of the tomatoes whole or in large pieces. They should be well peeled and cored. The net contents of a No. 3 can should not weigh less than 32 ounces.

Prior to 1890 all unit operations in the canning of tomatoes were done by hand. Between 1890 and 1900 the tomato scalder, cyclone, and "merry-go-round" peeling tables were put into use. In the early 1920s the juice extractor was developed and tomato juice came onto the market. In the 1930s homogenization and flash pasteurization were significant improvements in the processing of tomato juice. Stewed tomatoes were a reality in the 1940s. In the 1950s lye and flame peeling were significant contributions to the history of the tomato-processing industry. In the 1960s acid and sweeteners were permitted as food additives for canning tomatoes. In the late 1960's mechanical harvesting became a reality in much of the industry. Along with mechanical harvesting came bulk systems of handling tomatoes from the field to the factory with water unloading systems perfected first in Ohio. All these changes had to wait for the improved cultivars that were firm fleshed, thick walled, and uniform in ripening. Also, in the 1960's significant new products, such as, diced, quartered, crushed, stewed and sliced tomatoes, frozen sliced tomatoes, and many styles of tomato cocktail juices made their appearance. In the 1970's many styles of tomato sauces were introduced with pizza sauce leading the parade. Today over 20 types of tomato flavored sauces and many styles of ketchup are on the market

TABLE 1.2—WORLD PRODUCTION OF PROCESSING TOMATOES
(1,000 Metric Tons)

Country	1975	1980	1985	1990
United States	7,715	5,646	6,525	9,307
Italy	1,575	3,083	3,785	3,850
Greece	979	1,500	1,180	1,150
Turkey	520	600	1,100	1,500
Spain	827	499	819	1,134
Portugal	800	454	716	760
Canada	350	379	476	580
France	280	416	392	340
Taiwan	223	491	362	182
Israel	163	166	257	300
Mexico	210	220	250	365

TABLE 1.3 — STATISTICAL SUMMARY OF TOMATOES FOR PROCESSING IN THE U.S.

Year	Acres	Yield/Acre	Production/Ton	Price/Ton
1940	385,930	5.39	2,080,100	11.73
1941	460,450	6.09	2,802,200	15.06
1942	601,200	5.27	3,166,800	19.70
1943	551,650	4.80	2,645,600	26.14
1944	581,180	5.45	3,169,900	27.22
1945	546,750	4.91	2,689,200	27.58
1946	579,590	6.09	3,528,600	30.03
1947	511,370	6.34	3,242,800	28.63
1948	400,850	7.27	2,913,500	27.71
1949	358,700	7.34	2,633,700	23.51
1950	359,620	7.60	2,733,860	25.30
1951	423,830	10.06	4,267,070	31.70
1952	375,900	9.18	3,452,000	29.40
1953	297,300	10.88	3,234,910	27.50
1954	268,550	10.05	2,697,690	24.40
1955	330,800	9.91	3,277,990	24.90
1956	346,780	13.2	4,570,700	25.60
1957	305,200	10.9	3,314,500	26.20
1958	345,750	12.4	4,287,400	25.40
1959	292,130	12.0	3,508,800	24.40
1960	282,850	14.2	4,013,500	26.10
1961	301,850	14.0	4,247,700	29.70
1962	326,700	16.5	5,377,000	28.40
1963	248,100	16.4	4,070,600	26.70
1964	270,100	16.9	4,561,000	30.70
1965	255,200	17.6	4,482,200	37.10
1966	300.100	15.5	4,660,600	36.89
1967	326,100	15.8	5,164,300	42.80
1968	366,100	18.8	6,965,900	40.20
1969	266,900	18.4	4,897,700	34.70
1970	245,100	20.6	5,059,000	34.00
1971	258,100	21.4	5,515,600	35.80
1972	265,000	21.9	5,803,700	36.20
1973	295,100	20.1	5,934,600	42.00
1974	337,700	20.8	7,019,700	64.50
1975	384,300	22.1	8,503,800	63.20
1976	309,000	21.0	6,471,800	58.00
1977	346,700	22.4	7,779,200	64.10
1978	295,600	21.5	6,367,700	64.20
1979	312,000	23.5	7,329,500	67.60
1980	263,000	23.6	6,210,600	61.00
1981	253,900	22.5	5,716,100	67.50
1982	295,300	24.7	7,299,000	71.60
1983	292,000	24.1	7,029,800	68.40
1984	291,900	26.3	7,681,200	67.40
1985	265,500	27.0	7,177,100	66.30
1986	252,300	29.3	7,398,500	63.90
1987	257,400	29.6	7,607,700	59.10
1988	274,900	27.0	7,409,900	60.70
1989	320,800	29.6	9,484,000	67.50
1990	359,700	28.3	10,181,300	
1991	362,700	28.4	10,312,520	

with more yet to come. In the 80's, tomato juice from concentrate became a meaningful product.

Sullivan estimates that the pack distribution for the US is as follows: Sauces equals 35%, Paste equals 18%, Canned Tomatoes equals 17%, Ketchup equals 15%, and Juice equals 15%. For Canada he estimates the pack distribution as follows: Tomato Juice equals 32%, Canned tomatoes equals 29%, Ketchup and Sauces equals 24%, and Paste equals 15%.

During the past 100 years the location of the production of tomatoes has changed drastically. In the early years, the industry was centered in Maryland; then it moved to Indiana; and at present California dominates the production areas. Ohio has become the leader in the Midwest, with New Jersey a major factor in the East. To best illustrate these changes during the past 25 years, the data presented in Fig. 1.2 and 1.3 show that over 400,000 acres were required for the production of tomatoes in the United States in the early 1960s. At the present time, more total tons are produced on 300,000 acres. California and Ohio are the only states showing an upward trend in total tons produced for processing. During this same period of time, average yield per acre has gone up in the United States from 7.0 tons to over 29 tons. California and Ohio have been the leading states in tomato yield per acre.

In terms of total tons available for processing during the past quarter century, 3,000,000 tons were produced at the beginning of this period, while today the amount is over 9,000,000 tons. As would be expected, California produces over 85% of the total, and Ohio over 6%; Indiana, New Jersey, Pennsylvania, Maryland, Virginia, and Michigan produce the rest.

World production of tomatoes for processing now stands at over 20 million tons with the US producing over 50%. The European Economic Community (Italy, France, Greece, Spain, and Portugal) produces nearly 6 million with South America (Argentina and Chile primarily), Mexico, Israel, Canada, and Taiwan making up most of the remaining tonnage.

The European Community (EC) subsidy program, with its high grower prices and processor subsidies, gives Italian and French tomato processors the incentive to expand output. While Greece's tomato processing industry benefits from Government subsidies and duty-free entry into the EC, processors in Spain and Portugal have lost a large percentage of their traditional export markets in the EC because of added competition resulting from the EC subsidy program.

REFERENCES

ALDRICH, N.W., JR. 1979. The love apple. Country Journal 79 (Aug.) 34-40.

BITTING, A.W. 1912. The canning of foods. U.S. Dept. Agric. Bull. *151.*

BRAYTON, GARY N. and J. FLINT PULSKAMP. 1991. Red and black all over-tomato industries wide supply fluctuations pit grower against processor profitability. Deloitt & Touche Special January Edition.

GAROYAN, LEON. 1989. Trends in the global processing tomato industry. The California Tomato Grower. 32 (5): 4-7.

JUDGE, A.I. 1914. A History of the Canning Industry. The Canning Trade., Baltimore, MD.

JUDGE, E.E. & SONS. 1982. The Almanac of the Canning, Freezing, Preserving Industries. Edward E. Judge & Sons, Westminster, MD.

MAY, E.C. 1937. The Canning Clan. Macmillan Co., New York.

MORRISON, G. 1938. Tomato varieties. Mich. State Coll. Spec. Bull. *290.*

RICK, C.M. 1978. The tomato. Sci. Am. *7,* 77-88; *16,* 148.

SULLIVAN, GLENN H. 1990. Organization, structure and trade in the north american tomato processing industry. The California Tomato Grower. 33 (6): 4-10, 26-28.

ZOLLINGER, DAVE. 1989. Zollinger reviews eventful and highly successful Season. California Tomato Grower. 32 (1): 4-6, 19.

CHAPTER 2

Tomato Culture and Production for Processing

Profitable agriculture depends on productive soils. It is generally thought that land just taken into cultivation will produce attractive yields without treatment of the soil other than tillage. However, the natural distribution of plants has shown that soil conditions influence the location of plants. Plants that do best on soils with a high lime content are not found under natural conditions on extremely acid soils. In other words, the plant found growing on the soil is best suited for that land (Hester and Shelton 1939).

Crop production depends on factors other than the soil, and these conditions must also be met for the economical production of a crop. Tomatoes are no exception. Such factors as plant vigor, insect control, and climatic conditions are important in their production (Hester and Shelton 1939). Fortunately, most of the variables in tomato production can be altered or controlled by man. Aside from the weather, the grower and his productive methods are the determining factors in realizing success in tomato propagation. The field selection and land preparation, as well as the care taken in planting and cultivation, are usually reflected in the harvested crop. The quality and yield that the grower obtains is, in some manner, indicative of each of these factors. For this reason, each deserves appropriate discussion.

FIELD SELECTION

The first consideration in tomato growing is the selection of the field, as good field selection reduces the likelihood of later problems (Anon. 1969B). The area selected should be relatively level, with good drainage. Level fields with uniform soil conditions are preferred to low, poorly drained fields with heavy soils. The tomato fruit will be subjected to its fair share of problems without adding to them by planting the crop where inadequate drainage or improper soil conditions exist; 87% of the growers in the Top-Ten Club in Ohio reported that they had good drainage characteristics (Anon. 1968). This serves as quite a meaningful indication of the importance of eliminating excessive moisture caused by water left standing in the field.

FIGURE 2.1. LAND LEVELING A FIELD OF TOMATOES

The size and shape of the field should be such as to require a minimum number of turns for the mechanical equipment (Angell *et al.* 1971). This is especially important when mechanical harvesting is employed. Row lengths of less than 600 ft seriously decrease harvester efficiency (Anon. 1969A). In addition, the slope should be level or, at most, gently rolling land leveling is practiced where practical today. There should be few to no stones and a minimum of large soil clods. Fields with high weed populations should be avoided. Weeds, particularly grasses, cause clogging, jamming, and require harvester stops. Fields with trashy residues such as corn stalks should be avoided. The shape of the field is not as critical for handpicking as it is for machine harvest operations. However, it is recommended that, whenever and wherever possible, long rows facilitate picking, harvesting, and removal of the fruit from the field.

Fields with uniform soil conditions, well drained sandy loam, and good wind protection are preferred for direct seeding. Rows of tomatoes can generally be oriented to minimize the effect of sand blasting. Wind breaks can prove helpful if the sand-blasting condition is serious. Strips of rye or oats planted in the field may also help combat the problem.

Finally, the tomato field selected should be well balanced with organic matter (Hester and Shelton 1939). If the soil is sand, it should have over 1.0% organic matter. Avoid following corn with tomatoes because of trash, high residual nitrogen, and possible herbicide residue. Atrazine residues from the previous year are particularly harmful to tomatoes. If the tomato field is located on a sandy loam, it should have more than 1.5% organic matter if a good crop is to be expected. This relationship for various soil textures is shown in Table 2.1.

TABLE 2.1. INTERPRETATION OF THE ORGANIC MATTER CONTENT OF SOILS

Quality Interpretation	% Organic Matter by Soil Type				
	Sand	Fine Sand	Sandy Loam	Fine Sandy Loam	Loam and Silt Loams
Poor	Less than 0.9	Less than 1.1	Less than 1.4	Less than 1.4	Less than 1.9
Fair	1.0 to 1.4	1.2 to 1.6	1.5 to 1.9	1.5 to 1.9	2.0 to 2.9
Good	More than 1.5	More than 1.7	More than 2.0	More than 2.0	More than 3.0

CLIMATE, GEOGRAPHY, AND SOIL SELECTION

Tomatoes are grown throughout the United States and in many regions of the world. The tomato is a warm-season plant reasonably resistant to heat and drought, and grows under a wide range of climatic and soil conditions. The tomato is not sensitive to day length, and sets fruit in day lengths varying from 7 to 19 hr. It requires 3 to 4 months from the time of seeding to produce the first ripe fruit. The tomato thrives best when the weather is clear and rather dry and the temperatures are uniformly moderate, 65° to 85°F(18° to 30°C). Plants are usually frozen at temperatures below 32°F(0°C) and the fruits do not increase in size at temperatures above 95°F(35°C). High temperatures accompanied by high humidity favor the development of foliage diseases. Hot, drying winds cause the flowers to drop (Anon. 1969B).

Tomatoes are grown on many kinds of soil, from sands to heavy clays. Where earliness is of great importance, as for an early crop in the northern United States and Canada, sandy or sandy loam soils are preferred. When large yields are important, as in the production of a crop for processing, loams, clay loams and silt loams are preferred to lighter soils, provided the growing season is long enough. In general, a deep loamy soil well supplied with lime, organic matter, and fertilizer is most nearly ideal (Keirns and Wittmeyer 1951). The different types of soil in which tomatoes have been grown along with their respective compositions are shown in Table 2.2 (Hester and Shelton 1939). Figure 2.1 illustrates the portions of sand, silt, clay, and humus that may be found in soils for different textural groups.

The soil should be slightly acidic, and should be limed, if necessary, to raise the pH to the ideal range of 6.0 to 6.5. Caution should be exercised when applying lime, as an excess can be just as serious as a deficiency.

When the soil is low in organic matter it becomes hard and crusts badly during the summer months. This may be corrected by applying manure or organic matter to loosen the soil, and by planting a green cover crop such as rye or rye grass on the plot the winter before the tomatoes are to be grown. This cover crop may be seeded in August or September and plowed or tilled under in April or early May (Keirns and Wittmeyer 1951).

TABLE 2.2. TEXTURAL RELATIONS OF VARIOUS SOIL TYPES

Texture	Definition
Sands	Less than 20% silt and clay; 80% sand
Find sand	50% of the sand as very fine sand and 50% fine sand
Sandy loams	20–50% silt and clay; 50–70% sand
Loams	20% or less clay, 50% silt, 30% or less sand
Silt loams	20% or less clay, 50% or more silt and 30% or less sand
Clay loams	20–30% clay, 20–50% silt; 20–60% sand
Clays	30% or more clay, 70% or less silt and sand

LAND PREPARATION

Good soil preparation is important in the successful culture of tomatoes (Pierce *et al.* 1963). Where fall plowing can be done without sacrificing well-established cover crops, it is desirable. Fall plowing promotes more

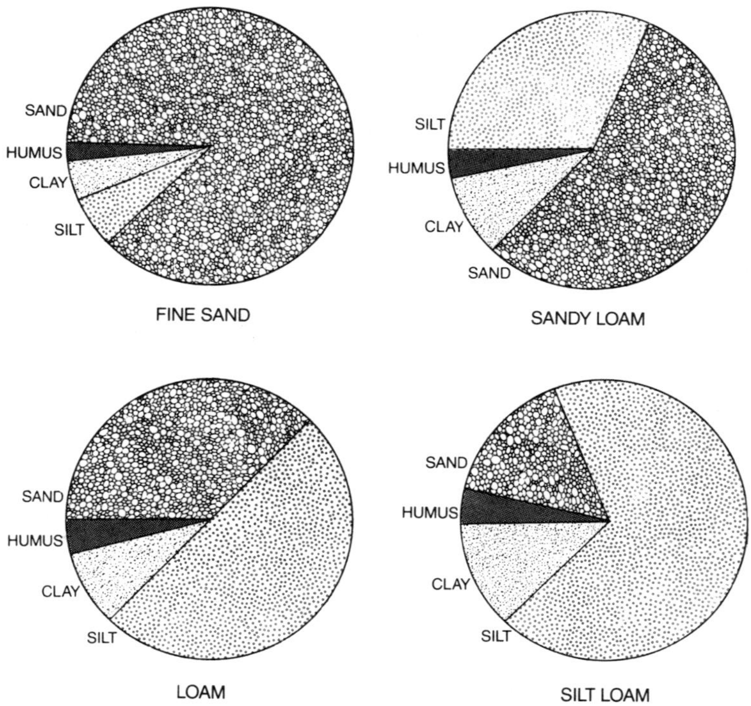

FIGURE 2.2 — DIFFERENT TYPES OF SOIL WITH THEIR RESPECTIVE COMPOSITIONS

thorough decay of roots and other organic matter in the soil. Further, as a result of alternate freezing and thawing, it leaves the soil in better physical condition (Beattie *et al.* 1942). If the sod crop is not a legume, 100 to 150 lb of ammonium nitrate or its equivalent should be broadcast just before plowing (Butler and Kerr 1952). Fall-plowed land should be left in the rough until spring, or sown to a winter cover crop that will not interfere with early spring preparation and planting. If left unplowed until spring, the land should be plowed as early as the soil is dry. No soil should be worked while wet. Heavy clay soils are especially subject to serious physical damage from tilling while too wet (Pierce *et al.* 1963).

Plowing should be done as deeply as the soil will permit, and the depth of plowing should be gradually increased by 0.5 in. each season until the soil is plowed at least 8 in. deep (Beattie *et al.* 1942). If the depth of plowing is gradually increased from year to year, the layer of fertile cropping soil can be deepened without affecting current crops (Pierce *et al.* 1963).

If a cover crop or sod is to be plowed under, disking is recommended before plowing, as this will hasten the decay of the material being turned under (Beattie *et al.* 1942). Preparation of the land after plowing should be more thorough than for general farm crops. Before setting the plants, the topsoil should be well pulverized to a depth of 3 to 4 in. (Beattie *et al.* 1942).

Soils having a hardpan or a layer of impervious clay 10 or 12 in. below the surface will be greatly improved for tomato production if the underlying soil is broken up without being brought to the surface. Deep tillage is accomplished by breaking the soil below ordinary plow depth. Plowing at the same depth year after year produces what is termed as "plow sole," and in time this becomes very hard. This condition can be corrected by an attachment to the plow that works in the bottom of the furrow and breaks the subsoil to a depth of 5 to 8 in. below the regular depth of plowing.

Manure, for best results, should be applied before plowing (Butler and Kerr 1952). Many growers prefer to apply stable or barn-lot manure to the crop preceding tomatoes rather than to the tomato crop. Others apply the manure to a cover crop of rye, wheat, or barley during the winter and then plow the manure under together with the cover crop, in ample time to properly prepare the land for setting the plants. When the manure is well decayed and of fine texture, 6 to 8 tons per acre may be applied broadcast after plowing, and thoroughly disked into the soil. Even 10 or 12 tons may be used without danger of adverse results, especially where the organic content of the soil is low and available plant food is not abundant. However, the cost of the manure in such large quantities may be a limiting factor (Beattie *et al.* 1942).

On soils that have been heavily manured during recent years or when the organic content of the soil is high, care should be exercised in the application of manure because of its tendency to produce a heavy vine growth at the expense of the set of fruit. When manure is applied before planting the tomatoes, the percentage of nitrogen in the commercial fertilizer is frequently reduced, or this element is omitted entirely (Beattie *et al.* 1942). Actually, heavy spring applications of manure should be avoided. Microorganisms feed on nitrogen while breaking down fibrous material. Microorganisms would, therefore, compete with tomatoes for available nitrogen early in the season. Also, excessive quantities of undecomposed straw manure may serve to accentuate a drought by drying out the soil and interfering with upward movement of soil moisture. During seasons of ample rainfall the decomposing manure may contribute to late growth and thus late maturity of the crop by the release of nitrate. This would have the same effect as late applications of nitrogen.

Beds are now used in many areas for growing tomatoes. They should be prepared in the Fall of the year and should be well shaped to help in surface drainage, particularly in clay loams or silt loam soils. If prepared in the Spring of the year one may find that it delays planting. The beds are generally spaced some 54 to 66 inches between centers to accommodate field and harvesting equipment, allowance for vine growth from different varieties, and whether using single or twin rows. The furrows should be some 8 to 10 inches deep depending on soil types. The single row of tomatoes should be planted in the center of the bed with the plants some 10 to 14 inches apart in the row. When planting twin rows the rows should be some 20 to 26 inches apart with the plants 12 to 16 inches apart in the row. If direct seeding, the clumps should be 9 to 12 inches apart with 2 to 3 plants per clump. Many of the beds are not so-called permanent beds and only a rotavator is used to loosen the soil in the Spring prior to seeding or planting. The bed should be as flat as possible and there should be no clods to interfere in harvesting. The beds make for more uniform ripening of fruit, allow for better drainage following heavy rains, and most reports indicate that increased yields are obtained with the bed system.

A last consideration in land preparation must involve long-range planning. Among other things, this involves a schedule of crop rotations. Soybeans, sugar beets, wheat, beans, and corn are popular yearly substitutes. Of course, corn may not be preferable if followed by tomatoes for mechanical harvesting, as it may leave undecayed stalks in the soil which could interfere with the harvester. Thorough land preparation and planning prior to the seeding or setting of tomatoes in the field is very important. No amount of cultivation after the plants have been set will take the place of adequate and thorough preparation before planting.

SOIL NUTRIENTS IMPORTANT TO THE TOMATO

The three major plant nutrients important in satisfactory development of the tomato are nitrogen, phosphorus, and potassium (potash). A number of minor nutrients are also important including calcium, magnesium, and sulfur, and the trace elements boron and manganese. In addition, there exist in the soil two important complexes from the standpoint of plant nutrition, namely, clay and organic matter.

Organic matter is extremely variable in the soil, depending on drainage and texture. For the most part, it is acid in nature, having a pH of about 3.5 when all the bases are removed. It analyzes about 5% nitrogen, which is slow to break down into nitrogen available to plants. In fact, in physical aspects, humus behaves not unlike clay in the soil since it is acid in nature, absorbs lime, potash, etc., and holds it available to plants. However, it is not subject to ready leaching from the soil by rain water (Hester and Shelton 1939).

Although lime (calcium and magnesium) and potash have a strong affinity for clay and organic matter, they have a stronger affinity for nitrates, sulfates and chlorides. Since these ions are soluble in water (unlike clay and organic matter), they cause the bases to leach from the light soil during rainy weather.

Nitrogen influences the quality of the tomato crop. There must be adequate nitrogen to produce sufficient foliage to protect the fruit from exposure to the hot sun. Furthermore, nitrogen greatly influences the date of maturity of the crop. If the crop has too much readily available nitrogen early in the season, it is likely to become too vegetative and to be too late in setting and maturing of the fruit. On light, sandy soils nitrogen from soluble sources may leach during rainy seasons and leave the crop with inadequate nitrogen. Yellow foliage or plants lacking sufficient nitrogen may then result. This condition in the soil is to be avoided whenever possible. Lastly, it is important to use good judgment in choosing the nitrogen compounds to be used on sandy soils. In addition, late applications of nitrogen should be avoided, as they may cause prolonged growth with late fruit and/or split sets. Nitrogen at excessive rates can have other effects. Limited research indicates high rates of nitrogen result in lower soluble solids and more blotchy ripening (gray wall), more yellow eye, more sprouted seed, more detinning problems, and poorer machinability (Zobel 1966). It is questionable whether the leaching of nitrogen is of tremendous importance during the growing seasons on heavy soils. It is perhaps of more importance to have the soil well cultivated and well limed so that microorganisms desirable for crop growth can function properly (Hester and Shelton 1939).

Phosphorus is of prime importance in the tomato fertility program. The importance of adequate phosphorus in the soil cannot be overemphasized. Phosphorus influences the quality of fruit in several ways. First, it stimulates vigorous root growth, which accounts for a better utilization of the nutrients in the soil. Second, it increases the efficiency of the plant by promoting a sturdy stem and healthy foliage. Zobel (1966), in a review of the literature on fertilization for *Mechanical Harvesting* reported, "Hepler found that phosphorus stimulated early growth to produce a larger number of blossoms earlier in the growth of the plant. MacGillivray found the composition of phosphorus-deficient leaves to have 0.105 to 0.162% phosphorus, while plants that contained sufficient phosphorus had leaves containing 0.35 to 0.56% phosphorus. The highest phosphorus content was in the top leaves and fruit. Kalin found that leaves containing more than 0.42% phosphorus did not respond to additions of phosphorus." However, there appears to be no particular constituent in the fruit directly influenced by phosphorus. Consequently, phosphorus fertilization increases yield. During a short growing season phosphorus in the fertilizer gives a greater increase in yield than in a long growing season, because the plant has a longer time to absorb the slowly available phosphorus from the soil. Generally, all soils carry a large reserve of phosphorus, but owing to certain constituents in the soil it becomes available very slowly (Hester and Shelton 1939).

The tomato plant absorbs and utilizes a large amount of potassium. According to Wilcox, potassium content of leaves is higher (3 to 4%) during the vegetative stage of the plant, and then declines during the fruiting period. The leaf composition should remain above 2% throughout the growth of the plant (Zobel 1966). In fact, more potassium is absorbed than any of the other minerals. The amount of potassium found in each ton of tomatoes varies from 5 to 6.5 lb. Accounting for the amount in the production of the plant, it can be readily seen that the production of a large crop consumes considerable potash. Assuming that one was producing only 10 tons of fruit per acre, it would require 2000 lb of mixed fertilizer analyzing 10% in available potash, or a 30-ton crop removes about 120 lb of potassium per acre in the fruit alone (Hester and Shelton 1939). However, it is not uncommon for a soil to carry from 30,000 to 60,000 lb of total potash per acre. It must be remembered, though, that only a fraction of this potash is available to the immediate crop. Therefore, under a system of crop production it becomes necessary to apply potash to the soil, sometimes in rather large amounts. Since this is the system under which tomatoes are produced, it behooves each grower to critically examine his method of fertilization to see if he is using the method that is yielding the most efficient production (Hester and Shelton 1939).

Potassium is important for stomatal movement in water regulation in plants. It is also required for carbohydrate metabolism and translocation, for nitrogen metabolism and protein synthesis, for regulating cell sap concentration, and as an enzyme activator. Potassium deficiency results in poor lycopene development in the fruit and in abscission of fruits as they approach maturity. Often, a heavy fruit load from a concentrated fruit may place such a stress on the plant that potassium deficiency symptoms occur (Kretchman *et al.* 1972). This deficiency frequently appears in the form of "yellow tops" about the plant. A frequent occurence in many fields is a very large set of fruit and an inadequate amount of potash to produce both the foliage and fruit.The foliage is often sacrificed, thus leaving the fruit exposed to the hot sun. When the weather becomes abnormally hot, the fruit is scalded or sunburned as it ripens and may spoil before it is ready to pick. Therefore, good-quality fruit depends on adequate and proper potash fertilization (Hester and Shelton 1939).

Calcium, magnesium, and sulfur are nutrients of minor importance. Calcium and magnesium can be purchased cheaply in the form of liming materials. It is difficult to measure the importance of calcium and magnesium on the quality of the fruit. Calcium and magnesium serve two functions in the soil: (1) they neutralize the acidity of the soil and (2) they serve as nutrients for the plant (Hester and Shelton 1939). Magnesium is essential to chlorophyll formation in the plant. Sulfur is essential to plant growth.

A number of trace elements are known to be necessary for plant growth. Some of these are iron, boron, manganese, copper, and zinc. Boron and manganese are needed only in small quantities by the plant.

SOIL TESTING

It can therefore be seen that soil and its ultimate composition are more than significant in determining the eventual success of growing tomatoes. For this reason, soil analyses carefully made from representative soil samples and properly interpreted can serve as useful guides in providing the required level of plant nutrients. Tomato growers who fail to have their soil analyzed are overlooking and important factor in the production of a profitable crop.

Soil testing will indicate what nutrients are necessary to add for proper plant growth and/or which nutrients are partially depleted or in short supply. These tests should include pH, magnesium, phosphorus, and potassium. Growers in sandy soils should pay particular attention to the need for lime whereas growers in heavy soils should pay particular attention to the phosphorus levels (Wittmeyer 1964). The measurement of pH will indicate whether and to what extent liming is desired to neutralize excess soil acidity and to raise the pH to an optimum level between 6.0 to 6.5.

LIMING

Land that is very acid in reaction (pH 5.0 or less) should be limed before planting tomatoes. Results obtained in experiments show that yields have been increased as much as 50% by liming very acid or calcium-deficient tomato land. Soil having a pH value of 6.0 to 6.5 is mildly acidic and is optimum for tomatoes. On soils with a pH of 5.0 or less, tomatoes are benefited markedly by an application of 1 or 2 tons of finely ground limestone. In soils deficient in magnesium, dolomitic limestone, which contains both calcium and magnesium lime, should be used. Some soils are naturally too alkaline for tomatoes. This is an important problem in California where whole areas or sometimes just spots in the field that are alkaline must be avoided or thoroughly leached before cropping (Porte 1952).

APPLICATION OF FERTILIZERS

As previously stated, which and how much fertilizer to apply should be a direct consequence of soil analysis. It has been found that tomatoes respond to liberal applications of fertilizer, but here again, as with lime, an excess can cause trouble (Keirns and Wittmeyer 1951). Most commercial fertilizers are combinations of three prime nutrients, with variable ratios of each nutrient. These are combinations given in percentages of nitrogen, phosphoric acid, and potash; or 100 lb of fertilizer with an analysis of 4−16−16 contains 4 lb of nitrogen (N), 16 lb of phosphoric acid (P_2O_5), and 16 lb of potash (K_2O). Fertilizers are also available that contain small amounts of iron, zinc, manganese, and other minor soil elements if it is found that the soil is deficient in these.

Applying the fertilizer before plowing continues to be the most popular single method of application. According to data compiled from the Top-Ten Club in Ohio, fertilizer usage by growers has increased steadily since 1947. The data in Table 2.3 summarize some of the changes over the years (Wittmeyer *et al.* 1971). They represent the basic fertilizer application made either before or at planting time (Wittmeyer 1971). Of the 82 growers, 35 used nitrogen as a side-dress application after the plants were established; 5 additional growers applied other fertilizer (but no nitrogen) as a side-dress application; 25 growers applied a soluble fertilizer with pesticide applications (Wittmeyer 1971).

TABLE 2.3. BASIC FERTILIZER APPLICATION

Fertilizer Constituents	lb per Acre				
	1955	1960	1965	1970	Suggested 1980
Nitrogen (N)	35	46	60	68	60
Phosphorus (P_2O_5)	106	147	155	167	175
Potassium (K_2O)	109	132	133	209	250
Total	250	325	348	444	485

It should be noted that tomatoes that are to be harvested by machine should be fertilized differently from those that are to be hand harvested. The uniformity of fruit maturity is greatly influenced by fertilization; thus, individual consideration must be given to each method. The principal difference is the rate and time of nitrogen application, since nitrogen influences the uniformity and rate of tomato maturity. The uniformity of fruit maturity is of prime importance in destructive harvesting; thus, late applications of nitrogen should not be made if the crop is to be mechanically harvested. In practice, the nitrogen supply should be depleted as the majority of fruit on the plant approaches maturity (Angell 1970).

Another consideration in fertilizer application is the method appropriate for direct-seeded or transplanted tomatoes. A broad generalization for each follows. It should always be remembered that no two growers' soils are exactly alike; consequently, each will require different rates of fertilizer nutrients as dictated by proper soil analysis. The total amount of nitrogen for the season will range between 30 and 100 lb per acre, depending on the cultivar of tomatoes grown, soil type, previous crops grown, and soil fertility level. The following examples illustrate two extreme situations. Intermediate conditions require intermediate rates of nitrogen.

1. Soils that are medium to heavy, with fertile land previously planted to pasture or legume cover, should use nitrogen at a rate of 30 lb per acre, applied before seeding. In addition, a band of fertilizer should be placed 1 to

TABLE 2.4 — SYMPTOMS OF NUTRITIONAL DEFICIENCIES*

Nutrient	Symptoms
Nitrogen	Plant may be stunted, light green foliage, older leaves turn chlorotic, plant may die.
Phosphorus	Stem, leaf veins and petioles turn reddish-purple.
Potassium	Older leaves appear chlorotic between veins, leaf margins may show burning and leaf roll.
Calcium	Young leaves become malformed, yellow to brown at margins.
Manganese	Young leaves appear mottled and chlorotic between veins.
Magnesium	Older leaves turn chlorotic between veins, young leaves may curl, brittle and dry up.
Iron	Young leaves develop chlorosis between veins.
Zinc	Leaves become chlorotic between veins.
Boron	Growing point turns yellow and dies. Leaves develop a blotchy appearance.
Molybdenum	Older leaves turn yellow and leaf margins curl up.
Copper	Leaves turn bluish green and curl up. Plant is stunted and chlorotic.
Sulfer	Older leaves turn light green. Stems may be woody and spindly.

*Taken from J. C. Watterson, *Tomato Diseases,* Petoseed Co., Inc.

2 in. directly below the seed. The band should not be placed more than 1 in. to the side of the seed because the first taproot goes straight down and may miss the fertilizer. The fertilizer and nitrogen should be applied just before planting. At the two to three true leaf stage (after thinning), a side-dress of 50 to 70 lb of nitrogen may be applied. Additional applications of nitrogen should not be made unless absolutely necessary. The remaining fertilizer should have been previously applied as a plow-down and/or a disk-in application.

2. Soils that are very light and sandy and low in fertility and organic material should have an application of 60 lb of nitrogen per acre. On such types of soil it may be desirable to use 25 to 50% of the nitrogen as a side-dressing, to be applied between the time the plants have two to three leaves and the time the flowers in the first cluster are open.

On many soils all the fertilizer may be applied before planting or as a side-dressing of N, or $N-P_2O_5-K_2O$ may be applied at the first cultivation. Additional nitrogen should not be applied thereafter except in sandy or light soils where heavy rains may produce leaching action. Should this be the case, 25 lb. of nitrogen can be added no later than 4 weeks after transplanting.

Regardless of whether tomatoes are to be direct-seeded or transplanted, or grown for machine harvesting or manual picking, the amount of fertilizer to be used on any given area or field cannot be subjectively measured; rather, it must be objectively measured. However, the fertilizer ratios (N−P−K) most generally used for tomatoes are 1−2−1, 1−2−2, 1−3−1, and 1−4−1. The 1−2−1 and 1−2−2 are used mainly on lighter soils when the tomatoes are grown in rotation with other cultivated crops. For sandy soils the 1−2−2 ratio is preferable to the 1−2−1 ratio. The 1−3−1 and 1−4−1 ratios are recommended mainly for tomatoes grown on loam, silt loam, and clay loam soils.

STARTER SOLUTION

Tomato plants suffer considerable shock when removed from the shelter of the nursery and transplanted in the field. Transplanting solution, or starter solution, provides available nutrients when and where they are most needed and stimulates root development, thereby assisting young plants to withstand shock and to become established in their new environment. This, in turn, will promote earlier maturity and heavier yield, both important considerations (Butler and Kerr 1952).

Starter solutions consist of solutions of water-soluble fertilizers or previously prepared liquid solutions that are generally high in phosphate content. Rates of 3 to 6 lb per 100 gal. of water have been suggested; 0.5 to 1 pint of the solution is applied around and in the soil where plants have been transplanted. The following starter solutions have been used in the past and exemplify common rates and types.

1. Ammo-Phos A and potassium nitrate: Dissolve 1 oz of Ammo-Phos A and 10 oz of potassium nitrate in 50 gal. of water (Hepler *et al.* 1950).
2. Diammonium phospate: Dissolve 3 lb of diammonium phosphate in 50 gal. of water (Hepler *et al.* 1950).
3. Ammo-Phos and potassium nitrate: Dissolve 2⅔ lb of Ammo-Phos plus 1⅓ lb of potassium nitrate in 50 gal. of water.
4. Commercial fertilizer (4−16−4): Dissolve 8 lb of 4−16−4 fertilizer in 50 gal. of water.
5. Commercial fertilizer (10−52−17; 10−55−10; 21−53−0): Dissolve 3 lb of the fertilizer in 50 gal. of water. A similar analysis of high-phosphorus fertilizer can be substituted (Anon. 1968).

When starter solutions were first introduced, trouble was often experienced because the insoluble portion plugged the water lines of the transplanter. This necessitated the preparation of a stock solution by suspending a bag of fertilizer in water, and then discarding the insoluble portion remaining in the bag. The introduction of complete forms has eliminated this difficulty. The required quantity of starter is merely dissolved and then added to the water in the tank on the transplanter (Butler and Kerr 1952).

FIGURE 2.3 — CULTIVATION OF TWIN ROW PLANTING OF TOMATOES

In addition to the incompletely and completely soluble forms, starter fertilizers may be purchased in liquid form. Although this form is certainly more convenient to use than the incompletely soluble form, it is, on the basis of plant food content, the most costly to use (Butler and Kerr 1952).

As a final word about starters, growers sometimes neglect to use transplanting solutions, or even water, in their transplanter, believing the soil to have ample moisture. Actually, if a soil is too wet to use transplanting solution, it is too wet to transplant tomatoes. When used, a certain amount of caution should be exercised. Excessive amounts of starter solution can do more harm than good as the growth of tomato plants can be stunted.

CULTIVARS

New varieties/cultivars are being developed rapidly to improve upon existing cultivars as to yield, quality, and cultivars more suitable for mechanical harvesting and handling systems. These new cultivars vary in shape from a flat round, to a true round, square round, oblong, pear, or oxheart and many variations in between. See Figure 2.4. Further, they vary widely in size and, hopefully, more disease and insect resistance. Their internal qualities vary widely, due in part to the needs and uses of the industry. The following are suggested guidelines to consider when developing and using new cultivars for processing.

1. Cultivars should be uniform in setting fruit and uniform in ripening with ability to set fruits over a wide range of temperture and climatic conditions.
2. Cultivars should be fully resistance to all tomato diseases, insects, and disorders.
3. New cultivars must be adaptable to mechanical harvesting and bulk handling.
4. All tomatoes for processing must be free from blossom end scars and cracking.
5. Tomatoes must be stemless when removed from the vine with stem scars less than ¼th of an inch in diameter. Further the stem scar should not brown during processing.
6. Tomatoes for peeling should be round to oval in shape, but shape may vary for juice or crushing and products manufacture.
7. Fruit size should be uniform with no fruits smaller than 50 grams and none larger than 90 grams.
8. Tomato total solids content should be in excess of 5.5% and preferably upwards to 8.5%.
9. Tomato soluble solids content (Brix value) should be in excess of 4.5% and preferable up to 7.5%.

FIGURE 2.4 — SOME SHAPES OF TOMATO CULTIVARS

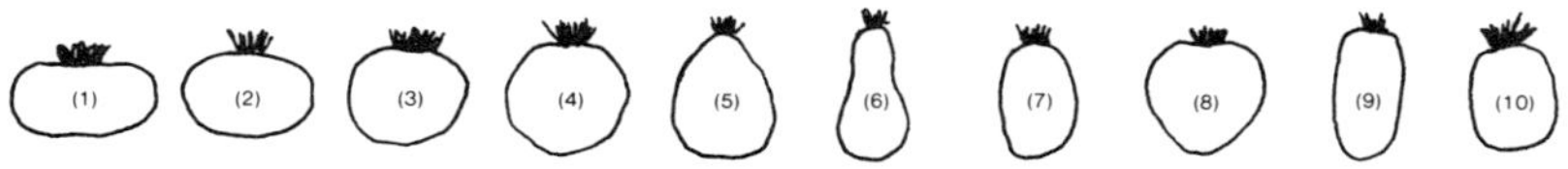

1. Flat
2. Oblate
3. Round
4. Square Round
5. Pear
6. Elongated Pear
7. Egg
8. Oxheart
9. Blocky Elongated
10. Blocky Round

---

10. Tomato water insoluble solids content should be in excess of 1% and increasing proportionally with total solids content.
11. Tomatoes should have a high acid (citric) content (Minimum of 0.35% and up to 0.55%).
12. Tomatoes should have a low pH content (Maximum of 4.4 and preferable all fruits with a pH of 4.2 or less).
13. Tomatoes should be high in Vitamin C content (in excess of 20 mg/100 grams).
14. Tomatoes for canning should have skin or peel that removes easily and completely without stripping. Also, tomatoes for canning should remain firm and whole (depending on style) after processing.
15. Tomatoes for juice manufacture should have a thick consistency (GOSUC value of 50 or more) after manufacture and the juice should not separate while in the can or jar during shelf life.
16. All tomatoes for processing should have a bright red glossy color after processing, regardless of the processed product.
17. All tomatoes should have typical tomato flavor before and after processing with no bitterness or stringent flavor.

The reader is referred to the appendix for a standard form when evaluating new or existing cultivars. It is also recommended that the reader take advantage of varietal work being conducted in each of the major tomato producing areas and obtain an up-to-date list of promising varieties for production and processing in that area. Further, the reader should study the varieties or cultivars being offered by the seed trade and request information on all of the above requirements plus yield and adaptability of their cultivars to any given region. Because of all of the above, no list of cultivars or varieties will be offered.

PLANTING

The time to set tomato plants or to direct-seed tomatoes in the field is governed by weather and soil conditions. Tomatoes grow best with average monthly temperatures of 70° to 75°F (21° to 24°C), but can be grown with average temperatures as low as 65°F (18°C) and as high as 80°F (27°C). Tomato growth is impaired by temperatures below 50°F (10°C), definite chilling injury occurring around 40°F (4°C). Temperatures that fall below 55°F (13°C) or rise above 95°F (35°C) for several hours when flowers are open at pollination time usually result in poor fruit set or no set (Porte 1952).

The question of when to plant depends on a number of factors: geographical and climatic conditions, transplanting or direct seeding, and mechanical or hand harvesting. Of these, temperature stability seems to be the most important factor. In California, where most of the crop is direct-seeded and mechanically harvested, seeding may start in January and continue into May. Harvesting may start in June and continue into October, thus scheduled plantings are essential. This is also important to ensure continuous delivery of quality fruit to the processor. It has been found that when soil temperatures reach 57°F (14°C) or more for 3 consecutive days, plantings may begin (Sims *et al.* 1968). In early plantings, seed development is slow. In later plantings, the seedlings grow more rapidly and often go from cotyledon to first true leaf stage in a day or two (Sims *et al.* 1968). The data in Table 2.4 show how temperature affects seedling emergence (Sims *et al.* 1968).

Consequently, the planting schedule may be based on calendar dates for mid- and late-season plantings. The number of days for seedlings to emerge to harvest are shown in Table 2.5 (Porte 1952). These data show that by allowing approximately 125 days for emergence to harvest, one can expect an orderly and continuous harvest, depending primarily on cultivar. In cooler climates, as in coastal valley areas, or in late plantings, the emergence to harvest day requirements may be over 130 days. The data in Figure 2.2 shows the relationship between date of planting and subsequent yield (Sims *et al.* 1968).

FIGURE 2.5 — EFFECT OF PLANTING DATES ON TOMATO YIELDS IN CALIFORNIA. SOLID BAR- RIPE. STRIPED BAR- GREEN.
From Sims et al. (1968)

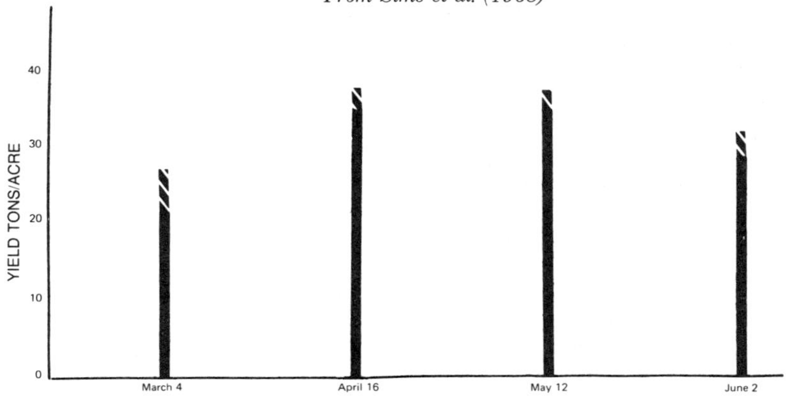

TABLE 2.5. EFFECT OF SOIL TEMPERATURE ON EMERGENCE

Average Soil Temperature (2-in. Depth)		
°F	°C	Days to Emerge
55	13	25
57	14	16
59	15	15
61	16	14
73	23	9
78	26	8
80	27	6

TABLE 2.6. EFFECT OF PLANTING DATE AND TIME OF EMERGENCE

Date Planted	Days to Emerge	Days from Emergence to Harvest	Date of Harvest	% Mature
March 4	25	124	July 31	80
April 16	14	123	Sept. 1	92
May 12	9	125	Sept. 23	93
June 2	6	126	Oct. 12	90

At the present time, Ohio is second to California in the acreage and total value of tomatoes for processing. Prior to 1960, Indiana was the second leading state, however, Indiana now grows about 1/10th the acreage prior to the early 1940's. Michigan has jumped into prominence as a tomato producing state, however, most of the tomatoes grown in Michigan are processed in Indiana or Ohio. Most of the acreage in the Mid-western states is from transplanted plants grown in the South or from plants grown in local greenhouses. About 15% of the acreage is direct-seeded in the Midwest, but not nearly to the extent that California acreage is direct seeded. Direct seeded plants may have the disadvantage of 10 to 15 days later in maturity over transplants. Very little irrigation is used in the Midwest or East as contrasted to California where nearly all the acreage is irrigated.

Direct-seeding of tomatoes in the Midwest is usually done from mid-April to early May. On light soils in warmer areas seeding can start earlier. On heavier soils in cooler areas where late harvests are desired, a delay until late April or the first week in May for direct-seeding is desirable. Transplants should be set as early as possible after any threat of frost. Growers are taking a chance with frost by planting late in April, but records indicate that growers are willing to do so in favor of a longer growing season.

Several precision planters for direct-seeding are now on the market that will equally space in the row the desired number of seeds (Johnson and Wilcox 1971) (Fig. 2.6). Research in clump versus single-seed planting indicates that clumps of two or three plants grow and develop as

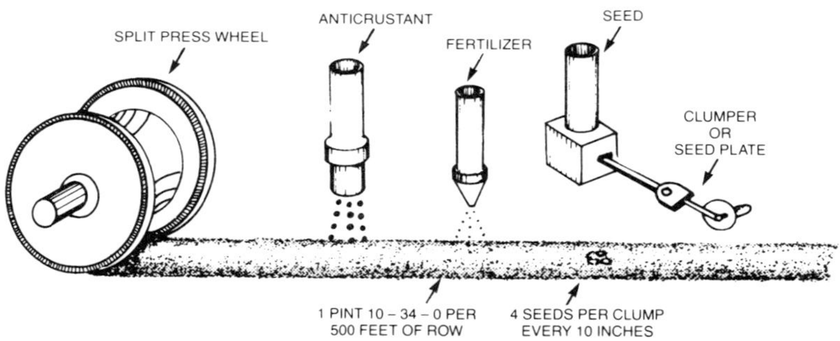

FIGURE 2.6 — EQUIPMENT FOR THE DIRECT-SEEDING OF TOMATOES
From Johnson and Wilcox (1971)

single plants with no sacrifice in yield (Johnson and Wilcox 1971). Therefore, it may be preferable to direct-seed in clumps of 4 to 7 seeds every 8 to 10 in. in rows on 5-6 ft centers. This results in a seeding rate of about 0.5 lb of seed per acre. The seed should be planted ½ to ¾ in. deep for optimum emergence. For single-row planting, populations of 10,000 to 20,000 plants per acre achieve uniform spacing. If twin row planting is desired, the rows should be spaced 16 to 24 in. apart on 6-ft centers, with seed clumps planted every 9 to 18 in. apart at a ¾ in. depth. Should thinning become necessary, the plants are thinned when they are in the two to three true leaf stage. This can be doen by hand or by mechanical thinners (Angell *et al.* 1971). After thinning, the field should be irrigated if the weather is dry and hot.

In addition, planters should be modified if necessary to allow for application of starter fertilizer and anticrust-resistant material as part of the seeding operation. The starter fertilizer can be applied in either a solid or liquid form while the anticrustant serves to facilitate uniform seedling emergence through crusty soil.

A better method of seeding is to pregerminate the seeds and mix with an anticrustant and source of nutrients for the seedlings (Anon. 1980). One mixture that has been successfully used is 1 bushel of Magamp (7–40–6, medium granules) with ½ oz tomato seeds. Use 3 to 4 quarts of water to moisten and mix. One bushel will plant about 600 clumps using ¼ cup per plug.

Transplanting is the process in which young, immature tomato plants are removed from their native localities and replanted where growth and eventual maturity are beneficial to grower and processor. Plants for transplanting in commercial operations are obtained from a number of sources, that is, from local greenhouses, from the South, ie., Georgia, or from the processor's own plantgrowing facilities.

FIGURE 2.7 - TWIN ROW PLANTING ON RAISED BEDS

Plants are subjected to severe shock when transplanted even under the most favorable conditions. Anything that can be done to assist in their reestablishment will be reflected in early plant development and in increased early yield. Only vigorous, well developed plants should be used. The plants should be stocky, compact, and reasonably uniform in size. Certified plants of suitable cultivars, free of nematodes, diseases, and insect injury, should always be used. If plants have to be held for several days because of unfavorable soil or weather conditions, a storage temperature of 50° to 55°F (10° to 13°C) is preferable (Angell 1970). They should be handled carefully and set in well-prepared land. Final preparation of the land should be done just before setting the plants. This usually consists of harrowing and leveling the soil until a smooth, even, and deep plant bed is made and weeds are destroyed (Porte 1952).

When possible, transplanting should be done either during the afternoon or on a still, cloudy day. Before removing the plants from the bed, the soil in which they are growing should be thoroughly watered. Pulled and shipped plants without soil on the roots should be protected from direct exposure to sunlight and from drying winds. A moisture covering of burlap or peat moss over the plants while they are waiting to be planted will help prevent excess wilting.

As suggested earlier, starter solutions high in phosphate should be applied at transplanting. A uniform stand and spacing are necessary for achieving uniform maturity and maximum yield of marketable fruit. There is a varietal response to spacing as influenced by type of plants and richness of soil; thus, the optimum spacing for the variety being grown should be used. There are, however, two basic plant types and suggested in-row spacings (Harbage 1971):

1. Small, compact vine: Transplant 12 to 16 in. apart with 7000 to 9000 plants per acre. The between-row spacing should approach 66 in.
2. Medium, large vine: Transplant 16 to 18 in. with 6000 to 8000 plants per acre. The between-row spacing should also approach 66 in.

It should be remembered that in determining the in-row spacing, plant and transplanting costs must be considered. Some growers may wish to transplant with populations approaching anywhere from 10,000 to 15,000 plants per acre. This would result in between-row spacings of around 60 in. and within-row spacing of 9 to 10 in. Some growers have had success with twin-row transplanting. This method requires populations of 12,000 to 15,000 plants per acre. Spacings between rows are then moved to 66 in. with spacing between rows on beds at 14 in. The within-row spacings are set farther apart at 12 to 15 in.

Transplanting or direct-seeding is a matter of economics, field selection, soil type, and locality. Some advantages of direct-seeding are: 1) reduced

FIGURE 2.8 TWIN ROW PLANTING AFTER THE PLANTS ARE FULLY GROWN

cost/acre; 2) less chance of introduction of diseases; 3) greater flexibility of planting time, variety selection, and plant population. Therefore, with present technology, the method that gives best results must be utilized.

CULTIVATION

The principal purpose of cultivating tomatoes is to eradicate weeds (Porte 1952). Certain weeds that are related to the tomato, such as the horsenettle, are carriers of tomato diseases and not only should be kept out of the tomato fields but also should be destroyed on adjacent land. Besides controlling weeds, the purpose of cultivations is to loosen soil that has become compacted, thereby conditioning it to receive and to absorb rainfall and to supply soil microorganisms with air so that they may thrive and liberate plant food for the crop (Butler and Kerr 1952). Thus, after emergence, the cultivation of soil toward the plants, or hilling, is beneficial in three ways (Butler and Kerr 1952):

1. Many small weeds close to the plant row will be smothered
2. Tomato plants will develop roots farther up the stem
3. Surplus moisture will not collect under tomato plants where it encourages disease, but tends to run away from the plants and to collect between the rows

The first cultivation may be fairly close to the recently set plants, but later cultivations should be shallower and farther from the stems. Shallow cultivations, 1 to 2 in. deep, should be given tomatoes, especially the first month after field planting. A healthy tomato plant has an enormous spread of feeder roots close to the surface as well as at considerable depth in the soil. When these surface roots are partially destroyed by cultivating too deep or too close to the plant, fruit production will usually be decreased significantly. Cultivation should not be continued when the plants have spread out in the rows; otherwise the vines and the growing fruit as well as the roots will be injured by the cultivating equipment (Porte 1952).

Cultivation either by machine or by hand hoeing should be only often enough to provide effective weed control (Pierce *et al.* 1963). Therefore, cultivation should be reduced to a minimum for the soil type, the crop, and the field or weed conditions. If one or two cultivations serve the purpose, then more are an unnecessary expense. Actually, excess cultivation increases loss of organic matter, impairs soil structure, and lowers soil moisture. Frequent cultivation may so deplete organic matter that the soil's water-holding capacity is materially impaired (Butler and Kerr 1952). It was once thought that cultivation conserved moisture. In reality, deep and frequent stirring of the soil may speed up water losses by exposing fresh amounts of soil. Also, the impairment of soil structure from unnecessary cultivation may be of more importance than moderate weed competition (Butler and Kerr 1952). Weeds can be controlled by the use of chemicals. Data from the various State Experiment Stations or Research Centers for the best-recommended chemicals for use in the tomato field for weed control should be obtained and used.

Finally, beds should be reshaped after cultivation. The beds should be kept level and free from clods and stone (Angell 1970). Whenever tomato vines are wet with dew or rain, they should be allowed to dry before cultivating, hoeing, or handling in any way. Such operations tend to spread the spores of various tomato diseases. Brushing against the vines while walking through the field when they are wet is likely to increase the spread of diseases. Timely cultivation can decrease necessary irrigation and be the first step in disease and insect control. After the crop reaches the stage where picking of the fruit begins, very little cultivation will be required.

WEED CONTROL

Weed control can be accomplished by a combination of mechanical cultivation and chemical control methods. In cultivation, weeds can be removed by various types of cultivators. Another method that is used extensively where seasonal farm help is available is hand removal of weeds by hoeing; however, chemical control is used more extensively today. Detailed herbicide application (rates, timing, and equipment) are specific for the types of weeds prominent. Specific recommendations are available from your Cooperative Extension Service or the Agricultural University in the area.

IRRIGATION

In sections of the country having an abundant rainfall during the growing season, tomatoes for canning and manufacturing may be grown without irrigation. During seasons of limited rainfall, or for short periods during which precipitation is below normal, irrigation gives a decided advantage in maintaining growth and obtaining good yields. In the East the overhead or sprinkler system is most commonly used, although on land that is level or only slightly sloping the furrow or flow method may be used. No irrigation is given late in the season after the crop has reached an advanced stage of maturity, for fear of causing fruit rots and cracks. In regions where crops are grown entirely under irrigation, the preparation of the land for planting, the width of rows, and other factors are made to conform to the approved methods of irrigation (Beattie *et al.* 1942).

Irrigation plays an important role in attaining uniform maturity, and it should be available for all the tomato acreage if needed. An adequate supply of water is necessary during the early plant growth, fruit set, and fruit enlargement periods. If water becomes a limiting factor during any one of these periods, optimum uniformity of fruit maturity cannot be achieved. Once the fruit has attained size, the only water required is for maintenance of the plant. Since rainfall cannot be controlled, it must be considered as a factor that can have significant effects on uniformity of fruit maturity and harvesting operations.

SUN-GARD

Several firms are now manufacturing and selling a material (Monterey CropWhite, Ortho Sun Shield) to be sprayed on tomato fields to protect the plant and fruit from sun burning. The material is a white, inert, nonabrasive powder, clay product mined near Bishop, CA. It is mixed with water at the rate of 60 to 150 lbs per acre in 30 to 200 gals of water along with an adjuvant at the rate of 2 lbs per 100 gallons to the diluted spray. The material may be applied from the ground or by air equipment. The tomato is covered with a very thin white coating which is easily removed during the washing of the tomatoes.

DISEASES

Tomato diseases are of two general types: parasitic and nonparasitic. Parasitic diseases are those caused by living organisms, chiefly bacteria and fungi, and by viruses. The parasitic group includes most of the common and serious tomato diseases (Barksdale *et al.* 1972). Nonparasitic diseases are due to unfavorable environmental conditions, such as excessive moisture or drought, extremes of temperatures, and lack or excess of certain mineral elements in the soil (Barksdale *et al.* 1972).

SYMPTOMS OF EARLY BLIGHT OF TOMATO CAUSED BY *Alternaria solani* (Ell. and G. Martin).

The fungus infects the young plant stem resulting in poor stand (UPPER). Advanced lesion on the mature stem (LOWER).
See description page 44.

Early development of black lesion, usually associated with yellowing (UPPER). Heavy infection of foliage and petiole (LOWER).
See description page 44.

SYMPTOMS OF EARLY BLIGHT (Continued)

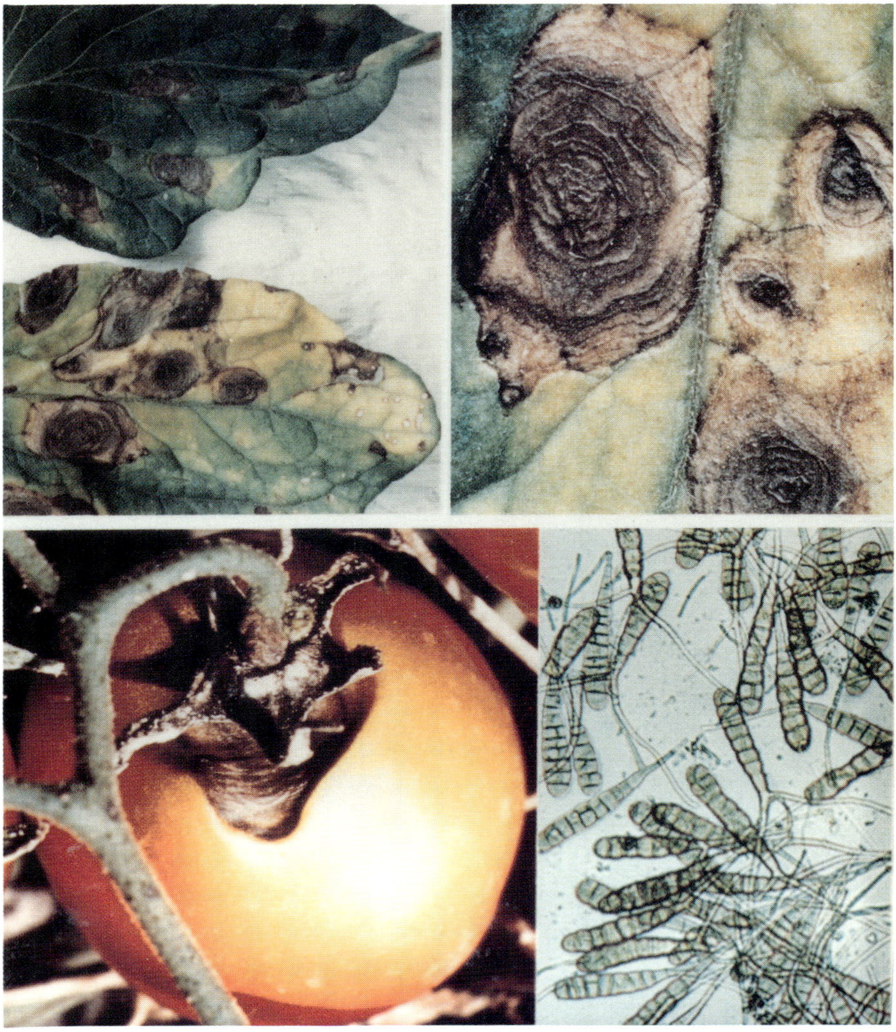

The lesions get larger forming concentric rings (UPPER). Fruit infection usually starts from the pedicel end (LOWER).
See description page 44.

The lesions clearly showing concentric rings with no white centers or specks (UPPER). The large conidia of *A. solani* (LOWER).
See description page 44.

EARLY BLIGHT

(Alternaria solani (Ell. and G. Martin)

Early blight occurs quite frequently in the Midwest. This fungus can cause stem canker or collar rot that greatly damages young seedlings and transplants in the field. These lesions are circular to elongated spots which occasionally show concentric markings. Large spots can occur on the stems, causing partial girdling, and stunting.

In the field, after plants are established, small, irregular, brown, dead spot usually first appear on the older leaves. The spots enlarge until they are one-fourth to one-half inch in diameter; as they enlarge, they commonly show ridged, concentric rings in a target pattern. These spots usually are surrounded by a diffuse yellow zone; when spotting is abundant, the entire leaf often is yellowed. Some spotting of the older leaves may appear early in the season, but the greatest injury usually occurs as the fruit begins to mature. If high temperatures and humidity occur at this time, much of the foliage frequently is killed before the end of the season.

Early blight also infects pedicel and may cause some blossom drop. On fruits, it can cause dark, leathery, sunken spots at point of attachment to the pedicel. These spots reach a considrable size and may show concentric markings like those on the leaf. The dark, dry decay extends to some depth into the flesh of the fruit. Infected fruits frequently drop; if they reach maturity, they are not fit for market or canning.

SEE PHOTOS PAGES 42-43

LATE BLIGHT

(*Phytopthora infestans* (Mont.) (Dby.)

Late blight, although not seen frequently on midwest tomatoes, can cause severe defoliation of foliage and destructive rot of fruits. At first, greenish-black, water-soaked patches appear on the older leaves. These spots enlarge rapidly, and, in moist weather, sometimes a white, downy growth of the fungus develops on their lower surfaces. The stems also may have water-soaked, brown areas similar to those on the leaves. Under cool nights and moderately warm days with abundance of moisture, the infection spreads so rapidly at times that almost all the foliage is affected, and the plants look as though they had been damaged by frost. Dry and hot weather conditions usually arrests the spread of the disease.

The first symptom on the fruit is a grayish-green, water-soaked spot. This spot enlarges until it may cover half the surface. The spot becomes brown and has a firm, corrugated surface that occasionally shows narrow, zonate markings. The margin of the spot may be somewhat indefinite, but usually it is slightly sunken where decayed and healthy tissues join. Under moist conditions, a white, downy growth of the fungus appears on the fruit.

SEE PHOTOS PAGES 46-47

SYMPTOMS OF LATE BLIGHT ON TOMATO CAUSED BY *Phytopthora infestans* (Mont.) DBy.

Irregular brownish lesions appear on the leaflet (UPPER) soon spread and kill the leaflet (LOWER).

See description page 45.

The white downy growth will be present under moist conditions (UPPER). Complete collapse of the leaf (LOWER).

See description page 45.

SYMPTOMS OF LATE BLIGHT (Continued)

The fungus will infect the flowers and shoot tip (UPPER) and soon completely defoliates the plant (LOWER).

See description page 45.

The initial grayish-green water soaked symptoms turn brown and hard with a rough surface (UPPER). The sporangia of the fungus (LOWER).

See description page 45.

SYMPTOMS OF SEPTORIA ON TOMATO CAUSED BY
Septoria lycopersici Spreg.

Very small circular lesions appear which soon will develop white centers (UPPER) and will spread rapidly to the rest of the foliage (LOWER).

See description page 50.

Portion of a leaflet or entire leaf may die (UPPER). In the middle of the white lesions minute black specks will develop (LOWER).

See description page 50.

SYMPTOMS OF SEPTORIA (Continued)

Close-up of a lesion shows black specks called pycnidia (UPPER). The photomicrograph of pycnidium shows the elongated conidia (LOWER).

See description page 50.

The flowers are infected by the fungus usually at a very advanced stage (UPPER). The lesions vary a great deal in size (LOWER).

See description page 50.

SEPTORIA LEAF SPOT

(*Septoria lycopersici* Spreg.)

Septoria leaf spot can be one of the most destructive tomato leaf diseases. The disease may occur in plants of any age, but it usually becomes most evident after plants have begun to set their fruit. First infection usually is found on the older leaves near the ground. Water-soaked spots, which often are scattered thickly over the leaf, are first noted. These spots soon become roughly circular and have gray centers surrounded by darker margins. Later the centers show tiny dark specks in which the spores of the fungus are produced. The spots are smaller and more numerous than those of early blight; usually they are one-sixteenth to one-eight inch in diameter. However, on larger leaves, the lesions may be bigger, and can be easily confused with early blight except for the presence of pycnidia (small fruiting-bodies) in the middle of the spots.

Severe infection causes the leaflet to completely dry out. When conditions favor infection, there is a progressive loss of foliage until only a few leaves are left at the top of the stem, and the fruits are exposed to sunscald. The fruits are rarely affected, but there may be spotting of the stem and blossoms.

SEE PHOTOS PAGES 48-49

BACTERIAL SPECK

(*Pseudomonas* (Syringae P.V.) *tomato* (Okabe))

Bacterial speck has become quite prevalent in midwest. On the leaves, dark brown spots, sometimes with a yellow halo cause twisting and distortion of leaves. Also causes discrete dark brown spots on stem. The general yellowing found with bacterial spot usually does not occur with speck.

Both speck and spot occur on the flower and can cause blossom drop. Field observations indicate flower abortion may be less with speck than spot.

Symptoms on fruit may help to differentiate spot and speck. Speck fruit lesions are smaller than spot lesions, and are surrounded on green fruit by a dark green halo. Spot lesions develop slightly deeper into the fruit whereas speck lesions are very superficial. Since both bacterial speck and bacterial spot may occur on the same fruit, the speck injury is frequently mistaken from small lesions of bacterial spot.

SEE PHOTOS PAGES 52-53

SYMPTOMS OF BACTERIAL SPECK OF TOMATO CAUSED BY *Pseudomonas syringae* pv. *tomato* (Okabe).

Dark, irregular lesions on young plants (UPPER). These lesions often distort the leaflet resulting in cupping of the leaflet (LOWER).

See description page 51.

Isolated lesions usually will have a yellow halo around them.

See description page 51.

SYMPTOMS OF BACTERIAL SPECK (Continued)

Flower infection and subsequent flower drop occurs after heavy infection of the foliage.

See description page 51.

Fruit lesions are black and small. Usually on the green fruit there will be a dark green halo around the lesion.

See description page 51.

EARLY (upper) AND ADVANCED (lower) SYMPTOMS OF TOMATO BACTERIAL SPOT CAUSED BY
Xanthomonas vesicatoria (Doidge) Dows.

Small water soaked lesions appear on under side of leaves (UPPER) which later form irregular dark spots with yellowing at the margin.(LOWER)
See description page 56.

Dark brown irregular spots appear on flower pedicel at very early stages (UPPER) which result in flower abortion (LOWER).
See description page 56.

SYMPTOMS OF BACTERIAL SPOT (Continued)

Small dark brown lesions on leaf peduncle and stem (UPPER). Later on they often coalesce to form larger lesions (LOWER).
See description page 56.

Small white lesions appear on green fruit with brown centers (UPPER) which form dark greasy spots without white margins (LOWER).
See description page 56.

BACTERIAL SPOT

(*Xanthomonas vesicatoria* (Doidge) Dows.)

Bacterial spot causes brown, water soaked, circular spots on leaves. They are rarely more than ⅛ inch in diameter The lesions often tend to cause necrosis (death) of leaf margins, but entire leaf can be infected. The bacterial spot lesions lack the concentric zoner found with early blight, and generally are darker in color. A yellowing is associated with the bacterial spot infections. Yellowing is usually seen not as a complete circle as in the case of bacterial speck, but as a path off to the side of a lesion or a group of lesions. The organism can cause severe blossom infection causing blossoms to drop. Stems of seedlings also may show spotting.

On the fruit, the early symptoms appear either as a white spot or a water-soaked spot. As the fruit matures, these spots become bigger and change to a dark, leathery, scab-like, slightly raised on the edge and sunken in the center and this white halo spot will disappear unlike in canker. The epidermis of the fruit finally ruptures and curls back from the center of the spot. This is often the most characteristic symptom of the disease on the fruit. The bacterial spots are seldom deeper than halfway through the outer fleshy layer of the tomato fruit.

SEE PHOTOS PAGES 54-55

BACTERIAL CANKER

(*Corynebacterium michiganese* (E.F. Sm.) Jensen)

Bacterial canker is not a frequently observed disease in the midwest; however, once when it occurs it can cause severe damage. There are two phases to this disease: the primary or vasuclar infection and secondary or foliage, flower and fruit infection. The primary infection usually starts on young plants in which case they may die in the field or show stunting and unilateral wilting. If the plant is heavily infected, the whole plant may be wilting and can be confused with bacterial wilt. Such plants when cut length-wise show a shiny reddish brown streak along the vascular tissue. Later, the whole stem may be infected, and the pith also becomes dark brown and the skin may break open to cause canker that gives the disease its name.

Many times in a field one can observe only the secondary phase of the disease. This can be characterized by a general blight appearance in the field. The infection spreads from the leaf margins to the internal tissue. The stem and pedicel can also be infected causing rusty flecking.

The most diagnostic feature of bacterial canker, however, is the presence of fruit lesions. They appear as snowy white raised spots, which can be rubbed off at the early stage. However, later on they are imbedded in the skin, and the center of the spot turns tan in color, slightly raised and will retain a distinct yellow halo around the spots, which will appear like a birds eye. It is possible that only one of these phases can be seen at a given time in a field.

SEE PHOTOS PAGES 58-59

SYMPTOMS OF TOMATO BACTERIAL CANKER CAUSED BY *Corynebacterium michiganense* (E.F. Sm.) Jensen

Sunken brownish lesion with a white halo on the cotyledons or on the first leaf (UPPER). Seedlings showing typical unilateral wilt (LOWER).

See description page 57.

Commercial fields showing wilting and killing (UPPER). Close up of wilted and healthy plants in the field (LOWER).

See description page 57.

SYMPTOMS OF BACTERIAL CANKER (Continued)

Leaflet showing unilateral wilt (UPPER). Heavy infection of the pedicel showing the sunken flecks (LOWER).

See description page 57.

The stem showing internal browning (UPPER) and typical birdseye spot on the fruit (LOWER).

See description page 57.

SYMPTOMS OF BACTERIAL WILT *(Pseudomonas solanacearum)* (E.F. Sm)

Healthy & Wilted plants (UPPER) the whole plant shows wilting at advanced stage (LOWER).
See description page 62.

Stem cut open to show the initiation of stem discoloration (UPPER) the whole pith area becomes brown at advanced stage (LOWER).
See description page 62.

SYMPTOMS OF SOUTHERN BLIGHT *(Sclerotium rolfsii)* (Sacc.)

Southern blight infected plants before planting (UPPER) closeup of infected plants showing white mycelium (LOWER).
See description page 62.

Infected plants from the field showing little or no growth (UPPER) mycelial growth in the field during wet conditions (LOWER).
See description page 62.

BACTERIAL WILT OR SOUTHERN BACTERIAL WILT
(*Pseudomonas solanacearum* E.F. Sm.)

Bacterial wilt or southern bacterial wilt is primarily a southern U.S. disease, but it occurs occasionally in the north when infected transplants are used.

The symptoms are rather rapid wilting and death of the entire plant unaccompanied by any yellowing or spotting of the leaves. If the stem of a wilted plant is cut across near the ground, the pith has a darkened, water-soaked appearance and there is a grayish, slimy exudate when the stem is pressed.

In later stages of the disease, decay of the pith, may cause extensive hollowing of the stem. These symptoms differ from those of fusarium and verticillium wilts, which do not cause sudden wilting or decay of the stems of older plants. Bacterial wilt causes no spotting of the fruits.

SEE PHOTOS PAGE 60

SOUTHERN BLIGHT OR SCLEROTIUM ROT
(*Sclerotium rolfsil* Sacc.)

Southern blight, or sclerotium rot occurs primarily in the southern United States where the organism attacks many crops. In the north, it is found sometimes in fields where southern-grown transplants have been used.

If infected plants are shipped from the south, careful examination will show a white mycelial growth in the plant crates. In the field severly infected plants usually die, however, slightly infected plants may show various wilt symptoms.

The first symptom on tomatoes is a general drooping of the leaves similar to bacterial or fusarium wilt. Wilting becomes more marked from day to day, and finally the plant dies without marked yellowing of the foliage. The stems show a brown decay of the outer tissues at the ground line. Frequently they are covered with a white fungus mat in which are embedded numerous small, light-brown bodies about the size of a cabbage or mustard seed. These are known as sclerotia and are characteristic of the disease.

The fungus also attacks the fruits where they touch the soil. It causes yellowed, slightly sunken areas that break open as the spots enlarge. The progress of the decay is rapid; the fruit soon collapses and is covered by the growth of the fungus.

SEE PHOTO PAGE 61

SYMPTOMS OF MAJOR MIDWEST TOMATO FRUIT ROTS.

Anthracnose (Colletotrichum) causes small saucer shaped lesions on ripe fruit (UPPER) which enlarge resulting in rotting of the fruit (LOWER).

See description page 63.

Black mold (Alternaria) causes irregular dark spots on ripe fruit (UPPER) which results in rotting at advanced stages (LOWER).

See description page 63.

ANTHRACNOSE

(*Colletotrichum phomoides* (Sacc.) Chester)

Anthracnose is the major fruit rot in the midwest. In the early stages, lesions are small, circular, slightly sunken, and water-soaked. Later they become darker and more depressed or develop concentric ring markings. Numerous dark pustules (fruiting structures) develop through the surface of the lesions. Under moist conditions, they become covered with cream to salmon-pink masses of fungus spores. In warm weather, the rot soon penetrates into the fruit and renders it worthless. In moist weather the conidia produced on the surface of the fruit are splashed to other fruits by rain or spread by pickers.

Fruits may be infected when green and small, but may show no evidence of spotting until they begin to ripen. The fruits become increasingly susceptible for development of lesions as they approach maturity. Fruits on partially defoliated plants seem to be quite susceptible to infection.

SEE PHOTO PAGE 64

BLACK MOLD (ALTERNARIA)

(*Alternaria tenus,* auct.)

Black mold, or alternaria rot, may occur wherever tomatoes are grown. The rot follows such fruit injuries as growth cracks, blossom-end rot, sunscald or anything which damages the fruit. The decayed area is brown to black in outward appearance. It may or may not have a definite margin; it has a flattened or slightly sunken surface. The lesions are firm; and the rot extends into the flesh of the fruit, producing a dark-brown to black, dry, corelike mass of decayed tissues. Dark-gray mold is often found in the cavities of the decayed tissues. Dense, velvety olive-green or black spore masses of the pathogen frequently grow over affected surfaces. Black mold, or alternaria rot, frequently develops from the stem scar in V-shaped lesions.

SEE PHOTOS PAGE 64

SYMPTOMS OF MAJOR MIDWEST TOMATO FRUIT ROTS.

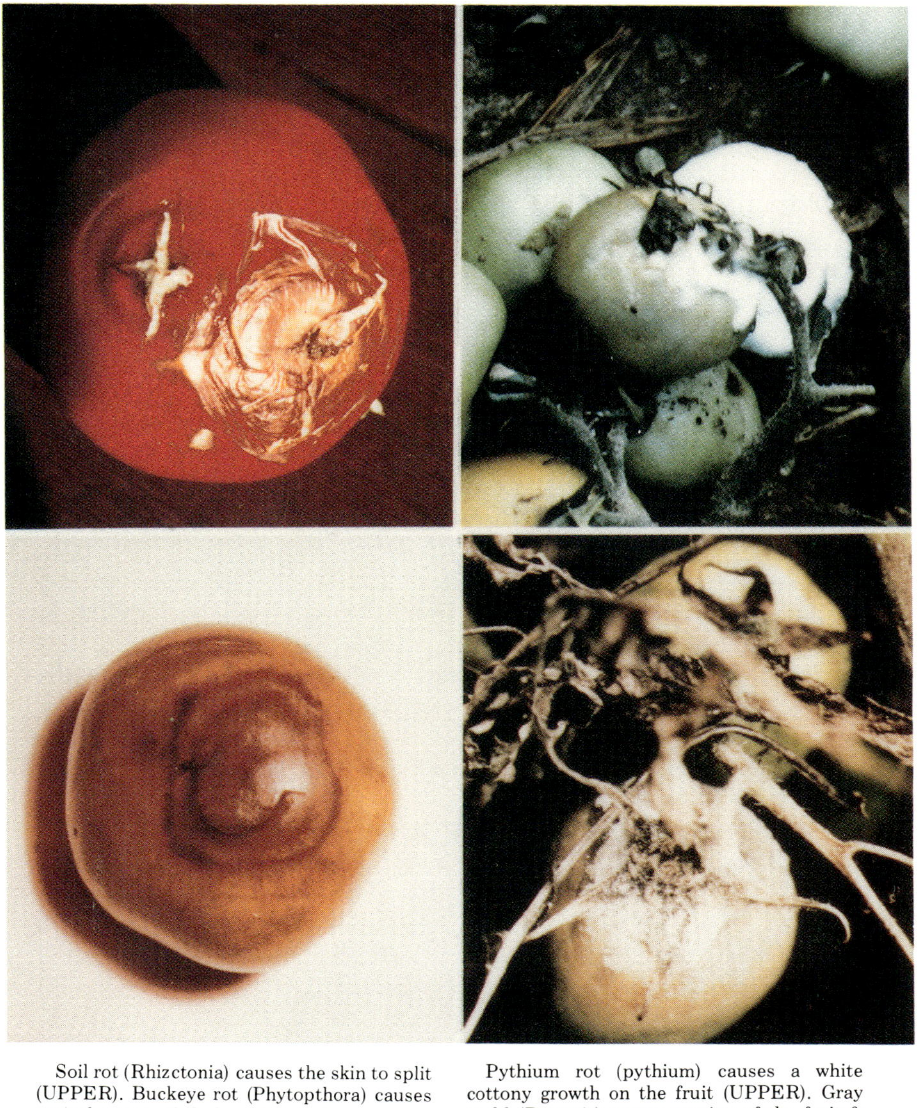

Soil rot (Rhizctonia) causes the skin to split (UPPER). Buckeye rot (Phytopthora) causes typical zones of dark and light brown areas (LOWER).

See description page 66.

Pythium rot (pythium) causes a white cottony growth on the fruit (UPPER). Gray mold (Botrytis) causes rotting of the fruit & pedicel (LOWER).

See description page 67.

SOIL ROT OR RHIZOCTONIA
(*Rhizoctonia solani* Kuehm)

Rhizoctonia or soil rot can occur in many areas. It develops on both green and ripe fruits. On green tomatoes, the first visible symptoms are small, circular, brown spots on the lower half of the fruits. Spots are one-fourth inch in diameter and larger usually showing very definite concentric-ring markings made up of alternating zones of light and dark-brown tissues. As the fruits ripen and the decayed areas enlarge, the concentric zones become less evident and may even disappear. On ripe fruits, the spots are moderately firm and reddish-brown, water-soaked areas. The definite margin, flattened surface, and, when present, the narrow, fairly regular concentric-zone markings, and the cracked skin on the decayed spots distinguish this disease readily from buckeye rot.

SEE PHOTOS PAGE 65

BUCKEYE ROT
(*Rhytophthora* sp.)

Buckeye rot occurs occasionally in the midwest, especially during periods of warm, wet weather. The rot is largely confined to fruits in contact with the soil; either green or ripe fruits may be infected. The first symptom is a grayish-green or brown, water-soaked spot that usually occurs where the fruit touches the soil. In warm weather the spot enlarges rapidly and may cover half, or more of the fruit. It may have no definite markings, but it usually has darker zonate bands. These markings give the disease the name "buckeye rot".

Fruits affected with late blight rot may have similar markings. The surface of the spot is firm and has a smooth and not sharply defined margin. This distinguishes buckeye rot from late blight whose spots have a roughened surface and are slightly sunken at the margin. Also buckeye rot differs from late blight in that the green fruits do not become soft for some time after infection and it affects only the fruits.

The fungus causing buckeye rot can penetrate the uninjured surface of the fruit, and infection occurs either where the fruit touches the soil or where soil is splashed on the fruit by rain.

SEE PHOTOS PAGE 65

PYTHIUM ROT
(*Pythium* sp.)

Pythium rot will sometimes cause serious losses during hot, wet weather. On mature-green tomatoes the rot appears as an extensive, firm water-soaked area that may eventually involve the entire fruit.

SEE PHOTOS PAGE 65

GRAY MOLD OR BOTRYTIS
(*Botrytis cinerea* Fr.)

Gray mold or botrytis has been a problem in greenhouse tomatoes for many years. Leaf infection can occur, but not too frequently. Infected leaves show typical light-tan or gray spots. The infected areas become covered by a growth of the fungus, and leaf collapses and withers.

On the fruits, water soaking and softening of the tissues at the point of infection are first noted. These spots are irregular in shape and may be an inch in diameter. Usually they are grayish or yellowish green with lighter margins. A darker gray growth of the fungus later develops on the surface of the fruit. Finally, the fruit is often destroyed by watery soft rot. The grayish mold may develop where the lesions have cracked or it may develop sparingly over the surface at the center of the more advanced spots.

SEE PHOTO PAGE 65

FIGURE 2.9 - PEAR VARIETY OF TOMATOES

FIGURE 2.10 - ROUND VARIETY OF TOMATOS

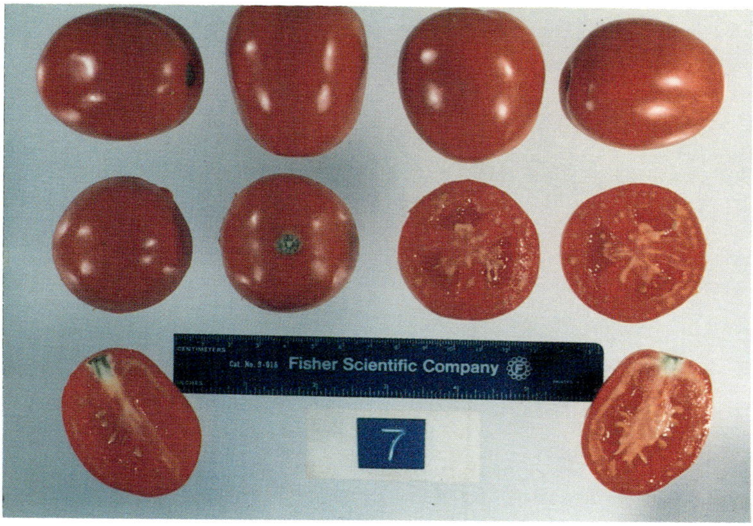

FIGURE 2.11 - NEW TOMATO CULTIVAR

FIGURE 2.12 - ETHREL TREATED FIELDS

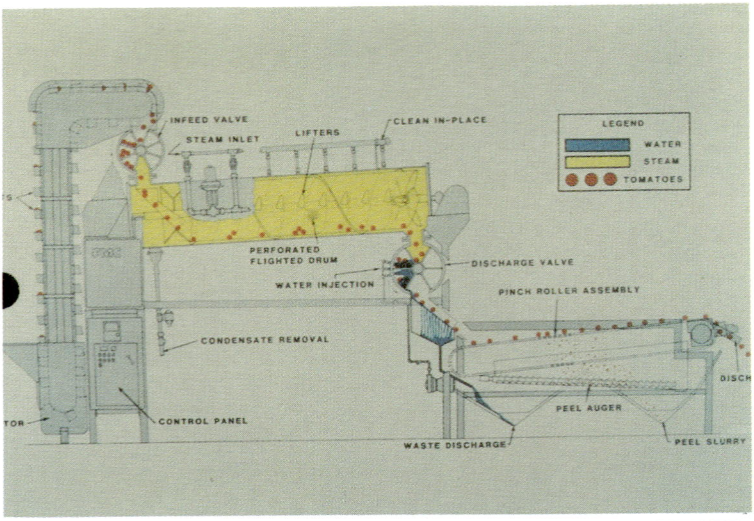

FIGURE 2.13 - FMC MODEL SP20 WET VACUUM STEAM PEELER

FIGURE 2.14 - SAMPLING TOMATOES FOR GRADE EVALUATION

FIGURE 2.15 - MECHANCIAL HARVESTING OF TOMATOES

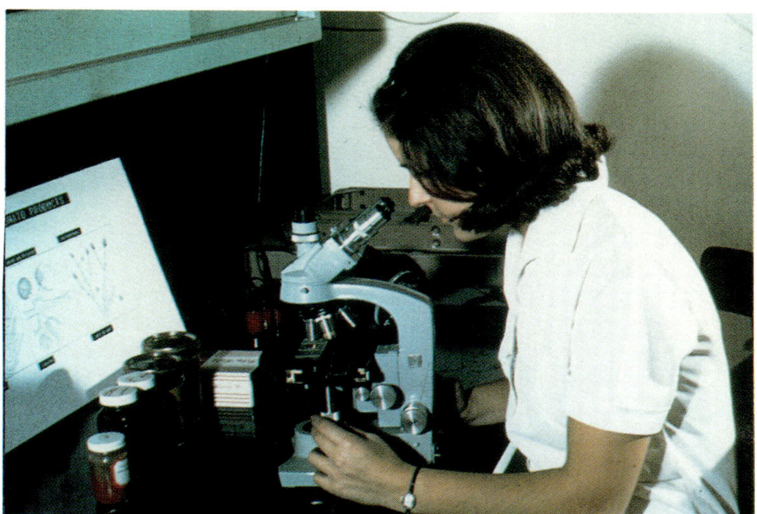

FIGURE 2.16 - MOLD COUNTING

Bacteria and fungi are microscopic organisms that obtain their food from the plant they attack or from decaying organic matter in the soil. They enter the plant through wounds and natural openings, or directly penetrate the epidermis. After entering the plant, they multiply within the plant tissues and produce specific symptoms, such as wilting of the plant, spotting or curling of the leaves, and decay of the fruits (Barksdale *et al.* 1972).

The bacteria affecting tomatoes are one-celled, rod-shaped organisms that multiply with great rapidity in the plant. They occur on the surface of diseased plants either as exudates or as a result of a breaking open of the diseased tissue and, when so exposed, are readily spread to other plants by splashing rain, insects, or field workers (Barksdale *et al.* 1972).

The fungi are more complex organisms with threadlike vegetative growth (mycelia) from which are usually produced various types of structures that bear seedlike bodies known as spores. In the presence of moisture spores can germinate and produce new infections. They are spread by wind, rain, drainage water, insects, and persons working among the plants (Sims *et al.* 1968).

Since bacteria and fungi are living organisms, they are destructive only when environmental conditions, particularly temperature and moisture, are favorable to their development and spread. Because tomatoes are grown under various environmental conditions, the disease of most importance in one region may be almost unknown in another (Barksdale *et al.* 1972).

Plant viruses are complex protein substances, that increase rapidly in the plant. The individual virus elements are too small to be visible except under the electron microscope. Viruses are highly infectious, and many of them are readily transmitted by any means that serve to introduce a minute amount of juice from a virus-infected plant into a light wound or abrasion in a healthy one (Barksdale *et al.* 1972).

The most common means of transmitting virus diseases are sucking insects, particularly aphids, and the brushing against, handling, or pruning first of diseased and then of healthy plants. Some viruses, such as those causing curly top and spotted wilt, however, are transmitted only by certain species of insects and are not spread by contact with the plants. Such diseases are serious only in regions where conditions permit the existence of the insect carriers in large numbers. A few viruses are transmitted in the seed of certain of their host plants, but such transmission is comparatively rare. Most viruses do not survive in the soil, but a few do. Certain viruses, though much alike in their chemical and physical properties, do not produce exactly the same symptoms on all species of plants. Such viruses are known as strains of a single virus (Barksdale *et al.* 1972).

Ordinarily, a diseased plant cannot be cured; therefore, control must be based on prevention of the disease and of its spread. Many tomato diseases are not readily controlled after they are once well established in the field or greenhouse, but it is often possible to limit their occurence by preventing infection from contaminated seed and soil or from weeds that carry disease-producing organisms or viruses. When disease-free plants are set in clean soil, the likelihood of serious loss reduced.

Some of the more common diseases which lower yield and quality of tomatoes are fusarium wilt, early blight, anthracnose, fruit rot, gray leaf spot, and late blight. All these diseases are caused by pathogenic fungi and can be controlled by following a disease-control program that includes the application of fungicidal sprays. Some diseases such as soft rot can be affected by cultural practices such as too much nitrogen application rates (Bartz *et al.* 1979). An effective disease-control program for tomatoes involves a number of steps (Angell *et al.* 1971). Although specific guidelines should be established for specific cases, there are a number of general rules.

1. All old tomato vines should be plowed under as soon as possible after harvest. Some of the fungi that cause diseases live from year to year on old tomato vines. Destroying these vines as soon as possible will reduce sources of infection for the next crop.
2. Use healthy, disease-free certified transplants. If direct-seeding is practiced, be sure to use certified disease-free seed.
3. Practice crop rotation. Tomatoes should be planted in fields in which potatoes, tomatoes, peppers, or eggplants have not been grown for 3 or 4 years.
4. Use disease-resistant varieties. This is the most effective way to control fusarium wilt.
5. Apply fungicides. All the tomato diseases previously mentioned with the exception of fusarium wilt can be controlled with a complete fungicide program. The most important point to remember with respect to fungicides is that their most important role is to prevent disease from getting started and spreading. They do not cure plants already infected. The important points which must be followed in an effective fungicide program are

    a. The proper use of given fungicides is necessary for controlling of specific diseases (see your Cooperative Extension Service for recommendations in your area).

    b. Make applications soon enough to protect plants before they are infected; on transplants, start the fungicidal sprays when crown fruit are about 1 in. in diameter or approximately 6 weeks after transplanting; on direct-seeded tomatoes, start sprays when plants are about at the similar stage of growth, as previously mentioned, or approximately 10 to 12 weeks after emergence.

    c. Make applications often enough to cover new growth and replace fungicide washed off by rain; a 7-day schedule is suggested with a total of at least eight or more applications during the season.

    d. Obtain complete coverage of entire plant, i.e., all leaves, stems, and fruits.

The data in Table 2.7 and Fig. 2.17 show the respective area and symptoms that are associated with some common tomato diseases.

TABLE 2.7 —SYMPTOMS FOR SOME OTHER TOMATO DISEASES

Disease	Part of Plant Affected	Causative Organism	Symptoms
Fusarium Wilt	Leaves, shoots	*Fusarium oxysporum* f. sp. *lycopersicum*	Yellowing, wilting, loss of leaves, dark brown discoloration, death of plant
Verticillium Wilt	Entire plant	*Verticillium albo-atrum* *Verticillium dahliae*	Yellowing, wilting of foliage
Damping-Off	Seed, seeding, stems, & roots	*Pythium alphanidermaum, P.ultimum, Phytophthora capsici, P. aprasitica, Rhizoctonia solani*	Seedling falls over & dies
Mosaic	Foliage	Virus	Mottled and distorted foliage
Aster yellows	Foliage	Virus	Stunted plant, purple & rolled leaves, death
Fruit Rots: Black mold, Soft Rot, Phoma Rot Buckeye Rot Cottony leak	Fruits	*Phthium sp phytophthora parasitica, Rhizoctonia solani, Alternaria alternati, Geotrichum candida, Phoma destruction, Erwina Carotovia*	Superficial flecking to brown black dry lesions, water soaked lesions, soft watery rot with foul odors, small sunken spots with concentric rings, soft water-soaked spots, white cottony growth

TABLE 2.8 — SOME PHYSIOLOGICAL DISORDERS OF TOMATOES

Name	Part of Plant Affected	Causual Agent	Symptoms
Blossom-End Rot	Fruit	Calcium deficiency,drought,and/or excessive ammonium nitrogern	Light tan lesions turning dark, sunken areas at blossom end of fruit
Blotchy Ripening/Gray wall	Fruit	High Nitrogen, low Potassium, high soil moisture, low light intensity, and other factors	Blotchy, brownish gray areas on green fruit. Vascular discoloration
Catface and cracking	Fruit	Environment	Concentric cracking of fruit on some cultivars
Sunscald	Fruit	Lack of shade	White, wrinkled flattened area
Blossom Drop	Blossoms	Adequate moisture, excess nitrogen	Loss of blossoms at time flowers fully developed
Puff	Fruit	Excess nitrogen, high temperature	Angular fruits, light in weight, locules not well filled

INSECT CONTROL

A number of different kinds of insect pests attack the tomato plant. Some of these are important pests of the tomato every year in some parts of the country and periodically in other tomato growing areas (Beattie *et al.* 1942). Thus, insects can be a serious problem. During seedling stage, stands can be damaged severely. Insects that attack seedlings are flea beetles, darkling ground beetles, cutworms, and occasionally, thrips, spring tails, vegetable weevils, garden centipedes, crickets, grasshoppers, earwigs, psyllids, seed corn maggots, aphids, and wireworms. Additional problem pests of seedling tomatoes may be birds, rabbits, squirrels, and other animals (Sims *et al.* 1968).

Later in the growing season, fruit and vine damaging insects become the problem. Tomato fruit worms, hornworms, stinkbugs, pinworms, army

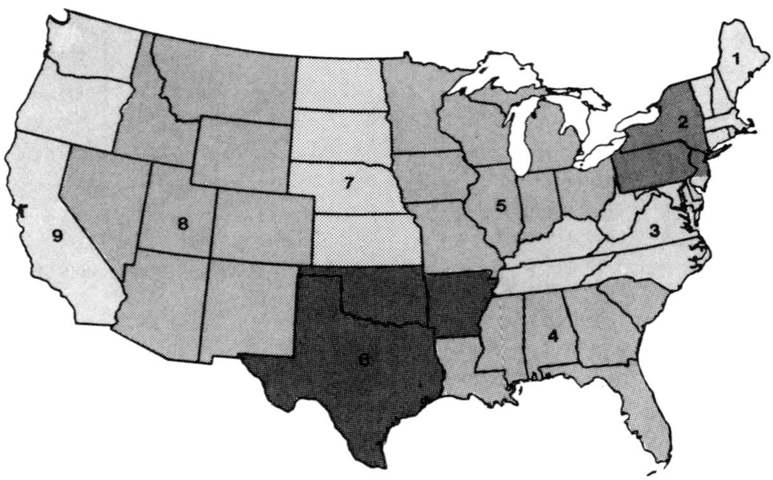

FIGURE 2.17 Regions of United States where tomato diseases are prominent. Numbers indicate the regions where serious and occasional losses occur. G. denotes possibility of disease occurring in any regional greenhouse. *From Barksdale et al. (1972).*

Fusarium Wilt (2,3,4,5,6,7,8,9,G)
Verticillium Wilt (1,2,8,9,G)
Bacterial Wilt (2,3,4,5,6)
Damping-Off (wherever seedlings)
Early Blight (1,2,3,4,5,6,7,9)
Late Blight (1,2,3,4,5,6,9,G)
Septoria Blight (1,2,3,4,5,6,7)
Southern Blight (3,4,6)
Tobacco Mosaic and Mottle Mosaics (1,2,3,4,5,6,7,8,9)
Cucumber Mosaic (occasionally everywhere)
Aster Yellows (5,6,7,8)
Anthracnose (2,3,5,6,7)
Buckeye Rot (1,2,3,4,5,6,9,G)
Bacterial Rot (2,3,4,5,6,7)
Nailhead Spot (3,4,6)
Gray Mold (4,G)
Ghost Spot (occasional)
Root Knot (3,4,6,8,9,G)
Blossom-end Rot (everywhere)
Cracking (everywhere)
Sunscald (in open fields)
Catfacing (wherever grown)
2,4-D Injury (wherever grown)
Graywall (wherever grown)
Blossom Drop (wherever grown)
Air Pollution (smog areas)
Fruit Pox (4,5,6,9)

worms, russet mites, loopers, and leaf miners can develop into damaging populations if not controlled. Usually mite and worm control is on a preventive rather than a corrective basis (Sims *et al.* 1968).

Occasionally, crickets, slugs, snails, and in certain areas, potato tuber worms may attack the crop. Late fields should be checked for insect activity, a point frequently overlooked (Sims *et al.*1968).

Among all the pests that can attack and damage tomatoes, the principal insects that are most bothersome are cutworms, flea beetles, aphids, hornworms, grasshoppers, and wireworms. With the possible exception of cutworms, none of these is usually troublesome in all tomato producing areas. Regionally, however, injury may be sufficient to justify control measures (Butler and Kerr 1952).

The following list of insects and characteristics of each may assist in their identification and control.

Aphids

These soft-bodied insects, which suck plant juices, are usually found on the underside of leaves. Aphids are often called plant lice. When plants are small and insects numerous, injury may be sufficient to retard development. The greatest injury, however, is the spread of virus diseases, particularly cucumber mosaic, from weeds and other plants to tomatoes (Beattie *et al.* 1942). A generation of aphids may be produced every 10-14 days and each generation may contain from 50 to 100 young.

Cutworms

Newly set tomato plants are often cut off at or near ground level by smooth, plump caterpillars. Cutworms feed during the night, and hide just below the surface of the soil during the day (Beattie *et al.* 1942).

Flea Beetles

These small, jet black beetles, which are about pinhead in size, eat small circular holes through leaves of newly transplanted tomatoes. Injury may be severe enough to defoliate plants and retard growth. Beetles jump when disturbed (Beattie *et al.* 1942).

Grasshoppers

Occasionally, when food is scarce during a dry period, grasshoppers migrate from adjoining road allowances, headlands, or sod crops to tomato fields where they eat both leaves and fruit (Beattie *et al.* 1942).

Hornworms

The tomato hornworm is a green caterpillar, which is easily recognized by characteristic V-shaped markings on the sides, and a large horn on one end.

When fully grown, caterpillars may be 4 in. long and nearly 0.5 in. in diameter. If not controlled, one or two hornworms may defoliate an entire plant (Beattie *et al.* 1942).

Wireworm

Wireworms, the slender cylindrical larvae of the chick beetle, infest the soil and attack plant roots. They may be yellow, straw, or reddish brown in color.

Lastly, there is the *Drosophila* problem. This insect can pose such problems that it and its control are given separate and special attention in Chapter 23.

PREPARING FOR HARVEST

The result of careful seeding, planting, cultivation, and spraying is the harvesting of a high-tonnage quality crop. However, before the start of harvesting, there should be a great deal of preparation; without this, countless hours of previous operations can, in the final analysis, be a meaningless waste. It therefore becomes mandatory for the grower and the processor to work together and plan for an upcoming period that is frequently hectic. The following list is worthy of thought.

1. Determine equipment needed for all steps from harvesting to delivery to the cannery. Purchase or build the equipment that you do not have and repair all other equipment (Angell 1970).
2. Determine the labor needed and make all arrangements for obtaining dependable labor. Train each worker for the job he is to do. Explain and show him how the work is to be done and also emphasize the importance of his job (Angell 1970). Arrive at methods of payment and a satisfactory pay scale.
3. Remove large weeds (rageweed and jimsonweed) from the fields. Prepare headlands at ends of fields to facilitate turning of harvesters and movement of trailers and trucks (Angell 1970).
4. Before harvest, loading areas should be smooth and large enough to maneuver forklift and trailers. Loading areas should have easy access and be located to minimize travel time to trailers.
5. The row length, yield, and field condition will determine the number of trailers needed. The field should be split if the rows are too long, or control loading areas should be considered.
6. Determine type of defects sorters will find. Before the sorters work on the machine, show them the defects that they will encounter and explain which and how much are to be discarded.
7. About 80% of the labor on a tomato harvester is sorting; therefore, it is vitally important that the grower develop efficient crews (Fletcher *et al.* 1971).

Finally, the time of harvest depends on the anticipated yield and the ultimate percentage of mature red ripe fruit. The percentage of mature red ripe fruit can be somewhat controlled by using Ethephon as a plant regulator. When applied to tomatoes it results in an increase of ethylene to trigger the ripening action of mature green fruits. Users indicate that mature green fruit will turn red in 7 to 18 days after proper application. The secret is to know that the green fruits are mature. one suggestion is to cut green fruits and note whether the seed cavities are filled with gelatinous pulp and that the seed coat is tan or brownish in color. Mature green fruits also show a change in color from green to light green or white appearance. Other factors that should be considered when using ethephon are the conditions of the plant. It should not be stressed for proper response to ethephon. If the temperature is too cold, the rate of application must be increased from the normal 1.5 pints (0.4 lbs ethephone) to the acre when the temperature is at 85 °F or above and up to 6½ pints (1.6 lb of ethephone) if the temperature is down to 60 °F. No application should be made if the temperature is below 65 °F. The ideal application rate is 3¼ pints per acre and applied when the temperature is in the 75 to 85 °F range with the temperature increasing. Further, at the time of application, the temperature should be rising rather than falling for success. Also, reports indicate that if the temperature is above 90 °F. no application of ethephone should be made as the plant is already under stress, too much leaf damage will occur, and the ripening response will not be favorable. Its most important to make a uniform and adequate coverage of the fruits and vines to promote fruit ripening and initiate aging and senescence of the leaves. Obviously, there are varietal differences and other effects when using ethephone, however, its the best tool available for aiding the ripening process of tomatoes.

REFERENCES

AL-SHAIBANI, A.M.H. and GREIG, J.K. 1979. Effects of stage of maturity, storage, and cultivar on some quality attributes of tomatoes. J. Am. Soc. Hortic. Sci. *104* (6) 880–882.

ANGELL, F.F. 1970. Production of Tomatoes for Mechanical Harvesting, Suggestions for the 1970 Tomato Season. Dep. Hortic. Mimeo, Univ. of Md., College Park.

ANGELL, F.F. *et al.* 1971. Growing tomatoes for mechanical harvesting. Md. Processors Rep. *17* (1) 3.

ANON. 1968. Practices followed by Ohio growers in 1963 top ten tomato club. Canning Trade *86* (20) 18–20.

ANON. 1969A. Bulk handling of tomatoes. Annu. Agric. Issue, Harvest *4*, 14–17.

ANON. 1969B. Fruit and Vegetable Facts and Pointers, 2nd Edition. United Fresh Fruit and Vegetables Assoc., Washington, DC.

ANON. 1980. Ohio guidelines for production of machine-harvested tomatoes—1980. Ohio State Univ. Coop. Ext. Serv. Bull. *647*.

ANON. 1980. Tomato Disease Handbook. Ohio Food Processors Association.

BARKSDALE, T.N., GOOD, J.M. and DANIELSON, LL. 1972. Tomato diseases and their control. U.S. Dep Agric., Agri. Handbook *203*, 1-20.

BARTZ, J.A., GERALDSON, G.M. and CRILL, J.P. 1979. Nitrogen nutrition of tomato plants and susceptibility of the fruit to bacterial soft rot. Phytopathology *69* (2) 163−166.

BEATTIE, J.H., BEATTIE, W.R. and DOOLITTLE, S.P. 1942. Production of tomatoes for canning and manufacturing. Farmers Bull. *1901*.

BETANCOURT, L.A., STEVENS, M.A. and KADER, A.A. 1977. Accumulation and loss of sugars and reduced ascorbic acid in attached and detached tomato fruits. J. Am. Soc. Hortic. Sci. *102* (6) 721−723.

BURG, S.P. 1973. Ethylene in plant growth. Proc. Natl. Acad. Sci. *70*, 591−597.

BURG, S.P. and BURG, E.A. 1969. Interaction of ethylene, oxygen and carbon dioxide in the control of fruit ripening. Qual. Plant. Mater. Veg. *19* (1−3) 185−200.

BUTLER, A.N.L. and KERR, E.A. 1952. Tomatoes for processing. Ont. Dep. Agric., Toronto, Bull. *491*.

CHIU, T.F. and BOULD, C. 1976. Effects of shortage of calcium and other cations on ^{45}Ca mobility, growth and nutritional disorders of tomato plants (Lycopersicon esculentum). J. Sci. Food Agric. *27*, 969−977.

DESAI, N. and CHISM, G.W. 1978. Changes in cytokinin activity in the ripening tomato fruit. J. Food Sci. *43*, 1324−1326.

FLETCHER, R.F., HEPLER, R.W., FERRETTI, P.A. and DAUM, D.R. 1971. Production of Tomatoes for Mechanical Harvesting: Suggested Practices and Procedures for the 1971 Season. Dep. Hortic., Pa. State Coll., University Park, PA.

GANMORE-NEUMANN, R. and KAFKAFI, U. 1980. Root temperature and percentage NO_3^-/NH_4^+ effect on tomato development. II. Nutrients composition of tomato plants. Agron. J. *72*, 762−766.

HARBAGE, R.P. 1971. An Engineering Evaluation of the Mechanized Production of Processed Tomatoes and Pickling Cucumbers in Ohio. Ohio Agric. Res. Dev. Cent., Wooster. Ohio.

HEPLER, J.R., RICHARDS, M.C., ELLIS, E.E. and CONKLIN, J.G. 1950. Tomatoes for New Hampshire. Univ. of New Hampshire Ext. Circ. *299*.

HESTER, J.B. and SHELTON, F.A. 1939. The soil side of tomato growing. Campbell Soup Co., Dep. Agric. Res., Camden, NJ. Bull. *1*.

HOWLETT, F.S. and KRETCHMAN, D.W. 1966. Test for nitrate nitrogen in tomato plants. Hortic. Mimeo. Ser. *327*.

JOHNSON, P.E. and WILCOX, G.E. 1971. Tomato Seeding for Commercial Production. Univ. of Purdue, Lafayette, IN.

KEIRNS, V.E. and WITTMEYER, E.C. 1951. Tomatoes in the Home Garden. Ohio Agric. Ext. Serv., Columbus.

KRETCHMAN, D.W. and SHORT, T.H. 1973. Considerations for using ethrel on tomatoes for processing in 1973. Dep. Agric. Eng., Ohio Agric. Res. Dev. Cent., Wooster, Ser. *397*.

KRETCHMAN, D.W. *et al.* 1972. Culture and physiology of tomatoes for processing. Ohio State Univ. and Ohio Agric. Res. Dev. Cent., Wooster. Mimeo Rep. *3*.

KRETCHMAN, DALE W. and E.C. WITTMEYER 1983. Twin Row Culture for Tomatoes for Processing. Am. Veg Grower 31 (8): pg 6-7.

LUH, B.S., UKAI, N. and CHUNG, J.I. 1973. Effects of nitrogen nutrition and day temperature on composition, color and nitrate in tomato fruit. J. Food Sci. *38*, 29-33.

LYON, W.F., FARLEY, J.D., GORSKE, S.F., ALBAN, E.K., WITTMEYER, E.C. and BROOKS, W.M. 1981. Pest control in vegetables for commercial growers. Ohio State Univ. Coop. Ext. Bull. *672*.

McKEEN, C.D. 1972. Tomato disease. Can. Dep. Agric. Publ. *1479*.

PIERCE, L.C. *et al.* 1963. Tomatoes. Iowa State Univ. Pamphl. *299*.

PORTE, W.S. 1952. Commercial production of tomatoes. Farmers Bull. *2045*.

SAKIYAMA, R. and STEVENS, M.A. 1976. Organic acid accumulation in attached and detached tomato fruits. J. Am. Soc. Hortic. Sci. *101* (4) 394–396.

SIMS, W.L., ZOBEL, M.P. and KING, R.C. 1968. Mechanized Growing and Harvesting of Processing Tomatoes. Univ. of Calif., Davis, Ext. Serv. *AXT-232*.

SIMS, W.L., ZOBEL, M.P., MAY, D.M., MULLEN, R.J. and OSTERLI, P.P. 1979. Mechanized growing and harvesting of processing tomatoes. Univ. of Calif., Div. Agric. Sci. Leafl. *2686*.

TAHA, A.A. and KRETCHMAN, D.W. 1980. Effect of daminozide and ethephon on transplant quality, plant growth and development, and yield of processing tomatoes. J. Am. Soc. Hortic. Sci. *105* (5) 705–709.

WATTERSON, J.C. 1985. Tomato Diseases. Petoseed Co. Inc.

WILLIAMS, J.W. and SISTRUNK, W.A. 1979. Effects of cultivar, irrigation, ethephon, and harvest date on the yield and quality of processing tomatoes. J. Am. Soc. Hortic. Sci. *104* (4) 435–439.

WITTMEYER, E.C. 1964. Practices followed by Ohio growers in 1963 Top Ten Tomato Club. Canning Trade (April) 18.

WITTMEYER, E.C. 1971. Summary of Practices Followed by Growers in the 1970 Ohio Top Ten Club. Ohio State Univ. Dep. Hortic., Columbus, Feb. 5.

WITTMEYER, E.C., JANSON, B.F. and GOLEMAN, D.L. 1971. Growing Tomatoes. Ohio State Univ. Agric. Ext. Serv., Columbus. Bull. *376*.

ZOBEL, M.B. 1966. Mechanization of tomato production. Proc. Natl. Conf. Tomatoes. Dept. Hortic., Purdue Univ., National Food Processors Assoc., Lafayette, IN. Dec. 1976.

CHAPTER 3
Genetics In Breeding of Processing Tomatoes

William L. George, Jr.[1] *and Stanley A. Berry*[2]

The tomato belongs to the nightshade or Solanaceae family and the genus *Lycopersicon*. The genus consists of relatively few species of annual or short-lived perennial herbaceous plants. The tomato is a warm-season crop, but despite its susceptibility to frost, it can be grown successfully from the equator to latitudes as far north as Fort Norman, Canada (65°N).

The tomato is normally highly self-pollinated, but there are also varying rates of natural cross-pollination which depend on the ecological conditions. Rick and Butler (1956) observed that the rates of cross-pollination in Peru, the center of origin of the genus, was much higher than rates obtained in California.

The center of origin of the genus *Lycopersicon* is a narrow elongated strip extending from northern Chile on the south to Ecuador on the north and reaching inland from the Pacific Ocean as much as 200 miles, but usually not more than 100 miles, and also including the Galapagos Islands. Generally, workers agree that the center of domestication of the cultivated tomato, *L. esculentum*, was Mexico. The indigenous tomatoes of these areas have provided a wealth of genetic diversity for tomato breeding programs.

The tomato was introduced to the Old World in the sixteenth century. Although it was used continuously in Italy since its first appearance there, superstitions concerning its poisonous qualities effectively suppressed its use elsewhere until well into the nineteenth century. It was not until about 1835 that the tomato became generally cultivated in the United States, and even at that time there was considerable bias against its use as food. The tomato growing industry began making rapid strides in the latter half of the nineteenth century and still greater in first half of the twentieth century.

[1] University of Illinois.
[2] The Ohio State University and the Ohio Agricultural Research and Development Center.

CLASSIFICATION AND CROSSING RELATIONSHIPS OF TOMATO

The genus *Lycopersicon* is divided into two subgenera; *Eulycopersicon*, the red-fruited species, and *Eriopersicon*, the green-fruited species. The species of these two subgenera are presented in Table 3.1.

TABLE 3.1. THE SPECIES OF THE GENUS *Lycopersicon*

Subgenera	Species	Common name	Somatic Chromosome Number
Eulycopersicon	*L. esculentum*	Common tomato	24
(red-fruited)	*L. pimpinellifolium*	Currant tomato	24
Eriopersicon	*L. cheesmanii*	Wild species	24
(green-fruited)	*L. glandulosum*	Wild species	24
	L. hirsutum	Wild species	24
	L. peruvianum	Wild species	24

Eulycopersicon, the red-fruited species, includes all the cultivated forms, and intraspecific crosses succeed readily in this group. Interspecific crosses between *L. esculentum* and *L. pimpinellifolium* have been extensively investigated by many workers and it has been the common experience that no barrier of any consequence exists to the hybridization of these species, or, in fact, to the production of progeny of the hybrid or to gene recombination in the hybrid.

Bailey (1949) classified *L. esculentum* into five botanical varieties: cv. *commune*, common tomato; cv. *cerasiforme*, cherry tomato; cv. *pyriforme*, pear tomato; cv. *grandifolium*, potato-leaved tomato; and cv. *validum*, upright tomato.

Both members of the red-fruited subgenus, *L. esculentum* and *L. pimpinellifolium*, are compatible with the wild species of the subgenus *Eriopersicon* when the latter functions as the male parent. Most forms of *Eriopersicon* are self-incompatible, and this condition is transmitted to all hybrids with *L. esculentum*. Crosses between *L. esculentum* and *L. peruvianum* set fruits readily if *L. esculentum* is used as the pistillate parent, but the embryos abort. This reproductive developmental barrier to obtaining hybrids was overcome by utilization of embryo culture (Smith 1944). Thus, valuable disease-resistant traits of *L. peruvianum* were made available for incorporation into commercial *L. esculentum* types.

Intergeneric crosses have succeeded with two species of *Solanum*: *S. lycopersicoides* and *S. pennellii*. The F_1 hybrid of *L. esculentum* and *S. lycopersicoides* is sterile. Hybridizations of *L. esculentum* with *S. pennellii* yield viable hybrids when *L. esculentum* is the female parent (Rick 1960).

THE TOMATO GENETICS COOPERATIVE

The importance of tomato as a world food crop has stimulated considerable research in the areas of genetics and breeding. Its characteristics of ease of culture and self-fertilizing mode of reproduction are advantages in genetic studies. Additionally, the organization of research workers with common interest in tomato genetics into the Tomato Genetics Cooperative in 1951 has contributed greatly to the progress achieved. In 1973, this group comprised 314 members from throughout the world.

Several gene lists have been published with brief descriptions of the known genes of tomato (Clayberg *et al.* 1971, 1973). At present 787 genes have been described and 256 of these have been assigned among the 12 linkage groups of tomato. These, of course, constitute only the genes controlling relatively simply inherited traits. Considerable genetic information exists concerning several important horticulture characters under polygenic control.

METHODS IN TOMATO BREEDING

The general pattern of the breeding program appropriate to a particular species is determined in part by the reproductive or mating system of the species. The cultivated tomato is a self-pollinated species. All varieties are highly inbred populations with no significant genetic diversity within a variety. It has generally been accepted that self-pollinating plants such as tomatoes, after they have become stabilized, do not change their genetic constitution to any great extent (Kerr 1969).

The tomato has perfect flowers. Crossing of tomatoes is accomplished by emasculation of the flowers of the seed parent as they begin to open but before pollen is shed. To emasculate a flower the stamens are removed with forceps, either alone or with the petals. Pollen is collected on the tip of a sterile forcep from the pollen parent and placed on the stigma of flowers of the seed parent. The pollination is labeled by recording the numbers of the seed and pollen parents and date on a tag, which is attached to the flower.

Normally bagging of the flower to prevent contamination is not necessary unless cross-pollination is suspected. When the fruit is ripe, the seeds are extracted, fermented naturally or digested with acid, washed, dried, and packaged. Selfing (backcrossing) plants of this first generation cross (F_1) gives the F_2 generation, or crossing the F_1 to one of the parents gives the backcross (BC) generation.

The choice of parental materials needed to achieve certain breeding goals is extremely important. It depends primarily on the availability of the genes required. It is advantageous to locate genes in horticulturally adapted varieties whenever possible. The most efficient systems of developing new varieties involve the use of "elite" germ plasm found in improved varieties, rather than of unadapted "exotic" varieties. However, often the desired genes are found only in primitive varieties or wild relatives of tomato.

The development of new varieties utilizing exotic gene pools takes considerable time. In using elite plasm, new variety developments usually takes 5 to 7 years, whereas, for example, the development of a new variety that incorporates a new gene for resistance from a wild species could take 20 years or more.

The time required to develop a new variety relates in part to the complexity of the plant breeding program. Complexity increases rapidly as the number of gene pairs contrasting two parents increases (Table 3.2). Stevens (1973) has determined that from cross-segregating for 21 genes with additive effects, a perfect F_2 population of tomatoes (one in which each phenotype possible occurs at least once) would require the growing of over 420,000 acres of tomatoes.

TABLE 3.2. KINDS OF PHENOTYPES POSSIBLE IN F_2 FROM PARENTS DIFFERING BY n ALLELIC PAIRS

Number of Allelic Pairs	Number of Phenotypes in F_2 Assuming: Full Dominance	Additive Genic Effects
1	2	3
2	4	9
3	8	27
4	16	81
10	1,024	59,049
21	2,097,152	10,460,353,203
n	2^n	3^n

Source: Allard (1960).

The basic breeding used following hybridization includes pedigree and backcross methods. These methods are used to handle the segregating populations and are based on the fact that selfing, or backcrossing to a homozygous parent, leads to homozygosity (uniformity). The speed with which a population reaches homozygosity upon self-fertilization is rapid. Continued self-pollination causes an increase in homozygosity by one-half per generation. It is evident that self-pollination, excluding special genetic situations, rapidly reduces any population to uniformity even when large numbers of heterozygous gene pairs are present initially (Fig. 3.1). Inbreeding is thus a powerful tool in plant breeding.

The pedigree method is widely used by tomato breeders. In this system records are kept of the ancestry or pedigree of each of the progenies (families). Selection is based on productivity and other horticultural characteristics of single plants or progenies. After the F_2 generation, selection is practiced within and between families. Considerable detailed record keeping is required, and the number of selections under test can quickly become burdensome. Once a degree of uniformity is reached, selections are placed in comprehensive evaluation trials.

The backcross method is particularly useful in transferring specific genes into an established variety which is deficient in one or a few characteristics. It is very effective in transferring genes for disease resistance. Continued backcrosses are made to the desirable parent, and selection is practiced for the characters being transferred from the donor parent. Essentially, the new variety will be the same as the original parent except for its improved characteristics. This is an advantage in that the new variety need not be extensively evaluated. It has the disadvantage in that the new variety will only have the specific improvements transferred. Many programs utilize a combination of the pedigree and backcross methods. The backcross system is used early in the program to transfer specific genes, but while sufficient heterozygosity exists pedigree selection is undertaken to improve other horticultural characters.

Testing of advanced breeding lines is an integral but costly part of a tomato breeding program. Early evaluation is primarily observational in nature without detailed yield or quality measurements. Here the skill or art of the breeder in selection is revealed.

Final testing of advanced lines involves replicated trials and grower trials. In replicated trials, and when possible in grower trials, detailed yield and quality data are collected. These trials should simulate the commercial growing situation as nearly as possible. In evaluating lines for machine harvest, often a one-time handpicked harvest is carried out to simulate a machine. Wherever possible it is most ideal to have the availability of a machine for complete performance evaluation of varieties and advanced lines.

Fig. 3.1. Generations of self-pollination.
From Allard (1960).

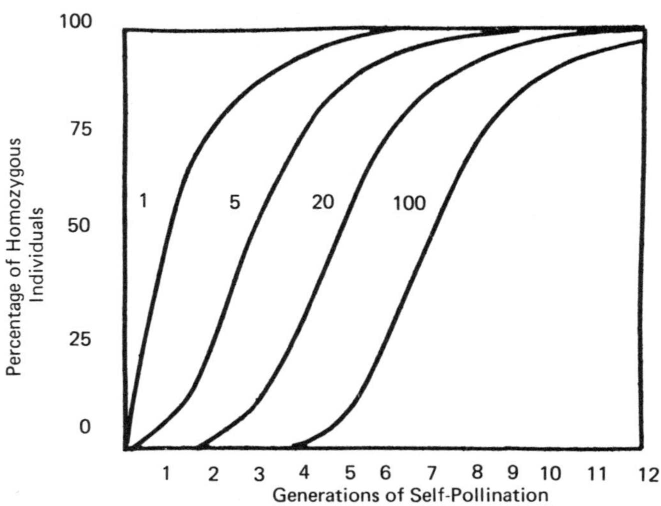

The procedures by which new varieties are released to farmers vary widely among public institutions and private industry. Generally, for experiment stations, if the decision to release is affirmative, the new variety is named and breeder seed produced and distributed.

GENERAL BREEDING OBJECTIVES

The objectives of any tomato breeding program depend on many factors. For example, method of culture (whether field or greenhouse grown), harvesting method (hand or machine), and product use (whether fresh or processed) all affect breeding objectives. More specifically, the use of the processed product, that is, as juice, paste, or whole-packed, also influences crop breeding. Equally important is the region or area of the world for which the cultivar is developed. It is interesting to note that many varieties have broad adaptability, whereas others have a narrow range of adaptability. It is difficult to define specific characteristics that account for this. In general, broad adaptability relates to the ability of cultivars to grow, flower, and fruit under environmental stresses of low and/or high temperatures, drought conditions, and disease.

The major breakthrough of the late 1960s in the production of processing tomatoes has been the advent of once-over machine harvest. The mechanization program in California, which is essentially 100% mechanized, is a milestone in horticulture crop production. Its success can be attributed largely to the close coordination of the varietal and the machine development programs in an area of extremely favorable climate for tomato production. In the eastern United States progress toward mechanization of harvest has been much slower, primarily because of less favorable climatic conditions during critical periods in the culture as well as harvest of the crop.

The breeding and development of tomato varieties for machine harvest adapted to any region is complex. It is not a function that can be performed by one person. In addition to the plant breeder, efforts by many specialists, including the geneticist, pathologist, entomologist, physiologist, biochemist, agricultural engineer, cultural specialist, and food technologist, as well as the cooperation of the grower and food processor, are fundamental to a successful program. Increasing efforts in developing interdisciplinary programs—the so-called team approach—will be necessary in the future, particularly as higher levels of crop performance and fruit quality are achieved (Younkin 1965).

In the next few pages the specific breeding requirements and problems related to processing-tomato improvement will be reviewed.

SPECIFIC BREEDING IMPROVEMENT OF PROCESSING TOMATOES

Plant Habit

Vine size as a criterion of suitability of processing tomatoes differs from that for the multihandpick type of fresh market tomato in that processing types are designed for once-over machine harvest or a maximum of two to three handpicks. Critical attributes of the machine-harvest type are determinate vine governed by gene *sp* with modifiers conditioning small vine size, dependable concentrated fruit set, and uniform ripening.

Vine expression can be radically affected by moisture and fertilization. Within limits, high levels of moisture in particular tend to result in correspondingly larger vine development, with a corresponding loss in fruit concentration and delayed fruit ripening (Wight *et al.* 1962).

High plant populations are a prerequisite for mechanized production (Hanna 1966; Wilcox 1970). This, then, based on economics and practicality, makes a growing system utilizing field direct-seeding advantageous (Sullivan and Wilcox 1971). California, with its long growing season and irrigation facilities, has been able to convert completely to direct-seeding. However, the eastern regions of the United States, with a shorter growing season and dependence on sporadic rainfall for germination moisture, have been able to utilize direct-seeding only to a limited extent. Generally direct-seeding, where feasible, must be carried out during early spring in cold soils, so that for greater effectiveness varieties are needed that are able to germinate and emerge at low temperature. Germ plasm (Berry 1969; Smith and Millett 1964) is available with heritable characteristics conditioning good germination at temperatures as low as 50° to 53°F (10° to 12°C) and progress is being made to incorporate these factors into commercial cultivars.

Earliness

Kerr (1973) has divided the development of the tomato plant as related to earliness into several periods. Briefly, these periods can be summarized as (1) seed germination and seedling emergence in cold soils; (2) seedling growth to first flowering; (3) first flowering to fruit setting; (4) fruit set to first ripe; and (5) first ripe fruit to peak production. These many components of earliness are under complex genetic control with considerable environmental influences (Honma *et al.* 1963; Powers *et al.* 1950).

Earliness as an objective for processing-tomato improvement differs from that of fresh market-tomato improvement. For the latter, earliness is critical for economically desirable seasonal marketing demands. For processing-tomato improvement, the objective of earliness, as far as it concerns the grower, advances, extends and makes more manageable the tomato harvest. Likewise, for the processor it means more efficient in-plant operation due to more even delivery of the crop to the plant.

Earliness takes on greater significance with the recent change over to mechanical harvesting. In contrast to multipick hand harvest, mechanical harvest involves a once-over destructive harvest. Thus, not only is the harvest season delayed until the crop is fully mature, but once under way, delivery patterns of the crop peak higher than under a multipick hand harvest regime; this can create delivery gluts and inefficiencies at the processing plant. By the use of varieties with a succession of maturities, particularly early varieties, this undesirable peak of delivery can be reduced and spread out (Fig. 3.2). The increasingly demanding conditions of employment being stipulated by present-day labor forces in tomato harvest and in-plant operations is making dependable and uniform harvest a requirement for profitable production.

Concentrated Fruit Set

For once-over machine harvest, concentrated fruit setting and maturity are important considerations. Plant growth habit, concentrated flowering, fruit setting ability, concentrated maturity, and ability of ripe fruit to store on the vine are factors that determine efficiency of machine harvest. In varietal breeding the self-pruning, *sp*, types with small plant size have been most widely utilized.

There are many sources of plant material that have variations in concentrated plant habit and higher degrees of earliness than presently available in commercial varieties. In general, these types have been of limited use due to low yield potential and deficiencies in quality, such as small fruit size, lack of firmness, crack susceptibility, low solids, and poor color.

Vine Storage of Fruit

The ability of tomato fruits to ripen and hold on the vine to achieve maximum fruit concentration to a large degree depends on climatic conditions. In the wet, humid conditions of the eastern United States resistance to fruit cracking is essential; this is now characteristic of many varieties, and several genes are involved (Armstrong and Thompson 1967; Reynard 1960). Work has been carried out to determine the degree of resistance of

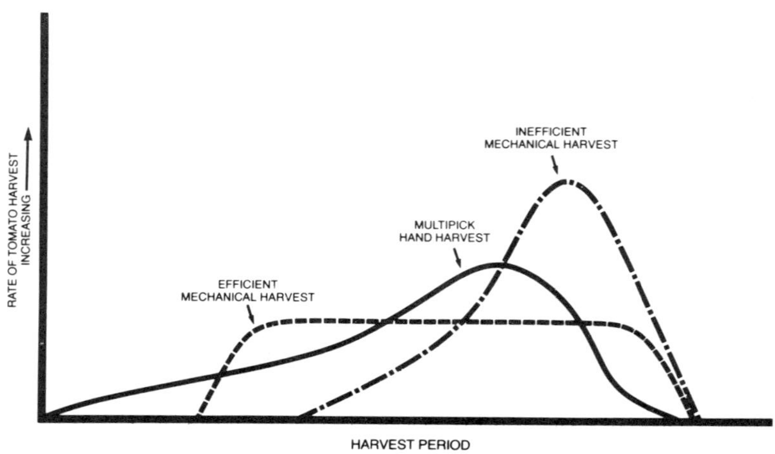

Fig. 3.2. Relationship between harvest date and method of harvest.

tomato fruit skin to puncture in relation to fruit cracking (Voisey *et al.* 1970).

Several fruit rots affect the tomato. Probably the most important and extensively studied are the various species of the fungus *Colletotricum* causing anthracnose. Genetic resistance and screening techniques are available (Barksdale 1969, 1970); however, it has not been possible to incorporate such resistance into commercial varieties.

Fruit firmness has been found to be under simple genetic control (El Sayed *et al.* 1966). Devices for measuring fruit firmness objectively have been reported (Hutchings *et al.* 1962).

Harvestability

For efficient mechanical harvest, vine size is important in its effect on the machine as to ease of vine pickup and effectiveness of fruit removal. Genetic variations have been noted in the force required to separate fruit from vine. During vine pickup, fruit should separate with ease and without the pedicel and calyx. Fruit separation (shatter) before harvest is undesirable. Stem retention is being overcome by use of the jointles pedical *j-2* gene (Reynard 1961).

Damage to fruit from machine handling depends on its firmness as well as its shape and size. Many variations occur in fruit shape and size, and these are under genetic control (Yeager 1937). Varieties with elongated fruits tend to be more resistant to damage from machine harvest than the round types; however, the exacting quality requirements of the industry dictate continued use of round varieties.

Many of the characteristics related to plant growing and harvesting are under complex genetic control and are difficult to evaluate. In these instances the tomato breeder is faced with major challenges in screening, incorporating, and maintaining these characteristics in a breeding program. Constant selection pressure must be maintained in the breeding program, and since the environment plays such an important role, this is often a major problem.

Resistance to Parasitic Disorders

In many instances diseases of various kinds are major factors in limiting the production of tomatoes. Their effects on yield reduction and product quality vary depending on the specific disease and crop region. Significant contributions have been made in breeding for disease resistance. The list of known and utilized genes for resistance is long and extensive. Only a few of the more important utilized by the tomato breeder will be reviewed.

Many of the types of genetic resistance to diseases used in breeding are governed by single dominant genes. Several cultivars now have resistance to fusarium wilt and verticillium wilt, controlled by genes *I* and *Ve*, respectively. Many cultivars have resistance to gray leaf spot caused by *Stem-*

phylium solani, which can be a destructive disease in humid regions. The fungus diseases late blight *(Phytophthora infestans)* and early blight *(Alternaria solani)* occur to some extent in most tomato growing areas. At times crop losses from these diseases are substantial. Resistance to late blight is available and is being utilized. Breeding for resistance to early blight has proved more difficult. Cultivars such as Manalucie, Southland, and Floradel seem to have some field tolerance.

Bacterial wilt *(Pseudomonas solanacearum)* and bacterial canker *(Corynebacterium michiganese)* are two important tomato diseases caused by bacteria. Bacterial wilt-resistant fresh market cultivars have recently been released (Henderson and Jenkins 1972), as has a breeding line with resistance to bacterial canker (John *et al.* 1973).

Parasitic nematodes often seriously infest tomatoes. Severe infestations increase the severity of the various wilt diseases. Resistance governed by gene *Mi* is available in commercial tomato cultivars to the root-knot nematode (*Meloidogyne* spp).

There are many virus diseases that affect tomatoes. The effect on productivity of many viruses is often difficult to measure. In many instances yield and fruit quality are seriously reduced. One of the most extensively studied viruses is tobacco mosaic virus (TMV). TMV is an important disease of tomatoes grown in greenhouses and is important in certain field-growing regions. Internal browning of tomato fruits, a fruit disorder, has often been associated with plants infested with TMV. Three major genes for tolerance or resistance are available, *Tm*, *Tm-2*, and $Tm\text{-}2^a$ (Alexander and Oakes 1971). These varieties are resistant to the several strains of TMV found in Ohio.

Leaf mold *(Cladosporium fulvum)* is a destructive disease of greenhouse tomatoes and can be of importance in field-grown tomatoes. Several dominant genes for resistance are available; however, variety breeding has been difficult because of the highly mutable characteristics of this fungus (Kerr *et al.* 1971).

Research on the identification and utilization of resistance to insects in horticultural crops has been minimal. Recently, however, progress has been made to identify germ plasm of various crops carrying resistance to various insects. Tomato germ plasms have been identified which have tolerance or resistance to spider mite (Stoner 1970), potato aphid (Stoner *et al.* 1968), leaf miner (Webb *et al.* 1971), tobacco flea beatle (Gentile and Stoner 1968) and greenhouse whitefly (Gentile *et al.* 1968).

Resistance to Nonparasitic Disorders

Several physiological disorders affecting fruit quality are a problem in tomatoes. Various environmental factors, nutrients, and diseases have been implicated in the expressions of disorders, such as blossom-end-rot, graywall (blotchy ripening), and internal browning. Genetic variability in susceptibility of various lines and varieties to these disorders has been noted.

Several studies have reported or suggested relationships between genetics and plant nutrition. Studies of the genetics and physiology of iron uptake (Wann and Hills 1972), boron uptake (Wall and Androus 1962), and ammonium tolerance (Maynard *et al.* 1966) suggest considerable variation in nutrient uptake and utilization in tomatoes. Development and improvement of proper screening methods in this area could improve the growth and quality of tomatoes.

Excesses of atmospheric pollutants are creating production problems in some tomato growing regions. Genetic variability suggesting tolerance or resistance to various air pollutants has been located in commercial cultivars (Clayberg 1971; Gentile *et al.* 1971). Future situations may place increased emphasis on modifying the plant through breeding to alleviate the effects of pollution of various kinds.

Raw Product Improvement

The concept of quality in vegetables is often rather vague and difficult to measure objectively. The processing-tomato breeder, in cooperation with the food technologist, has long been concerned with raw product quality. Because much of the tomato breeding work has been oriented to the production problems of growers, improvement of the characteristics of productivity and disease resistance has taken precedence over quality.

Assays for the measurement of several components of quality are not generally deemed difficult. Considerable attention has been given to the measurement and evaluation of soluble solids, pH, titratible acidity, and color (Porter 1960; Thompson *et al.* 1962). More recently, with the development of instrumentation and analytical methods, progress has been made on the determination of the constituents of flavor (Stevens 1970, 1972A).

The apparent simplicity of the study of the characteristics of quality obscures its complexity (McCollum 1970). Genetic studies of gross characteristics, such as total acidity and solids, have often indicated multigenic control (Ibarbia and Lambeth 1969; Lower and Thompson 1967). Conversely, study of the genetic control of individual components has indicated that differences in concentration levels are simply inherited. Stevens (1972B) has shown that differences in citrate and malate concentrations of tomato fruits from widely different sources are controlled by a single gene for each compound. Dominance was found for high concentration of citrate and low concentration of malate. Thus, to effectively breed for improvement of a characteristic of quality that depends on a large number of constituents it is necessary to separate the characteristic into its components and investigate their genetic control separately (Stevens 1973).

To complicate the problem, influences of the environment and stage of maturity affect phenotypic expression. This causes difficulties in separating genetic and nongenetic variations and places considerable importance on sampling error in a breeding program.

The development of varieties adapted to efficient machine harvest has had an adverse effect on quality (Stevens 1973). For example, several of the characteristics considered essential for mechanical harvest, such as fruit and vine characteristics, are in opposition to the characteristics needed for high solids. High solids in tomato fruits is associated with large, indeterminate vines with low fruit-to-leaf ratio and dispersed fruit set, late maturity, low yield, and small fruit size. Higher soluble solids in the fruit of the indeterminate genotype versus the determinate has been studied (Emery and Munger 1970).

Fruit color as a component of quality has been a major concern of the breeder. Color is important to the grower as it affects grade, and to the processor as it affects product appearance and ultimately salability (Younkin 1965). Considerable effort is under way to increase lycopene, the red pigment, and thus color by the incorporation of the crimson gene, og^c and the high pigment gene, *hp*. The simple inheritance of these characters makes incorporation by backcrossing not difficult. However, the crimson gene lowers β-carotene content and consequently reduces the nutritional value of tomato by lowering vitamin A. High pigment, *hp*, in combination with og^c, restores or increases β-carotene and improves fruit color. This dictates the use of both genes in the development of acceptable cultivars. The main difficulty is overcoming the deleterious effects of *hp* on plant characteristics (Thompson *et al.* 1962, 1967).

Several fruit characteristics in addition to the quality components mentioned influence the processed product. For whole-packed tomatoes intensity and uniformity of peeled fruit color and size of core are important. Size of stem scar and core are partly associated with fruit size. Small scar and core are advantageous in the elimination of the coring operation for whole-pack tomatoes. Progress has been made in selection for small core in small fruit types.

FUTURE CHALLENGES IN TOMATO BREEDING

Tomato breeding programs are complex. A major challenge involves increasing breeding efficiency by developing simplified screening methods and reducing the generation times required with present breeding procedures. The effective use of seedling screens, or seedling markers, increases the efficiency of handling large populations (Kerr 1965). The use of F_1 hybrids of processing tomatoes still offers several advantages, such as uniformity, earliness, and efficiency in creating new gene combinations. However, at present seed costs remain prohibitive. Progress is being made in utilizing some of the male-sterile genes to circumvent the emasculation process. Increased efforts are needed to devise an economical pollen-transfer system.

The production of haploids by anther culture in tobacco and the progress in this direction in tomato (Sharp *et al.* 1971) has potential for reducing the generation time to reach homozygosity required with present inbreeding methods.

Considerable progress has been made with agronomic crops in utilization of higher levels of photosynthetic efficiency, whereas horticultural crops have received little attention (Zelitch 1971). Many opportunities exist in this area. However, the basic concepts of yield as an all-exclusive term must be reevaluated and its component parts studied separately. Here, the biochemist and plant physiologist working closely with the breeder might make significant contributions. Black *et al.* (1969) have outlined some plant attributes associated with greater efficiency and indicate that genetic manipulation of some of the systems involved appears to be feasible. These traits are summarized as follows: (1) increased growth and vigor at high light intensities; (2) increased growth and vigor as temperatures rise above 59° to 68°F (15° to 20°C); (3) no inhibition of growth at normal oxygen concentrations and perhaps less inhibition of growth by the oxygen normally produced during photosynthesis; (4) effective storage of energy and substrates for metabolic processes based on minimum losses of reduced carbon; (5) a C_4 fixation cycle, the advantage of which is hypothesized to be higher rates of CO_2 uptake; and (6) low threshold of CO_2 utilization. Major factors conditioning these responses are morphology, capacity to extract nutrients or moisture from the soil, and differential responses to temperature (Black *et al.* 1969).

Hybrid varieties used for commercial plantings are being used more extensively. Approximately one-third of the tomato acreage in California alone is planted with hybrids and in the Midwest, although hybrid utilization has been minor, their use has begun to increase. Hybrid development involves identifying elite inbred lines and crossing the chosen lines to obtain the best possible combination, as determined by extensive trials. The first generation crosses between pairs of inbred lines are used for commercial planting. Hybrid seed production is a labor intensive manual operation and as such it is not economically viable to produce hybrid seed in the United States where high labor costs would make its production uneconomic. However, it is feasible in countries with low labor costs. Hybrids allow the more rapid utilization of dominant traits, such as disease resistance, in conventionally developed elite germplasm, as well as exploitation of heterosis for quantitative traits such as improved earliness, yield and quality attributes.

Genetic engineering techniques promise to further aid in the breeding of improved processing tomato varieties by providing potential for probing with new technologies more precise and efficient than traditional breeding methods. These also could provide a means for achieving heretofore unattainable levels of improvement in tomato yield quality and disease and pest resistance (Nevin, 1987).

Progress is being made utilizing such technologies as represented by: tissue culture, somaclonal variation *in vitro* selection, regeneration of protoplasts, recombinant transformation techniques, restriction fragment length polymorphism (RFLP) mapping and a host of additional new related emerging biotechnologies. Conventional breeding aided by genetic engineering techniques holds much promise for aiding in the development of improved processing tomato varieties.

REGULATION OF PLANT BREEDING

Many important contributions have been made by plant breeders in maintaining and improving the world food supply. The development of improved new cultivars has been an integral part of specialized agriculture and has contributed to abundant food supplies.

Recently, however, concern has been expressed by well-informed scientists and nonprofessionals alike regarding plant breeding activities. In certain instances critics have unjustifiably made the breeder into an ogre working to create an agricultural disaster. Several factors have brought about this unsettling situation. The major areas of concern involve the uniformity of germ plasms dictated by the utilization of certain breeding methodologies, levels of naturally occurring toxicants affecting food safety, and changes in nutritive value in food crops.

In 1970 the corn crop in the United States was severely damaged by an epidemic of a new race of Southern corn leaf blight. This was in part because a single source of cytoplasm was used in developing most of the corn hybrids. Genetic uniformity suggested genetic vulnerability. This prompted a study by a committee under the auspices of the National Academy of Sciences. Its report, entitled *Genetic Vulnerability of Major Crops* (Horsfall 1972), documents the genetic uniformity of many U.S. plants. This uniformity takes many forms, from the uniformity for all genes in individuals of vegetatively propagated crops to uniformity based on a single gene or single cytoplasm.

The tomato is self-pollinating and, as previously stated, all cultivars are highly inbred populations with no significant genetic diversity within a cultivars. Actually, the tomato industry is highly dependent on the relatively few genes already mentioned, namely, those associated with plant habit, disease resistance, and fruit color. In may cases the vulnerability that might be associated with these genes is unknown. Where vulnerability exists, for example, susceptibility of the *I* gene to race 2 of fusarium wilt in certain parts of the world, new resistance genes have been found and utilized. There are other examples of resistance loss; but there are also many of resistance that has remained stable. Resistance lost is publicized; stable resistance rarely receives more notice than the first announcement of its finding (Van der Plank 1968). The N.A.S. committee recommended greater use of exotic germ plasm to develop cultivars resistant to disease and insects, and to broaden the genetic base. But the questions then arise: (1) Are we thus inadvertently increasing the levels of natural toxicants in food crops (Kehr 1973) and (2) What changes, if any, either positive or negative, are occurring in the nutritive value of food crops?

In recognition of potential health hazards the Food and Drug Administration has proposed and published in the Federal Register (Anon. 1970, 1971) regulations effective June 25, 1971, which state that any new cultivar of food plant, before release, must be evaluated for significant alterations in values of food nutrients and increases in levels of toxicants. A 20% change in nutrient value and/or a 10% increase in toxicants are considered significant. Many questions remain unanswered as to what nutrients or toxicants are involved, standard levels, and so on, but it is imperative that all breeders of food crops be aware of these developments.

Breeding programs of food crops are becoming more and more complex and costly. Tomato breeding programs are no exception. In the future, increased emphasis on cooperative research inputs by many scientists from diverse disciplines is needed if tomato breeding programs are to meet the demands of growers, processors, and consumers.

REFERENCES

ALEXANDER, L.J. and OAKES, G.L. 1971. Ohio M-R9 and Ohio M-R12: Two new tomato varieties resistant to five Ohio strains of TMV. Ohio Agric. Res. Dev. Cent., Res. Bull. *1045*.

ALLARD, R.W. 1960. Principles of Plant Breeding. John Wiley & Sons, New York.

ANON. 1970. Eligibility of substances for classification as generally recognized as safe in food. Fed. Reg. *35*, 237.

ANON. 1971. Definitions and procedural interpretive regulations. Fed. Reg. *36*, 123.

ARMSTRONG, R.J. and THOMPSON, A.E. 1967. A diallel analysis of tomato fruit cracking. Proc. Am. Soc. Hortic. Sci. *91*, 505–513.

BAILEY, L.H. 1949. Manual of Cultivated Plants, 2nd Edition. Macmillan Co., New York.

BARKSDALE, T.H. 1969. Resistance of tomato seedlings to early blight. Phytopathology *59*, 443–446.

BARKSDALE, T.H. 1970. Resistance to anthracnose in tomato introductions. Plant Dis. Rep. *54*, 32–34.

BARKSDALE, T.H. and KOCH, E.J. 1969. Methods of testing tomatoes for anthracnose resistance. Phytopathology *59*, 1373–1376.

BERRY, S.Z. 1969. Germinating response of the tomato at high temperature. Hortic. Sci. *4*, 218–219.

BERRY, S.Z. and GOULD W.A. 1973. Ohio 2070, Ohio 2170, and Ohio 2470—Early, high quality, mechanically harvestable, whole-pack processing tomatoes. Ohio Agric. Res. Dev. Cent., Res. Circ. *195*.

BERRY, S.Z. and GOULD, W.A. 1979. New tomato variety for machine harvest. Ohio Rep. *64* (2) 22–23.

BERRY, S.Z. and GOULD, W.A. 1981. Evaluation of processing tomato breeding lines and cultivars for mechanical harvesting and quality in 1980. Ohio Agric. Res. Dev. Cent. Hortic. Ser. *494*.

BERRY, S.Z. and M.R. UDDIN. 1991. (In Press). Chapter 15. Breeding tomato for quality and processing attributes. *IN* Genetic Improvement of Tomato (ed. by Prof. Kalloo) pages 197-206. Springer-Verlag, Berlin, Hamburg.

BLACK, C.C., CHEN, T.M. and BROWN. R.H. 1969. Biochemical basis for plant competition. Weed Sci. *17*, 338-344.

BUESCHER, R.W. 1977. Factors affecting ethephon-induced red color development in harvested fruits of the rim tomato mutant. HortScience *12* (4) 315–316.

BUESCHER, R.W. and DOHERTY, J.H. 1978. Color development and carotenoid levels in rim and nor tomatoes as influenced by ethephon, light and oxygen. J. Food Sci. *43* (6) 1816–1818.

BUESCHER, R.W. and TIGCHELAAR, E.C. 1977. Utilization of nor tomato hybrids for extending storage-life and improving processed quality. Lebensm. Wiss. Technol. *10* (2) 111–113.

CLAYBERG, C.D. 1971. Screening tomatoes for ozone resistance. HortScience *6*, 396–397.

CLAYBERG, C.D. *et al.* 1971. Report of the gene list committee. Tomato Genet. Coop. Rep. *21*, 2–9.

CLAYBERG, C.D. *et al.* 1973. Report of the gene list committee. Tomato Genet. Coop Rep. *23*, 3–8.

EL SAYED, M.N.K., ERICKSON, H.T. and TOMES, M.L. 1966. Inheritance of tomato fruit firmness. Proc. Am. Soc. Hortic. Sci. *89*, 523–527.

EMERY, G.C. and MUNGER, H.M. 1970. Effects of inherited differences in growth habit on fruit size and soluble solids in tomato. J. Am. Soc. Hortic. Sci. *95*, 410–412.

GENTILE, A.G., FEDER, W.A., YOUNG, R.E. and SANTNER, Z. 1971. Susceptibility of *Lycopersicon* spp. to ozone injury. J. Am. Soc. Hortic. Sci. *96*, 94–96.

GENTILE, A.G. and STONER, A.K. 1968. Resistance in *Lycopersicon* spp. to the tobacco flea beetle. J. Econ. Entomol. *61*, 1347–1349.

GENTILE, A.G., WEBB, R.E. and STONER, A.K. 1968. Resistance in *Lycopersicon* and *Solanum* to greenhouse whiteflies. J. Econ. Entomol. *61*, 1355–1357.

HANNA, G.C. 1966. The development of tomato varieties for mechanical harvesting. Proc. Natl. Conf. Mech. Tom. Prod. Natl. Food Processors Assoc., Dept. Hortic., Purdue Univ. and National Food Processors Assoc., Lafayette, IN.

HENDERSON, W.R. and JENKINS, S.F. 1972. Venus and Saturn, two new tomato varieties combining desirable horticultural features with southern bacterial wilt resistance. N.C. Agric. Exp. Stn. Bull. *444*.

HONMA, S., WITTWER, S.H. and PHATAK, S.C. 1963. Flowering and earliness in the tomato; inheritance of associated characteristics. J. Hered. *54*, 212–218.

HORSFALL, H.G. 1972. Genetic Vulnerability of Major Crops. Natl. Acad. Sci. Washington, DC.

HUTCHINGS, I.J., JOHN, C.A., PREND, J and WYATT, C.C. 1962. An instrument for the measurement of the force required for the separation of cucumber fruit from the peduncle and for the measurement of the firmness of tomato fruit. Proc. Am. Soc. Hortic. Sci. *81*, 487–492.

IBARBIA, E.A. and LAMBETH, V.N. 1969. Inheritance of soluble solids in a large/small-fruited tomato cross. J. Am. Soc. Hortic. Sci. *94* 496–498.

JOHN, C.A., EMMATTY, D.A. and THYR, B.D. 1973. Release of bacterial canker resistant tomato breeding line, H 2990. Tomato Genet. Coop. Rep. *23*.

KEHR, A.E. 1973. Naturally-occurring toxicants and nutritive value in food crops: The challenge to plant breeders. HortScience *8*, 4–5.

KERR, E.A. 1965. Identification of high-pigment, *hp*, tomatoes in the seedling stage. Can. J. Plant Sci. *45*, 104.

KERR, E.A. 1969. Do tomato cultivars deteriorate after they are introduced. Rep. Hortic. Res. Inst. Ont. 75–80.

KERR, E.A. 1973. Breeding for earliness and concentrated maturity in tomato. Tomato Breeders Roundtable, Denver, Mimeo Rep.

KERR, E.A., PATRICK, Z.A. and BAILEY, D.L. 1971. Resistance in tomato species to new races of leaf mold (*Cladosporium fluvum* CKE). Hortic. Res. *11*, 84–92.

LOWER, R.L. and THOMPSON, A.E. 1967. Inheritance of acidity and solids content of small-fruited tomatoes. Proc. Am. Soc. Hortic. Sci. *91*, 486–494.

MAYNARD, D.N., BARKER, A.V. and LACHMAN, W.H. 1966. Variation among tomato lines with respect to ammonium tolerance. HortScience *1*, 17–18.

McCOLLUM, J.P. 1970. Plant constituents as they affect quality in vegetables. HortScience *5*, 99.

NEVIN, D.J. 1987. Why tomato biotechnology? A potential to accelerate the Applications. pp. 3-14. *IN* Tomato Biotechnology. A.R. Liss, Inc.

NITSCH, J.P. and NITSCH, C. 1969. Haploid plants from pollen grains. Science *163,* 85-87.

PORTER, D.R. 1960. Quality criteria and their evaluation in a breeding program for processing type tomatoes. Proc. Plant Sci. Symp., Campbell Soup Co., Camden, NJ, 137–150.

POWERS, L., LOCKE, L.F. and GARRETT, J.C. 1950. Partitioning method of genetic analysis applied to quantitative characters of tomato crosses. U.S. Dep. Agric. Tech. Bull. *998*.

REYNARD, G.B. 1960. Breeding tomatoes for resistance to fruit cracking. Proc. Plant Sci. Symp., Campbell Soup Co., Camden, NJ, 93–112.

REYNARD, G.B. 1961. A new source of the j_2 gene governing jointless pedicel in tomato. Science *134*, 2102.

RICK, C.M. 1960. Hybridization between *Lycopersicon esculentum* and *Solanum pennellii*: Phylogenetic and cytogenetic significance. Proc. Natl. Acad. Sci. *46*, 78–82.

RICK, C.M. and BUTLER, L. 1956. Cytogenetics of the tomato. Adv. Genet. *8*, 267–382.

SHARP, W.R., DOUGALL, D.K. and PADDOCK, E.F. 1971. Haploid plantlets and callus from immature pollen grains of *Nicotiana* and *Lycopersicon*. Bull. Torrey Bot. Club *98*, 219–222.

SMITH, P.G. 1944. Embryo culture of a tomato species hybrid. Proc. Am. Soc. Hortic. Sci. *44*, 413–416.

SMITH, P.G. and MILLETT, A.H. 1964. Germinating and sprouting responses of the tomato at low temperatures. Proc. Am. Soc. Hortic. Sci. *84*, 480–484.

STEVENS, M.A. 1970. Inheritance and flavor contribution of 2-isobutylthiazole, methyl salicylate and eugenol in tomatoes. J. Am. Soc. Hortic. Sci. *95*, 9–13.

STEVENS, M.A. 1972A. Relationships between components contributing to quality variation among tomato lines. J. Am. Soc. Hortic. Sci. *97*, 70–73.

STEVENS, M.A. 1972B. Citrate and malate concentrations in tomato fruits: Genetic control and maturational effects. J. Am. Soc. Hortic. Sci. *97*, 655–658.

STEVENS, M.A. 1973. The influence of multiple quality requirements on the plant breeder. HortScience *8*, 110–112.

STONER, A.K. 1970. Selecting tomatoes resistant to spider mites. J. Am. Soc. Hortic. Sci. *95*, 78–80.

STONER, A.K., WEBB, R.E. and GENTILE, A.G. 1968. Reaction of tomato varieties and breeding lines to aphids. HortScience *3*, 77.

SULLIVAN, G.H. and WILCOX, G.E. 1971. Costs for direct seeding and transplanting of tomatoes for processing. HortScience *6*, 479–480.

THOMPSON, A.E., HEPLER, R.W. and KERR, E.A. 1962. Clarification of the inheritance of high total carotenoid pigments in the tomato. Proc. Am. Soc. Hortic. Sci. *81*, 434–442.

THOMPSON, A.E., HEPLER, R.W., LOWER, R.L. and McCOLLUM, J.P. 1962. Characterization of tomato varieties and strains for constituents of fruit quality. Univ. Ill. Agric. Exp. Stn. Bull. *685*.

THOMPSON, A.E., TOMES, M.L., ERICKSON, H.T., WANN, E.V. and ARMSTRONG, R.J. 1967. Inheritance of crimson fruit color in tomatoes. Proc. Am. Soc. Hortic. Sci. *91*, 495–504.

VAN DER PLANK, J.E. 1968. Disease Resistance in Plants. Academic Press, New York.

VOISEY, P.W., LYALL, L.H. and KLOEK, M. 1970. Tomato skin strength—its measurement and relation to cracking. J. Am. Soc. Hortic. Sci. *95*, 485–488.

WALL, J.R. and ANDRUS, C.F. 1962. The inheritance and physiology of boron response in the tomato. Am. J. Bot. *49*, 758–762.

WANN, E.V. and HILLS, W.A. 1972. Inheritance of iron transport in tomato stock T3238. Tomato Genet. Coop. Rep. *22*, 28.

WEBB, R.E., STONER, A.K. and GENTILE, A.G. 1971. Resistance to leaf miners in Lycopersicon accessions. J. Am. Soc. Hortic. Sci. *96*, 65–67.

WIGHT, J.R., LINGLE, J.C., FLOCKER, W.J. and LEONARD, S.J. 1962. The effects of irrigation and nitrogen fertilization treatments on the yield, maturation and quality of canning tomatoes. Proc. Am. Soc. Hortic. Sci. *81*, 451–457.

WILCOX, G.E. 1970. Influence of row spacing and plant density on single harvest tomato yields. J. Am. Soc. Hortic. Sci. *95*, 435–437.

YEAGER, A.F. 1937. Studies on the inheritance and development of fruit size and shape in the tomato. J. Agric. Res. *55*, 141–152.

YOUNKIN, S.G. 1965. Preprocessing research problems from the viewpoint of the processor. Food Technol. *19*, 52–54.

ZELITCH, I. 1971. Photosynthesis, Photorespiration and Plant Productivity. Academic Press, New York.

CHAPTER 4

Tomato Harvesting, Systems and Methods

Harvesting is one of the most important aspects of crop production. Maximum yield and optimum quality are the goals of all phases of tomato culture.

The hand harvesting and handling of tomatoes common in the 1950s are as outdated today as using horse and buggy to haul fruit to the factory which was common in the 1930s.

Increased interest in the mechanical harvesting of tomatoes began in the 1960s. Some of the reasons for mechanical harvesting are:

1. Lack of labor to hand pick at reasonable cost,
2. Need for handling the crop mechanically and automatically,
3. Coordination of field and factory operations into efficient high-speed production systems,
4. Interest within segments of the industry to design, manufacture, and sell the new and needed equipment,
5. Development of new varieties or cultivars that:
   a. Are adapted to once-over harvest methods,
   b. Permit vine storage of fruit in the field without appreciable deterioration of quality for a reasonable length of time, and
6. Changes in cultural methods (Gould 1967).

Mechanical harvesting of tomatoes is now a reality. In California nearly all the crop is machine harvested, with an increasing percentage of the crop in the East and Midwest mechanically harvested. California's growing season is from 250 to 300 days (Angell 1970; Anon. 1967A; Hollis 1970) while the East and Midwest is 150 to 190 days.

California's long growing season enables farmers to use direct-seed method. Studies conducted in both the East and West have shown that high plant populations obtained by the direct-seed method favor the uniform plant growth and fruit development necessary for mechanical harvesting to be economical. It has also been shown that the transplant method, common in the East and Midwest, does not always provide uniform plant growth. A direct-seeding variety that will mature earlier than existing varieties is needed in these areas (Angell 1970; Anon. 1967A; Hollis 1970).

FIGURE 4.1. HAND HARVESTING INTO 33 lb. HAMPERS AND STACKING TOMATOES ON TRUCK FOR DELIVERY TO CANNERY.

For mechanical harvesting, a cultivar should possess at least six major characteristics:

1. A fruit should mature and ripen at the same time to permit top yields for a once-over harvest,
2. Plants should not have excessive foliage because vegetative growth interferes with the separation of fruit from the vine,
3. The tomato pedicel (stem) should be jointless so that the fruits are not punctured in handling,
4. Fruit should be firm and crack-resistant,
5. The tomato should have good vine-storage ability in the field after maturity, and
6. Fruit should be resistant to machine damage and should hold in a sound condition during transit (Gould *et al.* 1965A).

THE HARVESTER

Mechanical harvesters have passed through the experimental stage. Several machinery manufacturers and universities have developed equipment capable of mechanically harvesting tomatoes (Pearl 1962; Ries *et al.* 1960; Stout and Ries 1959). Although the operation and construction details are different for each machine, all are based on a "once-over" principle in which the entire plant is cut and carried over the harvester, where the fruit is then removed. Thus, the grower cannot return for any fruit unripe at the time of harvest. As a result, greater than 85% of the tomato field must ripen at the same time, or the grower sacrifices a large part of his crop (Anon. 1967B).

All mechanical tomato harvesters have four basic components:

1. Pick up mechanism,
2. Fruit and vine separating area,
3. Hand sorting area, and
4. Discharge or container-loading mechanism

When these components are properly designed and adjusted, the tomato harvester demonstrates very acceptable harvesting results.

Operation of the Harvester

The harvester cuts the vine at or slightly below ground surface. The vines, together with any loose fruit that may have fallen to the ground, are gathered into the machine's feed conveyor by the counter rotation of the pickup disks and convoluted belts. Loose fruit and dirt clods that do not drop through the slotted chain are discharged onto separate sorting conveyors. Provision for hand sorters or electronic dirt and color sorters on each conveyor ensures the recovery of all good fruit. Rejected fruit and dirt clods are discharged to the ground. Fruit-laden vines, meanwhile, are transferred from the feed conveyor to a reciprocating mechanism that begins a shaking action, causing the fruit to separate from the vine. As the fruit separates, it is transferred to a conveyor located directly below the shaking section. From this lower conveyor, the fruit is routed and distributed onto sorting belts located on each side of the machine, where culls and otherwise unacceptable fruit are removed by sorters. Many machines today are equipped with electronic color sorters, adjustable for any level of sorting, to eliminate hand sorting. Acceptable product continues its routing to a common discharge conveyor. The spent vines meanwhile are discharged onto the field behind the machine.

FIGURE 4.2. MECHANICAL HARVESTING AND LOADING OF TOMATOES FOR DELIVERY TO CANNERY.

The harvester is a high-capacity machine. Its production depends on the variety, potential yield, field conditions, and growing practices. Yields of 25 or more tons per hour have been reported under ideal conditions.

The harvester operator must control machine speed down rows to enable the crew to work at maximum efficiency and to avoid over-sorting. Down-the-row speed is a variable built into each harvester allowing the operator to keep fruit loss per acre at a minimum. Thus, maximum recovery of the fruit is obtained only by correct operation of the machine.

When to harvest

The optimum time to harvest, of course, is bsed on the maximum percentage of red ripe usable fruit. Some firms estimate their date of harvest by counting the number of days after full bloom (60 days plus or minus depending on the variations among different cultivars, maximum and/or minimum temperatures, rainfall, etc.) followed by actual fruit count of a given size, that is, 1 inch in diameter and projecting ahead for some 40 days, plus or minus, as indicated above. Each of the growing areas, that is, types of soil, irrigation practices, stress conditions, date of planting, etc. may, also, contribute to factors that affect the number of days to ripening and harvest. Some research workers have suggestd the use of the heat unit system as used for other crops, although little published data is available probably due to the wide range of maturities of the many cultivars in use at the present time. Of course, it is not always possible to operate the harvester every day during the harvest season and this creates a serious problem with over-mature fruit, decay, etc. It is important that the harvest time be planned using long range weather data, etheral, and good management practices. Should the field be ready for harvest and other factors prevent harvest as scheduled a new material has been released to protect the fruit from sun-burning. The material is sprayed on the field and leaves the fruit white and prevents much transpiration, therefore, water is conserved and the company states that there is an increase in quality and tonnage. The material is sold under the name of SUN-GUARD and is available from the Sun Guard Chemical Co. in Frenso, CA.

Importance of Sorting

Probably the most important concern in the actual operation of the harvesters is the fruit-sorting crew or the operation of the electronic color sorter on the machine. These persons should be trained in how and what to sort. Oversorting or elimination of good fruit is costly in terms of lost fruit (profit) and in terms of inefficient use of the harvester. Under-sorting also may be costly to the grower because the rejected loads may have to be re-sorted, as well as to the processor, again due to re-sorting expenses. It has been proved that women are more efficient than men for sorting. The sorting crew must be trained to recognize and to remove defects, e.g., sunburn, mold, defective fruits (insect, disease, growth cracks, etc.), and unacceptably colored fruits. Further, they must be instructed in the removal of all trash, clods of dirt, and green fruits. Probably the most critical factor for success with sorters is a well-trained supervisor who should be alert and highly responsible.

The speed of the machine must be varied because the crew will generally work at a steady rate. The supervisor must coordinate the down-the-row speed of the harvester with the quality of the fruit in order to obtain maximum crop yield and to fully utilize the sorting crew. This coordination problem is one reason for considering a central sort system, and only a limited crew on the harvester. Central sort can occur at the factory and eliminate large crews riding the harvester.

Advantages of a central sort system include adequate supervision, better working conditions (temperature, humidity, lighting conditions), updated safety program for the worker, easier shift operation (including rest periods and meal breaks) and, most important, greater efficiency (Gould 1967).

PROBLEMS WITH MECHANICAL HARVESTING

With modern methods of mechanically harvesting tomatoes, problems and concerns by the different segments of the industry have been accentuated. These problems include soil contamination, microbiological contamination, and loss of product quality.

Soil Contamination

Soil on machine-harvested tomatoes has been, is, and will continue to be a problem. Soil is prevalent on the fruit as a result of the method of severing or removing the plant from the ground as it enters the machine.

Soil is present in two forms: as a smear on the surface of the fruit and as the clods delivered into the bins with the tomatoes. The 1961 National Food Processors Association survey shows that about 8 lb of soil were present in each ton of fruit delivered to the plant. In 1966, the average soil load for each ton was 37½ lb. Of this soild load, 4.1 lb were present as a smear on the fruit. Soil in clod form averaged 33.4 lb per ton of fruit (Denny *et al.* 1961; Denny 1962; Denny and DeCamp 1962; Gould 1967; Gould *et al.* 1963; Olsen *et al.* 1966).

Soil smears on unbroken fruit may be removed by vigorous washing. Soil embedded in the tissues of broken tomatoes cannot always be removed in this manner. Such embedded, spore-bearing soil has been the cause of spoilage in canned tomatoes, in juice, and in light-weight purees and sauces (Anon. 1967).

Another aspect of the soil problem involves payment for soil at the same rate as payment for tomatoes. Though this may not seem serious, reports from California have shown loads of tomatoes with soil in the bulk bins of over 5%. This is equivalent to 100 lb of soil per ton of fruit, or for a typical plant running 100 tons of fruit per day, 5 tons of soil. This soil causes the additional problems of clogging the waste disposal system and additional washing requirements for fruit cleaning at the factory. It has been indicated that three times as much water for washing may be required as for handpicked tomatoes (Anon. 1967; Gould *et al.* 1965; Olson *et al.* 1966).

TABLE 4.1. — SIGNIFICANCE OF SOIL (MUD AND DIRT)

% Soil	lb/Ton	lb/Day/100 Tons
0.1	2	200
0.5	10	1000
1.0	20	2000
5.0	100	10000
10.0	200	20000

Microbiological Contamination

Soil in intimate contact with tomatoes during the harvesting operation can create a serious problem for the processor because soil contains many common spoilage microorganisms; 25% of all field soil samples have toxic organisms, 18% of which have been typed and found to contain *Clostridium botulinum* organisms (Gould 1971).

These figures prompted the National Food Processors Association (Denny *et al.* 1961; Denny 1962; Denny and DeCamp 1962) to theorize that a substantial increase in microbiological organisms would be present on the tomatoes at the time of arrival at the processing plant. Of major concern was the possibility that increased numbers of spore-forming bacteria might necessitate a drastic change in the thermal processing time and temperature presently used for tomatoes and tomato products.

Preliminary work conducted by the National Food Processors Association indicated a tenfold increase in spore counts for handpicked tomatoes that were cracked or broken, compared to handpicked tomatoes in sound condition (Anon. 1961). The NFPA also reported an increase in spore count for tomatoes mechanically harvested compared to handpicked tomatoes, as shown in Table 4.2. The soil may be a more serious bacteriological problem if the soil on the fruit is wet (Denny 1962).

TABLE 4.2 — AVERAGE ACID-TOLERANT, HEAT-RESISTANT AEROBIC SPORE COUNTS ON WASH WATER FROM TOMATOES

	No Fresh Cracks/g Tomato	Fresh Cracks/g Tomato
Machine picked	700	1150
Handpicked	30	30

Loss of Product Quality

The third problem created by the mechanical harvesting of tomatoes concerns the quality of fruit entering the factory. It is assumed that all the ripe, overripe, and cull fruits are separated from the tomato plants during the operation. Consequently, a wide range of quality is available because all the fruit is harvested at one time. The amount of sorting will determine the range of quality (color and defects) of the harvested fruit. Crushing or bruising of the fruit due to mechanical handling from the conveyor to the shaker, to the loaders, and into the containers will also help determine the quality of the raw product received by the processor. The amount of damaged and bruised fruit varies depending on the handling system used, including the size of containers, and whether it is wet or dry. Research data clearly show less damage for handpicked fruit placed in smaller containers than for machine-picked fruit placed in dry bulk containers (Gould 1971).

Damage to the fruit can be greatly reduced by:

1. Operation of the machines at slower down-the-row rates,
2. Correct use of side elevator or lowerator to transfer tomatoes from harvester to container,
3. Dropping fruit into tanks containing water (breakage is reduced by as much as 20 to 30% depending on variety and maturity,
4. Depth of fruit in container, and
5. Hauling shorter distances from field to factory. NFPA study has shown that fruit damage ranges from 20% for hauls up to 50 miles to 46% damage for hauls up to 150 miles (Olson *et al.* 1966; Gould 1967).

COST OF MECHANICAL HARVESTING

Machine harvesting of tomatoes has been a reality since the early 1960's with California 100% machine harvest. Most of Ohio is 100%, but other areas still hand pick portions of their crop.

A harvester will harvest from 3.5 to 5 acres of tomatoes a day depending on the yield per acre. 120 to 130 man hours are required to hand pick an acre of tomatoes requiring some 7½ man hours per ton of fruit.

There are significant advantages for both growers and processors from the adoption of mechanical harvesting of tomatoes. First, there is a great savings in labor costs. Secondly, the machine can and usually will work around the clock, weather permitting. Thirdly, Color can be better controlled, at least, when using electronic color sorters on the harvesters. Fourthly, machine harvesting is much more economical than hand harvesting and if everything is properly conducted, the machine can be ammortized in 3 years or less. Fifthly, machine harvesting lends itself to bulk handling of tomatoes and generally, with less damage to the harvested fruit.

Costs of growing, harvesting and handling of tomatoes vary quite widely.

FIGURE 4.3. BULK WAGON LOADS OF TOMATOES, NOTE GATES FOR UNLOADING.

FIGURE 4.4. BULK LOAD OF TOMATOES IN PLASTIC TANKS.

FIGURE 4.5. UNLOADING TOMATOES AT FACTORY. NOTE TRUCK ON 12-15° ANGLE AND USE OF 3-5" HOSE TO FLUSH TOMATOES OUT OF TRUCK.

FMC Corporation has published two charts that should serve as helpful guides for cost per ton to harvest based on tons harvested per hour (Fig. 4.7) and acres harvested per hour based on various row spacing and down-the-row speed (Fig. 4.8).

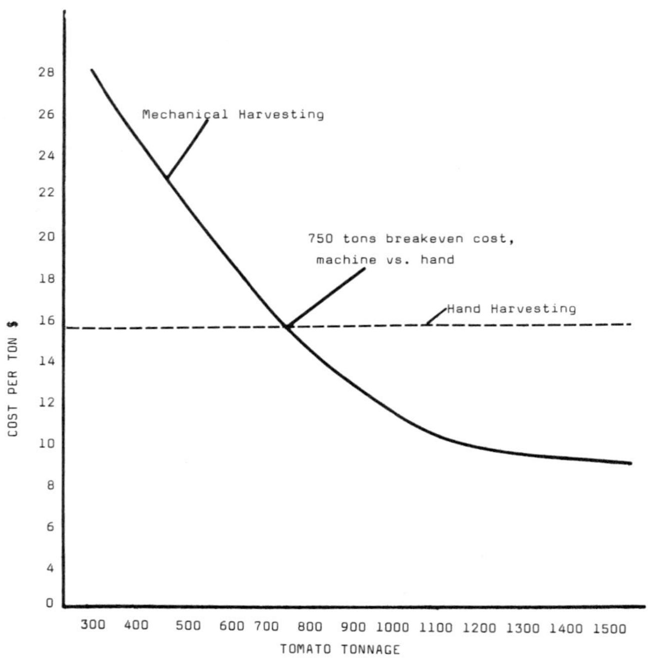

FIGURE 4.6. OUTPUT VERSUS COST PER TON, HAND AND MACHINE HARVEST, OHIO, 1970. FITTED CURVE, 95% CONFIDENCE.
From Wright (1970).

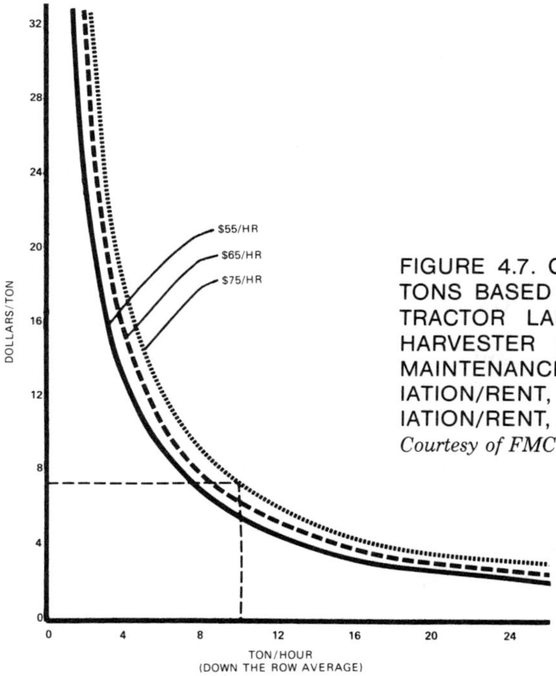

FIGURE 4.7. CALCULATING COST PER TONS BASED ON HARVESTER LABOR, TRACTOR LABOR, FORKLIFT LABOR, HARVESTER DEPRECIATION, REPAIR-MAINTENANCE TRAILER DEPRECIATION/RENT, FORKLIFT DEPRECIATION/RENT, AND SUPERVISION.
Courtesy of FMC Corporation

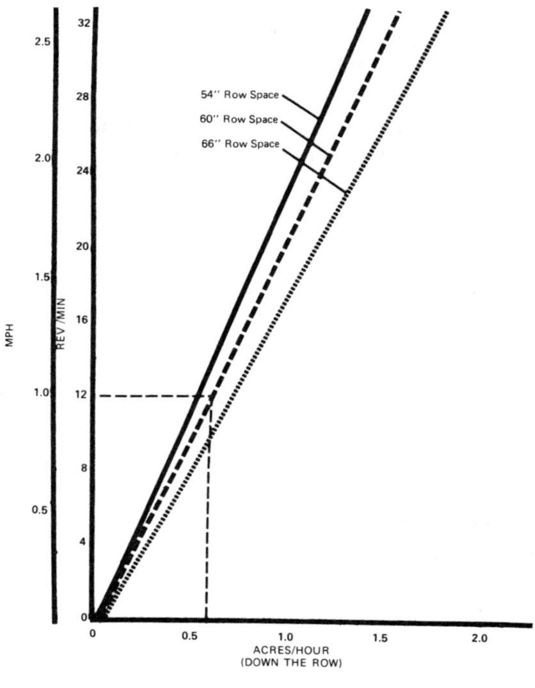

FIGURE 4.8. CALCULATING ACRES PER HOUR "DOWN THE ROW" WHEN REVOLUTIONS PER MINUTE OF STEERING TIRE IS KNOWN.
Courtesy of FMC Corporation.

REFERENCES

ANGELL, F.F. 1968. Tractor-Mounted Tomato Harvester. Univ. of Maryland, Md. Processors Rep. *14* (2).

ANGELL, F.F. 1970. Production of Tomatoes for Mechanical Harvesting: Suggested Practices and Procedures for the 1970 Season. Presented at Univ. of Maryland Hortic. Tomato Commodity Day, Easton, MD.

ANON. 1961. Procedures for Determining Comparative Spore Counts of Mechanically and Hand Picked Tomatoes. Natl. Food Proc. Assoc., Washington, DC.

ANON. 1965A. Mechanical harvesting in Colorado. Colorado State Univ., Colo. Ext. Serv. *4* (3).

ANON. 1965B. Costs of mechanical tomato harvesting compared to hand harvesting. Univ. Calif. Agric. Ext. Serv.

ANON 1967A. The Mechanization of Tomato Production. The Canning Trade, Baltimore, Maryland.

ANON. 1967B. Machines that pick tomatoes. Rohm Hass Rep.

DENNY, C.D. 1962. NCA bacterial spore counts on hand vs. machine harvested tomatoes. Proc. Raw Products Session, 55th Anual Conv., National Food Processors Association. Inform. Lett. 198

DENNY, C.D. and DECAMP, R.A. 1962. Determination on hand picked vs. machine picked tomatoes: 1961 data from Michigan. Res. Rep., *1–62*, Oct., Natl. Food Proc. Assoc., Washington, DC.

DENNY, C.D., REED, J.M. and DECAMP, R.A. 1961. Bacteriological determination on hand picked vs. machine picked tomatoes: 1961 data from Michigan. Res. Rep. 2-61, Oct., Natl. Food. Proc. Assoc., Washington, DC.

GOULD, W.A. 1967. Problems and Concerns with Modern Methods of Harvesting and Handling of Tomatoes. Ohio Agric. Res. Dev. Cent, Columbus. March. The Canning Trade, Baltimore, Maryland.

GOULD, W.A. 1971. Tomato Processor Quality Control Technologist Handbook. Ohio State Univ. Dep. Hortic., Columbus.

GOULD, W.A., BASH, W., YINGST, D., GEISMAN, J.R. and BROWN, W.N. 1963. Handling and holding studies of mechanically harvested tomatoes. Res. Progr. Rep. Fruit Veg. Process. Technol. Div., Ohio State Univ. Dep Hortic., Jan.

GOULD, W.A., BROWN, W.N., BASH, W.D. and GEISMAN, J.R. 1965A. Automated tomato harvesting. Ohio Rep. Res. Dev. Ohio Agric. Exp. Stn., Wooster (March–April).

GOULD, W.A., BROWN, W.N., BASH, W.D. and GEISMAN, J.R. 1965B. Tomato harvesting and handling updated. Ohio State Univ. Dep. Hortic. Ohio Agric. Exp. Stn., Ohio Rep. *50* (2).

HOLLIS, W.L. 1970. Status of mechanical harvesting of tomatoes in the East. Presented at Univ. of Md. Hortic. Tomato Commodity Day, Easton, MD., Jan. 27, Mimeo Rep.

MARLOWE, GEORGE A., JR., W.N. BROWN and W.A. GOULD. 1970. The Fruit Size Intercept Method of Predicting the Harvest Date and Yield of Field Tomatoes. OARDC Dept. of Hort. Mimeograph Series No. 288.

OLSON, N.A., ROSE, W.W., EICHNER, R.L. and MERCER, W.A. 1966. Biological and bacteriological factors in mechanical harvesting fruits and vegetables for processing. Presented at Am. Soc. Agric. Eng. 1966 Winter Meet., Chicago, Dec. 6–9.

PEARL, R.C. 1962. Mechanical harvesting experiments in California during 1961. Proc. Raw Products Session, 55th Annu. Conv., Natl. Food Proc. Assoc., Inform Lett. 1981.

RIES, S.K., STOUT, B.A., BEDFORD, C.L. and AUSTIN, M.E. 1960. A summary of 1960 mechanical tomato harvesting research at Michigan State Univ. Unpublished data. East Lansing, MI.

RIES, S.K., STOUT, B.A., BEDFORD, C.L. and AUSTIN, M.E. 1961. Mechanical harvesting and bulk handling tests with processing tomatoes. Q. Bull. Mich. Agric. Exp. Stn., Mich. State Univ., East Lansing, *44* (2).

STOUT, B.A. and RIES, S.K. 1959. A progress report on the Michigan State Univ. mechanical tomato harvester. Dep. Hortic., Dep. Agric. Eng. Mich. State. Univ., East Lansing. Mimeo Rep.

SULLIVAN, G.H. and UYESHIRE, R.Y. 1970. Cost analysis of mechanical harvesting and bulk handling tomatoes for processing in the Midwest. Purdue Univ. Agric. Exp. Stn. Res. Bull. *869*, Lafayette IN., Dec.

TRAUB, L.G., WRIGHT, P.L. and STEELE, H.L. 1971. An economic study: Hand versus mechanical harvest of tomatoes. Ohio Coop. Ext. Serv. Rep. Sept.–Oct., 67–69.

WRIGHT, P.L. 1970. The latest on machine harvesting of processing tomatoes in Ohio. Ohio Ext. Serv. Mimeo Rep.

ZOBEL, M.P. and PARSONS, P.S. 1969. Machine Harvest Costs Tomato–1968. Lolo County, Agric. Ext. Serv. Univ. of Calif. (Mar.).

CHAPTER 5
Tomato Handling

The quality of raw tomatoes cannot be improved after harvest. However, it can be maintained or decreased depending on the harvesting methods used, the handling of the raw product during harvesting, and the holding methods used. Destruction of quality can be ascertained in several ways: cracking; drosophila fly egg contamination; or the numbers of vegetative bacteria, spores, yeasts, and mold (see Part III).

Tomatoes are transported from the field to the processing plant in either hampers, lug boxes, plastic boxes, or bulk containers.

HAMPERS

The hamper, or five-eighths bushel, was the most widely used container in the Midwest and East before the 1970s. The advantages of the hamper center around its size and shape. Because of its small size, it can be handled conveniently and can be palletized for easy transport. Due to its shape, smaller at the bottom than at the top, the hamper prevents tight stacking, allowing air to circulate between stacked containers; its shape also permits space to exist between containers, allowing loads to be dusted or sprayed fairly uniformly.

The principal disadvantage of the hamper is the crushing of fruit in the bottom of the container. Loading the hampers onto trucks may be difficult because some method of bracing them is generally needed to secure the load while in transit. The small base of the hamper increases the possibility of tipping when being palletized. Because of the flexible construction material used in making hampers, their useful life is very short, generally less than 3 years.

LUG OR FIELD BOXES

The lug or field box is a wooden, rectangular box varying in capacity from 40 to 50 lb of tomatoes. The standard lug box has inside measurements of $7\frac{3}{4}$ in. deep, 14 in. wide, and $21\frac{7}{16}$ in. long. Most of the boxes have a $\frac{3}{4}$ in. cleat on the top edge to prevent damage to the fruit when the boxes are stacked. The boxes are easy to handle, can be stacked tightly for transporting, and are often palletized to facilitate unloading by machine. The tight-stacking characteristic of the box also facilitates loading onto trucks because the trucks do not need sideboards. Lug boxes can be used many times and, if they are kept clean and in good repair, they are expected to have a useful life of 5 to 7 years. The main disadvantages of lug boxes are the amount of crushed fruit due to overfilling, and the crushing of fruit at the bottom. The boxes stack tightly, thus reducing the air flow around the boxes, which may lead to retention of field heat, insect contamination, or bacterial and mold buildup. Tight stacking also may reduce the effectiveness of a dusting operation.

PLASTIC BOXES

The plastic box container holds 40 to 50 lb of tomatoes and can be handled like the lug box. The advantages of this container are that it is easily washed and can be chlorinated. The filled plastic box can be dipped into bacteriostatic and/or detergent solutions. Also, it is about half the weight of the standard lug box and can be nested.

BULK CONTAINERS

The development of machines and systems for mechanical harvesting of tomatoes has created a need for an economical and quality-retaining method of handling the fruit. The bulk system represents another step toward complete mechanization of the crop, which should eventually free the grower from the problem related to hand labor. Successful introduction of bulk handling, with its savings in time, money, and equipment, can do much to keep the tomato business in its present areas (Walson 1969). In recent years the industry has switched to almost all bulk handling.

FIGURE 5.1. HAND HARVESTING CONTAINERS. LEFT TO RIGHT- HAMPER, MIDWEST LUG, PLASTIC BOX, AND CALIFORNIA LUG.

FIGURE 5.2. BULK BOXES FOR HANDLING TOMATOES.

The Bulk Box

Bulk boxes holding from ¼ to more than a ton of fruit are used for handling tomatoes. Their use for holding tomatoes depends on cultivar, grade, hold time, intended use, and whether the fruit is held dry or in water.

Research workers at Michigan State University evaluated different sizes of bulk boxes for handling tomatoes. The bulk boxes used were 45 in. long, 43 in. wide, and 8, 12 or 16 in. deep, holding 320, 486, and 634 lb of tomatoes, respectively. Tomatoes are loaded directly from the harvester conveyor into the bulk boxes. After harvesting, the bulk boxes are loaded by forklift truck and hauled to the processing plant.

It has been found that as the depth of the box increases, fruit injury, expressed as cracked fruit, also increases. There was also a greater amount of injury when the machine dumped the fruit than when it was dumped by hand from lugs into bulk boxes, as is shown in Table 5.1. Since such a small percentage of crushed fruit resulted from most treatments, it would seem feasible to use bulk boxes that are about 12 in. deep for handling tomatoes (Ries *et al.* 1961).

No appreciable differences were found among wood, steel, and wire mesh bulk boxes as far as injury to the fruit is concerned. The cost of construction, durability, and ease of cleaning will determine the ultimate material for box construction (Stout and Ries 1959).

Water Tanks

Bulk tanks filled with varying amounts of water have been given much consideration. MacGillivray *et al.* (1958) made a bulk handling test in California of transporting tomatoes in water and found that the fruit devel-

TABLE 5.1 — THE INJURY OCCURRING DUE TO HARVEST METHODS INTO THREE DEPTHS OF BOXES

Depth of Bulk Box (in.)	Harvest Method		
	Hand	Machine	Average
8	11.1	22.1	16.6
12	12.8	30.2	21.5
16	13.0	33.0	23.0
Average	12.3	28.4	

[a]Injury expressed as percentage of fruit cracks over 1 in. long.

oped numerous small cracks. Five years later these studies were continued to test whether there were advantages outweighing the disadvantage of cracking. It was concluded that dry bulk handling of tomatoes would save $1 per ton of fruit and that water handling would offer no savings. Cooler and Kramer (1960) stated that in the water treatments the tomatoes appeared to be well cushioned from shock and showed little damage. After tomatoes were in water for 48 hr, there was some evidence of bursting. Research workers at Michigan State University (Ries *et al.* 1961) stated that water handling was no better than dry handling and that the quality was appreciably lower if the tomatoes were held longer than 24 hr. Leiss (1962) reported that tomatoes could be held in water and/or water—washing compound solutions for periods of 24 hr without significant quality deterioration.

Advantages of Water Handling. Water offers six main advantages over dry handling for a bulk system.

1. The water serves as a cushion for depositing the tomatoes from the harvester into the containers, as well as during transportation from the field to the factory, thus preventing cracking and bruising.
2. The soak time begins when the tomatoes are deposited into the water containers.
3. The drosophila problem is nonexistent except for the fruit floating on the top of the water in the tanks. National Food Processors Association experiments have shown that chlorine-washed tomatoes were less attractive to fruit flies (Mercer *et al.* 1967).
4. Mold can be controlled by water handling, since antimycotic or other chemical agents can be added to the water for control of the mold growth. MacGillivray *et al.* (1958) observed that mold count values increased over time with dry bulk-handling methods, but not with water-handling methods.
5. The quality of the fruit may improve with water handling depending on the variety, maturity of the fruit, and the amount of organisms entering the tanks. Tests made on canned whole tomatoes to determine whether

quality was lowered by mechanical harvesting and bulk handling of the raw product showed that drained weight of the processed tomatoes is the best quality factor on which compairsons can be made. Gould, Leiss, and Yingst (Gould and Leiss 1962; Leiss 1962; Yingst 1964) reported few differences in the drained weight of processed tomatoes regardless of the type of handling system when the water-hold period is short (up to 12 hr following harvest). However, as the hold periods increased above 24 hr, there was a greater loss in drained weight from tomatoes handled in water tanks than from those held in a dry condition or given the dip treatment (the tomatoes are immersed in a chlorine and water solution while the container is filled during harvest and then drained) (Gould *et al.* 1965).

6. Addition of detergents to the holding water promotes a better washing operation, and the addition of chlorine or chlorine dioxide gives good bacterial control (Cooler and Kramer 1960). The use of these chemicals in water-holding solutions or as dip treatments has reduced the spore counts greatly. Several groups of researchers reported (Bash and Gould 1962A,B; Gould *et al.* 1963; Gould and Leiss 1962; Leiss 1962; Yingst 1964) on the effectiveness of using chlorine solution for handling mechanically harvested tomatoes in water tanks. Two of these investigations (Gould *et al.* 1963; Leiss 1962) showed that chlorine solutions reduced the spore counts on tomatoes to almost zero. Bash (1964) found that chlorine-containing solutions of 550 ppm and 1000 ppm reduced the bacterial counts from tomatoes held continuously in water tanks up to 48 hr.

It is possible to improve quality over dry-handling systems by holding the fruit in water if the water temperature is low, 50° to 75°F (10° to 24°C), and if the time does not exceed 12 hr (Gould 1971; O'Brien *et al.* 1963).

Disadvantages of Water Handling. The water-holding system has some problems. The disadvantage most often cited is the shortage of water in some areas. Other criticisms are the added weight during hauling, the method of weighing fruit for payment to the grower, and the lack of an economical container. This latter point has been overcome by use of the steel hoppers (Gould *et al.* 1965).

Research workers at The Ohio State University, in cooperation with the Chase Foundry Manufacturing Company, Columbus, Ohio, have designed a steel container for use with harvesters. The tank holds 400 lb of fruit and 20 gal. of water. The welded steel hoppers can be handled with a forklift truck, and can be stacked up to three high. Each tank also has an inside splash seal at the top to reduce spillage during transit and to support a lid (Gould *et al.* 1965).

Bulk Trailers

To increase the efficiency of harvest mechanization, a bulk transport or trailer system has been introduced in the East and Midwest in the late 1960's as a means of handling the raw product from the harvesting machine to the processing plant. This system involves harvesting of fruit directly into a bulk trailer (capacity up to 14,000 lb), eliminating the use of the pallet bin (800-lb capacity) and its supporting equipment, including forklifts, field pallet trailers, and extra labor. With the trailer system, fruits are loaded to a depth of 30 inches or more.

The major advantage of the bulk trailer system is the reduced cost per ton of transporting the raw product, $4.59 for bins and $1.86 for trailers (Anon. 1971). An additional advantage is the easy unloading of the fruit. Most factories now use 3 to 6 in. hose to wash the tomatoes from the trailer directly into the flumes conveying them to the factory. The advantage is that soil adhering to the fruit becomes loosened and more easily removed in the final washing operation.

FIGURE 5.3. WATER TANKS FOR HANDLING MACHINE-HARVESTED TOMATOES.

REFERENCES

ANON. 1971. Bulk transport handling evaluated. Calif. Tomato Grower *14* (3) 8, 10.

AUSTIN, M.E., STOUT, B.A. and RIES, S.K. 1963. A Summary of 1963 Mechanical Tomato Harvesting and Handling Research. Dep. Hortic. and Agric. Eng. Mich. State Univ., East Lansing.

BASH, W.D. 1964. Effects of handling and holding practices on the aerobic heat resistant bacterial spore population of mechanically harvested tomatoes. Ph.D. Dissertation. Ohio State Univ., Columbus.

BASH, W.D. and GOULD, W.A. 1962A. Handling and holding studies of mechanically harvested tomatoes—pH. Res. Progr. Rep., Dep. Hortic., Fruit and Veg. Process., Technol. Div., Ohio State Univ., Columbus.

BASH, W.D. and GOULD, W.A. 1962B. Handling and holding studies of mechanically harvested tomatoes—Spore counts. Res. Progr. Rep., Dept. Hortic,. Fruit and Veg. Process., Technol. Div., Ohio State Univ., Columbus.

COOLER, F.W. and KRAMER, A. 1960. Water handling of tomatoes. Md. Processor's Rep. *6* (4). Univ. of Maryland. College Park.

GOULD, W.A. 1971. Tomato Processor Quality Control Technologist Handbook. Dep. Hortic., Fruit and Veg. Process., Technol. Div., Ohio State Univ., Columbus.

GOULD, W.A., BASH, W., YINGST, D., GEISMAN, J.R. and BROWN, W.N. 1963. Handling and holding studies of mechanically harvested tomatoes. Res. Progr. Rep. Dep. Hortic., Fruit and Veg. Process., Technol. Div., Ohio State Univ., Columbus. (Jan.).

GOULD, W.A., BROWN, W.N., BASH, W.D. and GEISMAN, J.R. 1965. Automated tomato harvesting. Ohio Rep. Res. Dev. Ohio Agric. Exp. Stn.,Wooster. March–April.

GOULD, W.A., DAVIS, R.B., KRANTZ, R., JR. and HEALY, N.C. 1956. A study of the factors affecting the grade relationship of fresh and processed vegetables: I. Canned tomatoes. Ohio Agric. Exp. Stn. Res. Bull. *781*, Wooster.

GOULD, W.A. and LEISS, R. 1962. Experiences on water holding of tomatoes. Proc. Raw Prod. Session, 55th Annu. Conv. Natl. Food Proc. Assoc., Inform. Lett. *1981*

LEISS, R.S. 1962. Effects of water handling and holding practices prior to processing on tomato quality. Master's Thesis. Ohio State Univ., Columbus.

MacGILLIVRAY, J.H., CLEMENTS, L.J. and YORK, G. 1958. Bulk handling of canning tomatoes. Unpublished data, Univ. of Calif., Davis.

MERCER, W.A., OLSON, N.A., ROSE, W.W. and EICHNER, R.L. 1967. Handling, Washing, and Utilization of Mechanically Harvested Tomatoes. Natl. Food Proc. Assoc., Berkeley, CA.

O'BRIEN, M., YORK, G.K., MacGILLIVARY, J.H. and LEONARD, S.J. 1963. Bulk handling of canning tomatoes. Food Technol. *17* (8) 96.

OLSON, N.A. 1965. Damage comparison—Tomatoes harvested in boxes and bins. Natl. Food Proc. Assoc. Res. Found. Final Rep. *R.F. 101,* Berkeley, CA.

RIES, S.K., STOUT, B.A., BEDFORD, C.L. and AUSTIN, M.E. 1960A. A Summary of 1960 Mechanical Tomato Harvesting Research. Dep. Hortic., Agric. Res., Food Sci. Mich. State Univ., East Lansing.

RIES, S.K., STOUT, B.A., BEDFORD, C.L. and AUSTIN, M.E. 1960B. Mechanical Tomato Harvesting and Handling Research. Dep. Hortic., Agric. Eng., Food Sci. Mich. State Univ., East Lansing.

RIES, S.K., STOUT, B.A., BEDFORD, C.L. and AUSTIN, M.E. 1961. Mechanical harvesting and bulk handling tests with processing tomatoes. Mich. Agric. Exp. Stn., Q. Bull. *44* (2), East Lansing.

STOUT, B.A. and RIES, S.K. 1959. Mechanical tomato harvester. Progr. Rep. Dep. Agric. Eng., Hortic., Mich. State Univ., East Lansing.

STOUT, B.A., RIES, S.K., BEDFORD, C.L., and AUSTIN, M.E. 1962. A Summary of 1962 Mechanical Tomato Harvesting and Handling Research. Dep. Agric. Eng., Hortic., Mich. State Univ., East Lansing.

WALSON, R.L. 1969. Tomato growing revolution in California. Harvest *4*, 11.

YINGST, D.E. 1964. Effect of various handling and holding practices of mechanically harvested tomatoes prior to processing on tomato quality. Master's Thesis. Ohio State Univ., Columbus.

CHAPTER 6
Tomato Grading

HISTORY AND DEVELOPMENT OF U.S. GRADES

The U.S. Standards for Canning Tomatoes were first issued in 1926, as a result of investigations started in 1923. Shortly after the issuance of these standards, canners began to contract for tomatoes on the basis of U.S. grades. In 1933, the U.S. standards for tomatoes for manufacture of strained products were issued. Since 1933, two things are of particular interest: less than 15% of tomatoes were graded then, while today over 90% are graded; and the percentage of culls has decreased from 8 to 9% in the early years to nearly 2% recently.

A discussion of federal–state grading methods was held at the 1950 NCA annual meeting. Many problems of the grading system existing today are similar or the same as in 1950. The problems cited at that meeting include: (1) inexperienced graders; (2) lack of supervision; (3) lack of uniformity of grade interpretation; (4) ambiguity of rule definition; and (5) poor sampling.

At the present time, four basic systems of grading tomatoes are used in this country. The first method segregates tomatoes into three grades, as set forth in the U.S. standards of grades for tomatoes. The second method segregates tomatoes into four grades, as recommended by the food technologists at the Ohio Agricultural Experiment Station in 1952. With the third system, employed in California, tomatoes are segregated into two grades and an Agtron E instrument is used to determine color. The fourth system, the new USDA system, uses a tomato colorimeter to objectively evaluate color and a four-way classification to subjectively evaluate defects.

SAMPLING

All grading systems are based on obtaining a random and representative sample. Proper sampling is the most important prerequisite in the final grade determination. Failure to obtain a uniform, representative sample results in inequities to both the canner and the grower. This problem is particularly evident when it is not possible, or at least not common practice, to sample the entire load. Such loads, when graded by a licensed inspector, must be marked "restricted" due to an inability to secure samples that are readily accessible. Such restricted sampling does not provide a desirable measurement of total quality not any meaningful grade classification.

Two typical, unrestricted sampling rates for processors in California and Mid-America are shown in Table 6.1. California law requires that approximately one-half of the number of bins in the sample be selected below the top layer, and the balance of the sample bins may be taken from the top layer in the load. In bulk sampling one-half of the probes in the sample shall be from the side of the bulk unit. The balance may be from the middle.

Most samples are taken from lug boxes or hampers. With the increase in use of bulk-handling systems capable of holding 400 to 1000 lb or more of tomatoes, sampling, especially representative sampling, presents a problem. To help solve this problem, California has established rates based either on the number of bins or bulk boxes or on the number of tons in the load. If the number of bins is from 1 to 20, two 50-lb samples should be drawn; for each additional 20 bins or portion thereof, 1 additional sample should be drawn. If the tomatoes are sampled in bulk, sampling is based on the number of tons in the load. If the number of tons is from 1 to 10, two 50-lb samples should be drawn; for each additional 10 tons or portion thereof, 1 additional sample should be drawn. Half the sample should be selected from the side of the load and the balance from the middle.

Presently three methods of sampling bulk-handling systems are being used: the scoop shovel method, the flume system, and the Davis sampler (Curtner 1972).

In the scoop shovel method, a shovel is used to scoop the sample from the upper 2 ft of tomatoes. A major disadvantage of this method is the inability to obtain a representative sample.

In the flume system the product is inloaded into a water flume and sampled while conveyed into the plant. A good random sample is obtained, but, in many instances, the load has been almost completely processed before the final grade is obtained. Unless a holding tank is used to retain the tomatoes until grading is completed, the advantages of grading to control processed quality are lost. However, it is a very fair system of sampling.

The Davis sampler is a mechanical sampling device commonly known as the "Yuba City Sampler." It appears to be the most reliable method presently available for representative sampling of bulk loads. The sampler is electronically and hydraulically operated by means of push buttons. A probe capable of sampling product up to 40 in. deep allows vertical sampling, and can be stopped at any particular point. When a section of the load has been probed, the load can be moved forward or the sampler can be moved across the load, and other areas can be probed. This is done as often as necessary to obtain the required sample size, as discussed above. One disadvantage of this method is its high cost. Also a number of cut tomatoes must be disregarded in defect scoring during the grading process.

TABLE 6.1 — SAMPLING RATES OF TOMATOES FOR GRADING

California		Ohio			
				Bulk Sampling	
No. of Bins	No. of Samples[a]	No. of Containers	No. of Samples	Tons	No. Samples
1–20	2	10–49	2	1–10	2
21–40	3	50–149	3	11–20	3
41–60	4	150–299	5	21–30	4
For each additional 20 bins or portion thereof select one additional sample		300–499	7	For each additional 10 tons or portion thereof select one additional sample.	
		500–749	9		
		750–999	10		
		More than 999	12		

[a]50-lb sample.

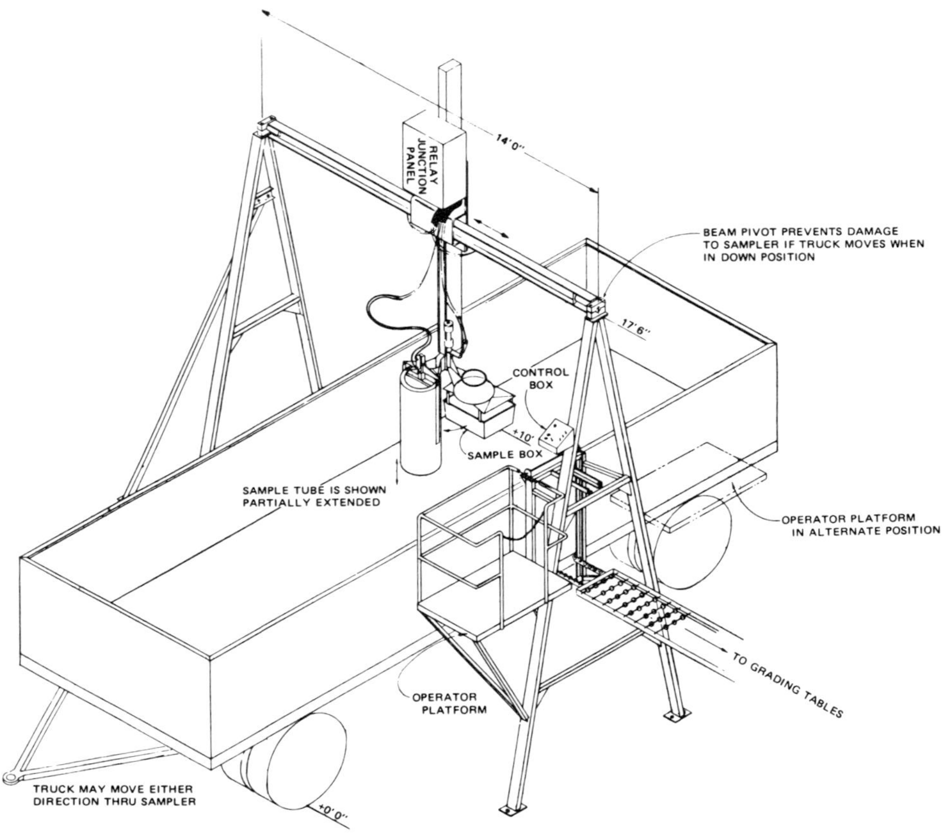

FIGURE 6.1 THE UC-YUBA BULK TOMATO SAMPLER.
Courtesy of Yuba Steel Products Company.

INSPECTORS AND INSPECTIONS

The application of grade standards requires the services of private or official inspectors. The inspector must be honest, tactful, respected, adequately trained, and always on the job. In addition, he should be checked very carefully for color blindness and color sensitivity. The federal–state inspection agencies conduct annual schools for training the inspectors.

The actual inspection and grading of a load of tomatoes requires a representative sample and a grading table. Until recently, stationary grading tables were used to sort and segregate tomatoes into proper categories. Most such tables were constructed with four compartments to facilitate tomato

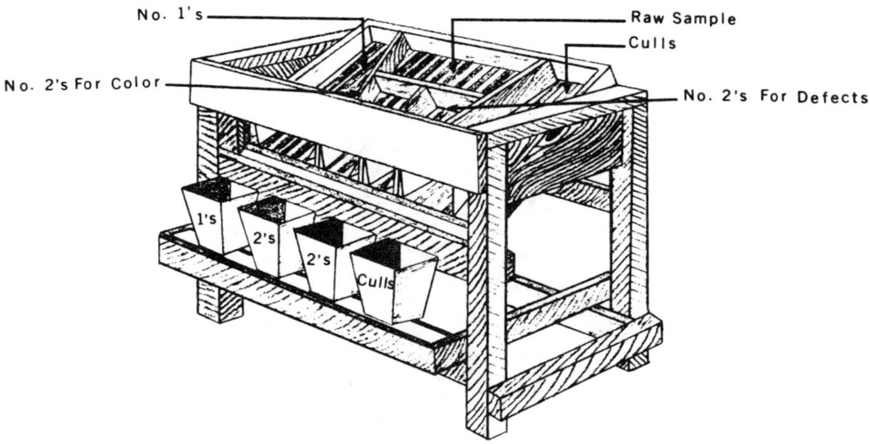

FIGURE 6.2. TILTING TABLE FOR GRADING AND INSPECTING TOMATOES.

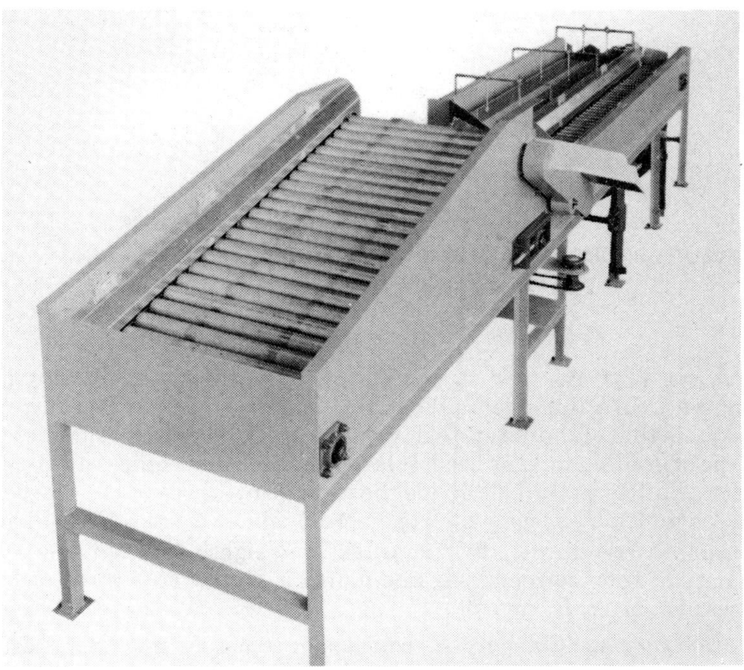

FIGURE 6.3. USDA DUMPING AND CONTINUOUS GRADING BELT FOR INSPECTION AND GRADING.
Courtesy of F.H. Langsenkamp Company.

FIGURE 6.4. UC YUBA CITY SAMPLER TAKING A CORE FROM A LOAD OF TOMATOES.

classification. With the new standards for tomato grade evaluation proposed by the USDA (included later in this chapter), sampling is at a rate almost double that of the old system. Because of the increase in sample size, a new type of grading table is used. It is mechanized and equipped with belt and rollers. Baffles on the belt divide the sample into three parts. The belt's speed is controlled by a foot switch operated by the inspector. Five-eighths of the sample is returned to the truck, one-eighth is used for color determination, and the remaining one-fourth is graded into three or four categories, either A, B, or cull.

(1) *Category A.* "Category A" consists of tomatoes free from worms present and feeding, worm injury that has penetrated through the

outer wall, free from mold or decay, and requiring 5% or less trim for other defects. These are essentially sound tomatoes which normally would be expected to pass over a cannery trim belt without being handled.

(2) *Category B.* "Category B" those free from worms, worm injury that penetrates through the outer wall and mold or decay and requiring more than an estimated 5 percent but not more than 20 percent trim for other defects.

(3) *Category C.* "Category C" those free from worms, worm injury that penetrated through the outer wall but affected by White Mold, Black Mold, Anthracnose or similar diseases, which combined with any other defects, if present, would not waste more than 20 percent, buy weight, of the individual tomato.

(4) *Culls.* "Culls" are those tomatoes which require more than an estimated 20 percent trim or totally green fruit.

The belt conveys the divided and graded sample to a container, usually the lug box or hamper in which the fruit was initially contained. Each category can then be weighed and percentages calculated for percent usable as specified in the "U.S. Standards for Grade Evaluation of Tomatoes for Processing." The sample for color determination is washed and extracted through a 4' mesh screen and a portion is evaluated using a colorimeter. This method provides a more objective indication of the quality of each load of tomatoes than other grading methods. Many contracts are now based on the specifications of this grading system to induce growers to deliver better quality.

GRADING PLATFORMS

Grading platforms should be located in buildings constructed to facilitate the grading of raw product. Modern facilities include such items as restrooms, lockers, storage areas, and instrument rooms. Grading platforms should meet the following requirements.

1. The platform should be situated to allow trucks or wagons to approach either side of the platform. This increases the number of loads which can be handled and provides a minimum of wasted time.
2. The floor space requirements should be estimated under current conditions and then increased 25% or more. This allows for increased volume in the future, or for space to install color-evaluating equipment or sampling devices.
3. The height of the floor will vary with the type of transportation vehicle used, but should be low enough to be at bed level with the vehicle.

4. Floor drains should be present to facilitate washing.
5. A tilt-top desk for comfortable stand-up writing should be available.
6. A sink with water tap and hose should be provided for washing down the equipment.
7. The roof should not extend past the edge of the platform, to prevent damage to high loads or trucks. Roof spouts should be provided for drainage in the event of sampling and grading during rain. A drop canvas or similar feature should be included to close off the platform on cold or windy days.

In addition, artificial lights should be provided above the grading tables. These lights should duplicate daylight of a moderately overcast north sky (7500°K). The lights should be designed to provide illumination on the grading table of 60 to 80 foot-candles when the units are 9 ft from the floor. Artificial lights provide standardization and uniformity of light quality on the inspection table. These lights also permit grading 24 hr per day.

GRADE STANDARDS

On February 15, 1972, the USDA issued notice that the grade standards for tomatoes were being considered for revision. These new standards, U.S.

FIGURE 6.5. USDA INSPECTORS GRADING TOMATOES ON CONTINUOUS BELT

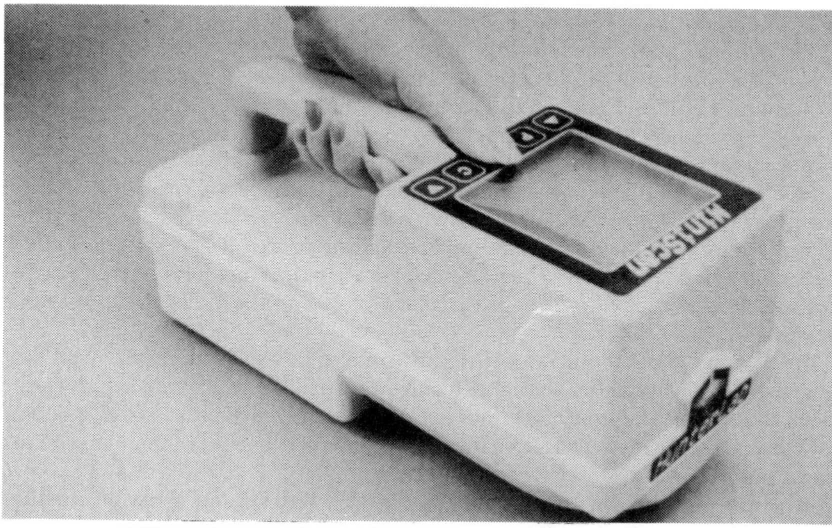

FIGURE 6.6. HunterLab's MiniScan portable hand-held color measurement system.

Standards for Grade Evaluation of Tomatoes for Processing, are to replace the former two-part classification, U.S. Standards for Canning tomatoes and U.S. Standards for Grades of Tomatoes for Manufacture of Strained Tomato Products. They provide a simpler classification of defects whereby percentage usable tomatoes can be calculated, and an optional color determination by color instrument.

The proposal, as it appeared in the "Federal Register," is cited as follows.

Grade Standards for Tomatoes

Effective March 1, 1973 new U.S. Standards for Grades of Tomatoes for Processing are to replace the former two-part classification, U.S. Standards for Canning Tomatoes and U.S. Standards for Grades of Tomatoes for Manufacture of Strained Tomato Products. This Standard was further updated on July 11, 1983. They provide a simpler classification of defects, whereby per cent usable tomatoes can be calculated, and an optional color determination by color instrument.

The U.S. standards provide an inspection procedure for determining the quality of raw tomatoes for processing based on two factors:

(1) classification of defects into various categories; and,
(2) optional color determination by use of either a color instrument or visual evaluation of the fruit.

The following are the principal changes in the new standards.

(1) The number of grade categories will be increased from 3 to 4. "Category B" tomatoes will be required to be free from mold or decay and a "Category C" will be provided for tomatoes affected by mold or decay, including Anthracnose, to not more than 10% by weight of the individual tomato.

(2) "Category A" tomatoes will be required to be firm, meaning that the tomato is not so water soaked, soft, shriveled, or puffy that it will lose more than 10% of its weight during the peeling or washing process. "Categories B and C" tomatoes will be required to be fairly firm. Fairly firm means that the tomato is not so soft that it will lose more than 20% of its weight during the processing process.

(3) "Category A" tomatoes will be required to be free from mechanical damage, meaning when more than one locule is exposed or when causing a waste of more than 10%. "Category B" tomatoes will be required to be free from mechanical damage when more than two locules are exposed or when causing a waste of more than 20% of the individual tomato.

(4) "Free from stems over 1 in. in length" will remain a requirement in "Category A." However, stems over 1 in. in length but not over 3 in. are permitted in "Categories B and C."

(5) The calculation of per cent usable and per cent waste will change to reflect the addition of one category as follows: Total weight of A's + 85% of B's + 75% of C's = % usable.

The standards, as revised, are as follows:

TABLE 6.2. U.S. STANDARDS FOR GRADES OF TOMATOES FOR PROCESSING (3-1-73)

Factor	Category A	Category B	Category C	Culls
Firmness	Firm	Fairly firm	Fairly firm	Water soaked, soft, shriveled, or puffy over 20% waste
Any worm attached	Free from	Free from	Free from	Affected tomatoes classed as Culls
Freezing	Free from	Free from	Free from	
Worm injury	Free from	Free from	Free from	
Anthracnose	Free from	Free from	Not more than 2	

General

The standards contained in this subpart apply to an inspection procedure for determining the quality of tomatoes for processing based on two factors:

(1) color measurement by use of a photoelectric instrument (USDA Tomato Colorimeter) or subjective visual color evaluation of individual fruit; and

(2) classification of defects.

Calculation of percentages shall be on the basis of weight.

Categories

"Category A" consists of tomatoes which meet the following requirements:

(a) Basic requirements:

(1) Firm; and,

(2) Color:

Color measurement by use of an electronic colorimeter instrument or subjective visual color evaluation of individual fruit*.

(b) Free from:

(1) Any worm attached;

(2) Worm injury;

(3) Freezing;

(4) Stems over 1 in. in length;

(5) Mechanical damage when more than one locule is exposed or when causing a loss of more than 10% by weight of the tomato;

(6) Mold or decay; and,

(7) Any other defect or combination of defects, the removal of which in the preparation for processing causes a loss of not more than 10% by weight of the tomato.

p. 51.3312 Category B.

"Category B" consists of tomatoes which meet the following requirements:

(a) Basic requirements:

(1) Fairly firm; and,

(2) Color:

Color measurement by use of an electronic colorimeter instrument or subjective visual color evaluation of individual fruit*.

(b) Free from:

(1) Any worm attached;

(2) Worm injury;

(3) Freezing;

(4) Stems over 3 in. in length;

(5) Mechanical damage when more than two locules are exposed or when causing a loss of more than 20% by weight of the tomato;

(6) Mold or decay; and,

(7) Any other defect or combination of defects, the removal of which in the preparation for processing causes a loss of more than 20% by weight of the tomato.

p. 51.3313 Category C

"Category C" consists of tomatoes which meet the following requirements:

(a) Basic requirements:

(1) Fairly firm; and,

*See notes next page.

(2) Color:
Color measurement by use of an electronic colorimeter instrument or subjective visual color evaluation of individual fruit*.

(b) Free from:
(1) Any worm attached;
(2) Worm injury;
(3) Freezing;
(4) Stems over 3 in. in length;
(5) Anthracnose when more than two spots or aggregating more than a circle 3/8 in. in diameter; and,
(6) Other mold or decay, or a combination of other defects including mold or decay, the removal of which in the preparation for processing causes a loss of more than 20% by weight of the individual tomato; including therein not more than 10% resulting from mold or decay.

Culls

p. 51.3314 Culls.
"Culls" are tomatoes which fail to meet the requirements of Category C and includes tomatoes which, when color evaluation is determined by means of an electronic instrument includes tomatoes which are completely green.

Percent Usable
p. 51.3315 Percent usable.

"Percent usable" is a calculation of total weights of tomatoes in Category A, plus 85% of the weight of tomatoes in Category B, plus 75% of the weight of tomatoes in Category C.

---

*(a) The electronic color evaluation shall be the color value of a composite raw juice sample. The equipment used in such evaluation shall be properly calibrated, and the type of device and procedures utilized shall be specified in grower processor contracts. The composite raw juice sample shall be extracted from tomatoes representative of the lot; Provided, that each tomato from which the juice is extracted must show a definite change in surface color from green to tannish-yellow, pink, red, or combination thereof.

(b) Visual analysis requires that each tomato be "fairly well colored" which means that at least two-thirds of the flesh of the tomato has good red color; Provided, that a tomato having flesh of a lighter shade of red shall be considered as "fairly well colored" if a sufficient amount of the flesh has a red color equivalent to that of a tomato with two thirds good red color.

TABLE 6.3 — DEFECT CLASSIFICATION GUIDE (p. 51.3326)

Defect	Waste 10%	Waste 20%
Sunburn (thin superficial type)	When extending more than 3/4 in. from stem scar, and more than 3/4 of the circumference of a 2½-in. tomato	When extending more than 1 in. from stem scar, and around the circumference of a 2½-in. tomato
Sunburn (type which penetrates outer wall)	When extending more than 3/4 in. from stem scar, and to more than ½ of the circumference of a 2½-in. tomato	When extending more than 3/4 in. from stem scar, and around the circumference of a 2½-in. tomato
Worms and wormy injury	Tomatoes with worms attached or with worm injury that has penetrated through the outer wall, or attached cocoons, shall be classed as "Culls." Worm on the fruit but not attached, and loose worms shall be ignored	
Insects	Grasshoppers, crickets, spiders, or other insects on the tomatoes shall be disregarded, but tomatoes injured by such insects shall be evaluated on a waste basis	
Growth cracks	Badly discolored cracks which are not affected by mold or decay shall be evaluated on a waste basis. Cracks affected by mold or decay which has penetrated the fleshy wall of the tomato shall be classed as "Category C," unless additional defects make them "Culls"	
Gray wall, virus mottling, cloudy spot, ghost spot, internal browning and irregular ripening	Fruit affected by such conditions shall not be handled on a waste basis. Presence of such factors shall be evaluated from the standpoint of their effect on color (See p. 51.3317)	
Blossom end rot	The initial stage of development, occurring as brown or silver discoloration of the skin, shall not be considered as decay. However, if the fleshy wall of the tomato is affected it shall be classed as decay	
Sunscald	Affected areas show a darkened, soft watery condition of the flesh or areas slightly sunken with a tough outer wall which has a whitish yellow appearance, shall be evaluated on a waste basis	
Freezing	Fruit affected by freezing injury develop a wide range of symptoms. Chief symptom of freezing injury is a glass or water-soaked appearance of the fruit. Tomatoes affected by any amount of freezing shall be classed as "Culls"	
Mold or decay	Tomatoes affected by mold or decay which has penetrated the flesh shall be classed as "Category C" or "Culls" depending upon the amount of waste	

Percent Waste

p. 51.3316 Percent waste.

"Percent waste" is a calculation of total weight of Culls, plus 15% of the weight of tomatoes in Category B, plus 25% of the weight of Category C.

Color Evaluation

p. 51.3317 Color evaluation.

Color shall be determined according to one of the following methods:

(a) Unless otherwise specified, the tomato color index (TCI) of a composite raw juice sample shall be not less than 63.0 as determined by means of a photoelectric instrument (USDA Tomato Colorimeter).

(1) The raw juice used for the color determination shall be extracted from a representative sample by means of a USDA approved extractor fitted with a 0.034-in. mesh screen juice attachment. (The extractor and the USDA Tomato Colorimeter are commercially available. Information on where they may be purchased and additional details concerning them, may be obtained from the Fresh Products Standardization and Inspection Branch, Fruit and Vegetable Division, Consumer and Marketing Service, U.S. Department of Agriculture, Washington, D.C. 20250);

(2) Each tomato in the color sample must show a definite change in surface color from green to tannish-yellow, pink, red, or a combination thereof; or,

(b) Each tomato shall be "fairly well colored."

(1) "Fairly well colored" means that at least two-thirds of the flesh of the tomato has good red color: *Provided*, That a tomato having flesh of a lighter shade of red shall be considered as "fairly well colored" if sufficient amount of the flesh has a red color equivalent to that of a tomato with two-thirds good red color.

Extraneous Material

p. 51.3318 Extraneous material.

(a) Extraneous material is loose stems, vines, dirt, adhering dirt, stones, trash, and other foreign material.

(b) The amount of extraneous material in any lot may be specified in connection with these standards.

Definitions

p. 51.3319 Firm.

"Firm" means that the tomato is not water-soaked to the extent that it is soft, shriveled or puffy that it will lose more than 10% of its weight during the peeling or washing process.

p. 51.3320 Fairly firm.

"Fairly firm" means that the tomato is not water-soaked to the extent that it is so soft, shriveled or puffy that it will lose more than 20% of its weight during the peeling process.

p. 51.3321 Worm injury.

"Worm injury" means any worm injury that has penetrated through the outer wall of the tomato.

p. 51.3322 Mold or decay.

"Mold or decay" means breakdown, distintegration or fermentation of the flesh of the tomato caused by bacteria or fungi.

p. 51.3323 Freezing.

"Freezing" means that the tomato is frozen or shows evidence of having been frozen.

p. 51.3324 Green.

"Green" means that the surface of the tomato is completely green in color. The shade of green color may vary from light to dark.

p. 51.3325 Mechanical damage.

"Mechanical damage" means that the tomato is bruised, crushed, or ruptured.

The California standards for defects of tomatoes for canning are presented in Table 6.4.

TABLE 6.4 — CALIFORNIA STANDARDS FOR DEFECTS OF TOMATOES FOR CANNING

Type of Defect	Tolerance for Processing	Description
Worm damage	0%	Damage penetrates the flesh Note: worm damage does not include clean open holes with no excreta present
Mold	10% of weight of tomato	Breakdown or watery appearance in the flesh, or any mycelium or spores of any type of mold fungus affecting the tomato
Rot	20% of tomato	Dry rot present
Sunburn	25% of skin or flesh	White or yellow color and flesh partially hardened
Sunscald	25% of flesh	Flesh is soft and watery due to heat damage
Growth cracks	25% of skin or flesh	Radiating cracks from the stem scar or around the shoulder with the stem scar approximately centered
Insect bites	25% of flesh	Skin of flesh punctured or chewed
Green or yellow color	25% of tomato	Evidence of green or yellow color at stem end and flesh may be partially hardened
Overripe	0%	Most of flesh sort or mushy due to overripeness
Shriveled	0%	Tomato is shriveled or rubbery
Frozen or frosted	0%	Glassy appearance on an aggregate area equal to that of ½-in. diameter circle
Gray wall	25% of tomato	Dark brown or black discoloration of the vascular bundles in wall of the tomato when halved through a cross section
Internal discoloration	25% of tomato	Whitish, greenish, or yellowish areas in the interior of the tomato
Seed sprouts	25% of tomato	Present in three or more cells or if any sprout exceeds 5⁄16 in. in any cell
Green	0%	No visible shade of red color on external surface

GRADE DETERMINATION BY OBJECTIVE COLOR MEASUREMENT

Subjective color determination of tomatoes by the human eye has certain perception limitations (see Chapter 14). Federal and state agencies have attempted to reduce these limitations. Colored disks and pictures of tomatoes are provided for visual comparison. Better lighting is now used over grading tables. In addition, inspectors rotate in order to better standardize results. Even with these improvements of subjective color measurement, an objective method is preferred. Thus, in 1949, a 4-year project was begun to measure objectively the relationship between the grades of raw and processed tomatoes by instruments. This research led to the development of various instruments for this purpose.

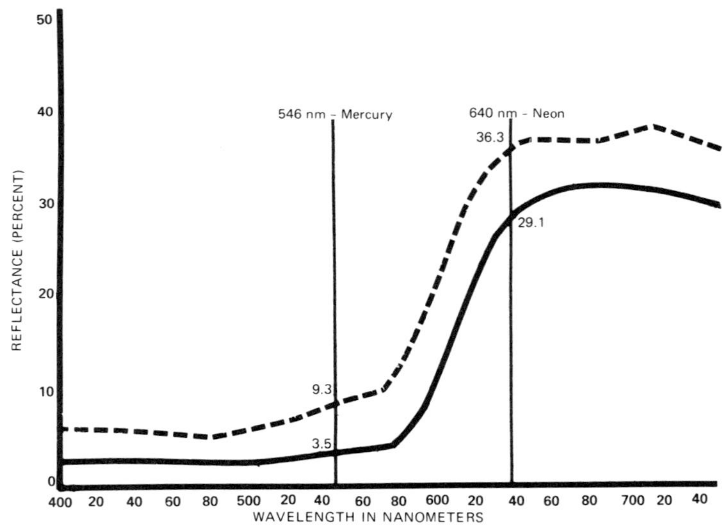

FIGURE 6.6. PRINCIPLE OF OPERATION OF THE AGTRON. TOMATO GRADING FORMULA:

$$G = 276 \frac{(X_G - 0.7)}{(X_R - 0.7)}$$

where G = Agtron grade, X_G = green reflectance (546 nm), and X_R = Red reflectance (640 nm). Therefore:

——, tomato A: $G = 276 \dfrac{(3.5 - 0.7)}{(29.1 - 0.7)} = 27.2$

– – –, tomato B: $G = 276 \dfrac{(9.3 - 0.7)}{(36.3 - 0.7)} = 66.6.$

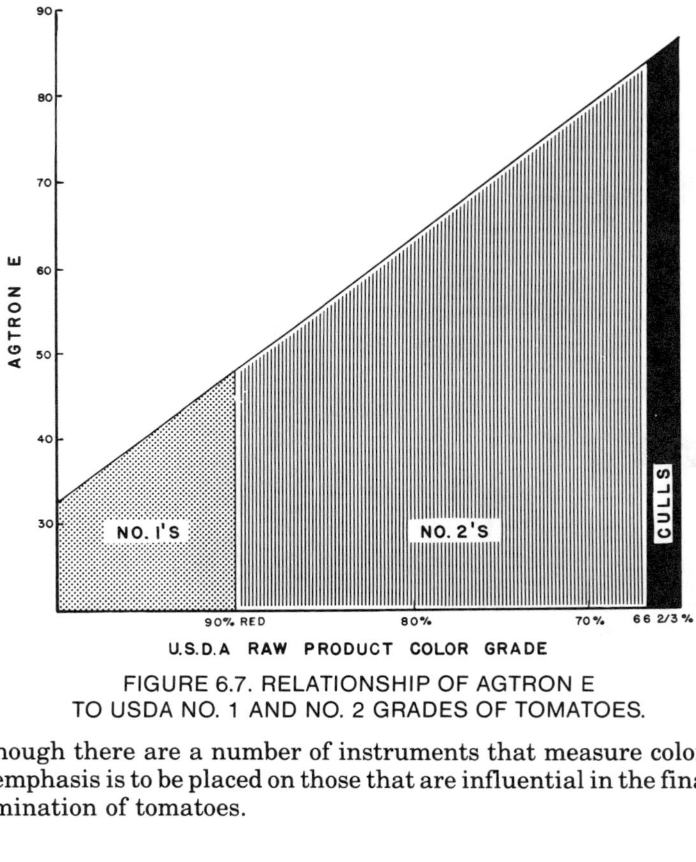

FIGURE 6.7. RELATIONSHIP OF AGTRON E TO USDA NO. 1 AND NO. 2 GRADES OF TOMATOES.

Although there are a number of instruments that measure color today, most emphasis is to be placed on those that are influential in the final grade determination of tomatoes.

Agtron Color Measurement

Agtron Corp. of Sparks, Nevada, have developed color-measuring devices that have gained wide acceptance in the field of tomato color measurement. These instruments are commonly known as Agtrons. Three of these models have enjoyed widespread use as a means of measuring tomato color.

Agtron Model E. The Agtron Model E is designed to measure the inside color of a tomato. The instrument, an abridged spectrophotometer, measures the spectral reflectance of tomatoes at two monochromatic wavelengths and provides a reading which is a dimensionless ratio of the two reflectances. It uses gas discharge tubes for illumination and incorporates selected glass filters to isolate individual spectrum lines. The spectrum lines used are the 546 nm line of mercury and the 640 nm line of neon. These lines were selected because they are in spectral regions which are critical to

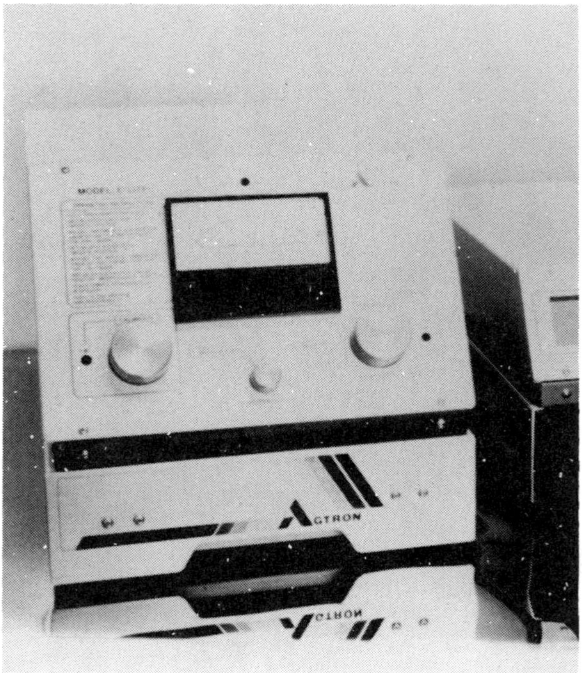

FIGURE 6.8.
AGTRON MODEL E-15TP.

tomato color. The ratio of reflectance at these two wavelengths correlates highly with visual evaluation of tomato color (Anon. 1960A).

In operating the Model E Agtron, the red and green zero levels are established by standardizing on special black disks. The relative sensitivity of the two circuits is then standardized on a red plastic standard, similar to "tomato red." The two tomato halves are placed in position and illuminated by a combination of the mercury vapor and neon gas discharge tubes. A red filter is used to isolate the 640 nm line of neon and the meter is adjusted to read 100. The filter is then changed to a green filter which isolates the 546 nm line of mercury, and the Agtron reading obtained is the color score assigned the product (Gould *et al.* 1954).

The Agtron Model E and a newer Model E-5 are used in California to score the inside color of tomatoes and to provide a basis for grade determination. The instrument is suited for use in California because the standards require that color be measured on the inside of the fruit.

Experiments conducted with the instrument at The Ohio State University have established lines for the categories of No. 1 and No. 2 quality for tomatoes as follows:

1. minimum No. 1: 48.0
2. minimum No. 2: 84.0

The procedure for operation of the Model E follows in two parts; the first deals with calibrating the instrument and the second with color measurement.

(1) Calibration

(a) Turn the meter switch ON and allow a 30-min. warm-up period.

(b) Open the drawer and place both black calibration discs in the supports. The discs must be clean and free from scratches. Center them beneath the photocells using the guide marks at the lower edge of the front panel.

(c) Put the Filter Lever in the Red position (right side). With the Red Zero control, adjust the meter needle on zero.

(d) Put the Filter Lever in the Green position (left side). With the Green Zero control, adjust the meter needle to zero again. This will not affect the previous adjustment of the Red Zero.

(e) Open the drawer, remove the black calibration discs, and put in the red calibration discs.

(f) Use the Standardize Control to set the meter on 46 (with the Filter Lever still in the Green position).

(g) Move the Filter Lever to the Red position, and, if the meter needle goes to 100 $\pm$ 2, the instrument is properly calibrated.

(2) Tomato Color Measurement

(a) Cut the tomato in half through a plane perpendicular to the stem-blossom axis. Avoid making a curved or rough cut surface; it is important that the surfaces be smooth and flat.

(b) Support the two halves in the drawer by means of the two horizontal wires and clips below the wires. Center them beneath the photocells, using the guide marks at the lower edge of the front panel.

(c) Put the Filter Lever in the Red position and set the meter on 100 with the Standardize control.

(d) Move the Filter Lever to the Green position and read the meter. This indicates the color grade. Riper tomatoes are indicated by a low numerical meter reading; greener ones by higher readings (Anon. 1960A).

Agtron Model F. The Agtron Model F is also an abridged spectrophotometer in which a single filter is employed for isolating a selected monchromatic line of the light source (Gould *et al.* 1954). The Model F was originally developed for the USDA to measure the color of extracted pulp or raw tomatoes establishing a color grade based on the USDA Strained Tomato Products Standards (Anon. 1960B). The instrument can be obtained with any one of four light sources and four filters: red, green, blue, or yellow. A gas discharge tube is employed as the light source for illuminating the sample. For the red measurement, the 640 nm line of neon is used; for the green measurement, the 546 nm line of mercury is used; for the blue measurement the 436 nm line of mercury is used; and, for the yellow measurement the 585 nm line of neon is used. The green illuminating light source and the green filter are used for the evaluation of liquid or pureed tomato products (Gould *et al.* 1954). Since tomato ripening is accompanied by a color change from green to red, the values obtained provided an indication of fruit maturity.

The Model F Agtron has a meter calibrated from 0 to 100. The amplifier circuit is designed to independently adjust the zero, or null point, and circuit sensitivity. Two different reference materials are used to standardize the instrument. One of these materials usually has a color slightly darker than the darkest sample to be controlled, and the other has a color slightly lighter than the lightest sample. Then, by means of separate controls, the instru-

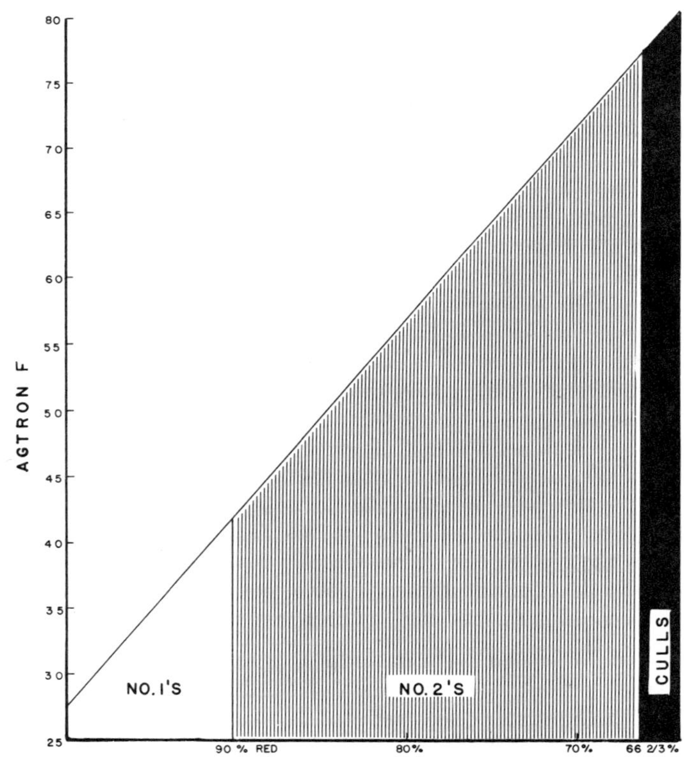

FIGURE 6.9. RELATIONSIP OF AGTRON F TO USDA NO. 1 AND NO. 2 GRADES
OF TOMATOES.

ment is standardized so that it reads zero on the dark standard and 100 on the light standard. In this way, the range of color between the two standards can be spread over the full 100 points on the meter scale. Thus, a manufacturer can establish specific tolerances for the various grades of a given product and establish close color tolerances.

The instrument is used primarily as a stationary, single-sample reading instrument, i.e., the color measurement is made by placing a sample of the product in a standard sample cup. The cup is then placed in a recessed opening in the top of the instrument and its reflectance is measured. However, at The Ohio State University, the instrument has been utilized for continuous color-measuring of tomato juice. So used, it is placed in the line and the juice is pumped through a continuous cell. By this method, all the product is measured and evaluated continuously rather than periodically.

Data indicate that continuous color readings will be of significant value because they allow the manufacturer to know at all times the color of his product. With this knowledge, he can make any necessary adjustment of the raw product, i.e., sorting, trimming, blending, or other processing variables, to keep the product within the desired color ranges.

The USDA, when testing the Agtron unit as a stationary single-sample reading instrument on samples of raw tomato juice, reported a correlation coefficient of 0.92 with the Hunter Color and Color Difference Meter *(L, a, b)*. Similar results have been obtained at The Ohio State University on raw tomato juice with the Agtron Model F and the Hunter Color and Color Difference Meter *(L, a, b)* giving a correlation coefficient of 0.94. Moreover, excellent relationships have been obtained between the Model E Agtron (official cut-surface instrument used in California for tomato grading) and the Model F Agtron (for liquid or puree grading).

Procedure for the color measurement of liquid or puréed tomato products with the Agtron Model F is made in 3 steps.

(1) Standardization

(a) Switch instrument on and allow 45-min warm-up period.

(b) Place red standardizing cup in recessed well and turn standardizing knob (right side) until the meter reads 70.

(c) Remove the red standardizing cup and replace with black standardizing disc. Turn zero adjustment knob (left side) until meter reads zero.

(d) Repeat steps b and c until no variance is noted from the initial settings.

(e) Keep recessed well covered when not measuring color. Handle discs in such a manner as to avoid scratching and periodically clean them in a mild detergent solution.

(2) Sample preparation (suggested by the California Department of Agriculture).

(a) Remove 8½ lb of fruit from the inspection sample taken in conjunction with and in addition to each normal 50-lb inspection sample.

(b) Wash the sample, if necessary, and dry.

(c) Place into gallon blender container, cover with blender lid connected to vacuum hose. Start vacuum pump and when vacuum gauge reaches 27, start the blender for 5 seconds. When blender stops, remove container from blender base (do not break vacuum). Turn upside down and shake once or twice. Return container to blender base.

(d) Blend for one min.

(e) When blender stops, remove lid and insert 14 mesh wire screen into the blender.

(f) Using ladle, remove 175 ml of sample and empty into Petri dish and read color on Agtron.

(3) Color Measurement

(a) If the reading exceeds 39, redeaerate 400 ml of the sample and reread on the Agtron.

(b) If the reading is 39 or less, the load can be certified.

Agtron E-10-TP Tomato Products Analyzer is a Special Applications Abridged Spectrophotometer.

The Agtron E-10-TP instrument is designed for simplicity of operation for grading of tomato products, providing quick, accurate and consistently reliable readings.

The Agtron E-10-TP is a microprocessor-based instrument, ruggedly designed for use in production as well as quality control environments. Simplified sample preparation, automatic calibration procedure, and LCD readout make the E-10-TP easy to use.

The Agtron E-10-TP uses regulated gas discharge tubes for illuminating the sample; condensing optics and narrow bandpass filters for isolating the monochromatic lines from the light sources; photosensors, solid state electronics, and metering circuitry for measuring the relative spectral reflectance of a sample; and a microprocessor for controlling function and processing information. Tomato products analysis is made by placing the product in the sample cup, placing the sample cup in the drawer and sliding the drawer into the instrument.

The E-10-TP is designed to measure relative spectral characteristics. The reflected light is passed through a narrow band pass filter and focused onto the photodiode sensor. The photodiode provides a signal proportional to the amount of monochromatic light reflected from the sample, which is processed and read out on the digital display. The spectral mode and monochromatic wave lengths employed are: Red and Green. The use of narrow band light and narrow band pass optical filters provides a high degree of consistency between units that gives dependable accuracy for many years.

The following features and benefits are standard on the E-10-TP:

Automatic Calibration: Automatically calibrates each time the sample drawer is open.

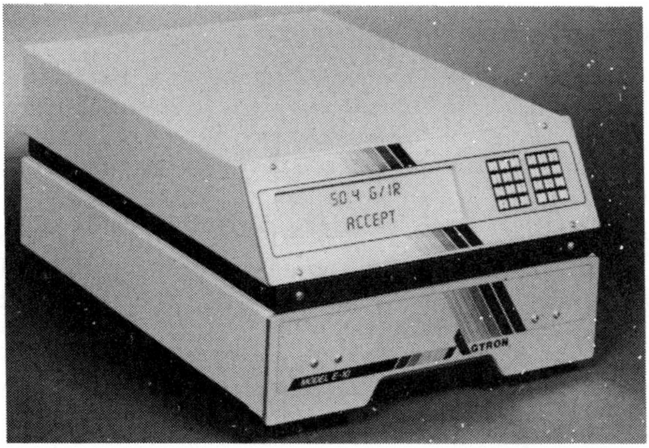

FIGURE 6.10
AGTRON MODEL E-10
FOR GRADING CUT SURFACE
TOMATO AND PULP.

Large View Area: Large viewer area of approximately 12 square inches (4" diameter) provides a well integrated assessment of tomato products.

Long Lamp Life: Agtron Gas Lamps have a useful life span of 3-5 years in continuous operation.

Super High Resolution and Accuracy: Resolution of 1 part in 1000 and repeatability within 0.2%.

Simple Sample Preparation: Puree the tomato to a fine uniform particle size, fill the petri dish to the scribed line (approximately ½ full) and test.

Printer Compatibility: 9600 Baud Serial port allows interfacing with most IBM compatible printers for hard copy of test data.

Green/Red Operation: Green/Red ratio operation dramatically enhances resolution of degree of tomato products analysis.

Monochromatic Operation: The operator can select a basic red or green analysis of product.

Calibration Routine: Software contained bulk calibration routine allows the user to perform a periodic laboratory level calibration, eliminating long and/or short term instrument drift problems and maintains factory calibration accuracy.

Dedicated Application Function: Dedicated application function keys are designed for converting Agtron readings to customer score points, and for inputting individual "accept" and "reject" product sample limits. Product identification and unit codes may also be inputted.

Other Features: Large character, easy-to-read liquid crystal display; single number display of product Test results; display prompts in plain English; standard RS-232 serial port for easy connection to printer or computer (9600 baud).

Hunter Color Measurement

At the request of the USDA, a photoelectric, tristimulus colorimeter was developed by the Hunter Associates Laboratory to measure tomato color. An instrument of this type was desired because a number of different pigments combine to produce the color of a tomato. As such, they could not be measured by any short-cut method such as an absorption of a single pigment at a single wavelength (Hunter 1961).

USDA Hunterlab D6 Tomato Colorimeter. The D6 uses the source–filtered–phototube combinations and optical unit of the color difference meter developed by Hunter (1961). Above the optical unit there is a sliding carriage with openings for a cup of tomato puree (on left) and a ceramic standard (on right) approximately "tomato color." Two beams of incident light strike the specimens at 45° through a window sealed to prevent leakage. The light measured for color is that reflected perpendicularly downward. This reflected light is collected by a light pipe through a diffusing window. The window serves to distribute the light uniformly to three phototubes below the light pipe.

These phototubes are embedded in a thermostitically controlled aluminum block. On the bottom of the light pipe and in front of the phototubes are the tristimulus filters. These filters adjust the spectral character of the

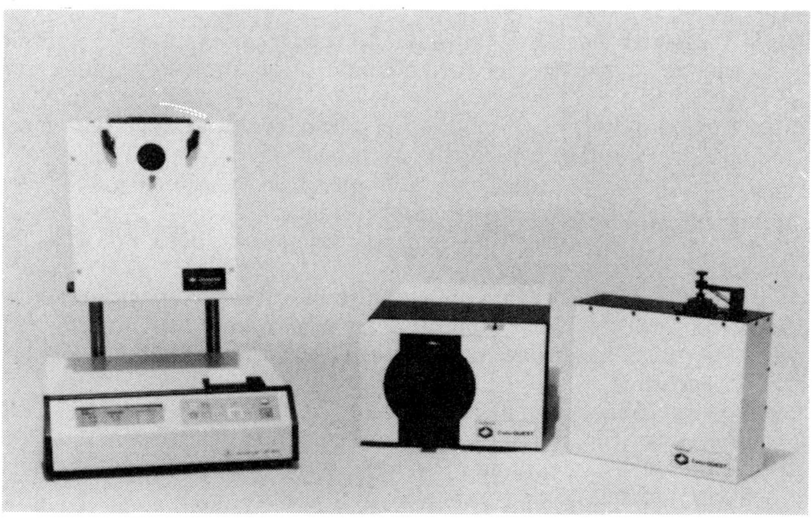

FIGURE 6.11. HunterLab's DP-9000 Color Measurement System.

source-filter-phototube combinations to simulate $\bar{x}$, $\bar{y}$, and $\bar{z}$ for CIE source C.

Signals from these three phototubes are taken to the measurement circuits of the instrument. The circuit for tomato color is on the right side of the instrument panel. The circuit for standardizing adjustments with the tomato color ceramic tile is located on the left side of the instrument panel.

Signals are directed to the TC (tomato color) or the CDM (color difference meter) scales by a relay. This relay is operated by a push switch at the right end of the specimen carriage. Thus, the TC scales are connected when the tomato puree is over the optical unit; the CDM scales are connected when the tomato standard is over the optical unit.

When color values assigned the tomato standard are set on the CDM dials inside, and when the tomato standard is over the optical unit, the standardizing knobs can be used to adjust the reference current and load resistors. This balances the CDM circuit for the standard.

Color measurement with the D6 colorimeter is made by placing puréed or extracted tomatoes, as explained earlier in 51.3316 (Color Evaluation), U.S. Standards for Grade Evaluation of Tomatoes for Processing, above the optical unit. The procedure for standardization and color measurement with the Tomato Colorimeter Hunterlab D6 follows.

(1) Warm-up and preparation for operation

(a) Have instrument power line permanently connected. Leave voltage regulator on continuously; the instrument lamp will burn at half voltage and it need be turned on full voltage only 5 min before starting measurements. If the lamp starts cold, it must be warmed up to 20 to 30 min.

(b) Turn instrument lamp on by flipping switch at bottom of front panel and allow 5 min to warm up.

(c) Center the galvanometer. Hold down on push switch to the left of the meter and turn centering knob on the right to bring the galvanometer needle to zero.

(2) Standardization

(a) Remove lock screws on front panel and pull panel down.

(b) Turn inside *L*, *a*, and *b* dials to the values given on the tomato color standard (*L*-25.6, *a*-27.6, *b*-12.4). These values should always be set on the dials; if necessary, tighten the screw lock on each of the dials to lock in these numbers.

(c) Replace front panel and lock screws.

(d) Have carriage to the left with standard tomato color tile in position.

(e) Turn switch on front panel to *L*. Adjust the left bottom standardizing knob to make the galvanometer read zero. Lock the knob in place by turning knob lock clockwise.

(f) Turn switch to *a*. Set galvanometer to zero with *a* standardizing knob. Lock in place.

(g) Turn switch to *TC*. Set galvanometer to zero with *TC* standardizing knob. Lock in place.

(3) Color Measurement

(a) Fill clean dry plastic sample cup with sample (to within ¼″ of top) and set on carriage.

(b) Move carriage to right so sample is over light beam and relay switch is depressed.

(c) Return pointer switch to *L*. Turn bottom knob on TC side of panel until galvanometer balances at zero. Then turn pointer switch to 21.6 and rebalance galvanometer with 21.6 knob.

(d) Turn pointer switch to *a*, balance with *a* knob on right side of panel. Turn pointer switch to 21.6 and rebalance with 21.6 knob.

(e) Turn pointer switch to *TC*, balance with *TC* knob on right side of panel. Turn pointer switch to 21.6 and rebalance with 21.6 knob.

(f) The tomato color is read from the dial beside the large TC knob after making the last setting.

(g) Check galvanometer frequently during operation of colorimeter and adjust it to zero with the push switch in.

(h) Move the standard into position and check the adjustment of the standardizing knobs at least once every 5 min. If the galvanometer deflects from zero by two scale divisions before restandardization, the previous TC readings are probably incorrect and should be redone.

(i) Clean plastic cup with cool water and dry with soft cloth (plastic will scratch easily). Clean standards with mild soap and warm water. Dry with soft cloth.

TABLE 6.5 — RELATIONSHIP OF COLOR INSTRUMENT VALUES TO QUALITY

Instrument	Unacceptable	Acceptable
AGRTON E5F	>44	43 – 25
Hunter a/b	<1.90	1.91 – 3.50
TCI Colorimeter	< 62	63 – 80

Hunterlab D6D Tomato Colorimeter.. Recently, a modified version of the D6 colorimeter has been designed, the D6D Tomato Colorimeter. It is unique in that it converts into a general purpose colorimeter by plugging a printed circuit board into the instrument. Thus, the unit can be used as a tomato colorimeter for 3 to 4 months of the year, and as a general purpose color and color difference meter for the remaining months of the year.

Relationship between color values obtained from the USDA Tomato Colorimeter and the Agtron E-5 are shown in Table 6.5. A value of 63 on the Tomato Colorimeter is equal to a value of 39 on the Agtron E-5. Values greater than 63 on the Tomato Colorimeter or less than 39 on the E-5 are desired for high colored fruit. Recent studies in Ohio have shown an Agtron

E-5 value of 43 is the cut-off value for acceptable quality of tomatoes for processing.

Firmness

Firmness of the tomato is an attribute of quality given consideration in the grade standards. However, the evaluation of firmness in these standards does not necessarily reflect differences among varieties of cultivars or differences due to handling efficiency on softening of the fruit. Several instruments have been described for measuring firmness (Sobotha *et al.* 1972). The Pressure-Load (P-L) Meter appears to be a nondestructive instrument to measure relative changes in compression of the fruits. Changes in pectic substances are known to be related to the softness of the fruit within a given variety. Differences in maturity by correlating quantitative changes in pectic substances with firmness have been confirmed with this instrument. Firmness is of great significance to the peeled tomato packer as it may directly affect drained weight and wholeness of the tomato after canning.

REFERENCES

ANON. 1960A. Agtron, Operating Principles of Model E Agtron. Magnuson Engineers, San Jose, CA.

ANON. 1960B. Technical Description. Model F Agtron. Magnuson Engineers, San Jose, CA.

BILTON, D.E. and O'BRIEN, M. 1976. A vacuum blending method for preparing tomato samples for color determination. Trans. ASAE Pap. *74-6525*, 386–389.

CURTNER, N.B. 1972. Tomato grading including sampling systems, methods and personnel. Paper presented at Ohio Canners Food Processors Assoc., Annu. Conf. Mar. 14–15, Fed.-State Spvr. Ohio Dep. Agric., Columbus.

GODDARD, W.B., O'BRIEN, M., LORENZEN, C. and WILLIAMS, D.W. 1975. Development of criteria for mechanization of grading processing tomatoes. Trans. ASAE, 190–193.

GOULD, W.A. 1982. Raw tomato color evaluation. Ohio State Univ., Columbus, Res. Circ. *271*, 18–20.

GOULD, W.A., DAVIS, R.B. and MAVIS, J.O. 1954. Color grading agricultural products. Symp. Color of Transparent and Translucent Products. Presented at Wash. Natl. Meet., Am. Soc. Testing Mater., Washington, DC, Feb. 3.

HUNTER, R.S. 1961. Direct-reading tomato colorimeter. J. Opt. Soc. Am. *51*, (5) 553–554.

KADER, A.A., MORRIS, L.L. and CHEN, P. 1978. Evaluation of two objective methods and a subjective rating scale for measuring tomato fruit firmness. J. Am. Soc. Hortic. Sci. *103* (1) 70–73.

KADER, A.A., STEVENS, M.A., ALBRIGHT-HOLTON, M., MORRIS, L.L. and ALGAZI, M. 1977. Effect of fruit ripeness when picked on flavor and composition in fresh market tomatoes. J. Am. Soc. Hortic. Sci *102* (6) 724–731.

SOBOTHA, F.E., WATADA, A.E. and DIENER, R.G. 1972. Effectiveness of the pressure-lead meter in measuring firmness of tomato fruit. Hortic. Sci. *7* (1) 34–36.

STENGEL, R.F. 1972. How ripe is a tomato? Design News, Ideas for Solving Design Problems (Jan.).

CHAPTER 7
Preparation of Tomatoes for Processing

As indicated in Chapter 5, tomatoes are primarily delivered to the processing factory in 10 to 25 ton loads. The tomatoes are unloaded by flushing the loads with water. The water, in most cases, is recycled water. Most firms today only use this water for unloading the tomatoes and it is used as long as possible. The tomatoes are washed into holding flumes (up to 50 tons) prior to being taken to the plant. In some cases the tomatoes are lifted (elevated) out of this first flume, thus keeping this unloading water separate from other flume systems. This first flume system may accumulate considerable sand, dirt, mud, etc. during a given period of time depending, of course, on the harvesting conditions. This first flume should be designed with a drag conveyor to constantly remove the sand, dirt, mud, etc. during the operation and of course, at the end of a given run.

DRY SORT

The first unit operation in preparing tomatoes for processing is dry sorting. The purpose of a dry sort is to remove gross contamination (material other than Tomatoes–MOT) and defective fruit (green, decomposed, or unfit) which would otherwise contaminate wash waters. A wire mesh conveyor or roller belt, which allows loose materials to be separated from the tomatoes, helps to accomplish this. The belt should be of variable speed (15 to 30 fpm) and have sufficient sorting area (two inspectors per ton of fruit) to permit adequate sorting. It should also be supplied with artificial daylight illumination to allow 24-hr operation (Gould 1965).

Grading of Size

Due to the variability in size of tomato varieties, various equipment for sorting has been developed. Generally, diverging-belt graders or cloverleaf drums with varying hole sizes are used. An advantage of size grading includes uniform fill for whole pack portion control. The newer varieties have a tendency toward smaller, more uniform size. Therefore, the need for size grading as a unit operation may be lessened in the future.

WASHING

The major unit opertion in the processing of tomatoes is washing the raw tomato. This operation is essential for the removal of "soil" that is, any substance foreign to the fruit. Types of soil present on tomatoes may include spray residues, microorganisms, dirt, mold, rodents, and drosophila eggs and larvae. Soil must be removed because any residue present in the canned product may cause it to be classed as adulterated, according to the Federal Food, Drug and Cosmetic Act of 1938 as amended.

Studies conducted at The Ohio State University showed that soil can be removed from tomatoes by specific washing practices (Gould 1965; Gould *et al.* 1959). The washing operation involves two phases: the soak period and the spray rinse period.

The Soak Period

Soaking of tomatoes is accomplished by using flumes to convey the fruit from the receiving–dumping area to the factory, or by using a soak tank. If flumes are used, they should be of sufficient size (depth and width with water volume and pressure) to allow fruits to move without clogging the system under maximum capacity. The flume should be designed with both

FIGURE 7.1. FLUME AND ROTARY WASHING TOMATOES.

air and steam pipes in the bottom of the flume to permit agitation of the water. Gould *et al.* (1959) found that agitation of the tomatoes during the soaking period aids in the removal of drosophila eggs and larvae and other contaminants.

The steam inlets should be sufficient to raise the water temperature to 130°F (54°C). A properly sized flume will allow fruits to be held for a period up to 3 min. The obvious purpose of this hold period is to loosen the soil, if any, on the fruit prior to the final wash. Gould *et al.* (1959) found that a temperature of 130°F (54°C) and an exposure of 3 min were effective in removing drosophila eggs and larvae, and resulted in an average reduction in eggs and larvae of 48%, as shown in Fig. 7.2. A greater reduction was obtained at 120°F (49°C) for 5 min; however, holding for this length of time may be too costly and time consuming. At 140°F (60°C) for 3 min, consider-

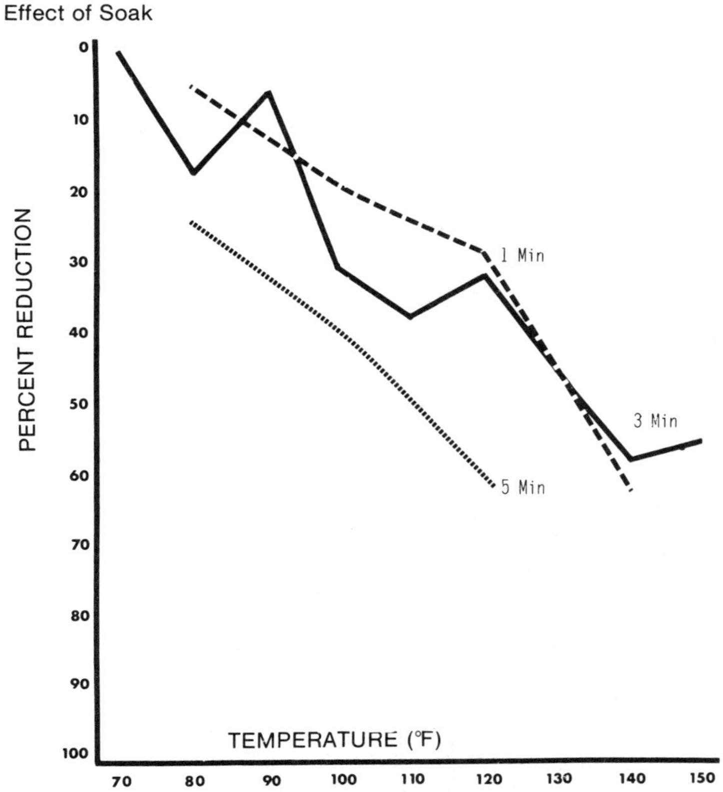

FIGURE 7.2. EFFECT OF SOAK TIME AND TEMPERATURE ON DROSOPHILA EGG REDUCTION.

able slipping of the peel occurred. Therefore, soaking at 130°F (54°C) for 3 min appears to be most nearly ideal. It is important that the flume water constantly be changed to prevent recontamination of the tomatoes.

The flume should be designed to permit the use of detergents, when and if needed. The detergent should be a low-foaming compound with a strongly alkaline solution (pH 11 to 12). The National Canners Association reported that detergent solutions having neutral or slightly alkaline pH gave poor egg removal performance. Gould *et al.* (1959) indicated that the count of drosophila eggs and larvae was reduced 86% when detergents formulated with alkali to a pH of 10 to 12 were added to the soak tank in a concentration of 0.25% by weight and at a temperature of 130°F (54°C).

Lye solution may also be used in washing tomatoes to remove drosophila eggs and larvae. Twigg and Gullette (1965) reported that soaking tomatoes in a 0.5% lye solution for 3 min at 130°F (54°C) was effective for this purpose.

A detergent or a wetting agent maybe added to this solution, or caustic soda already combined with a wetting agent may be used. The wetting agent may increase the effectiveness of the wash and will facilitate removal of residual caustic from the fruit in subsequent washings. They also showed that with sufficient rinsing there was no carryover of lye, and the pH and total acidity of the tomatoes were not affected.

El-Ashwah (1963) studied the effect of using detergents in tomato washing on the viability and thermal resistance of *Bacillus thermoacidurans* spores, the causative agent of flat sour spoilage in canned tomatoes and tomato juice. He reported that washing tomatoes with detergent solutions at 130°F (54°C) for 3 min resulted in a reduction of D values (time required under certain conditions to reduce 90% of the surviving spores) of between 17 and 25%.

Sodium hypochlorite solution may be proportioned into the soak water to control possible thermophilic buildup. Gould *et al.* (1959) reported that a residual chlorine content of 6 to 8 parts per million maintained at all times in the soak water was sufficient to prevent this.

If the flume is not used or is not of sufficient size to allow proper soaking of tomatoes, a soak tank is recommended to allow up to a 3 min hold prior to rinsing. The soak tank should be fitted with an overflow, steam for heating the water to 130°F (54°C) and/or air under pressure for agitation.

The Spray Rinse Period

Following the soak, a thorough rinsing of the fruit is necessary. The tomatoes are conveyed in a single layer using a Shufflo feeder or a roller conveyor. The conveyor should be of variable-speed design to permit a minimum of 2½ turns of the tomato during the rinsing operation. The water must be clean.

Spray nozzles located above the fruit are used for rinsing. Water from the spray should not flow into the soak tank because it would dilute the detergent and chlorine concentration of the soak water. Nozzle type and distance of the nozzle from the fruit are important variables to consider. Gould *et al.* (1959) showed that the Fullcone type giving a square spray pattern is adequate. This nozzle (3/8 GG 18 SQ—Spraying Systems Company) produces the proper particle size to accomplish the desired washing. The number and height of the nozzles are important to obtain optimum coverage and impact at the fruit surface; 9 square spray nozzles at a height of 7 in. above the rollers give optimum coverage on a 40-in. conveyor and optimum spray impact at the fruit surface.

In the rinsing operation, pressure is the most important factor. Impact, coverage, particle size, and volume of water used depend upon the pressure.

FIGURE 7.3. FULLCONE-TYPE SPRAY NOZZLE.
Courtesy of Spraying Systems Company.

Data presented in Fig. 7.5 show that as water pressure increases to a maximum of 150 psi, the effectiveness of drosophila egg reduction increases. However, at 150 psi, there is also an increase in the amount of tomato waste. Waste and water volume are too large to permit practical application. A pressure of 130 psi is most nearly ideal. At this pressure, the Fullcone nozzle delivers approximately 6.5 gal. of water per min. Furthermore, when placed 7 in. above the rollers, each nozzle covers 1 ft^2 of area (Gould *et al.* 1959).

Thus, when the soak operation and high-pressure rinse operation are combined, a great reduction in drosophila eggs and larvae results. Gould *et al.* (1959) found that when detergents were added to the soak tank the reduction in drosophila eggs and larvae was 86%, as is shown in Fig. 7.6.

FINAL SORTING AND TRIMMING

The purpose of sorting and trimming is the removal of off-color and defective fruit or parts (rotten areas, mold portions, insect damage, or sunscald). If the washed tomatoes are carefully sorted so that only the large perfect fruit goes to the peeler or scalder, the rotten fruit to the dump, and the small and misshapen fruit to the pulping line, then sorting will be a profitable investment. Only the best tomatoes should be used for canning. Unevenly ripened and overripe fruits should be used only for pulping. Tomatoes with a small amount of rot or green area may often be trimmed and canned as extra or standard grade (Cruess 1958). Early work by Howard showed that rot in excess of 0.5 to 1% may cause high mold counts in the finished products. The number of sorters and trimmers depends on the dry sort, the washing and rinsing, and the quality of the raw product.

For efficiency and for adequate sorting, roller conveyors, which turn the tomatoes, as they travel are used as sorting belts. The roller conveyor turns the fruit over in front of the sorters and is now in standard use in large factories. A central elevated conveyor belt is sometimes provided for removing defective tomatoes for separate trimming, and all waste materials are placed in chutes beside the picking tables. Sorters must be able to identify a product that should be completely discarded and that which is capable of being trimmed.

Methods of sorting and trimming vary. An effective practice is to use the first two or three sorters to remove absolute waste, while the middle and end sorters concentrate on removing partly defective material by trimming. Some managers prefer to carry out sorting and trimming on the same sorting table, providing the sorters with special trimming knives for cutting out bad tissue. If this method is employed, it is advisable to utilize some operators for sorting only, so that the trimmers do not miss any defective tomatoes.

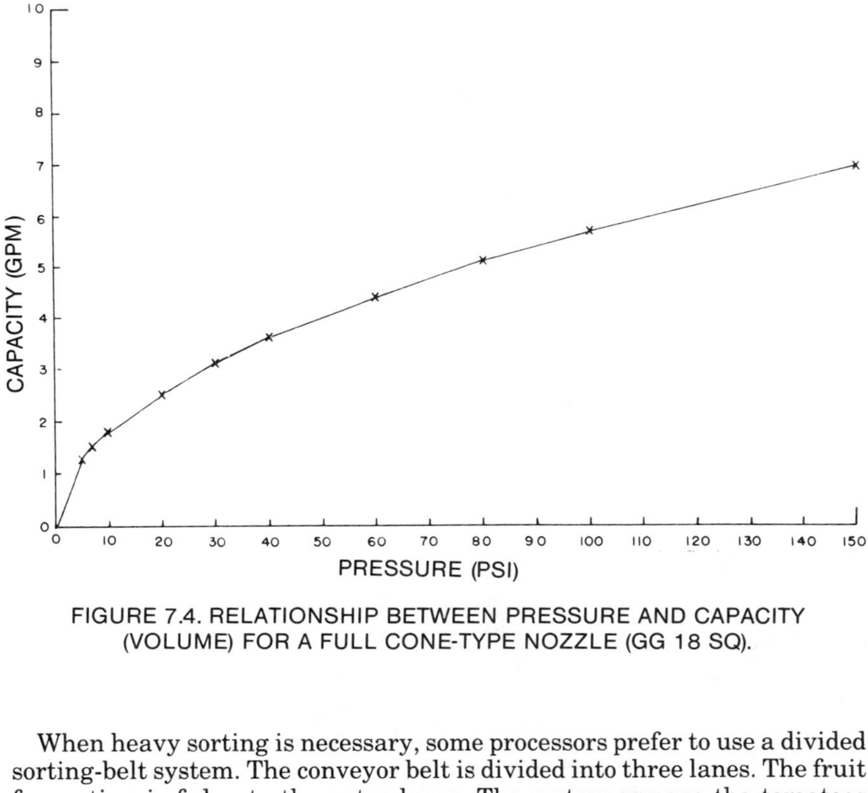

FIGURE 7.4. RELATIONSHIP BETWEEN PRESSURE AND CAPACITY (VOLUME) FOR A FULL CONE-TYPE NOZZLE (GG 18 SQ).

When heavy sorting is necessary, some processors prefer to use a divided sorting-belt system. The conveyor belt is divided into three lanes. The fruit for sorting is fed onto the outer lanes. The sorters remove the tomatoes individually and trim if necessary. The completely sound tomatoes are placed in the central lane for transfer to the canning line. This method ensures that every tomato is handled by either sorters or trimmers. It is claimed to reduce the possibility that defective material will be missed. The material in the center lane should receive a final inspection to ensure that unfit fruit is not thrown into the lane by mistake. The system is not ideal when the fruit is of high quality, when it has been selected in the field, or when it is subjected to dry sorting before entering the plant.

To ensure efficient sorting, it is preferable that operators do not perform the dual operations of sorting and trimming. Each operation should be treated individually. On each line, it is advisable to employ persons carrying out a single operation (Goose and Binsted 1964).

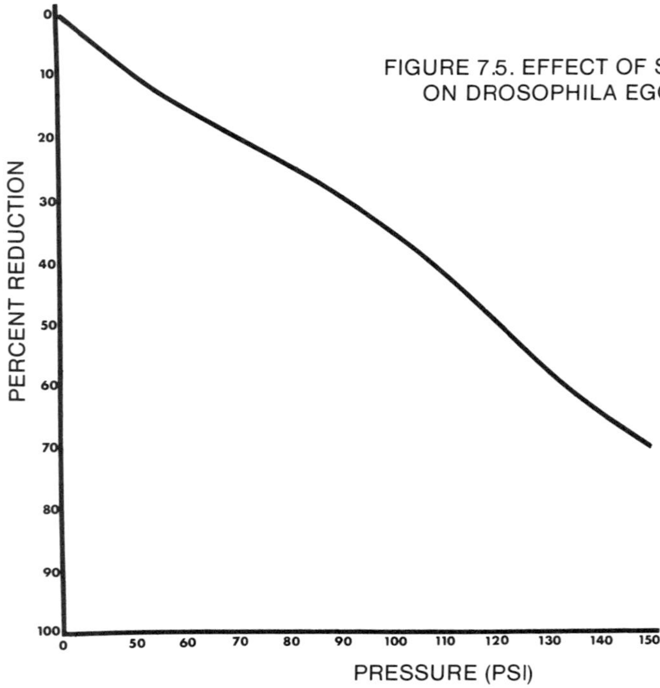

FIGURE 7.5. EFFECT OF SPARY PRESSURE ON DROSOPHILA EGG REDUCTION.

FIGURE 7.7. SORTING TOMATOES PRIOR TO TRIMMING.

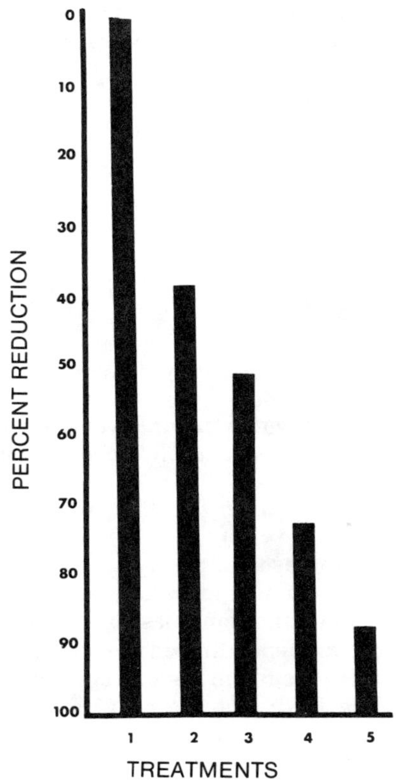

FIGURE 7.6. EFFECT OF COMBINATIONS OF SPECIFIC CHEMICAL AND PHYSICAL FACTORS ON DROSOPHILA EGG AND LARVAE REMOVAL. TREATMENT EXPLANATION: 1. Cold water regular sprays, no agitation (Check Lot). 2. Cold water, high pressure sprays. 3. Hot water, regular sprays. 4. Hot water, high pressure sprays. 5. Hot water, high pressure sprays, 0.25% detergent.

CORING

One of the major problems confronting processors is the removal of cores, if present, to meet minimum quality standards. Extreme care should be exercised to remove only the core tissue. If excess flesh tissue is removed, low drained weight and low wholeness scores may result (Gould 1953). With many of the new varieties, core removal is not necessary because little or no core exists. Generally, a stem scar greater than 0.25 in. in diameter necessitates coring of the tomato. The tomato may be cored by hand or machine.

Hand Coring

Formerly, the Smiley Tomato Knife was most commonly used for hand coring. Today most firms prefer the Mark Lowe or Boucher Tomato Spoon.

The Smiley knife has a curved blade of stainless steel (Fig. 7.8). A straight knife should never be used. Hand coring of tomatoes is more costly than machine coring and is more wasteful of fruit, because some corers and trimmers have a tendency to slice off the top of a tomato while removing the core, or they may chop off the top of the fruit, removing only part of the core.

Machine Coring

Machine coring of tomatoes is done by a small machine, the Hydrout, manufactured by Magnuson Corp., Reno, Nevada. The Hydrout is a water-powered coring and trimming machine. It consists of three essential parts:

1. A blade that varies in design according to the product or size of core to be removed
2. A water turbine wheel spun by a jet of water (100 to 200 psi of water pressure with each machine consuming from 0.6 to 1.2 gal per min of water)
3. A rubber diaphragm that controls the depth of cut

Water, as a source of power, keeps the Hydrout clean, washes away the core and trimmed material, and prevents mechanical breakdown.

In coring tomatoes, each Hydrout has a reported capacity of up to 700 lb per hr, depending on the operator and size of the fruit. To control production it is possible to install counters and pay the operator on a piecework basis. The Hydrouts are easy to install and may be located along trimming, sorting, and/or peeling conveyors. It is recommended that the Hydrout be located after the tomatoes have been selected for canning, prior to peeling. It was found that coring is easier before peeling and that the peel would begin to slip away before the trimmers had touched the fruit. In other words, the skin had been broken by coring before peeling, and, in the case of those fruits that were solid and free from defects, the skin came off without the use of a knife. Gould (1953) found that production capacity was increased by at least 25% and the same level of quality maintained when coring with the Hydrout and using the same number of peelers and corers. He also reported no significant difference in the finished product quality between the tomatoes cored with the Hydrout machine versus the conventional method, as shown in Table 7.1.

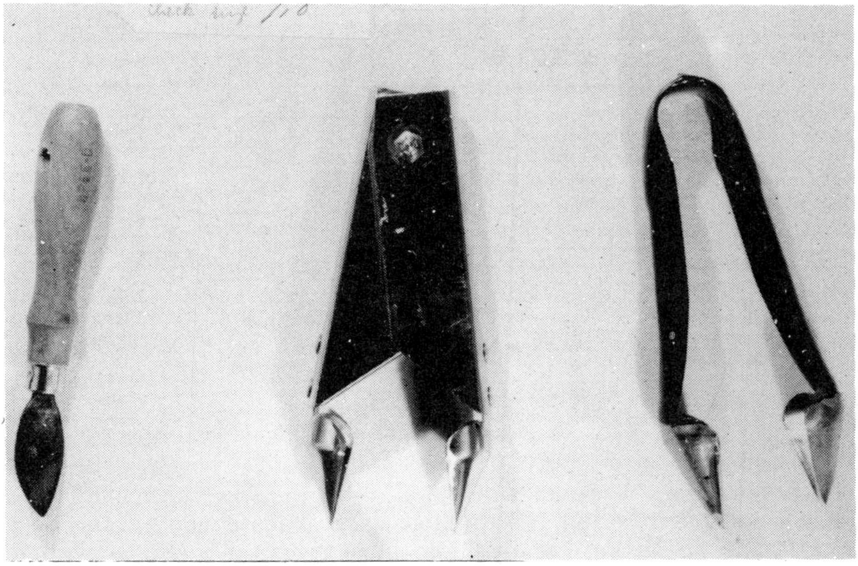

FIGURE 7.8. CORING KNIFE AND SPOONS. LEFT TO RIGHT- SIMLEY TOMATO KNIFE, MARK LOWE TOMATO SPOON, AND BOUCHER TOMATO SPOON.

FIGURE 7.9. HYDROUT CORING AND HYTAB COUNTERS FOR USE IN PREPARING TOMATOES FOR CANNING.

TABLE 7.1. EFFECT OF CORING METHOD ON QUALITY OF CANNED TOMATOES

Quality Attributes	Coring Method	
	Hand Cored	Hydrout Cored
No. of cans graded[a]	288	288
Total score	93.4	93.1
Drained weight score	18.0	18.0
Color score	28.2	28.1
Wholeness score	17.7	17.6
Absence of defects	29.5	29.4
U.S. grades	A	A

[a]These cans of tomatoes were graded by the USDA using the scoring system as outlined in the U.S. Standards for Grades of Canned Tomatoes (Aug. 1, 1946).

PEELING

Peeling for canned tomatoes is a high-cost operation in terms of labor. When coupled with coring, it accounts for approximately 60% of the total labor cost for processing tomatoes. The efficiency of the peeling operation influences the quality of the finished product (Schulte 1965).

Three major tomato-peeling methods are currently used. The cost of equipment and the volume of fruit determine choice of method.

Steam Peeling

Tomatoes are scalded in live steam long enough to loosen the skin but not so long that the pulp and flesh become soft or that the tomatoes are thoroughly heated (Cruess 1958). The scalding is accomplished by conveying the tomatoes through a live steam chamber. Steam peeling was widely used in the past. The steam scalder is an enclosed sheet-metal box filled with sprays of live steam. The tomatoes are exposed to live steam from 30 to 60 sec depending on variety, fruit size, and the stage of maturity or ripeness of the tomatoes. As they emerge from the scalder they are subjected to sprays of cold water to crack the skin.

For an efficient tomato-peeling operation, it is essential that the tomato skin be completely loosened. Light scalding normally requires considerable use of a peeling knife to remove the peel completely. This increases labor cost, adds to production time losses, gives undesirable appearance, and generally leaves some peel on the finished product. On the other hand, overscalding must be avoided, because it softens the tomatoes, resulting in excessive product losses and poor quality.

A good steam scalder operating at 208° to 212°F (98° to 100°C) is considered satisfactory. Sometimes closed steam coils are added to the scalder

FIGURE 7.10. B-3 Stepeel After HC Magnuscrubber.

above the tomatoes to increase heat in the box. A well-designed scalder of this type, when used with dry 125-psi steam, can attain 300°F (149°C). Obviously, high temperatures will result in shorter heating time (Anon. Undated). Presently, high pressures with resulting high temperatures are used.

A system developed by Savi Antonio of Italy, operates by immersing tomatoes in hot water agitated by steam for a few seconds followed by discharging into a hopper where the tomato temperature is reduced by a cold water spray. The sudden drop in temperature helps loosen the peel without loss of other tissues. A conveyor moves the fruit out of the hopper to a distribution chute, which separates the tomatoes into multilanes with an overhead scoring knife making a shallow slit in the peel of each tomato Next, the tomatoes go through an open lane that closes in segments to grip the peel and slip the tomato through the scored slit in the peel.

Lye Peeling

Chemical peeling of fruits and vegetables has been used in the food-processing industry for many years. The first commercial use of lye solution to remove skins from food product was in the production of hominy. The use of lye peeling in the canning industry is more recent; peaches and apricots are early examples. There has been a dramatic increase in the use of lye

FIGURE 7.11. MECHANICAL NONCAUSTIC TOMATO PEELER.

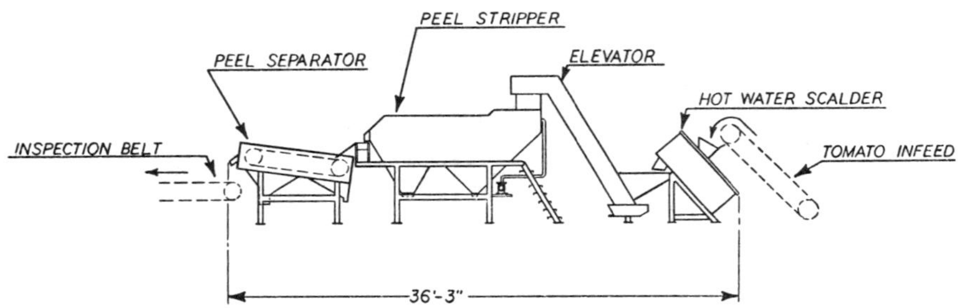

peeling of many fruits and vegetables due to its economy, simplicity, and labor-saving advantages (Lucas 1967).

The active agent in lye peeling is a water solution of caustic soda. Caustic soda (sodium hydroxide, NaOH) is often referred to as lye (Anon. 1964). Lye peeling is made possible because caustic soda is able to attack cuticular tissue and cause its dissolution. Caustic soda is selective in its action, rapidly attacking and breaking down the outer tissues of fruits and vegetables while leaving the inner flesh untouched. The broken-down outer tissue, soft and gelatinous, is easily washed away with a water spray (Anon. 1964).

Equipment for lye peeling of tomatoes consists of a means of immersing the product in a hot lye bath or of spraying lye on the product for a definite period of time; provision for draining the excess lye from the product; and a means of washing away the lye and disintegrated peel. Equipment includes the draper-type scalder, Fox lye-spray scalder, mill-wheel scalder, and the combination FMC lye-filming and pressure steam peeler.

The concentration of the caustic soda varies from 16 to 20% depending on temperature and the addition of wetting agent. Suitable wetting agents added to the lye peeling solutions give more uniform peeling, either in less time or with lower concentration of lye, and decrease the amount of water required to remove residual lye. Two of the wetting agents approved by FDA for use in lye peeling and detergent washing are sodium 2-ethylhexyl sulfate and sodium mono- and dimethylnaphthalene sulfonates. Both are anionic surfactants supplied under various trade names (Tergitol anionic 08 from Union Carbide and Chemical Co. or Faspeel from Wyandotte Chemical Co.).

TABLE 7.2. DENSITY AND CAUSTIC SODA CONTENT OF CAUSTIC SODA SOLUTIONS AT 60°F (16°C)

% NaOH	Specific Gravity 60/60°F	NaOH (g/Liter)	Total Wt of Solution (lb/gal.)
1	1.012	10.11	8.44
2	1.023	20.46	8.53
3	1.034	31.02	8.62
4	1.045	41.80	8.71
5	1.056	52.80	8.80
6	1.067	64.02	8.90
7	1.079	75.53	9.00
8	1.090	87.20	9.09
9	1.101	99.09	9.18
10	1.112	111.20	9.27
15	1.167	175.05	9.73
20	1.223	244.60	10.20
25	1.278	319.50	10.65
30	1.332	399.60	11.10
35	1.384	484.40	11.54
40	1.434	573.60	11.96
45	1.483	667.35	12.36
50	1.530	765.00	12.76

TABLE 7.3. — PREPARATION OF CAUSTIC SOLUTIONS

Caustic (% by wt)	Oz Caustic Required per Gal. Water	Lb Caustic Required per 10 Gal.	Temperature Rise		Gal. of Final Caustic Solution at 60°F(16°C)
			°F	°C	
1	1.4	0.9	5	−15.0	0.997
2	2.9	1.7	10	−12.2	0.997
3	4.3	2.7	15	−9.4	0.996
4	5.0	3.6	20	−6.7	0.996
5	7.2	4.5	25	−3.9	0.996
6	8.8	5.5	31	−0.6	0.996
7	10.4	6.5	37	2.8	0.997
8	12.0	7.5	42	5.6	0.997
9	13.6	8.5	48	8.9	0.998
10	15.4	9.6	54	12.2	0.999
15	24.0	15.0	86	30.0	1.007
20	33.6	21.0	113	45.0	1.021
25	44.0	27.5	139	59.4	1.042
30	54.9	34.3	160	71.1	1.071
35	66.6	41.6	177	80.6	1.111
40	78.9	49.3	185	85.0	1.161
45	91.8	57.4	191	88.3	1.224
50	105.3	65.8	191	88.3	1.306

FIGURE 7.12. ELECTRONIC SORTING FOR COLOR AND DEFECTS GREATLY IMPROVES THE EFFICIENCY AND COSTS OF OPERATION.

Schulte (1965) successfully peeled tomatoes using 190° to 200°F (88° to 93°C) and 18% caustic soda with an immersion time of 25 sec. He found that the use of a wetting agent with the caustic solution facilitated the action of the solution and that the peel loss was reduced by addition of the wetting agent (El-Ashwah 1963).

Schultz and Smith (1968) successfully peeled tomatoes using 16 to 20% caustic soda with 0.3% Faspeel added at a temperature of 190 to210°F (88 to 99°C) with an immersion time of 20 to 30 sec. They found that the wetting agent reduced the caustic soda concentration required to perform the peeling operation. Lucas (1969) found that the wetting agents Tergitol and Faspeel at a concentration of 0.3% were effective in reducing tomato peel loss when used in conjunction with 20% caustic solution. He also found that the application of wetting agents to the peeling solution increased the tomato drained weight scores.

The temperature of the lye solution varies from 190° to 210°F (88° to 99°C) with an exposure time from 20 to 30 sec. The fruits are allowed a 45- to 60-sec reaction time and then cold water is sprayed to remove the digested peel or skin.

Floros and Chinnan have explained the chemical peeling process as complex reaction involving diffusion and chemical reactions. They state that once the caustic solution comes in contact with the surface of the fruit, it dissolves the epiculticular waxes, penetrates the epidermis, and diffuses through the skin into the fruit. Inside the fruit, the NaOH reacts with macromolecules and organic acids in the cytoplasm, middle lamella and cell wall, and, as a result, separation of the skin takes place. Above a certain baseline (concentration of greater than 2 Molar NaOH), the diffusivity was independent of concentration, which verifies that the theory of Fickian diffusion holds true and applies to chemical peeling. The temperature dependence of the diffusivity was of the Arrhenius type above a certain temperature (50 degree C.). At lower temperatures (e.g., 30 degree C), diffusion did not take place due to inhibition by the solid state epiculticular waxes present at the surface.

Lye peeling may cause neutralization of fruit acids due to carryover of the lye from the peeling solution. Furthermore, the FDA regulation 21 CFR 121.0191 requires that the use of wetting agents by "followed by rinsing with potable water to remove, to the extent possible, residues of the chemicals."

National Food Processors Association experiments have shown that lye carry-over can be reduced or prevented by flooding the peeled and cored tomatoes with clean water. Abundant water is necesary to remove caustic from the surfaces and tissues of the tomato. Two rinsing tanks may be used, the tomatoes being spray washed as they leave each tank. Some processors also follow the rinse with up to a 10% citric acid dip. Citric acid can be added to the rinse water also. The addition of citric acid to an immersion tank is a practical method of controlling lye carryover on tomatoes. The pH of the rinse should be between 5 and 6 (Anon. 1969).

Graham and co-workers developed a rubber disc system for removal of the peel after caustic or steam peeling. The discs are 1/32 inch in thickness and 4¼ inch in diameter mounted ⅞ inch apart on steel shafts. The shafts are parallel spaced 3 inches apart, thus providing 1¼ inch overlap between adjacent rows of staggered discs. The discs are driven at 400 rpm. Workers at The Ohio State University found that the system operated more effectively if it was on a 10°incline. The advantages of the system are that it uses very little water (5 gallons per ton of fruit) and a near perfect job of peeling and adhering stem removal.

Infrared Peeling

Infrared is defined as those rays lying just beyond the red end of the visible spectrum. The wavelengths are longer than those of visible light but

FIGURE 7.14. FMC CA 20 TOMATO PEELER (Left)
PR 20 TOMATO PEEL REMOVER (Right)
Courtesy of FMC Corporation

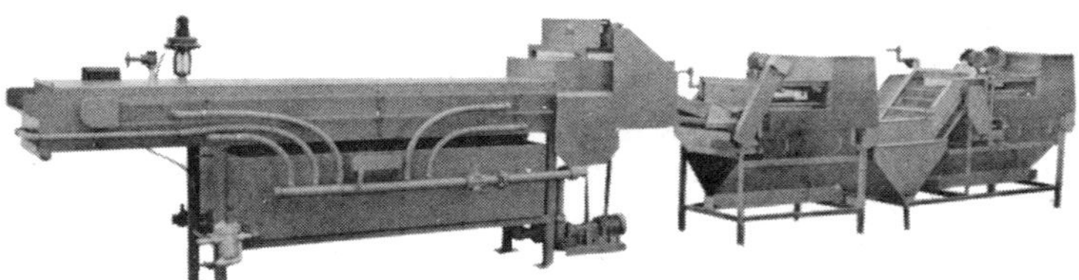

FIGURE 7.15. LEADER/FOX CAUSTIC PEELER

FIGURE 7.16. DISCHARGE END OF LEADER/FOX PEELER AND REACTION BELT PRIOR TO SCRUBBING THE PEEL OFF.

FIGURE 7.17. MAGNUSON TOMATO WASHER AND SCRUBBER (MAGNUSON MODEL HC MAGNUSCRUBBER).
Courtesy of Magnuson Corporation

shorter than radio waves. The infrared region of the electromagnetic spectrum extends from wavelengths of about 0.75 to 1000 μm.

Infrared radiation as a source of heat has been studied for a number of applications in the food industry (Lafferty 1960).

A prototype model for infrared peeling was designed and used in the Pilot Plant Laboratory of the Department of Horticulture, The Ohio State University. This model consisted of four gas-fired red burners which expose the fruit to high temperatures (1500° to 1800°F) (816° to 982°C) while revolving on a spindle to allow more of the tomato surface to be directly exposed to the infrared heat. Also, the vertical angles of the burners can be adjusted to ensure maximum heat penetration of the skin. Exposure time varies from 4 to 20 sec, depending on variety, size of the fruit, and maturity. When the fruit is so exposed, only the epidermal cells are heated. The water contained in these cells is changed from liquid to steam and the peel "explodes" or is broken away from the tomato, and is then easily removed by a fine cold water spray wash or by slight rubbing action. Schulte (1965) found that peeling tomatoes with an infrared method had a peel loss of 5.27% while the steam method had a peel loss of 7.47%. The data in Table 7.4 illustrate the relative efficiency of the different peeling methods as compared with the usual method, live steam. Perhaps of more significance is the relation of treatment to quality determined in accordance with the USDA Standards for Grades of Canned Tomatoes. The data in Table 7.4, representing averages for 12 varieties of tomatoes, clearly point out the possibilities of infrared peeling.

TABLE 7.4. EFFECT OF PEELING METHOD AND SIZE OF TOMATOES ON EFFICIENCY AND QUALITY OF CANNED TOMATOES

	Peeling Method by Size of Fruit					
	Steam		Lye		Infrared	
	Small	Large	Small	Large	Small	Large
No. of fruit/25 lb	87.5	57.5	87.0	56.5	83.0	53.0
% core	5.2	3.2	4.5	2.8	4.7	3.2
% peel	7.8	7.0	2.9	2.4	5.7	5.0
lb peeled/hr	305	352	533	741	232	261
pH	4.28	4.30	4.30	4.32	4.28	4.31
Total acid	0.469	0.461	0.455	0.445	0.445	0.445
Drained weight (oz)	10.77	10.77	10.87	10.88	10.89	10.91
Drained weight (20)	17.3	17.3	17.9	17.9	17.9	17.7
Wholeness (20)	16.1	16.4	16.4	17.2	15.1	16.4
Color (30)	27.7	28.7	27.5	28.3	28.0	28.5
Absence of defects (30)	29.6	29.4	29.5	29.6	29.2	29.1
Total score	90.7	91.8	91.3	93.0	90.2	91.7
Grade	B	B	B	B	B	B

Food Machinery Corporation has developed two new systems for peeling tomatoes, a Wet Steam System as illustrated in Figure 7.18 and A Dry Vacuum System as show in Figure 7.19. These peelers eliminate the use of caustic and, of course, the resultant caustic disposal problem. The tomato peel is, thus, usable waste. The basic principle in both systems is high temperature peeling in pressurized continuous chambers followed by immediate water cooling upon discharge. The peel, if necessary, is further removed by passing over pinch rollers. The further advantage of the Dry-Vac System is the ability to collect the tomato vapors to add back to the tomato and tomato products.

FIGURE 7.18. FMC WET VACUUM SYSTEM FOR PEELING TOMATOES.
Actual Layout of Equipment.

Other Methods of Peeling

Cryogenic Scalding. Recently, a peeling process called cryogenic scalding was developed by the Agricultural Research Service of USDA. This method is designed to speed up the tomato peeling process for canning and to reduce the amount of flesh lost during peeling. The new process involves

FIGURE 7.19. FMC DRY VACUUM SYSTEM FOR PEELING TOMATOES. Actual layout.

the use of liquid nitrogen, liquid air, or Freon 12 to freeze the skin of tomatoes within a few seconds. After the skin is frozen, the tomatoes are rapidly thawed in tap water, leaving them in a loose-fitting skin sack which can be quickly and easily removed with minimum loss of flesh. Peeling is done by slitting the skin either mechanically or manually and allowing the fruit to slip out. Losses during peeling are reduced to about half of those incurred by the outdated conventional hot water process. The reduction in loss is due entirely to a smaller loss in the layers of flesh just beneath the skin. These layers are rich in color and are essential to maintaining color grades in processed products (U.S. Dep. Agric. 1970).

Peeling with Gas. When peeling with gas, the tomatoes are exposed for 6.5 to 7.5 sec to a high temperature, high velocity blast of spent combustion gases. During this time they are gently revolved to expose their entire surfaces. The tomato peeler, called Hydoff and developed by Magnuson Corporation, Sparks, NV., operates between 650° and 700° F (343° and 371°C). At this temperature and exposure time, the skin "blisters" and pulls away from the flesh. By forcing the hot gases across the tomato at 250 mph, only the skin is torn off and the important fleshy layers immediately below the skin, rich in red pigment and vitamin C, are left untouched (Anon. Undated).

INSPECTION

Final inspection of peeled tomatoes prior to canning is necessary to assess defects for grade classification.

The tolerances as presented in Table 7.5 are the maximum defects allowed for the respective grades according to U.S. Standards for Grades of Canned Tomatoes.

TABLE 7.5. CANNED TOMATOES — MAXIMUM DEFECTS

		In Cans of Less Than 2 lb Total Contents	In Cans of 2 or More lb Total Contents	In Cans of Any Size
Grades	Defect (Aggregate Area)	In Any Container	Equivalent Amount per lb of Contents of Any Container	Per lb of Total Contents of All Containers (Avg)
A and A Whole	Peel	2 in.^{2}	1 in.^{2}	½ in.^{2}
	Blemished areas	⅛ in.^{2}	1⁄16 in.^{2}	1⁄16 in.^{2}
	Discolored portions	½ in.^{2}	¼ in.^{2}	¼ in.^{2}
	Objectionable core material—practically none			
	Harmless plant material—not more than a trace			
B	Peel	3 in.^{2}	2 in.^{2}	1 in.^{2}
	Blemished areas	¼ in.^{2}	⅛ in.^{2}	⅛ in.^{2}
	Discolored portions	1 in.^{2}	½ in.^{2}	½ in.^{2}
	Objectionable core material—slight amount			
	Harmless plant material—slight amount			
C	Peel	no limit	no limit	1 in.^{2}
	Blemished areas	½ in.^{2}	¼ in.^{2}	¼ in.^{2}
	Discolored portions	1½ in.^{2}	¾ in.^{2}	¾ in.^{2}
	Objectionable core material—moderate amount			
	Harmless plant material—moderate amount			

REFERENCES

ANON. 1964. Lye Peeling of Fruit and Root Crops. Diamond Chemical Co., Cleveland, OH.

ANON. 1969. Lye peeling of tomatoes. Canning Trade (Sept. 1) 6.

ANON. (Undated.) The secret of modern tomato processing. Magnuson Engineers, San Jose, CA Mimeo Rep.

CARROAD, P.A. and ROSE, W.W. 1977. Water recycle improves disc cleaning of tomatoes. Food Technol. *31* (3) 92−96.

CH2M HILL. Properties of Wastes from Conventional Peeling Versus Dry Caustic Peeling of Peaches and Tomatoes. Study for Magnuson Engineers, San Jose, CA. Proj. *F8736.0.*

CRUESS, W.V. 1958. Commercial Fruit and Vegetable Products, 4th Edition. McGraw-Hill Book Co., New York.

EL-ASHWAH, E.T. 1963. Effect of detergents on the growth and thermal resistance of *Bacillus thermoacidurans*. Ph.D. Dissertation. Ohio State Univ., Columbus.

FLOROS, JOHN D. and MANJEET S. CHINNAN. 1988. Seven Factor Response Surface Optimization of a Double Stage Lye (NaOH) Peeling Process for Pimiento Peppers. J. Food Sc. 53(2): 631-638.

FLOROS, JOHN D. and MANJEET S. CHINNAN. 1988. Microstructural Changes During Steam Peeling of Fruits and Vegetables. J. of Food Sc. 53 (3): 849-853.

FLOROS, JOHN D. and MANJEET S. CHINNAN. 1989. Determining the Diffusivity of Sodium Hydroxide through Tomato and Capsicum Skins. J. of Food Eng. 9: 129-141.

FLOROS, JOHN D. and MANJEET S. CHINNAN. 1990. Diffusion Phenomena during Chemical (NaOH) Peeling of Tomatoes. J. Food Sc. 55 (2): 552-553.

FLOROS, JOHN D., HAZSEL Y. WETZSTEIN and MANJEET S. CHINNAN, 1987. Chemical (NaOH) Peeling as viewed by Scanning Electron Microscopy: Pimiento Peppers as a Case Study. 52 (5): 1312-1316 & 1320.

GOOSE, P.G. and BINSTED, R. 1964. Tomato Paste, Puree, Juice and Powder. Food Trade Press, London.

GOULD, W.A. 1953. Hydrout corer helps quality and saves labor in canning tomatoes, OSU tests show. Food Packer *34* (9) 30−32.

GOULD, W.A. 1965. Effect of processing factors on the quality of fruits and vegetables. *In* Food Quality: Effects of Production Practices and Processing. Adv. Sci., Washington, DC Pub. *77*, 57.

GOULD, W.A. 1975. Mass sorting of mechanically harvested tomatoes. Ohio Agric. Res. Dev. Cent. Res. Circ. *209.*

GOULD, W.A., GEISMAN, J.R. and SLEESMAN, J.R. 1959. A study of some of the physical and chemical factors affecting the efficiency of washing tomatoes. Ohio Agric. Exp. Stn. Res. Bull. *825.*

GRAHAM, R.P., HART, M.R., KROCHTA, J.M. and ROSE, W.W. 1974. Cleaning and Lye Peeling of Tomatoes Using Rotating Rubber Discs. West. Reg. Res. Cent., U.S. Dept. Agric., Natl. Food Proc. Assoc. D *2704,* April.

KADER, A.A., MORRIS, L.L., STEVENS, M.A. and ALBRIGHT-HOLTON, M. 1978. Composition and flavor quality of fresh market tomatoes as influenced by some postharvest handling procedures. J. Am. Soc. Hortic. Sci. *103* (1) 6–13.

LAFFERTY, L.E. 1960. Adoption of infra-red technique for the peeling of apples. M.Sc. Thesis. Ohio State Univ., Columbus.

LUCAS, L.L. 1967. Evaluation of lye peeling of tomatoes using high lye concentration and short time of exposure with wetting agents. Spec. Group Studies Rep., Dep. Hortic., Ohio State Univ., Columbus.

LUCAS, L.L. 1969. Variables affecting the efficiency of peeling tomato with caustic soda. Ph.D. Dissertation. Ohio State Univ., Columbus.

MAY, D., ZOBEL, M.P., BRENDLER, R.A. and PARSONS, P.S. 1970. Central sorting of cannery tomatoes. Calif. Agric. *1970* (Feb.) 5–6.

SCHULTE, W.A. 1965. Efficiency of chemical and physical tomato peeling systems and their effects on canned product quality. Ph.D. Dissertation. Ohio State Univ., Columbus.

SCHULTZ, R.A. and SMITH, G.A. 1968. Additives raise peeling efficiency. Food Eng. *40* (6) 95.

SULLIVAN, G.H. and WILCOX, G.E. 1974. Tomato mechanization stalls in Midwest. Am. Veg. Grower *9*, 14–15.

THOMAS, W.M., STANLEY, D.W. and ARNOTT, D.R. 1976. An evaluation of blanch, lye and freeze-heat methods for tomato peel removal. Can. Inst. Food Sci. Technol. J. *9* (3) 118–124.

TWIGG, B.A. and GULLETTE, L.O. 1965. Lye washing of tomatoes for *Drosophila* egg removal. Canning Trade (July 12) 18.

Part II — Processing

CHAPTER 8
Canning Tomatoes

FILLING

Accurately filling the cans with tomatoes is an important unit operation for the processor in relation to consumer satisfaction, the company's profit, and satisfaction of government standards.

According to the United States Standards for Grades of Canned Tomatoes, U.S. Grade A or U.S. Fancy canned tomatoes must have a drained weight not less than 66% of the capacity of the container; U.S. Grade B or U.S. Extra Standard must have a drained weight of not less than 58% of the capacity of the container, and U.S. Standard or Grade C canned tomatoes must have a drained weight of not less than 50% of the capacity of the container.

With increasing emphasis on drained weight and fill of containers, the processor must understand the performance of his fillers and maintain this performance within the desired control limits. The processor should be aware of overfill or underfill by the fillers, to avoid giving away the product on the one hand or cheating the consumers on the other.

At the end of the trimming belt, the tomatoes are packed into previously cleaned cans. Cans with enamel bodies and enameled ends are recommended for tomatoes.

Tomatoes are packed by hand and by machine. Fancy whole, evenly colored, large tomatoes are carefully packed by hand. The tomato juice may be from the peeled tomatoes or juice from similar tomatoes (Lopez 1987).

Most cans are filled by machine. In this case, peeled tomatoes are dumped into a hopper and a mixture of whole tomatoes and juice fills the cans automatically. In some machine packing, the tomatoes may be somewhat broken in filling if the filler is run at a high speed.

The solid-pack style consists of tomato only, i.e., the free juice has been drained from the peeled tomatoes. This style is usually packed by machine.

The California standard-pack canned tomatoes consist of a mixture of peeled tomatoes and tomato puree. This requires a special label statement (Lopez 1987).

PROCESSING OF TOMATOES

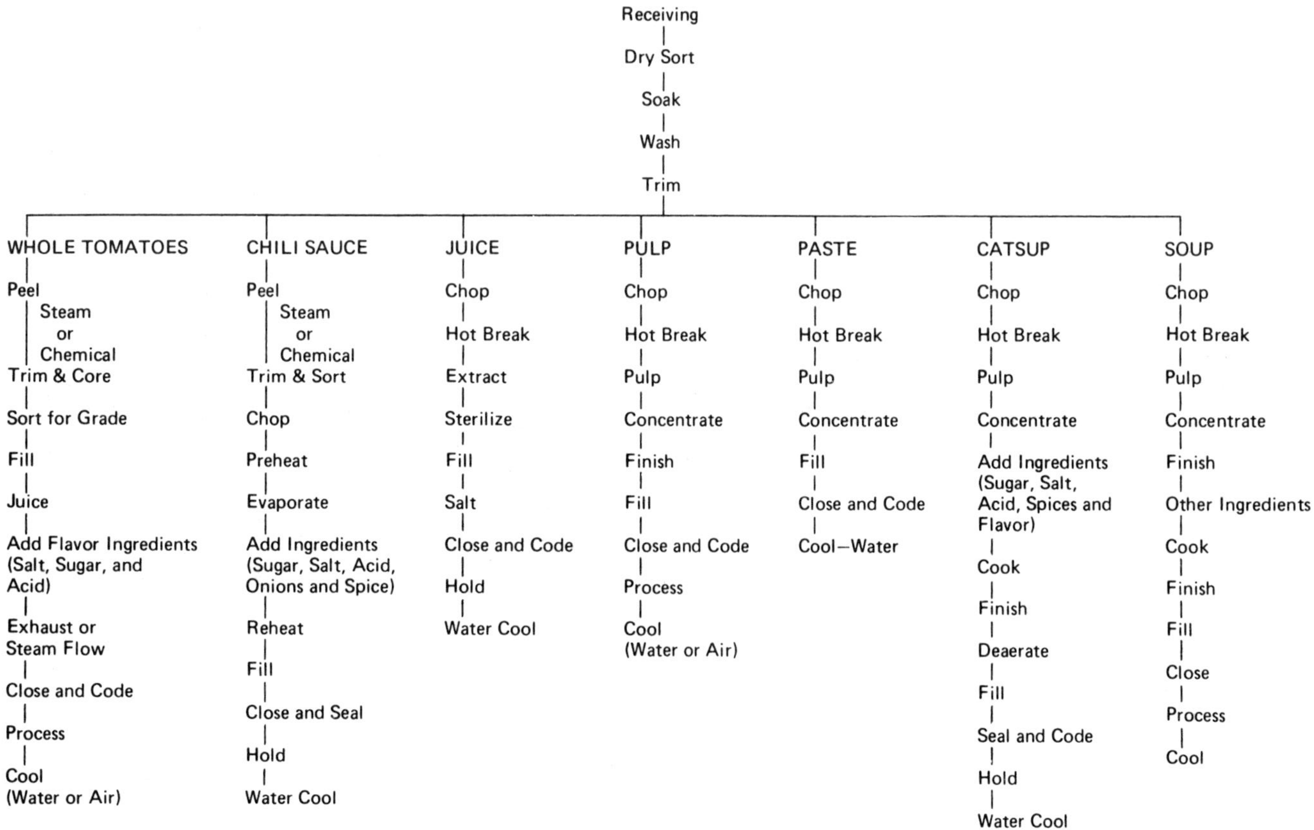

FIGURE 8.1. FMC HAND PACK FILLER FOR TOMATOES.

The quality of the final product is greatly influenced by the fill weight because the fill weight is highly significant relative to the drained weight. Schulte (1961) studies the relationship between fill weight and drained weight of canned tomatoes at five different fill weights in No. 303 cans. He reported a loss in the drained weight ranging from 8.5 to 11.9% after 4 months of storage. He also found that tomatoes could be packed at specified fill weights to yield predetermined drained weights of Grade A, Grade B, and Grade C. A processor could calculate the fill weight required to yield a specific grade quality drained weight if the fruit is of that quality in terms of wholeness, color, and absence of defects.

The minimum drained weights of drained tomatoes for commonly used containers are presented in Table 8.1. by grades. The nomograph in Fig. 8.2 shows dollar loss per hr from overfill based on cost of ingredients.

SALTING AND FIRMING

Firmness is a major factor in determining the quality of canned whole tomatoes. During advanced ripening, tomatoes have a tendency to become soft and to break up easily during processing. As a result, most manufacturers are obliged to pack some tomates that are excellent in flavor and color but which must be sold at lower prices because of lack of firmness (Kertesz *et al.* 1941).

Many investigators have reported on the use and effect of calcium salts for the firming of canned tomatoes (Appleman and Conrad 1927; Kertesz *et al.* 1940; LaCrone and Haber 1933; Siegal 1943). Jacobs (1951) explained the action of the calcium salts on the firming of products like tomatoes. He stated:

Fresh tomatoes, like all fresh fruits, contain pectin components which are relatively insoluble and which form a firm gel around the fibrous tissues of the tomato, thus preventing the collapse of the vegetable and in that way aiding in keeping it firm. When there is a breakdown of the cell structure, the pectin components are brought into contact with the enzymes of the food and the pectin is converted into pectic acid. This imparts less firmness to the tomato tissues than the original pectin, in turn causing collapse. The addition of calcium salts to tomatoes causes the formation of a calcium pectate gel which supports the tissues and protects the tomato against softening.

The FDA has approved the use of the following salts as firming agents: purified calcium chloride, calcium sulfate, calcium citrate, monocalcium phosphate, or any two or more of these in concentrations not to exceed 0.045% except for diced, wedges, and sliced, and the calcium level shall not exceed 0.08% calcium by weight in the finished canned tomatoes (1977). Added calcium salts must be properly declared on the label (Anon. 1971).

TABLE 8.1 - MINIMUM DRAINED WEIGHT BY CONTAINER SIZE TO COMPLY WITH USDA GRADES AND FDA 50% DRAINED WEIGHT SPECIFICATIONS

Container	Container Dimensions Diameter X Height	Minimum Drained Weight U. S. Grade A & B	U. S. Grade C
8 oz Tall	211 X 304	5.0	4.3
No. 300	300 X 407	8.8	7.6
No. 303	303 X 406	9.8	8.5
No. 2	307 X 409	11.9	10.3
No. 2½	401 X 411	17.3	14.9
No. 10	603 X 700	63.5	54.7

Methods of Use

Two methods of applying the calcium treatment have been demonstrated.

Salt Tablets. These are now commercially available in sizes suitable for the various can sizes. Each tablet contains anhydrous calcium chloride in addition to a proper amount of sodium chloride (common salt). It is stated that these tablets feed successfully in the usual machines. However, the hygroscopic nature of calcium chloride makes it necessary to package such tablets in small moisture-proof containers, which should be opened only as used. Without this precaution the tablets absorb moisture and disintegrate (Greenleaf, Undated).

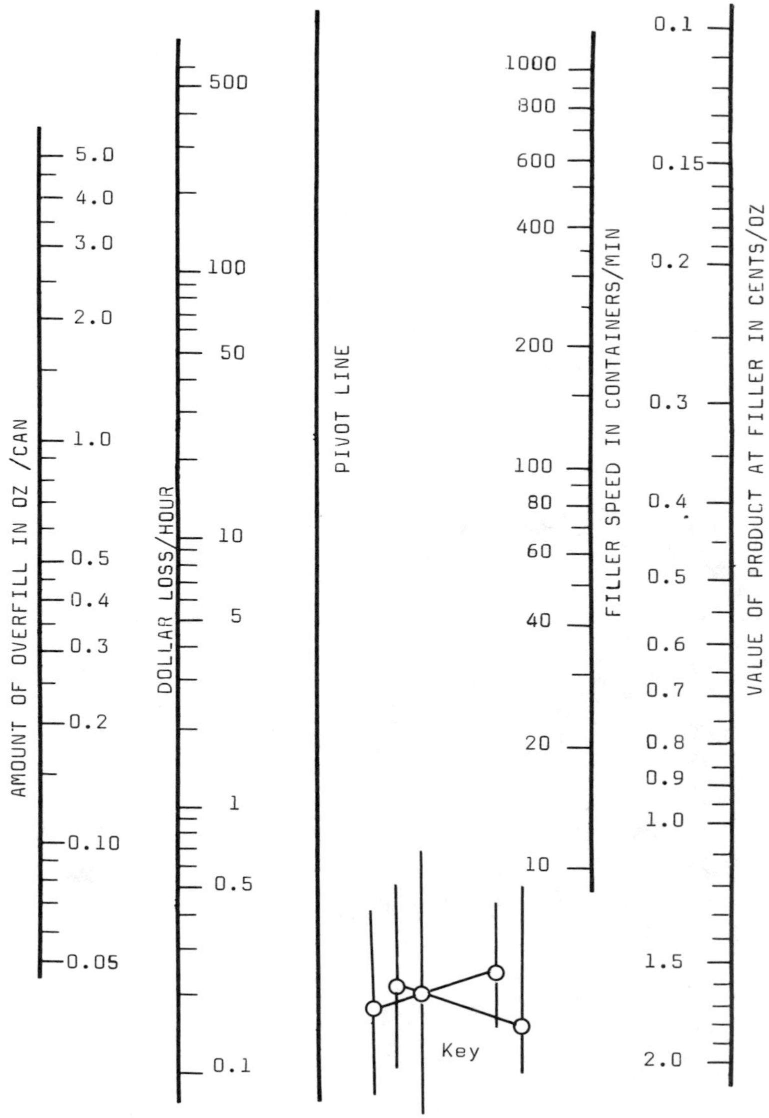

FIGURE 8.2. NOMOGRAPH FOR FILLER CONTROL. WITH THIS NOMOGRAPH AND STRAIGHT EDGE, YOU CAN COMPUTE DOLLAR LOSS PER HOUR FOR ANY SIUTATION.

Courtesy of Packaging Engineering (1963).

TABLES 8.2. QUANTITIES OF CALCIUM SALTS FOR CANNED TOMATOES BY CAN SIZE IN GRAINS

Can Size	Recommended Amount*[a]	Maximum Calcium Allowable in Grains as			
		Chloride†[a]	Sulfate†	Citrate†	Phosphate†
211 × 400	15–20	10.8	8.6	9.4	12.4
303 × 406	25–30	17.3	13.6	15.2	19.9
307 × 409	30–35	20.6	16.3	18.0	23.6
401 × 411	40–50	30.4	24.0	26.4	34.8
603 × 700	150–200	111.6	83.6	96.8	126.9

[a]Difference between * and † is NaCl.

Juice Solution. It is also possible to introduce sodium chloride-calcium chloride dissolved in the tomato juice. Preferably, this solution is placed in the can before the tomatoes are packed in, to avoid loss due to topping

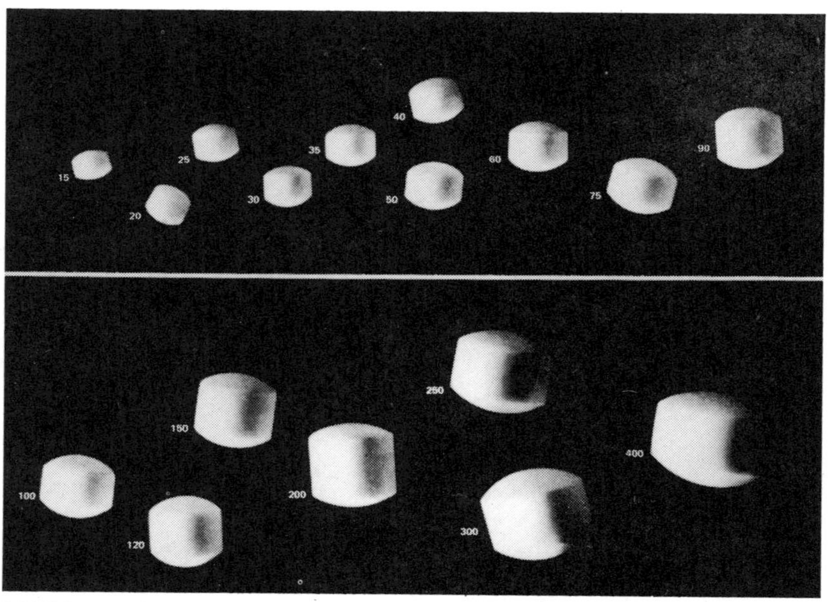

FIGURE 8.3. SALT TABLETS USED FOR CANNED TOMATOES AND TOMATO JUICE.
Courtesy of Morton Salt Company, Chicago, IL.

FIGURE 8.4. SALT TABLET DISPENSER.

(Greenleaf). Devices called Flocron and the Electro Portioner are used to dispense the tomato juice containing the flavoring and firming agents into the empty cans before the tomatoes are packed into them. The juice solution consists of 9 lb of U.S.P. calcium chloride (77% $CaCl_2$) and 27 lb of salt dissolved in 15 gal. of tomato juice. The amount of the juice solution injected into the cans depends on the size of the cans. It has been found that 5 ml of this juice solution produces the firming effect of tomatoes packed in No. 303 cans; the calcium content of these tomatoes is well below the maximum permitted by the standards for canned tomatoes (Siegal 1957).

Addition of calcium salts to tomatoes during canning improves the wholeness and drained weight of the canned tomatoes. Gould *et al.* (1956) found that tomatoes treated with sodium chloride had an average drained weight score of 17.31 and were rated U.S. Grade B, whereas tomatoes treated with

FIGURE 8.5. FILLED CANS OF TOMATOES BEFORE ADDING OF COVER JUICE AND CLOSURE.

FIGURE 8.6. GRANULAR SALT BEING DISPENSED INTO SIDE CAN.

sodium chloride-calcium chloride tablets (18.2 grains of NaCl and 11.8 grains of $CaCl_2$) had an average drained weight score of 18.35 and were rated U.S. Grade A.

Schulte (1961) indicated that canned tomatoes treated with sodium chloride tablets had significantly lower drained weight and lost 2% more weight than tomatoes treated with sodium chloride–calcium chloride tablets. Figure 8.2 represents the influence of NaCl and NaCl–$CaCl_2$ on the drained weight of canned tomatoes at five different fill weights in No. 303 cans.

EXHAUSTING

Because the product is generally cold when packed into cans, the filled cans must be heated. This is done by conveying the cans through an exhaust box or tunnel of steam. During this operation sufficient vacuum is obtained to prevent spoilage in the cans during storage (Lopez 1987). Tomatoes should be thoroughly exhausted, because solid-packed tomatoes heat very slowly (Cruess 1958). The center of the can should reach at least 130°F

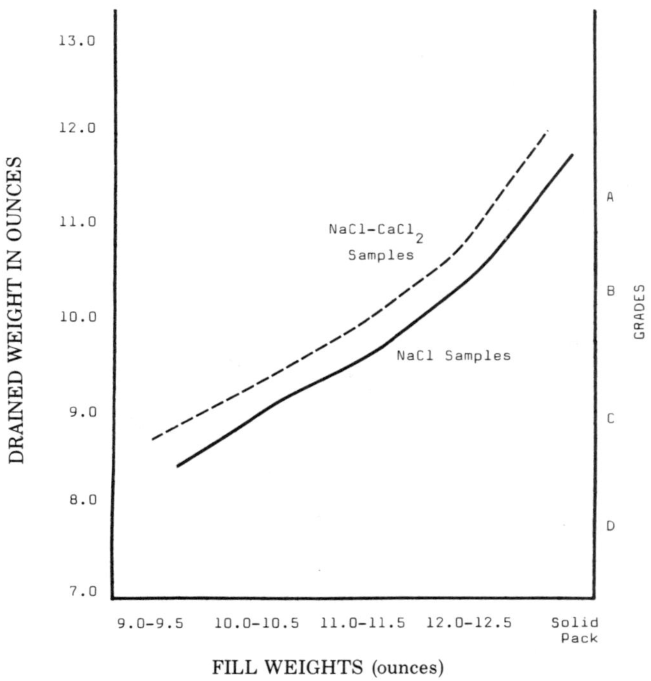

FIGURE 8.7. THE RELATIONSHIP OF FILL WEIGHTS TO DRAINED WEIGHTS AND GRADE OF CANNED TOMATOES PACKED WITH NaCl-$CaCl_2$ VERSUS SALT TABLETS.

(54°C) and the length of the exhaust should be adjusted to accomplish this temperature (Gould 1951). The following are minimum lengths of exhaust in live steam (1987):

No. 2 or No. 303 cans	3 min
No. 2½ and No. 3 cans	4 min
No. 10 cans	10 min

Too short an exhaust may cause springers or flippers through overfilling.

A satisfactory vacuum may be obtained by use of the steam-flow closing machine. In this operation, there is no need for exhausting the cans if the head space is carefully controlled to 5⁄16 in. for small cans and 6⁄16 in. for No. 10 cans (Lopez 1987) and the closure is preceeded by adding cover juice that is at least 190°F.

PROCESS TIME AND TEMPERATURE

The process time and temperature of canned tomatoes depend on the type of equipment used and the can size.

The agitating continuous cooker for tomatoes operates at 212°F (100°C) and has largely superseded the retort and open-still cooker. Table 8.3 shows

FIGURE 8.7. FMC ROTARY ATMOSPHERIC COOKER
Courtesy FMC Corporation

FIGURE 8.8 ATMOSPHERIC COOLER FOR TOMATOES.
Courtesy of FMC Corporation

process times and temperatures recommended for processing canned tomatoes.

A new means of sterilizing canned tomatoes is the high temperature, short time flame sterilization. This is a process in which the cans of tomatoes are first preheated in steam and then further heated by means of direct contact with the flame while rotating rapidly. After being held for sufficient time to ensure sterilization, the cans are cooled by means of water sprays in the continuous unit. (Leonard *et al.* 1975, 1976).

It is essential to check the center temperature of the processed cans in order to ascertain the efficiency of the process. Generally the center temperatures of the cans should be at least 180°F (82°C) when air cooling is used and 190°F (88°C) when water cooling is used. The center temperature is taken regularly during the day by plunging an armored, sharp-pointed thermometer into the centers of several cans taken directly from the outlet of the cooker (Cruess 1958).

TABLE 8.3. RECOMMENDED PROCESSING TIMES FOR CANNED TOMATOES BY CAN SIZE

	Still Retort, 212°F (100°C)		Agitating Cooker, 212°F (100°C)	
Can Size	Water Cool	Air Cool	Water Cool	Air Cool
303 × 406	45	35	14	9
307 × 409	45	35	14	9
401 × 411	55	45	18	13
603 × 700	100	80	25	20

COOLING

The canned tomatoes should be cooled completely and quickly after processing to a temperature of 100°F (37.6°C) to avoid "stack burning," which results in lower drained weight, browning of the color, and loss of flavor (Cruess 1958). Two methods are used for cooling the tomatoes after processing: water cooling and air cooling.

In water cooling, the cans are cooled in flowing water immediately following the process. The duration of the water cool is governed by the length of time required to reduce the can center temperature to 100°F. The water used for cooling should be chlorinated to 15 ppm residual chlorine to maintain a zero or low bacterial count.

Canners in regions with limited water supply may find it necessary to air cool canned tomatoes in ricks. It is claimed that water cooling is more effective in maintaining a higher percentage of drained weight than air cooling (Germain *et al.* 1939).

ACIDIFICATION

The tomato is considered an acid product with a pH value generally of 4.6 or less. Consequently, minimal sterilization processes are considered adequate. However, in recent years, the acidity of tomatoes has been decreasing, with an associated increase in product spoilage (Lopez 1971).

Control of flat-sour spoilage in canned whole tomatoes becomes more and more difficult as the pH value of the product becomes higher than 4.6 (i.e., less acid). Some high pH tomatoes increase the spoilage potential even with the best industry practice. If spoilage is to be prevented in high pH tomatoes, they must be processed to a point at which adverse changes result. Therefore, acidification of the product has been found to be the best solution (Lopez 1971).

The FDA has sanctioned the addition of edible organic acids for the purpose of acidification. Allowing these edible organic acidifying agents as ingredients in canned tomatoes enables tomato packers to more easily control the pH and other factors involved in spoilage.

The FDA Standard of Identity for canned tomatoes allows the use of any edible organic acid including in the list of substances generally recognized as safe (GRAS). When an edible organic acid is added, any nutritive sweetener in solid form may be added to compensate for any tartness resulting from the added acid (Anon. 1966). Edible organic acids include citric, malic, and fumaric acids. However, citric acid is the only one in general use.

FIGURE 8.9. 4 STAGE TOMATO JUICE EVAPORATORS
Courtesy of FENCO/Food Machinery Sales & Service

In addition to the possibility of increased spoilage from the use of high pH tomatoes, increased spoilage may also be due to the use of mechanically harvested tomatoes. Tomatoes harvested in this manner carry a higher load of microorganisms than handpicked ones. For these reasons, it is recommended that sufficient acid be added to increase the acidity to between pH values of 4.1 and 4.3.

The addition of calcium chloride in normally used proportions depresses pH value by approximately 0.1 (Lamb *et al.* 1962). When acidification is

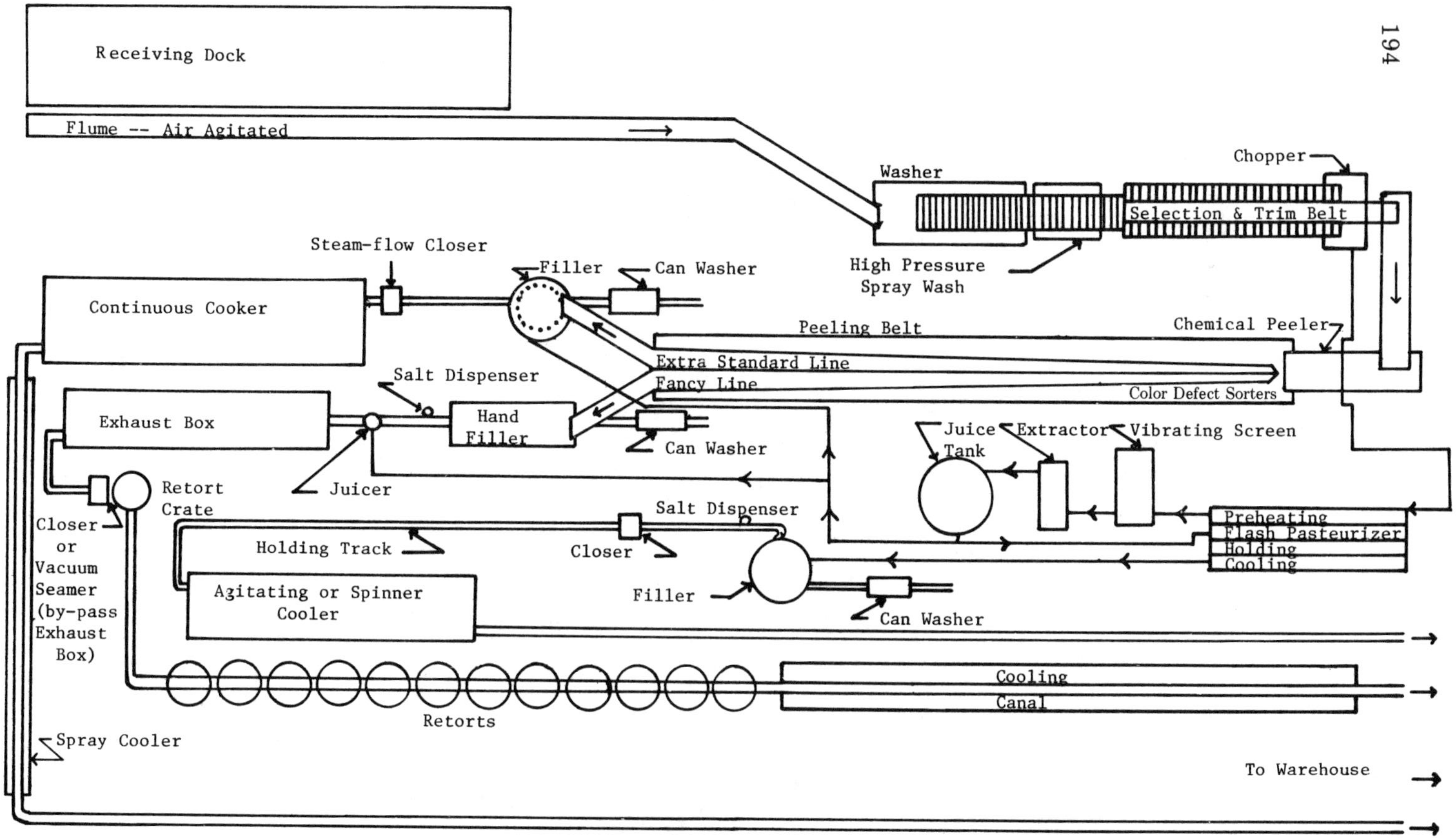

FIGURE 8.11. FLOW SHEET OF CANNED TOMATO (FANCY AND EXTRA STANDARD) AND TOMATO JUICE LINES.

used, the addition of an acid in sufficient quantity to depress pH value by another 0.2 of a pH unit is all that is usually necessary to prevent spoilage and assure the safety of the pack, assuming that the sterilization process is adequate (Lopez 1987). Experience indicated that 0.1% of added citric acid will lower the pH value by this amount for average tomatoes, or slightly more when the tomatoes are abnormally high in pH value.

The addition of 0.1% anhydrous citric acid requires 8 grains for No. 303 can, 14 grains for a No. 2½ can, and 50 grains for a No. 10 can.

TABLE 8.4. PERCENTAGE CITRIC ACID NEEDED TO OBTAIN DESIRED pH IN WHOLE PACK TOMATOES[a]

	% Citric Acid to Reduce pH to		
Starting pH	4.1	4.0	3.9
4.6	0.3	0.4	0.5
4.4	0.2	0.3	0.4
4.2	0.1	0.2	0.3
4.0	—	—	0.1

[a] Percentages and amounts listed are approximate since individual tomatoes will vary in the extent to which they are buffered and will, therefore, vary in their reaction to the addition of citric acid.

TABLE 8.5. CITRIC ACID ADDITION PER CAN[a]

Citric Acid	No. 303			No. 2½			No. 10		
	Grains	g	ml[b]	Grains	g	ml[b]	Grains	g	ml[b]
0.05	4	0.3	2	7	0.5	4	25	1.5	14
0.10	8	0.5	5	14	1.0	8	50	3.5	28
0.20	16	1.0	9	28	2.0	16	100	7.0	56
0.30	24	1.5	14	42	3.0	23	150	10.0	83
0.40	32	2.0	19	56	4.0	31	200	13.5	111
0.50	40	2.5	25	70	5.0	40	250	17.5	140

[a] Percentages and amounts listed are approximate since individual tomatoes will vary in the extent to which they are buffered and will, therefore, vary in their reaction to the addition of citric acid.
[b] Solution of 1 lb citric acid anhydrous dissolved in 1 gal. tomato juice.

Method of Application

Citric acid may be added by five methods:

1. Direction addition of granular anhydrous citric acid by means of a separate dispenser or with the salt tablet. This is the currently preferred method.

2. Addition of citric acid to the tomato juice as an optional packing medium. The amount to be added in oz per gal. of juice may be calculated by the following formula:

$$\frac{\text{Net weight of can contents}}{\text{Weight of tomato juice added per can}} \times \%\,\text{citric acid} \times 1.37$$

In the formula the percent citric acid is expressed as a decimal, i.e., 0.1 for 0.1%, etc. The disadvantage of this procedure is that the amount of tomato juice added per can is not uniform and this could result in a considerable variation in the final pH.

3. Addition of a dry mixture of citric acid and salt. Calcium chloride cannot be used because it causes the mixture to cake.

4. Dipping the tomatoes in a citric acid bath following peeling. Concentration of acid may be in the range of 1 to 2%. This may cause variation in pickup of acid depending on such factors as size of the tomatoes, immersion time, and decrease in concentration of the bath during use.

5. Addition of a measured quantity of a solution of acid in tomato juice to each empty can prior to packing. To be of value in reducing spoilage the citric acid must go into solution during processing of the cans (Anon. 1966).

Canned tomato packers may be in doubt sometimes as to whether or not to acidify the product. The following are some conditions that may make acidification desirable (Lopez 1971):

1. If tomatoes are too ripe
2. If tomatoes are too watery
3. If higher-yielding varieties are being used
4. If tomatoes are mechanically harvested
5. If some canned product spoilage has been observed in previous years.

OTHER PRODUCTS

Diced tomatoes, sliced tomatoes, and tomato wedges are new products finding considerable consumer acceptance. These new styles of packing tomatoes allow the tomato canner diversification in utilizing his raw materials, as not all tomatoes are suitable for canning as whole or as pieces of tomatoes. These new styles have one basic change in the Standard of Identity in that the amount of calcium added is not more than 0.1% of the weight of the finished food.

Stewed tomatoes, tomatoes and okra, and the canning of tomatoes with other vegetables offer the tomato canner other possible products to add to his line. As with tomatoes, these products should be acidified (0.1 to 0.2% citric acid), calcium chloride added at rate of 0.1%, and processed to a

TABLE 8.6 — TOMATO PROCESSING OPERATIONS & CRITICAL CONTROL POINTS

UNIT OPERATION	FUNCTION	SPECIFICATIONS	CRITICAL CONTROL POINT
Recieving	Quantity Quality	Usable Tonnage	Mud, Dirt, Soil, HEM Color,pH % Mold & Type, Brix, Acidity
Washing	Cleaning	All "soil" removed	1 Min. dwell time, 150 psi at nozzle, Violent agitation, Min. 130°F Temp., 2 % Concentration of detergents
Sorting	Removal of usable fruit	Maximum of 1% defective fruits	Agtron color of <45, max. of 10% mold count, no HEM
Peeling	Removal of all cores & peel	Max. of: 1/2 sq. in. of peel, 1/16 sq. in. blemished areas, 1/4 sq. in. of discolored portion	Min. of 15% caustic, min. of 190°F, min. Dwell Min. of 8 seconds
Filling	Fruit/Juice ratio, Salt, Acid + other Additives	Min. of 60% fruit Max. of 40% juice	Drained weight of 60% capicity of can, but can 90% full
Closure	Integrity of clousure Code (product, date, year)	Double seam lg. =W Double seam tk =S Body hook lg =BH Cover hook lg =CH	Within tolerances of can Co. specifications
Processing	Sterilization and cooling	Center Temp. of 190°F Chlorine of 5 ppm, cool to 105°F	Recorder and Thermometer in agreement
Labeling	Identification	0% defective	No smears, or cocked labels
Warehousing	Storage time & temperature	Maximum of 100°F Low RH, FIFO	No sweating of containers

center temperature of 190 °F. Many formulations exist for these products, but the following is a good starting point for stewed tomatoes:

Sliced, diced, or small whole tomatoes	70.0%
Tomato juice	27.5%
Sugar	1.0%
Salt, up to	0.5%
Dehydrated onions	0.5%
Dehydrated celery	0.2%
Dehydrated peppers	0.2%
Stewed Tomato Seasoning	0.1%

The onions, celery, and peppers should be diced and added precisely to each can using separate fillers, if possible. The sugar, salt and seasoning may be fixed with the tomato juice. The tomato juice should be heated to 190 °F. or above prior to adding to the can. All of these ingredients should be added first, prior to adding the tomatoes.

REFERENCES

ANDERSON, D. and MENDENHALL, V.T. 1978. A survey of tomatoes home-canned in Utah. J. Food Protect. *41* (7) 514–517.

ANON. 1966. Acidification of whole tomatoes. Canning Trade (Sept. 19) 10.

APPLEMAN, C.O. and CONRAD, C.M. 1927. The pectic constituents of tomatoes and their relation to the canned product. Md. Agric. Exp. Stn. Bull. *291*.

CRUESS, W.V. 1958. Commercial Fruit and Vegetable Products, 4th Edition. McGraw-Hill Book Co., New York.

DIXON, S., MARSHALL, R.B. and ROSS, J. 1963. Food processing method and apparatus. U.S. Pat. 3,096,181.

GERMAIN, L.G., HOY, H.A. and WEINER, L.G. 1939. Influence of method of cooling on the solidity of canned tomatoes. Am. Can Co. Res. Dep. Presented to Canning Probl. Conf. Natl. Food Proc. Assoc., Chicago, Jan.

GOULD, W.A. 1951. High quality tomatoes, juice, pulp step-by-step. Food Packer *32* (8) 28–29, 46.

GOULD, W.A. 1971. Tomato Processor Quality Control Technologist Handbook. Dep. Hortic., Ohio State Univ., Columbus.

GOULD, W.A., DAVIS, R.B., KRANTZ, F.A., JR. and HEALY, N.C. 1956. A study of some of the factors affecting the grade relationship of fresh and processed vegetables. I. Canned tomato. Ohio Agric. Exp. Stn. Res. Bull. *781*.

GREENLEAF, C.A. (Undated.) Suggestions concerning the use of calcium chlorine in canning tomatoes. Natl. Food Proc. Assoc. Res. Lab., Washington, DC Mimeo Rep.

GUTHEIL, R.A., PRICE, L.G. and SWANSON, B.G. 1980. pH, acidity, and vitamin C content of fresh and canned homegrown Washington tomatoes. J. Food Protect. *43* (5) 366–369.

JACOBS, M.B. 1951. Food and Food Products, Vol. 3, 2nd Edition. John Wiley & Sons, New York.

JOHANNESSEN, G.A. 1981. Solids anyone? Calif. Process. Tomatoes News Views *4* (2) 1–2.

JUDGE, E.E. & SONS. The Almanac of the Canning, Freezing, Preserving Industries, Edward E. Judge & Sons, Westminster, MD.

KERTESZ, Z.I., TOLMAN, T.G., LACONTI, J.D. and RUYLE, E.H. 1940. The use of calcium in the commercial canning of whole tomatoes. N.Y. State Agric. Exp. Stn. Tech. Bull. *252*.

LACRONE, F. and HABER, E.S. 1933. Changes in the pectin constituents of tomatoes in storage. Iowa State Coll. J. Sci. *7*, 467.

LAMB, F.C., LEWIS, L.D. and KIMBALL, J.R., JR. 1962. Factors affecting the pH of tomatoes. Natl. Food Processors Assoc. Res. Lab. Rep *61C-44*.

LEONARD, S., MARSH, G.L., MERSON, R.L., YORK G.K., BUHLERT, J.E., HEIL, J.R. and WOLCOTT, T. 1975. Chemical, physical and biological aspects of canned whole peeled tomatoes thermally processed by Stériflamme. J. Food Sci. *40*, 254–256.

LEONARD, S., MARSH, G.L., YORK, G.K., MERSON, R.L., HEIL, J.R., WOLCOTT, T. and ANSAR, A. 1976. Flame sterilization of some tomato products and fruits in 603 × 700 cans. J. Food Sci. *41* (4) 828–832.

LOPEZ, A. 1987. A Complete Course in Canning, 12th Edition. Canning Trade, Baltimore, MD.

LOPEZ, A. 1971. Updating developments in acidification of canned whole tomatoes. Canning Trade (Apr. 12) 8.

LOPEZ, A., COOLER, F.W. and WYATT, W.B. 1968. Acidification of canned mid-Alantic tomatoes. Canning Trade (Apr 1) 8.

LOPEZ, A. and WILLIAMS, H.L. 1981. Essential elements in fresh and canned tomatoes. J. Food Sci. *46*, 432–434.

POWERS, J.J. and GODWIN, D.R. 1978. pH of tomatoes canned at home in Georgia. J. Food Sci. *43* (4) 1053–1055.

SAPERS, G.M., PHILLIPS, J.G. and STONER, A.K. 1977. Tomato acidity and the safety of home canned tomatoes. HortScience *12* (3) 204–208.

SAPERS, G.M., PHILLIPS, J.G., TALLEY, F.B., PANASIUK, O. and CARRE, J. 1978. Acidulation of home canned tomatoes. J. Food Sci. *43* (4) 1049–1052.

SCHULTE, W.A. 1961. The relationship between fill weight and drained weight of canned tomatoes. Master's Thesis. Ohio State Univ., Columbus.

SIEGAL, A. 1943. A new calcium compound for firming tomatoes for canning. Canner/Packer (Feb. 13).

SIEGAL, M. 1957. More data on the salting of canning tomatoes. Canning Trade (Feb. 11) 7.

SKELTON, M. and CRAIG, J.A. 1978. Ascorbic acid content, pH and flavor characteristics of acidified home canned tomatoes. J. Food Sci. *43* (4) 1043–1045.

CHAPTER 9
Tomato Juice Manufacture

Tomato juice contributes significantly to man's nutrition in the diet as a breakfast juice or an appetizer served at any meal of the day. It contains from 15 to 25 mg/100 g of vitamin C and has four times the vitamin A content of orange juice (Gould 1971). Like orange juice it contains considerable basic ash, and on digestion leaves an alkaline residue. It is a good source of iron, manganese, and copper (Tressler and Joslyn 1971).

Tomato juice is defined by the Federal Food, Drug, and Cosmetic Act as "the unconcentrated liquid extracted from mature tomatoes of red or reddish varieties, with or without scalding followed by straining. In the extraction of such liquid, heat may be applied by any method which does not add water thereto. Such liquid is strained free from skins, seeds, and other coarse or hard substances, but carries finely divided insoluble solids from the flesh of the tomato. Such liquid may be homogenized, and may be seasoned with salt. When sealed in a container it is so processed by heat, before or after sealing, as to prevent spoilage" (Judge & Sons 1971).

Tomato juice was introduced in the middle 1920s. the idea for a commercially canned tomato juice is attributed to Elliott O. Grosvener of Tomato Products Company. In 1925, the first tomato juice was packed under factory conditions as a part of the regular manufacturing operations of that company (Cruess 1958). It was first distributed in significant commercial quantities in 1928 (Tressler and Joslyn 1971). Total pack of canned tomato juice and combination vegetable juices containing 70% or more tomato juice increased rapidly before World War II.

The attributes of quality in tomato juice, that is, flavor, color, consistency, and nutritive value, are greatly influenced by variety, climate, cultural practice in the field, harvest procedure, degree of ripeness at the time of harvest, length of storage before processing, washing and sorting, and each step of the processing procedure (Tressler and Joslyn 1971).

Tomatoes used for juice production should possess high color, rich flavor, and high total acidity (Cruess 1958).

PREPARATION FOR PROCESSING

In manufacturing tomato juice, pulp, and catsup, the tomatoes are subjected to the same operations as previously described in the preparation of tomatoes for canning in regard to dry sort, washing, final sorting, and trimming. Great care should be taken during the sorting and trimming operations to remove mold and rot. The inclusion of stems or unripe portions may impart unpleasant flavor characteristics to the final product (Cruess 1958).

CRUSHING OR CHOPPING

At the end of the trimming belt, the tomatoes drop into the chopper for crushing prior to juice extraction. The tomatoes are normally chopped to 0.040 to 0.060 inch prior to heating or breaking. They may be crushed under pressure, but this is not too efficient if the tomatoes are firm ripe.

Most workers agree that the "hot break" method produces a superior tomato juice product over the older "cold break" method. Some have described these two methods as being heated under 150°F for the "cold break" and over 170°F for the "hot break". Some feel that the hot break temperature should be over 200°F to be really effective.

In the "hot break" method the tomatoes are rapidly heated immediately following chopping or crushing. The advantage of the "hot break" is that a greater yield is obtained and a more viscous product that does not separate upon standing.

In the cold break process the tomato is chopped or crushed and then extracted following mild heating. This product has a more natural tomato color and it has a fresher tomato flavor. However, this product does not retain as much of its natural Vitamin C content and it may separate in the jar or the can.

It is generally agreed that hot break produces a better-quality juice with respect to cooked tomato flavor, and body. A heavier-bodied, more homogeneous juice is obtained by the hot break method because heat destroys the pectic enzymes and permits more efficient extraction of pectin. Luh and Daoud reported that the breakdown of pectic materials in tomato juice by enzymatic action yields a product of low consistency (Luh and Daoud 1970). The heat stability of pectic enzymes, when subjected to thermal treatment, is therefore most important. Pectinesterase is less stable when subjected to heat than polygalacturonase (Tressler and Joslyn 1971). It should be pointed out that the activity of the pectic enzymes is greatly accelerated as the temperature is increased to about 140° to 150°F (60° to 66°C). Beyond this point, the activity is retarded until inactivation is reached at a temperature of about 180°F (82°C) (Lopez 1987). The pectic enzymes cause a breakdown of the pectin resulting in a thin-bodied

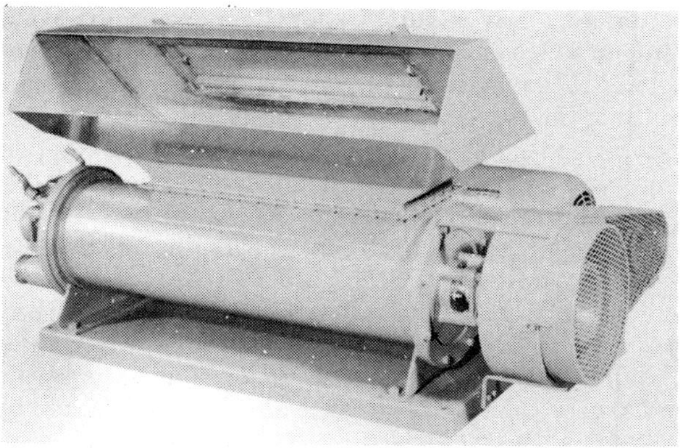

FIGURE 9.1.
TOMATO CHOPPER.
Courtesy of FMC Corporation.

product that separates readily. The temperatures should be raised to at least 180°F (82°C) for 15 sec (Lopez 1987).

The fastest means of enzyme inactivation is by steam injection. However, this technique is not recommended for tomato juice because it dilutes the juice with condensate and is prohibited by the Standards of Identity. Steam injection would retain at least 95% of the potential serum viscosity of the fresh tomatoes (Tressler and Joslyn 1971).

A very satisfactory enzyme inactivation can be achieved by heat treatment in a rotary coil tank (vertical or horizontal), followed by a heat exchanger and holding tube to achieve 220°F (104°C). The rotary coil tank, when run at the designed heating capacity of the system, will inactivate the enzymes fast enough to retain at least 90% of the potential serum viscosity in the original fresh tomato (Tressler and Joslyn 1971). A rotary coil tank procedure has the added advantage that the violent boiling occurring at the designed heat transfer capacity is an excellent means of deaeration. Air removal is important nutritionally because tomato juice containing dissolved or occluded air and processed at a high temperature will not retain all the original vitamin C (Tressler and Joslyn 1971).

Catelle has "developed a process of feeding whole tomatoes under vacuum, chopping, or mashing them in conditions where residual oxygen is reduced to a minimum– depending upon the degree of vacuum obtainable inside the feeder-tank and, then introducing the tomato in its liquidized state into a pressurized system or circuit designed for operation at whatever temperature represents the optimum for enzyme-deactivation; the introduction of freshly-mashed deaerated tomato into the circuit being such that it mingles with tomato already heated and circulating therein." I have seen this system in operation and the data indicate that the consistency of the final product is superior to conventional break systems, the color is "set" quicker and there is greater retention of the Vitamin C content as the entire system is under a vacuum or, at least, the residual oxygen is greatly reduced and held to a minimum.

Other types of heat exchangers may be suitable for the inactivation of the pectic enzymes found in tomatoes. Tubular heat exchangers with large tubes are commonly used in the tomato industry. In this type of equipment a major portion of the pectin is destroyed.

Many heat exchangers, including the tubular and swept-surface types, have the additional disadvantage that the crushed tomatoes contain dissolved and occluded air, which markedly reduces the ascorbic acid (vitamin C) during the heating treatment. This problem could be avoided by deaeration before heating (Tressler and Joslyn 1971).

The cold-break procedure, in which the fruit temperature is less than 140°F (60°C), is claimed to give a better-colored juice, particularly if the raw tomatoes are not fully colored throughout. Under these conditions a cold break may give a better-flavored juice. Vitamin C is also better retained by the cold-break procedure, since its destruction is accelerated by high temperatures when air is present (Tressler and Joslyn 1971).

Tomatoes that are extracted by the cold break procedure are usually first scalded to loosen the skins so that no tomato flesh will cling to them during extraction; failure to scald at this point reduces the yield of juice. Tomatoes are passed directly from the scalder over the inspection belt to a chopper and

FIGURE 9.2. FBR HOT BREAK SYSTEM
Courtesy FMC Corporation

then to the extractor (Lopez 1987). Quick processing of the extracted juice is necessary to produce high-quality tomato juice by cold-break procedure (Gould 1971).

EXTRACTION

Extraction of tomato juice may be accomplished by two main types of commercially available extractors: the screw type and the paddle type. Screw type extractors press the tomatoes between a screw and screen (Lopez 1987). The pressing action of the juice extractor consists of an expanding helix inside a tomato juice screen, in which tomato pulp is forced against the screen at continuing and increasing pressures (Tressler/Joslyn

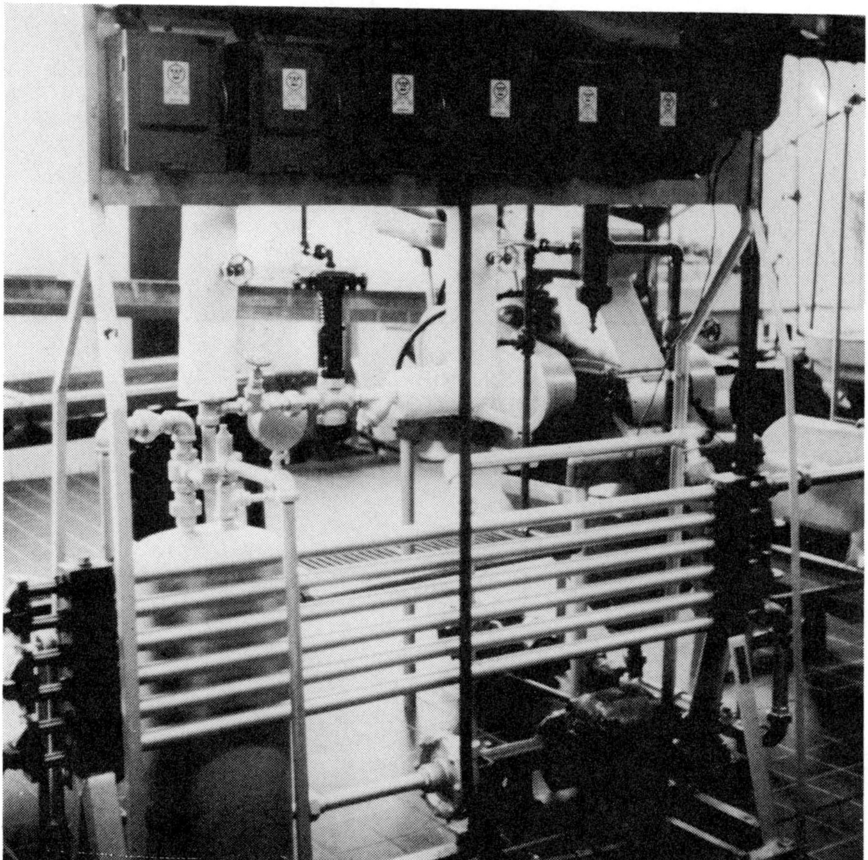

FIGURE 9.3. TUBULAR PREHEATER FOLLOWING CHOPPING PRIOR TO EXTRACTION.

1971). The holes in the screen vary but usually are about 0.02 to 0.03 in. in diameter (Gould 1971; Lopez 1987). This pressing action does not churn the product; therefore, very little air is incorporated into the expressed juice (Tressler and Joslyn 1971). Paddle-type extractors beat the tomato against the screen (Gould 1971). Some installations are equipped with a shaker screen ahead of the extractor where green areas, along with stems, cores, and other foreign material, can be removed. When shaker screens

Table 9.1 — RELATIONSHIP OF BRIX (SOLUBLE SOLIDS CONTENT) TO TITRATABLE ACIDITY (%CITRIC ACID) TO BRIX/ACID RATIOS FOR TOMATO JUICE

	Brix or Soluuble Solids Content						
% Acid	4.5	5.0	5.5	6.0	6.5	7.0	7.5
0.30	15.0	16.7	18.3	20.0	21.7	23.3	25.0
0.35	12.9	14.3	15.7	17.1	18.6	20.0	21.4
0.40	11.3	12.5	13.8	15.0	16.3	17.5	18.8
0.45	10.0	11.1	12.2	13.3	14.4	15.6	16.7
0.50	9.0	10.0	11.0	12.0	13.0	14.0	15.0
0.55	8.2	9.1	10.0	10.9	11.8	12.7	13.6
0.60	7.5	8.3	9.2	10.0	10.8	11.7	12.5
0.65	6.9	7.7	8.5	9.2	10.0	10.8	11.5
0.70	6.4	7.1	7.9	8.6	9.3	10.0	10.7
0.75	6.0	6.7	7.3	8.0	8.7	9.3	10.0

CHART 9.1. RELATIONSHIP OF VALUES TO FLAVOR OF TOMATO JUICE*

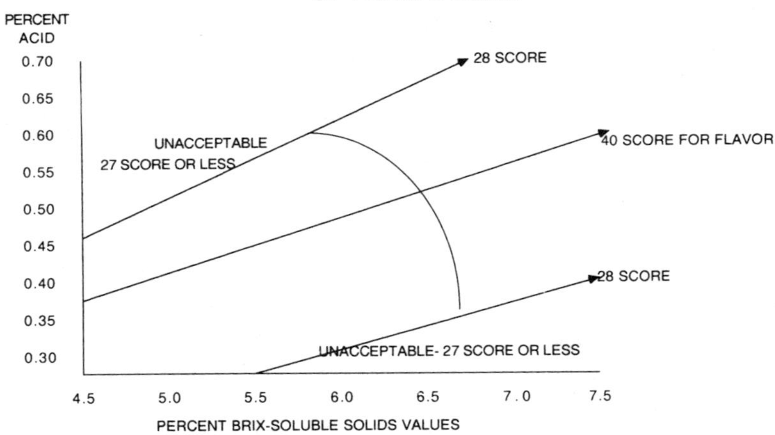

*These flavor scores are based on distinct canned tomato juice flavor, that is, tomato juice not affected by stems, leaves, crushed seeds, cores, immature tomatoes or the effects of improper trimming.

are used, some canners employ a higher temperature hot break and a tighter setting of the extractor.

Moyer *et al.* (1959) observed that the yield of tomato juice extracted from fresh tomatoes ranged from 29.4 to 91.5%, depending on the type of equipment used. They found that the pressing action of a screw-type juice extractor gave an average yield of 78.9%, whereas the beating action of a paddle pulper and a paddle finisher gave an average yield of 82.4%. Either of these types of extractors may be preset to deliver either a high or relatively low percentage of juice extraction. A high extraction would yield 3% skins and seeds and 97% juice. It is, however, commercially feasible to extract only 70 to 80% juice, a procedure that yields a very moist residual containing useful tomato materials, which can be reextracted for use in other tomato products. In some cases, this low extraction yield (70%) is desirable because the extracted juice will have a high percentage of soluble solid components, which improve flavor, and at the same time a lower percentage of insoluble solids, which tend to reduce the quality of the finished juice (Tressler and Joslyn 1971).

DEAERATION

Since heating tomato juice containing dissolved or occluded air impairs the retention of vitamin C, some canners employ deaerators in which the product is vacuum deaerated (Tressler and Joslyn 1971).

Ideally, deaeration should be applied as soon as possible after crushing the tomatoes because, from this point on, oxidation is rapid, particularly at high temperatures (Goose and Binsted 1964). For practical reasons, however, vacuum deaeration takes place immediately after extraction of the juice. Normally a 10° flash is sufficient to remove the dissolved and occluded air (Tressler and Joslyn 1971). If a hot-break procedure is used, the effectiveness of deaeration at this stage loses some of its advantages because oxida-

FIGURE 9.4. TOMATO JUICE TURBO EXTRACTOR.
Courtesy of FENCO/Food Machinery Sales & Service.

tion may already have become quite advanced as a result of natural agitation during the process (Goose and Binsted 1964). However, deaeration is still capable of averting serious loss of vitamin C in subsequent sterilization of the juice (Tressler and Joslyn 1971).

Once deaeration has been accomplished, it is important to engineer the processing line to guard against recurrence of aeration. This requires the use of adequately sealed pumps to prevent air from being incorporated into the product during a pumping action (Tressler and Joslyn 1971).

ACIDIFICATION

According to the new Standard of Identity change for Tomato Juice, it may be acidified with any safe and suitable organic acid. The most logical organic acid to use is citric acid as it is the natural acid of the tomato. The amount of citric acid to add to the tomato juice will vary with the makeup of the varieties being used. My recommendation is that sufficient acid should be added to balance the soluble solids content according to the data in Table 9.1 and Chart 9.1 to bring the product into the grade A classification for flavor. That is, if the soluble solids is 5.5% the acid content should be between 0.35 and 0.55%. If the soluble solids content should be as high as 6.5% then the acid content should be between 0.40 and 0.65%. The best flavor in either case will be near the mid point, that is, with 5.5% soluble solids the acid should be at 0.45 and for the 6.5% soluble solids juice the acid should be at 0.50%. All of this is predicated on the fact that there are no off-flavors present such as green tomato flavor, burn-on flavor, chemical residue flavor, etc. Acidification is an excellent way to improve tomato juice flavor and, of course, aid in processing requirements.

SALTING AND FILLING

Salt may be added to the extracted juice by direct dissolution in batch quantities, by use of tablets added to each can at the time of filling, or by injecting concentrated brine made by dissolving salt in tomato juice or serum (Tressler and Joslyn 1971). The standard practice of using salt tablets is preferred because it eliminates the need to accumulate the tomato juice in tanks, thus removing a complete operation from the production line.

TABLE 9.2. SALT TABLETS BY CAN SIZES FOR TOMATO JUICE

Can Size	Volume of Fill (oz)	Salt Tablet (Grains)
202 × 308	5¼	10–15
211 × 414	12	30–35
300 × 407	13½	35–40
307 × 409	19	50–60
404 × 700	46	120–150
603 × 700	96	250–300

The sodium chloride added to tomato juice will range from 0.5% to 1.25% by weight. The average sodium chloride content of commercial samples of tomato juice is 0.65% by weight. If using granular salt, it is recommended that 2 to 3 lb be added to 100 gal. of juice prior to sterilization. Salt tablets may be used according to Table 9.2.

Juice need not be salted, but the flavor of unsalted tomato juice will be most distinct as the natural acids are dominant. Unsalted tomato juice may find a natural market, at least, for the elderly and those on low sodium diets.

Filling machines are adjusted to give maximum fill which gives the best retention of quality and of vitamin C. In No. 10 (603 × 700) and No. 3 cylinder (404 × 700) cans, the net headspace after cooling should not be over 7/16 in. (10/16 in. to the top of the double seam) and in shorter cans not over 3/16 in. below the cover (6/16 in. gross headspace) for best results (Tressler and Joslyn 1971).

FIGURE 9.5. TOMATO JUICE 10 POCKET VOLUME FILLER.
Courtesy of Hirzel Canning Co.

CONTAINERS

Tomato juice is packed in cans made with plain, hot-dipped, or electrolytic tinplate bodies and enameled electrolytic tinplate ends, or in glass.

The U.S. Department of Agriculture has recently taken an interest in the amount of tin in canned tomato juice. The Codex Alimentarius appears to be directed toward a tolerance of 250 ppm of tin in canned products. At present there is no official tolerance on tin in tomato products. However, more tomato juice will be processed in enameled cans or glass in the future.

The empty cans should be kept as clean as possible during storage. The can-handling methods employed should be properly designed to prevent scratching or denting, particularly the flange of the can. Before filling, the cans should be sanitized with a relatively large volume of water at a minimum temperature of 180°F (82°C) to remove any possible dust or other foreign material.

HOMOGENIZATION

Tomato juice is sometimes homogenized before canning in machines similar to those used for milk and other dairy products. Homogenization retards or prevents settling of the solids and produces a thicker-bodied juice. It is generally used for cold-break juice. The juice is forced through narrow orifices at a pressure in the range of 1000 to 1500 psi and at a temperature of about 150°F (66°C) to break up the suspended solids (Goose and Binsted 1964). This step is a usual one when juice is packed in glass (Tressler and Joslyn 1971). Tomato juice may also be milled by using a comminutor or a Fitzpatrick Mill to assist in control of separation and product consistency.

THERMAL PROCESSING OF TOMATO JUICE

Although tomato juice is an acid product, it has been subject to frequent outbreaks of spoilage when conventional thermal processes for acid products have been employed. The spoilage is caused by heat-resistant strains of *Bacillus thermoacidurans* and is known as flat-sour spoilage.

Commercially canned tomato juice should be sufficiently heat-processed either before or after filling to prevent spoilage. The following methods are currently employed for processing tomato juice (Troy and Schenk 1960):

1. In-can processing
   a. Pressure processing in continuous agitating cookers
   b. Atmospheric processing in continuous agitating cookers
   c. Processing in boiling water with agitation
   d. Hot fill, followed by steam processing at atmospheric pressure
2. Bulk processing
   a. Flash sterilization followed by hot fill–hold–water cool;
   b. Hot fill–hold–air cool

In-Can Processing

Pressure Processing in Continuous Cookers. During the pressure processing in continuous cookers, the cans of juice are normally closed at about 185° to 190°F (85°to 88°C) processed while rotating under steam pressure at 240° to 250°F (116° to 121°C) followed by water cooling to approximately 100°F (38°C). This is a safe method for processing tomato juice to avoid flat-sour spoilage. The juice is sterilized after the can is closed and consequently is not exposed to possible recontamination. Beside the spoilage protection provided by this procedure, the equipment permits use of higher temperatures and shorter processing times. Due to agitation of the product, sterility can be attained in a shorter time. This method provides continuous mechanical handling of the can, thereby reducing labor and production costs.

Atmospheric Processing in Continuous Agitating Cookers. In atmospheric processing in continuous agitating cookers, the cans of juice are normally closed at about 200° to 205°F (93.3° to 96.1°C). They are then heated for 15 to 20 min while rotating in water near the boiling temperature, followed by water cooling. This method destroys low heat-resistant organisms in the juice, but does not provide any significant protection against the more heat-resistant spores of the flat-sour organism *B. thermoacidurans*.

Boiling Water Process. Some packers employ a boiling water process without agitation (still cook) for processing tomato juice. If heat-resistant organisms are present, this method will not give a sterile product unless extended process times are used. The process time varies with the can size and the initial temperature of the juice at the time the process is started.

Hot Fill Followed by Steam Processing at Atmospheric Pressure. In this method, the product is hot filled at 200° to 205°F (93.3° to 96.1°C), conveyed from 7.5 to 10 min on a metal mesh belt through a steam tunnel at atmospheric pressure, followed by water cooling. The belt is enclosed in sheet metal with perforated pipes running the length of the tunnel to distribute the steam. This method makes it possible to use production line speeds without going to processing kettles. However, this method is not adequate to destroy heat-resistant flat-sour spores.

Bulk Processing

Flash Sterilization Followed by Hot Fill–Hold–Water Cool. Next to pressure processing in a continuous agitating cooker, flash sterilization is the safest method employed in canning tomato juice to avoid flat-sour

FIGURE 9.7. THERMAL DEATH TIME CURVE FOR *BACILLUS THERMOACIDURANS*.

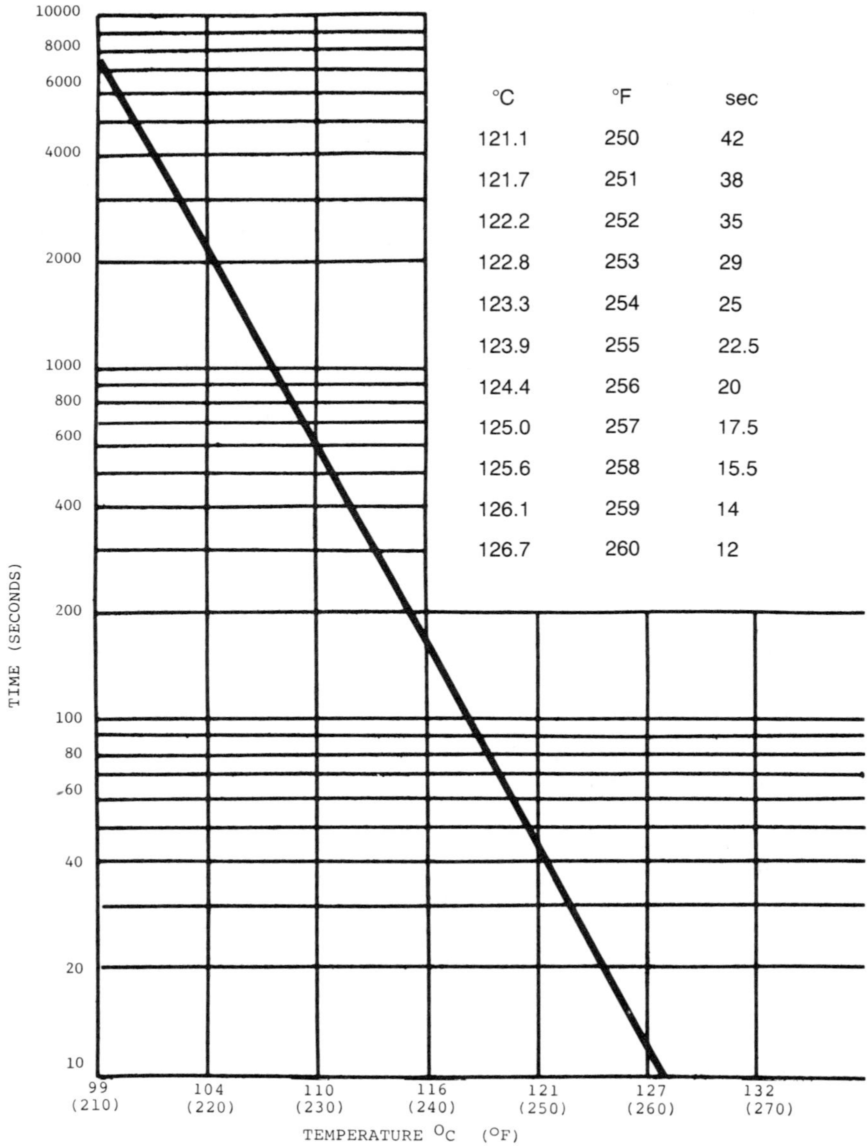

°C	°F	sec
121.1	250	42
121.7	251	38
122.2	252	35
122.8	253	29
123.3	254	25
123.9	255	22.5
124.4	256	20
125.0	257	17.5
125.6	258	15.5
126.1	259	14
126.7	260	12

TABLE 9.3. THERMAL DEATH TIMES FOR *Bacillus thermoaciduran*s IN TOMATO JUICE[a]

Temperature		
°F	°C	Time of 100% Destruction
180	82.3	5600 min
190	87.8	1525 min
200	93.3	440 min
210	98.9	117 min
212	100.0	90 min
220	104.4	32.5 min
230	110.0	9 min
235	112.8	4.7 min
240	115.6	2.5 min
245	118.3	1.2 min
250	121.1	0.7 min or 42 sec
255	123.9	22 sec
260	126.7	12 sec
265	129.4	6 sec
270	132.2	3 sec
280	137.8	1 sec
290	143.3	0.3 sec
300	148.9	0.07 sec
310	154.4	0.02 sec

[a]The data give the thermal death time for 100% destruction of spores of *Bacillus thermoacidurans* in tomato juice having a pH of 4.25 to 4.30. The data are based on a concentration of 400,000 spores per ml of tomato juice.

spoilage. The juice is presterilized by heating in continuous heat exchangers to temperatures substantially above the boiling point to completely destroy the spores of heat-resistant flat-sour organism which may be present in the juice. A common practice is to provide a sterilization time and temperature equivalent to about 0.7 min at 250°F (121°C), as shown in Fig. 9.1. The juice must be cooled below the boiling point before filling, but must still be hot enough to insure sterilization of the containers. A minimum closing temperature of 200°F (93°C), is suggested, after which the cans should be inverted and conveyed for a minimum of 3 min *at this temperature* prior to water cooling. Small cans, after filling, should be processed 5 to 10 min at 212°F (100°C), or conveyed in an atmosphere of steam for 5 to 10 min prior to water cooling to maintain the closing temperature and to counteract the rapid cooling that occurs in smaller cans.

Although this processing procedure minimizes the possibility of flat-sour spoilage, it does not necessarily guarantee a sterile canned product. Prevention of spoilage depends on control of possible recontamination and rigid adherence to good sanitation practices in all operations following juice presterilization.

FIGURE 9.8. TOMATO JUICE PASTEURIZER
Courtesy of FMC Corporation

Hot Fill–Hold–Cool. In the hot fill–hold–cool process, the tomato juice is heated to 200° to 205°F (93° to 96°C), filled at about 195° to 200°F (91° to 93°C), and closed. The cans are inverted, held for about 3 min, and then water cooled. This procedure often results in excessive spoilage, especially where the pH of the juice is 4.35 or higher. Low heat-resistant organisms may be destroyed by this process, but it has practically no effect against the heat-resistant spores of *B. thermoacidurans*. Use of this method is not recommended (Troy and Schenk 1960).

TOMATO JUICE FROM CONCENTRATE

Tomato Juice from concentrate is now a reality. The original research initiated in the early '60's showed that an excellent tomato juice could be manufactured from pulp or paste. However, this was not permitted except under permit until FDA established a Standard of Identity for Tomato Juice from Concentrate (July 1, 1985 (21 CFR 156.145)).

Tomato Juice from Concentrate (TJC) can be manufactured from tomato pulp or puree (8.0 to 24% soluble solids), or Tomato paste (24% or more soluble solids), or concentrated tomato juice (20 to 24% soluble solids) as long as the concentration upon diluting the food according to label directions it will not contain more than 5.0% by weight tomato soluble solids. The product should be finished by passing through a finisher with screens ranging from 0.020 to 0.030 inch prior to homogenization (500 to 2000 psi), seasoned with salt, acidified with any suitable organic acid, and preserved by heat (canning), refrigeration, or freezing.

Some consumers state that this new tomato juice from concentrate has superior quality. This may be because it is made from pulp or paste or tomato juice concentrate that was manufactured with extremely high temperature hot-break or that it was manufactured directly and not a by product from the peeling line. The consumption of this product is increasing and it is a credit to these manufacturers because the product does have consumer appeal.

NEW PRODUCTS

Many newly developed tomato juice based products with or without other vegetable juices are now appearing on the market. There is no Standard of Identity for these products. The flavors are controlled by the correct additions of spices, sugar, flavors, acids, and or other ingredients. The possibilities are unlimited. The process conditions are essentially similar to tomato juice manufacture except for the formulations. Hopefully, these new products will find their niche with consumers and help in the consumption of more tomato juice products.

REFERENCES

AMMU, K., RADHAKRISHNA, K., SUBRAMANIAN, V., SHARMA, T.R. and NATH, H. 1977. Storage behavior of freeze dried fruit juice powders. J. Food Technol. *12*, 541–554.

BECKER, R., MIERS, J.C., NUTTING, M.D., DIETRICH, W.C. and WAGNER, J.R. 1972. Consistency of tomato products. 7. Effects of acidification on cell walls and cell breakage. J. Food Sci. *37*, 118–125.

BECKER, R., WAGNER, J.R., MIERS, J.C., SANSHUCK, D.W. and DIETRICH, W.C. 1968. Consistency of tomato products. 3. Effects of pH adjustment during tomato juice preparation on pectin contents and characteristics. Food Technol. *22* (4) 503–505.

BROOKMAN, J.S.G. 1974. Mechanism of cell disintegration in a high pressure homogenizer. Biotechnol. Bioeng. *16*, 371–383.

CATELLE, Camillo. 1985. Plant for bringing about enzymedeactivation in fruit and vegetables: Tomatoes in particular. U.S. Patent No. 4, 543, 879.

CRANDALL, P.G. and NELSON, P.E. 1975. Effects of preparation and milling on consistency of tomato juice and puree. J. Food Sci. *40*, 710–713.

CRUESS, W.V. 1958. Commercial Fruit and Vegetable Products, 4th Edition. McGraw-Hill Book Co., New York.

DOUGHERTY, R.H. and NELSON, P.E. 1974. Effects of pH on quality of stored tomato juice. J. Food Sci. *39*, 254–256.

EBERTS, E.C. 1934. Method of producing juice beverage. U. S. Patent No. 2, 092,729.

FODA, Y.H. and McCOLLUM, J.P. 1970. Viscosity as affected by various constituents of tomato juice. J. Food Sci. *35*, 333–338.

FONSECA, H. and LUH, B.S. 1976. Effect of break temperature on quality of tomato juice reconstituted from frozen tomato concentrates. J. Food Sci. *41* (6) 1308–1311.

GERSCHENSON, L.N., BARTHOLOMAI, G.B. and CHIRIFE, J. 1979. Retention of volatiles during freeze drying of tomato juice. J. Food. Technol. *14*, 351–360.

GOOSE, P.G. and BINSTED, R. 1964. Tomato Paste, Puree, Juice and Powder. Food Trade Press, London.

GOULD, W.A. 1971. Tomato Processor Quality Control Technologist Handbook. Dep. Hortic., Ohio State Univ. Columbus.

GOULD, W.A. 1978. Quality evaluation of processed tomato juice. J. Agric. Food Chem. *26* (5) 1006–1010.

HAND, D.B., MOYER, J.C., RANSFORD, J.R., HENING, J.C. and WHITTENBERGER, R.T. 1955. Effect of processing conditions on the viscosity of tomato juice. Food Technol. *9* (5) 228–235.

HUHTANEN, C.N., NAGHSKI, J., CUSTER, C.S. and RUSSELL, R.W. 1976. Growth and toxin production by Clostridium botulinum in moldy tomato juice. Appl. Environ. Microbiol. *32* (5) 711–715.

JANORIA, M.P., THOMPSON, A.E. and RHODES, A.M. 1975. Inheritance and evaluation of alcohol insoluble solids of tomatoes as a secondary character in selection for juice viscosity. J. Am. Soc. Hortic. Sci. *100* (3) 219–221.

JOHNSON, J.H., GOULD, W.A., BADENHOP, A.F. and JOHNSON, R.M., JR. 1968. Quantitative comparison of isoamylol, pentanol, and 3-hexenol-1 in tomato juice. Varietal and harvest differences and processing effects. Argic. Food Chem. *16* (2) 255–258.

JUDGE, E.E. & SONS. The Almanac of the Canning, Freezing, Preserving Industries, Edward E. Judge & Sons, Westminster, MD.

LAMB, F.C. 1967. Relation between refractive index, specific gravity, and total solids of tomato juice, puree, and paste. J. Assoc. Off. Anal. Chem. *50* (6) 690–700.

LEE, Y.C., KIRK, J.R., BEDFORD, C.L. and HELDMAN, D.R. 1977. Kinetics and computer simulation of ascorbic acid stability of tomato juice as functions of temperature, pH and metal catalyst. J. Food Sci. *42* (3) 640–648.

LOPEZ A. 1987. A Complete Course in Canning, 12th Edition. Canning Trade, Baltimore.

LUH, B.S. and DAOUD, H.N. 1970. Pectin and pectin enzymes in VF-145 tomatoes. Univ. of Calif. Tomato Res. Prog. Rep. (Apr. 15).

McCOLLOCH, R.J. 1952. Determination of pectic substances and pectic enzymes in citrus and tomato products. U.S. Dep. Agric., Bur. Agric. Ind. Chem., Agric. Res. Admin. *AIC-337*.

McCREADY, R.M. and McCOMB, E.A. 1952. Extraction and determination of total pectic materials in fruits. Anal. Chem. *24* (12) 1986–1988.

MIERS, J.C., SANSHUCK, D.W., NUTTING, M.D. and WAGNER, J.R. 1970. Consistency of tomato products. 6. Effects of holding temperature and pH. Food Technol. *24* (12) 81–84.

MIERS, J.C., WAGNER, J.R. and SANSCHUCK, D.W. 1967. Consistency of tomato products. II. Effect of pH during extraction in tomato juice consistency. Food Technol. *21* (6) 117–120.

MOHR, W.P. 1976. Effect of difference in degree of ripeness on canned juice quality of "high colour" vs. normal colour tomatoes: A preliminary assessment. J. Inst. Can. Sci. Technol. Aliment. *9* (3) 167–170.

MOYER, J.C., HOLGATE, K.C., LABELLE, R.L. and HAND, D.B. 1951. Tomato juice line. Ind. Eng. Chem. *43*, 1874.

MOYER, J.C., ROBINSON, W., RANSFORD, J.R., LaBELLE, R.L. and HAND, D.B. 1959. Processing conditions affecting the yield of tomato juice. Food Technol. *13*, 270.

NELSON, P.E. and TRESSLER, D.K. 1980. Fruit and Vegetable Juice Processing Technology, 3rd Edition. AVI Publishing Co., Westport, CT.

SARAVACOS, G.D. 1970. Effect of temperature on viscosity of fruit juice and purees. J. Food Sci. *35*, 122–125.

SARAVACOS, G.D., ODA, Y. and MOYER, J.C. 1967. Tube viscometry of tomato juice and concentrates. Dep. Food Sci. Technol., N.Y. State Agric. Exp. Stn., Cornell Univ. Bull.

SHERKAT, F. and LUH, B.S. 1977. Effect of break temperature on quality of paste, reconstituted juice, and sauce made from M-32 tomatoes. J. Inst. Can. Sci. Technol. Aliment. *10* (2) 92–96.

SOC. PAMPRYL. 1964. Process of preservation of fruit juices and the apparatus for putting this process into practice. Fr. Pat. 1,406,168. (French)

STADTMAN, F.H., BUHLERT, J.E. and MARSH, G.L. 1977. Titratable acidity of tomato juice as affected by break procedure. J. Food Sci. *42* (2) 379–382.

TRESSLER, D.K. and JOSLYN, M.A. 1971. Fruit and Vegetable Juice Processing Technology. AVI Publishing Co., Westport, CT.

TROY, V.S. and SCHENK, A.M. 1960. Flat Sour Spoilage of Tomato Juice. Continental Can. Co., Chicago.

U.S. DEP. AGRIC. 1963. Composition of foods. U.S. Dep. Agric. Agric. Handb. *8*.

WAGNER, R. and MIERS, J. 1967. Consistency of tomato juice products. I. The effects of tomato enzyme inhibition by additives. Food Technol. *21* (6) 114–117.

WAGNER, J.R., MIERS, J.C., SANSHUCK, D.W. and BECKER, R. 1968. Consistency of tomato products. 4. Improvement of the acidified hot break process. Food Technol. *22* (11) 150–154.

WAGNER, J.R., MIERS, J.C., SANSHUCK, D.W. and BECKER, R. 1969. Consistency of tomato products. 5. Differentiation of extractive and enzyme inhibitory aspects of the acidified hot break process. Food Technol. *23* (2) 113–116.

WHITTENBERGER, R.T. and NUTTING, G.C. 1957. Effect of tomato cell structures on consistency of tomato juice. Food Technol. *11* (1) 19–22.

WHITTENBERGER, R.T. and NUTTING, G.C. 1958. High viscosity of cell wall suspensions prepared from tomato juice. Food Technol. *10* (8) 420–423.

YORK, G.K., O'BRIEN, M., TROMBROPOULOS, D., WINTER, F.H. and LEONARD, S.J. 1967. Relation of fruit damage to quality and consistency of tomato concentrates. Food Technol. *21* (1) 69–92.

CHAPTER 10
Tomato Pulp and Paste Manufacture

TOMATO PULP AND PUREE

The definition of tomato pulp or purée under the Food, Drug, and Cosmetic Act follows.

Section 53.20 Tomato purée, tomato pulp—Identity; labeling of optional ingredients.

(a) Tomato purée, tomato pulp, is the food prepared from one or any combination of two or all of the following optional ingredients:

(1) The liquid obtained from mature tomatoes of red or reddish varieties.

(2) The liquid obtained from the residue from preparing such tomatoes for canning, consisting of peelings and cores with or without such tomatoes or pieces thereof.

(3) The liquid obtained from the residue from partial extraction of juice from such tomatoes.

(4) Salt.

Such liquid is obtained by so straining such tomatoes or residue, with or without heating, as to exclude skins, seeds, and other coarse or hard substances. Prior to straining, food-grade hydrochloric acid may be added to the tomato material at a rate to obtain a pH no lower than 2.0 ± 0.2. Such acid is then neutralized with food-grade sodium hydroxide so that the treated tomato material is restored to a pH of 4.2 ± 0.2, prior to straining. It is concentrated and may be seasoned with salt (sodium chloride formed during acid neutralization shall be considered added salt). When sealed in a container it is so processed by heat, before or after sealing, as to prevent spoilage. It contains no less than 8.0%, but less than 24.0%, of natural tomato soluble solids, as determined by the following method:

Determine the refractive index of the clear serum obtained from the product, corrected for temperature, converting the resultant index to '% Sucrose' in accordance with the 'International Scale of Refractive Indices of Sucrose at

20°C.,' pages 828-30, Reference Tables 43.008 and 43.009 of the book 'Official Methods of Analysis of the Association of Official Agricultural Chemists,' 10th edition, 1965. If no salt has been added, this percent sucrose from the reference table shall be considered the percent of natural tomato soluble solids. If salt has been added, determine the percent of sodium chloride by the method prescribed on page 519, section 30.009, under 'Sodium Chloride-Official,' of said book. Subtract the percent of sodium chloride found from the percent of total soluble solids found and multiply the difference by 1.016. The product shall be considered the percent of natural tomato soluble solids.

(b) When optional ingredient (2) is present, in whole or in part, the label shall bear the statement 'Made from ___________' (or 'Made in Part from ___________,' as the case may be), 'Residual Tomato Material from Canning.' When optional ingredient (3) is present, in whole or in part, the label shall bear the statement 'Made from ___________' (or 'Made in Part from ___________,' as the case may be), 'Residual Tomato Material from Canning and from Partial Extraction of Juice.' Wherever the name 'Tomato Purée' or 'Tomato Pulp' appears on the label so conspicuously as to be easily seen under customary conditions of purchase, the statement or statements herein specified showing the optional ingredients present shall immediately and conspicuously precede or follow such name, without intervening written, printed, or graphic matter."

The USDA has available for voluntary use "U.S. Standards for Grades of Canned Tomato Pulp" based on a scoring system in which the attribute of color scores 50 points and the attribute of absence of defects scores 50 points.

Manufacture of Tomato Pulp

Tomato pulp is made from raw tomatoes first by separating the liquid and fleshy portions from the seeds, skins, cores, etc., and, second, by removing water from this cyclone juice until the concentrated product contains at least 8.0% salt-free tomato solids, as shown in Table 10.1.

TABLE 10.1. RELATIONSHIP OF DIRECT REFRACTOMETER READING FOR NATURAL TOMATO SOLIDS AND SPECIFIC GRAVITY FOR RAW JUICE, PULP, AND PASTE

Minimum values.

	Pulp-Puree		Paste	
Concentration	% Solids	Specific Gravity	% Solids	Specific Gravity
Raw juice	4.0	1.0163		
Light	8.0	1.0332	24.0	1.1165
Medium	10.1	1.0421	28.0	1.1365
Heavy	11.3	1.0472	32.0	1.1580
Extra heavy	15.0	1.0635	38.5	1.1765
Concentrated tomato juice	20.0	1.0965		

The raw tomatoes are first washed, sorted, and trimmed to remove all visible defects. This is a most important unit operation in the manufacture of pulp. The tomatoes are then conveyed either direct to a cyclone for the cold-break extraction, or to a chopper or crusher and then to a preheater, followed by cycloning in the hot-break procedure.

The chopped tomatoes, with or without preheating, are sometimes passed through a vibrating screen before going to the cyclone. Vibrating screens are used to remove defective areas, green portions, hard cores, and yellow sunburn, thereby producing better color and flavor in the finished product.

In the hot-break procedure, the preliminary heating given the tomatoes completely destroys the enzymes and protects the constituents of the tomato (especially pectin) from enzymatic change. The tomatoes are crushed with a minimum inclusion of air and quickly heated by direct contact with free flowing steam or under steam pressure or in an atmospheric tank with rotary coils for indirect heating to as high as 220°F (104°C). The yield of cyclone juice is greater than with the cold-break procedure.

In cold-break systems the tomatoes are crushed at temperatures less than 150°F (66°C) and then fall into a holding tank, where they remain static for periods ranging from seconds to many minutes. During this holding period, the enzymes liberated during crushing can catalyze the breakdown of pectins. This effect may be considered a function of the time of holding cold after breaking (Goose and Binsted 1964).

Concentration or evaporation is carried out in tanks with coils, or in vacuum pans. The tanks may be made of stainless steel or Monel metal. Aluminum should not be used as the acid can severely pit the tanks. Oxidation of aluminum into the finished product could also be a serious problem. The tanks should be easy to clean, and the coils should be of the same metal as the tanks (Lopez 1987).

The success of the concentration depends on a good steam supply under pressure, 100 psi or over, and steam traps. Coils operating under low pressure burn and cause the solids to stick on the surface, resulting in slow, if any, evaporation. This difficulty can be overcome to a certain extent by

FIGURE 10.1. ROTARY COILS FOR CONCENTRATION OF TOMATO PULP. *Courtesy of F.H. Langsenkamp Co.*

forcing circulation with a propeller stirring device. A hot coil causes rapid circulation which prevents burning or sticking. The foam on pulp is best broken by a blast of air from a small blower. A vacuum kettle works quickly, but because the pulp needs a high temperature for sterilization, there is no great advantage in using it for low concentrated pulp.

The steam should be turned on as soon as the coils are covered with pulp. To secure the best color and flavor, evaporation would be as rapid as possible. The evaporation time of a batch should preferably be less than 30 min. With proper operation of the steam traps on the coils and proper boiler pressure, a batch of 600 gal. can be concentrated to 1.035 specific gravity in 25 to 30 min. A steam pressure of 100 psi can be carried in the average steam coils.

One thousand gal. of 1.020 specific gravity cyclone or raw juice will give 574 gal. of 1.035 pulp, 501 gal. of 1.040 pulp, 444 gal. of 1.045 pulp, or 398 gal. of 1.050 pulp (Lopez 1987).

The cyclone juice is usually pumped to a storage tank and withdrawn from this to the evaporating kettles when needed. If the cyclone juice is left in the storage tank undisturbed for any length of time, the insoluble fiber will rise to the top, liquid withdrawn from the bottom will be watery, and the batches of pulp will vary in consistency. It is suggested that the storage tank be equipped with a steam coil so that the contents may be kept hot and mixed at all times. Some manufacturers also install a mechanical stirrer to ensure thorough mixing.

The older method consisted of evaporating the juice to one-half its original volume, or until it reached a certain consistency determined by its appearance when breaking from a spoon or the dipper. This method has been abandoned as unreliable and exceedingly variable as a basis on which to buy or sell. It is used only by those who pack pulp for their own use or by the few who purchase upon sample. A slight difference in specific gravity may mean a considerable difference in volume of raw juice for concentrating. The average specific gravity of pulp evaporated to one-half its volume is about 1.035; this is accepted as the standard for pulp. Pulp is now made to gravities of 1.040, 1.045, 1.050 and even higher in order to save cans and freight in shipment. Accordingly, the price is scaled on the basis of the tomato solids content.

After the pulp is evaporated to the desired point, it is run through finishing operation to remove any rough particles or fiber which were driven through by the pulper. At temperatures of 190°F (88°C) or higher, it is then filled into cans. Pulp filled at 190°F (88°C) and not cooled usually keeps, though not always. Pulp canned hot, given a 30-min cook in boiling water, and then cooled is more certain of keeping and will have a brighter appearance.

When the product is not processed in the cans, the filled containers, after leaving the closing machine, should be passed through a warm water spray to remove any pulp on the outside. Unless the cans are cooled carefully, loss in color, loss in flavor, or even stack burning may result.

Determination of Total Solids

Total solids in tomato pulp may be determined in various ways, which include drying in vacuum at 158°F (70°C), drying at atmospheric pressure at the temperature of boiling water, calculating from the specific gravity of the pulp, or using the index of refraction of the filtrate. The last is now the most common method. The direct refractometer reading, as shown in Table 10.1, is used to determine the amount of concentration for the various label declarations.

When using the refractometer the scale should be checked periodically by taking the reading of distilled water. Abbé refractometer readings of distilled water at several temperatures are shown in Table 10.2. The Abbé refractometer has been used to control the degree of concentration during the manufacture of tomato pulp. In addition to the refractive index scale, the instrument may also have a Brix scale, which shows the percent of soluble solids calculated as sugar.

TABLE 10.2. RELATIONSHIP BETWEEN TEMPERATURE AND REFRACTOMETER READINGS FOR DISTILLED WATER

Temperature		Abbé Refractometer Reading (n_D)
°F	°C	
68	20	1.3330
71.5	22	1.3328
75.0	24	1.3326
79.0	26	1.3324

The following procedure has been found to be rapid and satisfactory for factory control purposes. With a 10-ml pipette having a fairly large opening at the tip, remove about 10 ml of the hot pulp from the sampling cup. By placing the forefinger over the top of the pipette, the sample is held withing the pipette. To cool the sample, immerse the length of the pipette in a can of cold water for a few seconds. Take the pipette out of the water and wipe the outside dry with a towel. Allow the bottom portion of the sample in the pipette, which may have been diluted by contact with the cooling water, to flow from the bottom of the pipette. Use a drop or two from approximately the middle of the sample in the pipette to examine with the Abbé refractometer. The drop of sample, which comes in contact with the prisms, quickly assumes the temperature of the prisms and the reading may be made at once. When cleaning the prisms after the reading is taken, care should be taken to wash the surfaces with water and to remove all soluble solids and moisture without scratching the surfaces by using a soft linen cloth.

TOMATO PASTE

The definition for this product under the Federal Food, Drug, and Cosmetic Act follows.

Section 53.30 Tomato paste—Identity; labeling of optional ingredients.

(a) Tomato paste is the food prepared from one or any combination of two or all of the following optional ingredients:

(1) The liquid obtained from mature tomatoes of red or reddish varieties.

(2) The liquid obtained from the residue from preparing such tomatoes for canning, consisting of peelings and cores with or without such tomatoes or pieces thereof.

(3) The liquid obtained from the residue from partial extraction of juice from such tomatoes.

Such liquid is obtained by so straining such tomatoes or residue, with or without heating, as to exclude skins, seeds, and other coarse or hard substances. Prior to straining, food-grade hydrochloric acid may be added to the tomato material at a rate to obtain a pH no lower than 2.0 ± 0.2. Such acid is then neutralized with food-grade sodium hydroxide so that the treated tomato material is restored to a pH of 4.2 ± 0.2, prior to straining. It is concentrated and may be seasoned with one or more of the optional ingredients:

(4) Salt (sodium chloride formed during acid neutralizations shall be considered added salt).

(5) Spice.

(6) Flavoring.

It may contain, in such quantity as neutralizes a part of the tomato acids, the optional ingredient:

(7) Baking soda.

When sealed in a container it is so processed by heat, before or after sealing, as to prevent spoilage. It contains not less than 24.0% of natural tomato soluble solids, as determined by the method prescribed in the Standard for Tomato Purée.

(b) When optional ingredient (a) (2) is present, in whole or in part, the label shall bear the statement 'Made From __________' (or 'Made in Part From __________,' as the case may be) 'Residual Tomato Material from Canning.' When optional ingredient (a) (3) is present, in whole or in part, the label shall bear the statement 'Made From __________' (or 'Made in Part From __________,' as the case may be) 'Residual Tomato Material from Partial Extraction of Juice.' If both such ingredients are present, such statements may be combined in the statement 'Made From __________,' (or Made in Part From __________,' as the case may be) 'Residual Tomato Material from Canning and from Partial Extraction of Juice.' When optional ingredient (a) (4), (5) or (6) is present the label shall bear the statement or statements 'Salt added' or 'With added salt,' 'Spice added' or 'With added spice,' 'Flavoring added' or 'With added flavoring,' as the case may be. When optional ingredient (a) (7) is present, the label shall bear the statement 'Baking soda added.' If two or all of the optional ingredients (a) (4), (5), (6) and (7) are present, such statements may be combined; for example, 'Salt, spice, flavoring, and baking soda added.' In lieu of the word 'salt,' 'spice,' or 'flavoring' in such statement or statements, the common or usual name of such salt, spice, or flavoring may be used.

(c) Wherever the name 'Tomato Paste' appears on the label so conspicuously as to be easily seen under customary conditions of purchase, the statement or statements herein specified showing the optional ingredients present shall immediately and conspicuously precede or follow such name, without intervening written, printed, or graphic matter.

It should be noted that this definition does not permit the use of color added to tomato paste.

Manufacture of Tomato Paste

Commercially, this product nearly always contains salt and basil. In the heavier pastes, a portion of the acid is generally neutralized with sodium bicarbonate or sodium carbonate, preferably the former.

The packing of tomato paste differs from the packing of tomato pulp only in the degree to which the concentration is carried, as shown in Table 10.1. The preliminary steps are the same. In concentration, however, the vacuum pan is preferred to the open kettle because the cooking is done at a lower temperature, thus conserving color and flavor (Lopez 1987).

The concentration systems, always operating under reduced pressures and usually double or multiple effect, fall into two main types.

The batch type consists of a large-capacity preconcentrator and from three to five secondary vacuum-pan evaporators, which receive batch charges of partially concentrated paste from the preconcentrator.

The continuous concentration plant handles pulp in a continuous flow and discharges finished paste at almost any desired concentration.

Evaporation under partial vacuum takes place at a lower temperature than would occur if the pulp were boiled at atmospheric pressure. Consequently, the resulting paste retains most of the flavor and color of the fresh tomatoes.

In a batch-type system the pulp is drawn into the preconcentrator from the receiving tank. It is made to circulate through banks of vertical tubes, losing water rapidly without excessive frothing. The evaporator is started by means of steam but, when operating, heating is continued by utilizing the steam and vapors from the secondary operators to which it is connected. When the pulp has been concentrated to a solids content of about 12% it is transferred in batches to the steam-jacketed vacuum pans, known commonly as boules. The boules are fitted with mechanical agitators to keep the paste moving during heating and to prevent it from sticking and burning onto the walls of vessels. These pans are fitted with domed tops to allow the vapors to be drawn off through a central duct and then used to heat the preconcentrator, as mentioned above. Both the boules and the preconcentrator are maintained under reduced pressure by a vacuum pump and a barometric condenser.

As the paste concentrates, samples have to be taken for testing the solids content; special vapor locks are fitted to the sides of the vacuum pans for this purpose. Rapid determination of solids content is made during production by means of refractometers. When the required concentration is reached, the contents of each pan are drained. This system is known as an inverted double-effect system because the pulp, during concentration, flows in a direction opposite that of the heating medium. The capacity of a boule is lower than the preconcentrator because the boule must handle a product that progressively increases in viscosity while the preconcentrator carries a liquid.

Continuous evaporation systems tend to produce a more consistent product than batch-type systems, but they have not superseded them. Some manufacturers still prefer to concentrate their pulp in batch quantities. When small runs of different concentrations produced from different raw materials are desired, the batch system is more convenient.

One attractive feature of a continuous plant is its ease of cleaning. A cleaning solution, such as hot caustic, may be circulated through the equipment, following the path of the tomato product (Goose and Binsted 1964).

The acidity is increased at the same rate as the reduction in volume. Therefore, paste of high concentration becomes very acid. A few processors neutralize a part of the acid in order to improve both the flavor and the color. This is done by adding bicarbonate of soda to the fresh pulp; 16 oz of bicarbonate will neutralize the equivalent of 0.1% of acidity in 100 gal. of pulp. The amount of soda used will depend on the original acidity and the degree of neutralization desired (Lopez 1987).

If sodium bicarbonate or sodium carbonate is used, it should be dissolved in a small amount of water and added gradually during evaporation. Both cause foaming and must be carefully added, preferably to the hot cyclone juice prior to introducing the juice into the vacuum pan. The juice is then heated for a short time to drive off the carbon dioxide gas (Lopez 1987).

The amount of salt needed also depends on the concentration. Approximately 8 lb per 100 gal. of finished paste is necessary. Salt should be added near the end of the evaporation, allowing sufficient time for it to be thoroughly dissolved.

Basil may or may not be used. The basil is sometimes added as a single leaf placed on the bottom of each small can, or as oil of sweet basil. If the oil is used, it should be added to the paste shortly before completing the evaporation and should be well stirred in. Care should be used not to add too much of the strong-flavored oil. The oil should be reduced with alcohol or a vegetable oil before adding. Three to 5 oz of a 1% solution of the oil may be used for each 100 gal. of finished paste (Lopez 1987).

Filling

The temperature of the paste must be raised to at least 194°F (90°C) before it is packed into cans. This prevents the survival of any micro-organisms likely to bring about subsequent spoilage of the product. No further heating is given to the cans after closing; therefore, hot filling and immediate seaming are essential parts of the packing process (Goose and Binsted 1964). The cans should be well cooled before casing (Lopez 1987).

BULK STORAGE

Tomato products may be stored in bulk containers (55 gal. drums to more than 40,000 gal. tanks) by various aseptic storage, freezing, or acidified bulk storage. These methods allow the processor to store product during the harvest season and make various finished products on a year around basis. This allows the industry to compensate for market demands and save on transportation cost by producing finished product close to its distribution point.

FIGURE 10.2. BULK STORAGE FACILITY.

Aseptic storage is based on the sterilization of product on a continuous basis before pumping into a sterile storage container. In aseptic storage a continuous heat exchanger sterilization system is first sterilized with chemical sterilant, steam, or superheated water. For concentrates, scraped surface heat exchanger surfaces are required in order to limit fouling problems. Specialized pumps and valving systems are employed to eliminate the risk of contamination. Storage tanks are very difficult to sterilize. Sterilization is accomplished by using iodine or compounds such as perchloracetic acid. With iodine the entire tank must be flooded for a sufficient time with a concentration of iodine that will kill all microorganisms. Sterile nitrogen is used as the headspace on top of a filled tank. This is make by passing air through a bacteriological filter. This nitrogen blanket also forms a sufficient positive pressure to prevent the entry of spoilage microorganisms. This product may be removed for remanufacture into catsup, sauces, soup, or other specialty products packaged into smaller containers. Aseptically stored finished products can also be transfered into sterile aseptic containers that may be heat or hydrogen peroxide sterilized. Aseptic bulk storage affords the processor a good method for storing concentrates. This system is very expensive to initially install and complicated to run. This makes it prohibitive to many small processors. Another problem is that there is a risk of contamination unless skilled personnel operate the system.

Freezing storage is an alternative storage system used by large paste producers. Freezing storage enables the processor to concentrate to higher soluble solids than utilized with an aseptic storage system. Standard equipment can be used and no labor is needed to operate the storages except to maintain refrigeration equipment. A much lower initial investment is also needed which makes it appealing to the processor. The major problem with this system is the enormous cost of energy.

Acidified bulk storage is another type of storage system that has recently been developed. This system has, for the first time, enabled the storage of whole tomatoes for better processing any finished product needed. This method works if the pH of the product is lowered to about 1.3 and oxygen is excluded in the storage container. This is the first bulk storage system for tomatoes that has allowed the processor to store tomatoes. With this system whole tomatoes are mixed with tomato juice acidified with HCl to an equilibrium pH of 1.3. The product must be stored in a corrosion resistant tank with a headspace filled with nitrogen gas. When the finished product is needed, the product is neutralized with sodium hydroxide or sodium carbonate. This system promises to give whole tomato processors a storage tool that has complete product flexibility. Very little initial investment is needed to start such a system. The size of the containers depends on the processor desires.

REFERENCES

ANON. 1950. Tomato products pulp, paste, catsup, chili sauce and juice. Natl. Food Proc. Assoc. Res. Lab. Bull. *27-L,* Washington, D.C.

ATTIYATE, Y. 1979. Cold-storing tomato pulp. Food Eng. *51* (10) 104-105.

BASEL, Richard M. 1983. Current trends in bulk storage preservation of fruits and vegetables with emphasis on tomatoes. OH. Agr. Res. & Develop. Ctr. Spec. Circ. 107.

BUESCHER, R.W. and DOSTAL, H.C. 1974. Effects of low temperature storage of tomato fruits on acid-soluble nucleotides. J. Food Sci. *39,* 774–776.

DANZIGER, M.T., STEINBERG, M.P. and NELSON, A.I. 1970. Thermal browning of tomato solids as affected by concentration and inhibitors. J. Food Sci. *35,* 808–810.

FRŌLICH, M. 1970. Storage tank installation for sterilized fluid substances. Fr. Pat. 1,599,388. (French)

GIO, S.P.A. and SANSEPOLCRO, E. LLI BUITONI 1971. A process and plant for the aseptic storage of sterile liquids in closed containers. Pat. 1,235,621.

GOOSE, P.G. and BINSTED, R. 1964. Tomato Paste, Puree, Juice and Powder. Food Trade Press, London.

H.J. HEINZ CO. 1964. Improvements in or relating to a method and apparatus for maintaining sterilized perishable liquids in a sterile condition during the bulk transportation of the liquids in tanks. Pat. 949,712.

HARPER, J.C. 1960. Viscometric behavior in relation to evaporation of fruit purees. Food Technol. *14* (11) 557–561.

HARPER, J.C. and EL SAHRIGI, A.F. 1965. Viscometric behavior of tomato concentrates. J. Food Sci. *30* (3) 470–476.

HON, V.M.L., CHEN, C.S. and MARSAIOLI, A., JR. 1979. Computer simulation of dynamic behavior in vacuum evaporation of tomato paste. Trans. ASAE *22* (1) 215–218, 224.

HULSEY, R.G. 1971. Aseptic processing and bulk storage of pulped tomato: System design and development. M.S. Thesis submitted to Faculty of Purdue Univ., Lafayette, IN.

ISENBERG, F.M.R. 1979. Controlled atmosphere storage of fruits. Hortic. Rev. *1,* 380–394.

KIRK, J., DENNISON, D., KOKOCZKA, P. and HELDMAN, D. 1977. Degradation of ascorbic acid in a dehydrated food system. J. Food Sci. *42* (5) 1274–1279.

LEONARD, S., MARSH, G.L., BUHLERT, J.E., HEIL, J.R., WOLCOTT, T.K. and BIRNBAUM, D.G. 1977. Microbial stability and remanufacturing characteristics of high solids tomato concentrates. J. Food Process. Preserv. *38* (1) 191–206.

LOPEZ, A. 1987. A Complete Course in Canning, 12th Edition. Canning Trade, Baltimore.

LUH, B.S., LEONARD, S. and MARSH, G.L. 1958. Objective criteria for storage changes in tomato paste. Food Technol. *12* (7) 347–350.

LUSTER, C., III. 1978. A rapid and sensitive sterility monitoring technique for aseptically processed bulk tomato paste. J. Food Sci. *43*, 1046–1048.

MARSH, G.L., BUHLERT, J. and LEONARD, S. 1977. Effect of degree of concentration and of heat treatment on consistency of tomato pastes after dilution. J. Food Process. Preserv. *38*, 340–346.

MARSH, G.L., BUHLERT, J.E. and LEONARD, S.J. 1980. Effect of composition upon Bostwick consistency of tomato concentrate. J. Food Sci. *45* (3) 703–706, 710.

NELSON, P. 1977. Method and apparatus for aseptic bulk storage of apple sauce. U.S. Pat. 4,022,922. May 10.

NELSON, P.E. 1972. Aseptic storage and valving system. U.S. Pat. 3,678,955. July 25.

NELSON, P.E. 1973. Aseptic storage and valving system. U.S. Pat. 3,714,956. Feb. 6.

NELSON, P.E. and HOFF, J.E. 1969. Tomato volatiles: Effect of variety, processing and storage time. J. Food Sci. *34* (1) 53–57.

NELSON, P.E. and SULLIVAN, G.H. 1975. Method for processing and storing tomatoes. U.S. Pat. 3,873,753. Mar. 25.

NELSON, P.E., SULLIVAN, G.H. and HERON, J.R. 1974. Aseptic processing and bulk storage of fruit products. Proc. 4th Int. Congr. Food Sci. Technol. *4*, 5–10.

PERNOD & RICARD. 1962. Plant for storing fermentable liquids. Br. Pat. 892,357. Mar. 28.

RECHTSTEINER, S.A., NELSON, P.E. and DEBONTE, M. 1975A. Aseptic bulk material storage system and improved aseptic valve therefor. U.S. Pat. 3,918,678. Nov. 11.

RECHTSTEINER, S.A., NELSON, P.E. and DEBONTE, M. 1975B. Aseptic storage system for bulk materials and improved microbiological filter therefor. U.S. Pat. 3,918,942. Nov. 11.

RECHTSTEINER, S.A., NELSON, P.E. and HERON, J.R. 1975C. Method of aseptically connecting a fitting to an aseptic storage tank. U.S. Pat. 3,871,824. Mar. 18.

RECHTSTEINER, S.A., NELSON, P.E. and HERON, J.R. 1976. Method of filling, sampling and sealing as aseptic tank with sterile product without destroying asepsis of either the sterile product or the tank and its associated valves and fittings. U.S. Pat. 3,951,184, Apr. 20.

RIETZ, C.A. 1962. Method of processing foodstuffs. U.S. Patent No. 3,036, 921.

SCHULTZ, W.G., NEUMANN, H.J., SCHADE, J.E., MORGAN, J.P., HANNI, P.F., KATSUYAMA, A.M. and MAAGDENBERG, H.J. 1978. Commercial feasibility of recovering tomato processing residuals for food use. U.S. Environ. Protect. Agency, *EPA 600/2-78-202*.

SMIT, C.J.B. and NORTJE, B.K. 1958. Observations on the consistency of tomato paste. Food Technol. *12* (7) 356–358.

STIER, E.F., BALL, C.O. and MACLINN, W.A. 1956. Changes in pectic substances of tomatoes during storage. Food Technol. *10* (1) 39–43.

WALKER, ROBERT G. and GEORGE BOSY. 1970. Tomato product and method of making same. U.S. Patent No. 3,549,384.

CHAPTER 11
Tomato Catsup and Chili Sauce Manufacture

TOMATO CATSUP

The definition of this product under the Federal Food, Drug, and Cosmetic Act of 1938 follows (Anon. 1950).

Section 53.10 Catsup, Ketchup, Catchup—Identity; Label Statement of Optional Ingredients.

(a) Catsup, Ketchup, Catchup, is the food prepared from one or any combination of two or all of the following optional ingredients.

(1) The liquid obtained from mature tomatoes of red or reddish varieties.

(2) The liquid obtained from the residue from preparing such tomatoes for canning, consisting of peelings and cores with or without such tomatoes or pieces thereof.

(3) The liquid obtained from the residue from partial extraction of juice from such tomatoes.

Such liquid is obtained by so straining such tomatoes or residue, with or without heating, as to exclude skins, seeds, and other coarse or hard substances. Prior to straining, food-grade hydrochloric acid may be added to the tomato material at a rate to obtain a pH no lower than 2.0 ± 0.2. Such acid is then neutralized with food-grade sodium hydroxide so that the treated tomato material is restored to a pH of 4.2 ± 0.2, prior to straining. It is concentrated and seasoned with salt (sodium chloride formed during acid neutralization shall be considered added salt), a vinegar or vinegars, spices or flavorings or both, and onions or garlic or both and is sweetened with sugar or dextrose, or corn sirup (including dried corn sirup) or glucose sirup (including dried glucose sirup) or any mixture of these; provided that when the solids of corn sirup, or dried corn sirup, or glucose sirup, or dried glucose sirup (or any combination of these) used contains less than 58 percent by weight of reducing sugars calculated as anhydrous dextrose, then such corn sirup of glucose sirup shall be mixed with sugar or dextrose or both, in such quantity that the weight of the solids of such corn sirup or dried corn sirup or both, or glucose sirup, or dried glucose sirup or both, is not more than one-third of the weight of the solids of such mixture. When sealed in a container it is so processed by heat, before or after sealing, as to prevent spoilage.

(b) (1) For the purpose of this section, the term "Corn Sirup" means refined corn sirup (including dried corn sirup) the solids of which contain not less than 40 percent by weight of reducing sugars calculated as anhydrous dextrose.

(2) The term "glucose sirup" means a clarified, concentrated, aqueous solution of the products obtained by the incomplete hydrolysis of any edible starch. the solids of glucose sirup contain not less than 40 percent by weight of reducing sugars calculated as anhydrous dextrose. "Dried glucose sirup" means the product obtained by drying "glucose sirup."

(c) When optional ingredient (a) (2) is present, in whole or in part, the label shall bear the statement "Made From ____________" (or "Made in Part From ____________," as the case may be) "Residual Tomato Material from Canning." When optional ingredient (a) (3) is present in whole or in part, the label shall bear the statement "Made From ____________" (or "Made in Part From ____________," as the case may be) "Residual Tomato Material from Partial Extraction of Juice." If both such ingredients are present, such statement may be combined in the statement "Made From ____________" (or "Made in Part From ____________," as the case may be) "Residual Tomato Material from Canning and from Partial Extraction of Juice." Wherever the name "Catsup," "Ketchup," or "Catchup" appears on the label so conspicuously as to be easily seen under customary conditions of purchase, the statement or statements herein specified showing the optional ingredients present shall immediately and conspicuously precede or follow such name, without intervening written, printed, or graphic matter.

It will be noted that the definition of tomato catsup does not permit the use of artificial color, artificial preservatives, or added thickeners of any kind.

The USDA, through its Production and Marketing Administration, has available for voluntary use "U.S. Standards for Grades of Tomato Catsup." These standards are based on a scoring system where color accounts for 25 points; consistency, 25 points; absence of defects, 25 points; and flavor, 25 points.

Manufacturing Tomato Catsup

Catsup may be made directly from fresh cyclone juice or from concentrated pulp or bulk stored tomato paste. With the same quality of raw stock and the same care in manufacture, there are some advantages in making catsup from fresh tomatoes. The concentrated pulp may lose some of its color by holding, and the catsup made from the concentrated pulp is subjected to more heating than that made from fresh tomatoes.

Pulping. Following washing, sorting and trimming, the tomatoes are normally chopped, heated as in hot break described for juice, and put through a pulper or cyclone. The seeds, skins and fiber pass through the pulper, and the pulp and juice pass through the screen. Usually the screen openings are from 0.023 to 0.027 in. The liquid product is then pumped to the concentration tanks or continuous evaporators.

Constituents of Catsup. The constituents used in the manufacture of catsup, in addition to tomatoes, are sugar, vinegar, salt, onions, and spices. Generally, granulated cane sugar or beet sugar is used. Liquid sugar is also being used. The vinegar generally used is 100-grain distilled vinegar. The salt may be the grade known as "dairy salt," or a more refined grade. Among the spices commonly used in making catsup are cinnamon, cassia, cloves, allspice, pepper, cayenne pepper, ginger, mustard, and paprika. Spices may be used either in the form of whole spices, ground spices, or volatile spice oils.

There is a difference of opinion among catsup manufacturers about the relative merits of the different forms of spices. Whole spices are thought by some to produce a better flavor, somewhat more mild and pleasing. Ground spices, when used, should be secured from a reputable manufacturer because there is a possibility of adulteration of the use of low-grade spice material in preparing the ground product. The volatile spice oils, especially those of spices containing considerable tannin, are better to use when there is a possibility of discoloration due to the formation of iron tannate during the manufacture of catsup (Anon. 1950). Only high-grade or extracts should be used. The spices, either whole or ground, are generally placed in a bag and added to the batch at the beginning of the cook (Anon. 1950; Lopez 1987). When volatile spice oils are used they should be added shortly before finishing the catsup; otherwise, a large proportion of them may be carried off as vapors with the steam. In using spice oils, care is necessary

FIGURE 11.1. INSIDE VIEW OF TOMATO PULPER SHOWING PADDLE AND SCREEN ASSEMBLY.
Courtesy of F.H. Langsenkamp Co.

to avoid adding an excessive amount. The object should be to accentuate the tomato flavor rather than mask it. It is advantageous to use the spice oils in the form of an emulsion, in which the particles of oil are very finely divided, to permit the flavor to be uniformly distributed throughout the batch of catsup (Anon. 1950). A recent trend is the use of encapsulated spices or pre-prepared spice mixtures. These spice mixtures are formulated to a processor's specifications and are packaged in lots required per batch. While being more expensive than bulk spices, they are easy to add and do not require many spice oils and ingredients. The use of raw acetic acid is illegal.

The sugar may be added at any time during the manufacture of the catsup, but preferably during the latter part of the cooking. To avoid the sugar remaining undissolved in the batch, some manufacturers add the sugar gradually and scatter it over the surface of the cooking catsup.

Vinegar is always added a few minutes before finishing. The acetic acid of vinegar is volatile; thus, a large portion of it will evaporate if the vinegar is added at the beginning of the cooking.

Salt may be added at any time during the cooking; caution should be taken to ascertain that it is thoroughly dissolved and mixed with the product (Anon. 1950).

The addition of onions and garlic can be made with the spices or in a separate bag. They should be cooked for 20 to 30 min. Some catsup manufacturers chop these very fine, and then add them directly to the pulp depending on their partial removal by the finishing machine (Lopez 1987).

Formula. Nearly every manufacturer of catsup has a formula of his own which differs in some respect from those of other manufacturers. These differences are mainly in the quantities of spices or other flavoring agents used.

The formulas shown in Table 11.1, which are based on suggestions made to the Department of Defense by a committee of catsup manufacturers, may be used as a working basis for manufacturing the type of catsup desired. No.

TABLE 11.1. INGREDIENTS FOR MAKING 100 GAL. OF FINISHED PRODUCT

Ingredient	Formula No. 1	Formula No. 2	Formula No. 3	Formula No. 4
Cyclone juice (sp. gr. 1.020)	182 gal.	182 gal.	254 gal.	290 gal.
Sugar	60 lb	75 lb	118 lb	150 lb
Salt	13 lb	15 lb	20 lb	24 lb
Vinegar	4 gal.	5 gal.	6.3 gal.	8 gal.
Onions	Optional	Optional	27 lb	26 lb
Cloves	16 oz	16 oz	25 oz	21 oz
Cinnamon	16 oz	16 oz	25 oz	25 oz
Allspice	8 oz	8 oz	13 oz	—
Cayenne	4 oz	4 oz	6 oz	4 oz
Garlic	Optional	Optional	4 oz	4 oz

Source: Lopez (1987).

1 and No. 2 are for catsup with a light body, No. 3 for one with a medium body, and No. 4 for one with a heavy body. No. 1 and No. 2 are too light in consistency for most commercial use, and they are not recommended except for special purposes.

In these formulas, the specific gravity of the cyclone juice is assumed to be 1.020. When it runs higher than this, the amount of cyclone juice used can be reduced proportionally. The amount of cyclone juice in formulas No. 1 and No. 2 is equivalent to 100 gal. of 1.035 pulp, in formula No. 3 to 140 gal. of 1.035 pulp, and in formula No. 4 to 160 gal. of 1.035 pulp.

Table 11.2, listing the ingredients for 100 gal. of catsup, is reported to give a good-flavored catsup with a heavy body.

TABLE 11.2. INGREDIENTS FOR MAKING 100 GAL. OF FINISHED PRODUCT

Ingredient	Amount
Cyclone juice	240 gal.
Sugar	110 lb
Salt	24 lb
Saigon cassia (broken and sifted)	16 oz
Allspice, whole	14 oz
Cloves, headless, broken and sifted	12 oz
Mace, pinang, broken and sifted	7 oz
Onion, chopped	25 lb
Garlic	10 lb
Vinegar, 100-grain	7 gal.

Cooking. Catsup is cooked in kettles and tanks of almost every size. Sometimes the desire for quantity outweighs that for quality; thus, the batches are made too large or require too much time for completion. A number of cookers of about 250 gal. capacity or less are to be preferred to the larger sizes. Such kettles are quickly filled, emptied, and cleaned. A high steam pressure, from 90 to 120 psi, is the best preventive against burning and sticking on a kettle or coils. A high heat ensures circulation in the batch. If this is not possible, a small, high-speed propeller should be installed in the cooker. Coils or kettles that are not self-cleaning delay the work and require much attention. The evaporation of a batch should not take more than 45 min; if whole spices are used, it should not be less than 30 min. A long slow cook gives a flat soggy body, whereas one of less than 30 min may fail to extract the spices.

When the kettle is first heated, foaming may be present, especially when the pulp is made from fresh crushed tomatoes. Some varieties foam more than others. The use of compressed air is the best means to overcome the foaming; cottonseed or other oil will serve when air is not provided. Special antifoaming compounds are also available.

As soon as the batch is finished, the spice bag is removed, *if used*, and the catsup is run through the finishing machine, which removes all fiber and particles to give a smooth body. Finishers vary in screen openings from 0.033 to 0.040. The larger openings, generally, give more body to the catsup,

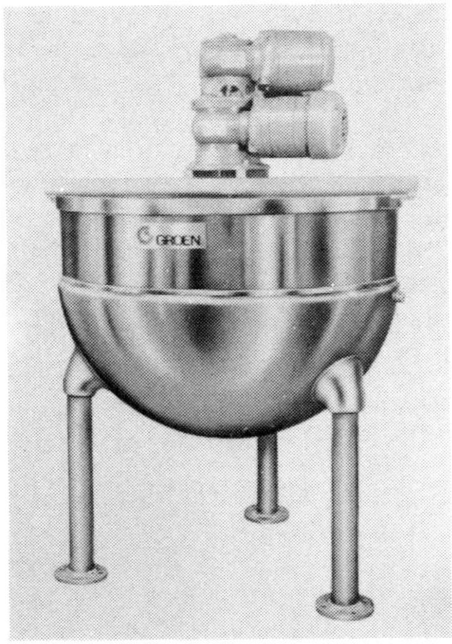

FIGURE 11.3. MODEL DA/1 JACKETED MIXING KETTLE.
Courtesy of Groen.

but it is not as smooth. From the finishing machine, the catsup is placed in a holding tank supplying the filling machine.

The thickness of bottled catsup is an important part of its quality. If it is too thick it will not pour well; if it is too thin, it is generally distasteful. Control of the consistency requires care and experience. Part of the thickness of properly made catsup is due to the pectin from the tomatoes. If the pulp has been carelessly made, there may be little, if any, of the natural pectin remaining. Some catsup manufacturers demand a hot break before the tomatoes are cycloned in order to retain the largest possible amount of pectin from the tomatoes. The hot-break method dissolves some of the mucilaginous material from the tomato seeds, again contributing to the final consistency.

Milling. The consistencies of tomato products have considerable variation depending on character, size, and proportion of the suspended particles and the serum viscosity (Kertesz and Loconti 1944). Further, the consistencies can also be altered by the extent of cell rupturing. Additional mechanical cell fragmentations by a piston-type homogenizer, a Waring Blendor, and a creamery homogenizer result in an increase in consistencies of tomato products (Hand *et al.* 1955; Whittenberger and Nutting, 1957; Becker *et al.* 1972; Crandall and Nelson 1975). Namely, milling changes spherical particles of tomato products to brush-heap type (elongated particles) and increases the consistencies of tomato products (Hand *et al.* 1955). Many types

FIGURE 11.2. MODEL NEM JACKETED MIXING KETTLE.
Courtesy of Groen.

of milling machines are currently used by the industry. Further, it is a known fact that the higher temperature in milling results in the higher consistency of tomato catsup, and the higher pressure in milling results in the higher consistency of tomato catsup.

Filling and Sterilizing. Catsup is packed in bottles of various sizes and in No. 10 cans. In filling the finished catsup into containers every effort should be made to keep the air out. Air in the finished product may result in unsightly air pockets or excessive headspace, or may endanger the desirable bright color of the product. The presence of air also interferes with certain types of vapor-seal closures. To overcome the air trapped in the catsup when it is passed through the filler, the finished catsup may be put through a deaerator.

Whether or not catsup should be processed after filling into bottles depends on the condition under which it is bottled. If the bottled product can be sealed at a temperature of at least 180°F (82°C), further heating is unnecessary because it may impair the color of the product (Anon. 1950).

The temperature of the catsup in the receiving tank feeding the filler should not fall below 200°F (93°C). It is suggested that the usual closing temperature be up to 190°F (88°C) as a measure of safety (Anon. 1950). With a closing temperature lower than 180°F (82°C), processing is generally advisable.

Cooling. No. 10 cans of catsup should be stacked in racks two cans deep and allowed to completely air cool before being cased or stacked in larger piles. This prevents injury to the flavor or stack burning, which results from remaining at a high temperature for a long time. Catsup which has been processed may be cooled a short time in water to reduce the temperature before casing or stacking (Lopez 1987).

"Black neck" in bottled catsup, a dark discoloration of the top layer apparently due to oxidation, formerly a rather frequent occurrence, is now seldom encountered. This may be accounted for, in part, by the improvement in closures, and also by increased use of deaeration before filling. When it does occur, it is usually associated with greater than normal headspace; this factor should therefore be carefully controlled.

Quality Control of Catsup. Catsup of uniform color, consistency, and flavor can be produced only by controlling the quality and amount of each ingredient used. The distinctive tomato flavor and consistency of the finished product depend largely on the tomato solids used in each batch. Approximately, one-third of the final acidity and sugar content is derived from the tomato solids content. Thus, any satisfactory method of control requires a knowledge of the amount of tomato solids in the cyclone juice or tomato pulp used in the manufacture of the catsup.

In the manufacture of catsup, factory control based solely on a uniform specific gravity of the finished product assures only that the percentage of total solids in the finished product is uniform. It does not provide uniformity in consistency, sweetness, acidity, or other characteristics of the product. Fortunately, the ratio of total solids to insoluble solids in whole tomato cyclone juice is fairly constant, as is the ratio of sugar to acid. About half of the total solids present in whole tomato cyclone juice is sugar. Therefore, if a batch of catsup is made from a given volume of cyclone juice of known concentration, the total amount of tomato solids and, approximately, the amounts of the various constituents of the tomato solids may be calculated.

It is convenient to plan each batch of catsup on the basis of a uniform volume of finished product. The amounts of salt, onions, and spices for the same volume of finished product are always the same. The proportion of sugar and acid in the raw stock used may vary somewhat in different varieties of tomatoes, and it may also change due to growing conditions. Slight changes in the amounts of added sugar and vinegar in the batch formula may be necessary from time to time.

The final concentration of the catsup batch may be controlled by determining the specific gravity of the catsup or by the percent of total solids corresponding to the Abbé refractometer reading of the filtrate. The latter method is simple and quick. It requires only a few drops of the liquid (Anon. 1950).

The results obtained by the refractometer reading represent the soluble solids present in catsup. For the various grades, the readings should be 25–29 for Grade C, 29–33 for Grade B and over 33 for Grade A.

Methods of Analysis. The methods of analysis used with catsup are similar to those given under tomato pulp. The methods given under tomato pulp for microscopic examination, insoluble solids, acidity, and salt are entirely applicable to the examination of catsup.

Consistency. The physical state of catsup, a mixture of liquid, insoluble tomato fibers, and coagulated flecks of pectin, does not lend itself to examination with the usual viscosimeter. Consistency refers to the tendency of the product to hold its liquid portion in suspension. One of the principal reasons for determining the consistency of catsup is to judge its rate of flow through the neck of the catsup bottle. The USDA recommends the use of the Bostwick Consistometer. Grades A and B catsups flow not more than 9 cm in 30 sec at 20°C; Grade C flows not more than 14 cm in 30 sec at 20°C in the Bostwick Consistometer.

CHILI SAUCE

Chili sauce is of the same general character as catsup but is made from peeled and cored tomatoes without removing the seeds. It contains more sugar and onions and sometimes is made hotter in flavor than catsup by the use of more cayenne pepper. There is a great variation among different manufacturers with respect to the method of treating the tomatoes. Usually large to medium-sized tomatoes are employed, separated from the small tomatoes used for making pulp and catsup. Some manufacturers of chili sauce place the peeled and cored tomatoes directly into the kettle and mix the other ingredients without any form of breaking. Other manufacturers have various method of breaking and crushing the tomatoes. Several crushers for this purpose are on the market and other means of breaking, such as meat choppers, meat cutters, and apple graters, are employed. Some convey the tomatoes from the peeling room to the kettle through a pump that breaks them up.

Because of the nature of the product there is not a method available for testing the concentration of chili sauce and determining the point at which the cooking should be stopped. The refractometer may be used as a method of controlling the concentration. The consistency of the product is always regulated by its appearance. The amount of cooking varies with different manufacturers, but, in general, 100 gal. of peeled and cored tomatoes yield from 40 to 45 gal. of chili sauce.

The amount of onions added to chili sauce is substantially larger than that used with catsup; some manufacturers use approximately twice as much. Large onions should be used because they are easier to peel and give less waste than smaller onions. They should be carefully peeled and finely chopped. Most processors today use dried onions.

The cooking and handling are the same as for catsup, but the finishing operation is eliminated.

Because of its lumpy condition, a chili sauce bottle is more difficult to fill than a catsup one. The bottle is also harder to seal because of its wide neck. The black rings in the top of the bottle are more conspicuous than is the case with catsup (Anon. 1950).

Chili sauce has not been defined; therefore, all ingredients must be stated on the label (Lopez 1987).

REFERENCES

ANON. 1950. Tomato products. Pulp, paste, catsup, chili sauce and juice. Natl. Food Proc. Assoc. Res. Lab. Bull. *28-L,* Washington, DC.

BECKER, R., MIERS, J.C., NUTTING, M.D., DIETRICH, W.C. and WAGNER, J.R. 1972. Consistency of tomato products. 7. Effects of acidification on cell walls and cell breakage. J. Food Sci. *37,* 118.

CRANDALL, P.G. and NELSON, P.E. 1975. Effect of preparation and milling on consistency of tomato juice and puree. J. Food Sci. *40,* 710.

HAND, D.B., MOYER, J.C., RANSFORD, J.R., HENING, J.C. and WHITTENBERGER, R.T. 1955. Effect of processing conditions on the viscosity of tomato juice. Food Technol. *9,* 288.

KERTESZ, Z.I. and LOCONTI, J.D. 1944. Factors determining the consistency of commercial canned tomato juice. N.Y. State Agric. Exp. Stn., Bull. *272.*

LOPEZ, A. 1987. A Complete Course in Canning, 12th Edition. Canning Trade, Baltimore.

MARCH, G.L., LEONARD, S.J. and BUHLERT, J.E. 1979. Yield and quality of catsup produced to a standard solids and consistency level. II. Influence of handling practices, break temperature and cultivar. J. Food Process. Preserv. *40* (3) 195-212.

MARSH, G.L., LEONARD, S.J., BUHLERT, J.E. and VILAS, M.R. 1979. Yield and quality of catsup produced to a standard solids and consistency level. I. Method of determining the amount of tomato solids required. J. Food Process. Preserv. *40* (3) 189-193.

WALKER, ROBERT G. and GEORGE BOSY. 1970. Tomato product and method of making same. U.S. Patent No. 3, 549, 384.

WHITTENBERGER, R.T. and NUTTING, G.C. 1957. Effect of tomato cell structures on consistency of tomato juice. Food Technol. *11,* 19.

CHAPTER 12

Tomato Soup

Most tomato soup packed today is produced from fresh tomatoes; however, increasing numbers of processors are considering manufacture of soup by using tomato paste. Among the reasons for using the paste are economics, uninterrupted yearly schedules, elimination of added part-time employees, and elimination of a perishable crop capable of creating a crisis for 2 to 3 months of the year.

However, many processors are resisting this transition for a number of reasons. Even though the switch is inevitable, as conceded by many, the chief objection to the paste is loss of fresh tomato flavor, duller and darker tomato colors, and variation in viscosity. Some processors have come very close to equaling the quality of the fresh tomato pack by using concentrates extracted from fresh tomatoes.

Tomatoes that have been initially prepared for soup preparation by crushing and hot-break heating are extracted to remove extraneous material such as stems, skins, and cores. The extracted rough pulp is then pumped into a finisher where completion of the pulping process takes place.

The finished pulp is then pumped to the pulp tank, where it is continuously agitated to ensure uniform solids distribution. The tomato pulp is now ready to be used exclusively, or in part, as the main ingredient of tomato soup.

The blending operation of the soup batch starts with precise measuring of the pulp into the blending kettle. The pulp is then heated to approximately 200°F (93°C); the heat is turned down, and agitation continued as the seasonings are added. Butter and vegetable oil are also added. Water may be added to give a specified quantity, and the batch is reheated to boiling and simmered for 1 to 2 min. Liquid spices, flavorings, and ascorbic acid are added. The batch is adjusted to a volume allowing for addition of thickening agents, usually wheat flour, starch, and/or cracker meal.

After addition of thickeners, the batch is adjusted to a final specified volume. Batch standardization is accomplished by a volume or weight measurement. Volume standardization, the more common method, involves use of a stick gauge or strap gauge. Weight standardization is more precise and controllable. Air is forced through the batch at a set depth within the kettle. As batch weight increases, the force required to push air through the batch increases at a proportional rate. The final batch volume responds to weight changes.

The standardized batch is now agitated to complete uniformity. The batch is then pumped to rotary piston fillers. The soup may be forced through a final screen to remove any starch lumps, seeds or other undissolved material prior to filling. Problems frequently occur with lumps due to the volume of the necessary thickeners and the viscosity of the soup. For this reason, a final screening or finishing is often necessary.

The soup product is then filled and closed at a temperature of 180°F (82°C) or above. The hermetically closed product is then processed in still retorts, hydrostatic cookers, or continuous agitating cookers.

After adequate processing, the sterilized product is cooled to an average temperature of 100°F.

FORMULATIONS

Most of the soup produced in the United States is condensed or concentrated, allowing the consumer to dilute the product with an equal volume of liquid. Another form produced is the ready-to-serve or single-strength variety. The ready-to-serve line has a convenience appeal because it does not require additional liquid. The vending industry makes widespread use of this type.

There are hundreds of different varieties of soup produced commercially, along with the unique home styles. Some of the more popular are tomato, noodle, vegetable, beef, poultry, and creamed styles. The demand for any of

TABLE 12.1. CONTINUOUS HIGH TEMPERATURE AGITATING COOKER.
Courtesy of FMC Corporation

these is usually dependent on regional preference. A particular variety may sell well in one area while the demand for it is nonexistent elsewhere.

Since the conception of canned soup, the market for it has steadily grown. A wholesome and nutritious food, it is enjoyed by all age groups.

A typical formula for tomato soup is shown in Table 12.1. The procedure for manufacture of tomato soup follows.

1. Weigh the required amount of prepared pulp and add to the blending kettle
2. Heat to 200°F (93°C) and add the sugar, salt, pepper, and ascorbic acid
3. Add the butter and vegetable oil and reheat to 200°C (93°C); let simmer for 2 min
4. Add the liquid spice tea prepared from the bay leaves, thyme, parsley, allspice, and cloves (cook the bay leaves, allspice and cloves in a bag or cheesecloth to extract flavor for 45 min)
5. Adjust volume to approximately 0.8 gal., leaving room for the thickeners; heat to 200° to 205°F (93° to 96°C)
6. With steam turned off and vigorous agitation, slowly add the flour–meal–water mixture.
7. Adjust volume to 1 gal. with water
8. Agitate to complete uniformity.
9. Finish the completed batch through a screen if necessary
10. Fill the soup into 211 × 400 plain body cans with "R" enamel ends
11. Omit step #1 and add paste plus 300 ml water for paste pack.
12. Fill to 10½ to 10¾ oz with a minimum fill temperature of 160°F (71°C)
13. Process time is 25 min at 240°F (116°C)
14. Cool cans to average of 100°F (38°C)
15. Yield is 11¾ cans per gal.
16. pH is 4.0 to 4.3
17. Viscosity is 10,000 to 14,000 cps (after setup)

TABLE 12.1 FORMULA FOR TOMATO SOUP

Ingredients	g/Gal.
Tomato pulp (or paste 36%, 220g)	1500
Wheat flour	205
Cracker meal	50
Sugar	90
Salt	45
Soybean oil	15
Butter	8
Ascorbic acid	0.3
Bay leaves, whole	10
Thyme, whole	10
Parsley, ground	10
Allspice, whole	12
Cloves, whole	12
White pepper	0.1

A typical formula for the manufacture of Cream of Mushroom soup is shown in Table 12.2. The procedure for manufacture of cream of mushroom soup follows. Blend the soup in two fractions: a dark fraction and a light fraction. Successful blending depends on the utilization of live steam to heat the soup, thus eliminating kettle scorch. The hold time on diced mushrooms is critical because oxidation is rapid if they are left exposed for any length of time.

TABLE 12.2 FORMULA FOR CREAM OF MUSHROOM SOUP

Ingredients	g/Gal.
Mushrooms, fresh 3/8"	280
Condensed milk, 12%	50
Cream, 44% butter fat	75
Soybean oil	180
Uncolored margarine	15
Salt	78
Sugar	45
Tomato puree 1.035	15
Wheat flour	160
Corn starch	80
Oleoresin of celery (or celery seed or celery salt)	0.03
White pepper	0.5
Mushroom concentrate (optional)	2.5

1. Dark fraction

a. Add the fresh diced mushrooms, salt, sugar, white pepper, and tomato puree to a separate kettle other than the final blending kettle. Agitate, heat to a boil, and let simmer for 3 min

b. Turn off the steam after this period and with adequate agitation, add *half* of the thickeners

c. Agitate to uniformity and turn off agitator

d. Let stand until ready to add to light fraction

2. Light fraction

a. Dilute the frozen cream one-to-one with water, add the margarine, and heat to 160°F (71°C)

b. Add the prepared cream, milk, vegetable oil, and margarine to a suitable hold tank; add 300 ml of water and vigorously agitate

c. Run this mixture through a Rapisonic at 200 psi or a homogenizer at around 1000 psi, forming an emulsion (the emulsion can be simulated for small batches with the aid of a Waring Blendor)

3. Blending procedure

a. Add the prepared emulsion to the kettle along with the remaining half of the thickeners; add about 100 ml of water and begin heating with live steam

b. Pump, bail, or pour in the prepared dark fraction after all the emulsion and flour-starch has been added

c. Add the celery flavoring
d. Adjust volume to 0.9 gal. and heat to 190°F (88°C) with live steam
e. Adjust the final volume to 1 gal. with water and agitate until uniform

4. Fill to 10½ to 10¾ oz in 211 × 400 plain body cans with "C" enamel ends
5. Process 65 min at 245°F (118°C)
6. Cool to 100°F (38°C)
7. Yield is 11¾ cans per gal.
8. Grams of mushroom per can after process is 12 to 16 g.

A typical formula for the manufacture of vegetable beef soup appears in Table 12.3. The procedure for manufacture of vegetable beef soup follows.

TABLE 12.3. FORMULA FOR VEGETABLE BEEF SOUP

Ingredients	g/Gal.
Bone stock	500
Meat stock	500
Beef, dice, ⅜ in. cooked (50–55% yield raw)	210
Potatoes, fresh, ⅜ in.	250
Carrots, fresh, ⅜ in.	220
Barley, blanched (dry wt, 55 g)	105
Tomato puree, 1.035 sp. gr.	40
Peas, Alaska	25
Salt	70
Sugar	10
Potato starch	35
Green beans, ¾ in.	30
Celery, fresh, ⅜ in.	25
Beef fat	15
Beef meat purée (fines, scrap)	14
Yeast extract	2
Hydrolyzed milk protein	1.5
Onion powder	1
Caramel color	6
Lactic acid, 50%	1.5
White pepper, ground	0.5
Garlic, ground	0.1
Shank bones	1000

1. Beef and bone portion

a. Cook the shank bones after crushing in a digestor or pressure cooker for about 1 hr at 5 to 8 lb pressure; cover the bones with 700 to 800 g of water; screen or centrifuge the stock after complete digestion; soluble solids should range from 2.5 to 3.5%; any beef fat rising to the surface can be skimmed and saved

b. The raw beef is cooked for 1 hr at a moderate boil in about 1000 g of water; fat rising to the surface may be skimmed and saved; solids will average about 1.5% for the meat stock or broth

c. Remove the meat after adequate cooking and allow to cool; trim and sort beef for gristle, cartilage, ligament, and connective tissues; save fines and scrap material for puree

d. Dice meat to ⅜ in.

e. Screen meat stock before using

2. Blending procedure

a. Add the fresh diced carrots and potatoes to the kettle. Add water sufficient to cover the steam jacket and blanch for 4 min; skim off any foam rising to the surface; drain and thoroughly rinse the vegetables after completion of the blanch

b. Add the meat and bone stocks to the kettle; heat to boiling

c. Add the blanched barley, beef fat, celery, beef meat puree, hydrolyzed milk protein (cm-1); heat to boiling

d. Add the condiments (salt, sugar, and MSG)

e. Add the tomato puree, peas, and green beans

f. Reheat to boiling and simmer for 1 min

g. Add the caramel coloring (liquid burnt sugar), lactic acid, and yeast extract

h. Adjust volume to 0.9 gal. and reheat to boiling; simmer for 1 min

i. Slowly pour in the screened potato starch–water mixture and agitate vigorously

j. Adjust volume to 1 gal. and agitate to uniformity

3. Fill in 211 × 400 full "C" enamel cap to 10¾ to 11 oz
4. Process 55 min at 245°F (118°C)
5. Washed drained weight of solids after processing is 5 to 6 oz
6. Grams of beef per can is 12 to 18 g
7. Yield is 11¾ cans per gal.
8. Minimum fill temperature is 160°C (71°C)

CHAPTER 13
Tomato Wastes

Wastes from a tomato processing plant are a problem for many tomato processors. The wastes include both liquid and solid wastes. The liquid waste includes the flume waters, the wash waters, chemical peeling solutions and peel waters, the clean up waters, cooling waters, and the personnel wash and toilet waters. The solid wastes include all the mud, dirt and sand brought in with the fruit; the tomato peel, seeds, and defective materials; and the human wastes. In addition, wastes from the packaging lines, worn out and replaced equipment and the office wastes are in need of proper disposal.

There is no one solution to tomato factory wastes. In the past one could accumulate the solid waste and have it taken to the trash dump. That is not possible today and wastes must be separated and handled differently. The mud, sand and dirt should be taken back to the farm for disposition. This waste can be dried and then spread evenly back on the soil.

The tomato peel, seeds, and defective or cull material, if any, can be utilized by drying and sold as animal feed. The seeds are a high source of good protein and of much interest to mink and even humans. The peel, if kept in a sanitary condition and acidified if chemically needed, should be useful as a fiber supplement. If peeling additives are used in chemical peeling, this waste product cannot be used for human consumption. The defective or trim out material can be composted and used as mulch or, at least, mixed with organic mulches and then used as a mulch.

The other solid material from the factory should be kept separate and sold to waste handlers. Systems are in place to take paper products once they are baled. Many times these can be recycled or reused. The glass, metal and plastic spent containers are recyclable as recyclers are in place to handle these materials for reuse or development of new products manufacture. Cartons and packaging materials can be baled and sold to paper mills for recycling. The worn out or replaced machinery can be sold for parts or to used equipment manufacturers for rebuilding.

The liquid wastes are a separate problem as are the human wastes. If the municipal system can handle these wastes at a nominal charge this is the best handling of these materials. If they can only handle the human wastes, that may be a plus and the easiest way to handle. If not, the firm may have to install its own septic system for disposal of human wastes.

The other liquid wastes can go to digestion pounds and once digested they can be recycled or reused in the plant following filtration and chlorination. The wastes, particularly from the chemical peeling factories are somewhat difficult to digest without neutralizing the alkalinity. In some cases the natural acid of the tomato may help in neutralizing this material. Once it has been digested most municipal systems will handle it. Another solution is to spray the liquid wastes back on the land as irrigation waters. It may be necessary to mill these wastes before attempting to spray them on the land to break up any tomato material rather than clog up nozzles. Spray Irrigation or Laniation, as it is sometimes called, has worked well in the past and should be acceptable in the future if the operator does not drain this material off the land before the organic material in this waste is naturally digested in the field and soil.

Waste from a tomato factory is inevitable, however by careful and respectful operation of the equipment, controlling water usage during the clean up, and elimination of the unusable fruit in the field much reduction in tomato factory waste can accrue.

Down the road one would hope that the Standards of Identity could be corrected to permit the use of the peel and seeds in all tomato products. Certainly they are acceptable in the fresh tomatoes and if used they would add much fiber and nutrition to many tomato products like juice, ketchup, sauces, and soups. Even permitting the peel on canned tomatoes would be a step forward.

My personal belief is that the time is right to make a move in the direction of utilizing peels and seeds from the tomato rather than having the worry about disposition through the development of new and innovative waste systems. Initiatives should be undertaken by the industry to change the Standards of Identity where applicable and to up-date the utilization of the whole tomato.

REFERENCES

GEISMAN, J.R. 1981. From Waste to Resource. Protein from Tomato Seeds. Ohio Report Nov-Dec. 92-94.

SCHULTZ, W.G., H.J. NEUMANN, J.E. SCHADE, J.P. MORGAN, P.F. HANNI, A.M. KATSUYAMA, and H.J. MAAGDENBERG. 1978. Commercial Feasibility of Recovering Tomato Processing Residuals for Food Use. USEPA R&D publication 600-2/78-202. 66 pages.

Part III — Technology

CHAPTER 14
Quality Assurance

Quality assurance is a program designed to provide confidence for management and the ultimate user, that is, the customer. It is the modern term for describing the control, the evaluation, and the auditing of a food processing system or operation.

A firm is in business to produce a product intended for sale to a customer fromwhich the firm hopes to make a profit. The key word is the customer. The customer is the one a firm must satisfy and it is the customer who ultimately established the level of quality the firm must manufacture to. The customer is management's guide to quality and this is what the firm must build its specifications and label requirements around. Only by having a planned program can a firm continue to succeed in supplying the customer with what they expect.

A large part of a quality assurance program is built around the control of quality. Quality control means to regulate to some standard or specification. It is usually associated with the production line, that is, the specific processes and operations. As used today, quality control is the tool for production worker to help him operate the given operation or line in conformance with the predetermined parameters for any given quality level. Quality assurance personnel must spend a large part of their time and talents in educating the line operator to work within the specifications and requirements set-forth by management. In the next chapter a more thorough discussion of this topic is presented.

Quality evaluation is also part of any modern quality assurance program. It is the modern term used to describe or appraise the worth of the products. It generally means taking a measurement of given attributes or characteristics of products in the laboratory. It is used to include the evaluation of all incoming materials, products in process, and or finished products. This aspect of quality assurance is discussed in detail in the following chapters by looking specifically at the various attributes or characteristics one at a time.

Another part of a quality assurance program is to audit or verify or examine the products or even the processes over time. It is a term used frequently with firms having multiple plants. It should be part of any quality assurance program to verify products in the warehouse as to quality, shelf life, etc. or products in distribution, and/or competitors products in the market place. Data should be accumulated and interpreted using statistical programs, such as, frequency distributions, histograms, and/or Analysis of Variance (ANOVA). An audit of quality should be looked at in the same manner as the financial records are audited. In many cases a 3rd party may be utilized to audit the quality of the products, the processes, and/or the quality assurance programs.

DEFINITION OF QUALITY

Quality makes a product what it is: it is the combination of attributes or characteristics of a product that have significance in determining the degree of acceptability of the product to a user, and that determines its value or worth. As used by the industry, it is a concept involving degree: degree of purity, strength, flavor, color, size, maturity, workmanship, condition, and any other distinctive attribute or characteristic of the product. Thus, the term quality, without being defined in terms of some standard, means little. The trade also uses the term to mean the finest product attainable. From past experience food processors have learned that high-quality products never fail to sell. This is true because the consumer recognizes the brands that maintain their quality at the standard set for that particular product. Repeat sales are the outgrowth of quality control practices.

Many large companies attain their position by the control of quality of the products they process. They do not process a product that is lacking in uniformity or one that will "just get by," but a uniform quality product that will continue to build their business. Perhaps more important than their success is management's knowledge, at all times, of what level of quality is being packed. Thus, a quality assurance program informs management not only of the quality and the condition of the products being packed but also of the industry trends.

STANDARDS FOR QUALITY

Legal Standards

Legal standards are those established by the federal, state or municipal agencies; they are generally mandatory. These standards are set up by law or regulation and represent the Federal Food, Drug, and Cosmetic Act, the various state or municipal minimum standards of quality. They are generally concerned with freedom from adulteration or contamination. They may involve insects, molds, yeasts, pesticides, etc., or maximum limits of food or chemical additives, or the establishment of specific conditions in processing to prevent contamination by extraneous materials.

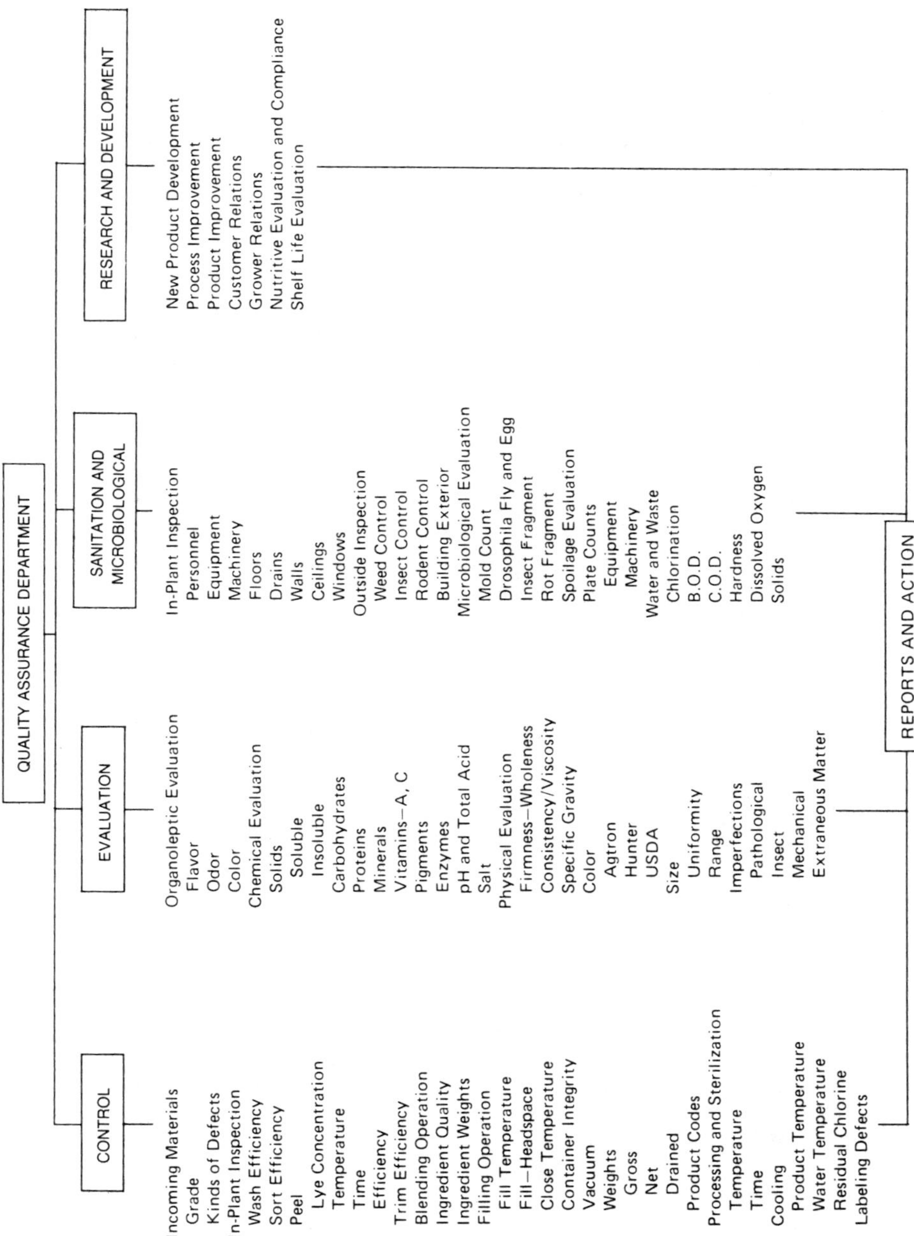

FUNCTIONS OF A QUALITY ASSURANCE DEPARTMENT.

Company or Voluntary Label Standards

Company or voluntary label standards represent standards established by various segments of the food industry. They generally are not known by the consumer except through his personal experience. These standards may become a trademark or symbol of product quality. Standards are used by private firms or supermarkets and tend to vary depending on the particular requirement for any given label.

Grade Standards

These are legal and voluntary representing levels of quality or grade established by the USDA. They are widely used and recognized by industry and consumers.

Industrial Standards

Industrial standards are attempts by an organized group to establish quality limits for given commodities. These become effective when adopted by marketing organizations or specific commodity groups.

Consumer Standards

Consumer standards are the consumers' requirements for a product and generally are based on their experience. Consumers are presently not too effective as a group, but individually they represent the day-to-day demands for any given product.

METHODS FOR DETERMINING QUALITY

Subjective Methods

This type of quality evaluation is based on the investigator's opinion, usually a physiological reaction resulting from past training, experience of the individual, influence of personal preference, powers of perception, etc. This evaluation is subjective because the individual is required to use a mental process to give his opinion on qualitative and quantitative values of the characteristic or characteristics under study. These methods usually involve the various sense organs and are called sensory methods. Examples include flavor, odor, color, tactile, or texture.

Objective Methods

Objective quality evaluation is based on observations that exclude the investigator's attitude. As recognized standard scientific tests, they are applicable to any sample of the product or products without reference to its previous history or ultimate use. They are representative of modern quality control because the human element is excluded. They can be divided into three general groups.

Physical Methods of Measurement. Physical methods of measurement are perhaps the quickest of the three and require the least training. They measure product attributes such as quality, size, texture, color, consistency, or imperfections, or process variables such as headspace, fill, drained weight, or vacuum. Instruments can generally be found or adapted for the physical evaluation of product quality.

Chemical Methods of Measurement. For most products, standard food analysis methods are used for quantitative evaluation of nutritive values and quality levels. However, chemical analysis is long, tedious, and expensive. As a result, the industry and other interested parties have developed methods that are termed quick tests, such as enzyme, solids, vitamins, pH, or acidity.

Microscopic Methods. Microscopic methods have excellent application in a quality control program, but require training to properly interpret the results. They can be divided in two general categories.

Adulteration and contamination tests indicate the presence of bacteria, yeast, mold, insect fragments, insect excreta, foreign materials, etc. The technologist must have the proper background to be able to differentiate the various types of adulteration and contamination which may be present in the products.

Differentiation between cell types, tissue types, and microorganisms of various stored foods are used to test for deficiency of fertilizer materials, stored food in the tissues of plant materials, microorganisms causing spoilage, and/or desirable fermentative organisms.

PURPOSES OF A QUALITY ASSURANCE PROGRAM

The primary objective of a quality *assurance* program is to obtain adequate information on all product factors or characteristics that affect the quality of a product. The intelligent interpretation of this information provides management with a quality index of the entire operation. The information serves as a management guide for the exact quality packed from a given quality of raw stock, or it may provide management with information necessary for the processing of a product to obtain a given quality.

Quality control also opens the door to research. Charles Kettering said, "Research is a high-hat work that scares a lot of people, but it needn't as it is rather simple. Essentially, it is nothing but a state of mind—a friendly, welcoming attitude toward change. This change may involve people, facilities, materials, and equipment." Mr. Kettering also said, "Research is something that if you don't do it till you have to, it's too late." An investment in the future, research must delve into quality *assurance* activities. In fact, quality assurance is the application of ideas and techniques derived from research and product development.

Some reasons for quality assurance are

1. Control over raw materials through setting of specifications
2. Improvement of product quality
3. Improvement of processing methods resulting in lower production costs and in greater profits
4. Standardization of the finished product according to level specifications
5. Increased order and better housekeeping in a sanitary plant
6. Greater consumer confidence in the uniformly high quality of a product

BASES OF A QUALITY ASSURANCE PROGRAM

Six facets must be carefully considered and planned for a successful quality assurance program. These include (1) organization, (2) personnel, (3) sampling, (4) standards and specifications, (5) measurement, and (6) interpretations.

Organization

The quality assurance department should be responsible to top management and report directly to them. It must necessarily provide other departments with specific information on the quality at the receiving platform, or on the line, or in the warehouse, but it does not control these departments. Management makes the decision between quality and quantity, not any one of the several departments of the company. However, the quality assurance department should be authorized to cooperate with production to maintain the desired standards during operations.

Personnel

The personnel in the quality assurance department vary with the products being packed, the size of the operation, and the amount of control desired by management. Some important qualifications necessary for the quality assurance technologist to fulfill his responsibilities are an ability to:

1. Be truthful in his reports, decisions, and, especially, analysis,
2. Possess salesmanship,
3. Speak the language of the industry and write intelligently,

4. Work with people,
5. Be alert and responsive to necessary changes,
6. Always be on the job,
7. Be well mannered and neat in appearance,
8. Give instruction to production supervisory employees concerning,
   a. What is to be done,
   b. How it is to be done, and
   c. Why it is to be done.

Samples and Sampling Plans

Probably the greatest limiting factor in the successful control of product quality is the sample for production evaluation. Briefly, the sample must be representative of the lot in question and must be random. Otherwise, the sample is useless for inspection or evaluation of product quality.

The new regulations of the USDA governing inspection and certification of processed fruits and vegetables and related products change the terminology and sampling procedures. Some of the new terms are defined below. Specific procedures, including tables for sampling plans and acceptance levels provide a guide for the quality assurance technologist.

Definitions

1. *Acceptance number:* the maximum number of deviants permitted in a sample to meet a specific requirement.

2. *Certificate of sampling:* a statement, written or printed, issued pursuant to the regulations, identifying officially drawn samples. It may include a description of condition of containers and the condition under which the processed product is stored.

3. *Deviant:* a sample unit affected by one or more deviation, or that varies in a specifically defined manner from the requirements of a standard, specification, or other inspection document.

4. *Lot:* for the purpose of charging fees and issuing certificates, a lot is any number of containers of the same size and type which contain a processed product of the same type and style, located in the same or adjacent warehouses, and available for inspection at any one time, provided that

   a. Processed products in separate piles which differ from each other as to grade or other factors may be deemed to be separate lots,

   b. Containers in a pile bearing an identification mark different from other containers of such processed product in that pile, if determined to be of lower grade or deficient in other factors, may be deemed to be a separate lot, or

   c. If the applicant requests more than one inspection certificate covering different portions of such processed product, the quantity of the product covered by each certificate shall be deemed to be a separate lot.

For the purpose of sampling and determining the grade of compliance with a specification, lot means each pile of containers of the same size and type containing a processed product of the same type and style which is separated from other piles in the same warehouse. Containers in the same pile bearing an identification mark different from other containers in that pile may be deemed to be a separate lot.

5. *Rejection number:* the minimum number of deviants in a sample that will cause a lot to fail a specific requirement.
6. *Sample:* any number of units to be used for inspection.
7. *Sample unit:* a container and/or its entire contents, a portion of the contents of a container or other unit of commodity, or a composite mixture of a product to be used for inspection.
8. *Sampling:* the act of selecting samples for the purpose of inspection under USDA regulation.
9. *Unofficially drawn sample:* any sample that has been selected by a person other than an inspector or licensed sampler, or by any other person not authorized by the Administrator of the USDA pursuant to the regulations.

The USDA inspection service shall be performed on the basis of the appropriate U.S. standards for grades of processed products, or any written specification or instruction which is approved by the Administration.

When inspection for quality is based on any U.S. grade standard that contains a scoring system, the grade to be assigned to a lot is the grade indicated by the average of the total scores of the sample units: provided further, that

1. Such sample complies with the applicable standards of quality promulgated under the Federal Food, Drug, and Cosmetic Act,
2. Such sample complies with the product description,
3. Such sample meets the indicated grade with respect to factors of quality which are not rated by score points,
4. With respect to those factors of quality which are rated by score points, each of the following requirements is met:
    a. None of the sample units falls more than *one grade* below the indicated grade because of any quality factor to which a limiting rule applies,
    b. None of the sample units falls more than *4 score points* below the minimum total score for the indicated grade,
    c. The number of sample units classified as deviants does not exceed the applicable acceptance number indicated in the sampling plans contained in the section of lot compliance (included in the following material) (a deviant, as used in this paragraph, means a sample unit that falls into the grade below the indicated grade but does not score more than 4 points below the minimum total score for the indicated grade).

5. If any of the provisions contained in the above paragraphs 3 and 4 of this section are not met the grade is determined by considering such provisions in connection with succedingly lower grades until the grade of the lot, if assignable, is established.

Sampling Plans

Samples are selected from each lot in the number indicated in the sampling plans. At the discretion of the inspection service, any comparable multiple sampling plan may be used if the number of sample units selected may be increased to the exact number of sample units indicated for any one of the larger sample sizes provided for in the appropriate plans.

Under the single sampling plans, if the number of deviants in the samples does not exceed the acceptance number prescribed for the sample size, the lot fails the requirement.

Under the multiple sampling plans, inspection commences with the smallest sample size indicated under the appropriate plan.

1. If the number of deviants in the sample does not exceed the acceptance number prescribed for that sample size, the lot meets the requirement.
2. If the number of deviants in the sample equals or exceeds the rejection number prescribed for that sample size, the lot fails the requirement.
3. If the number of deviants in the sample falls between the acceptance and rejection numbers of the plan, additional sample units are added to the sample so that the sample thus accumulated equals the next larger cumulative sample size in the plan. It may then be determined that the lot meets or fails the specific requirement by considering the cumulative sample and applying the procedures outlined in (1) and (2) of this paragraph or by considering successively larger samples accumulated in the same manner until the lot meets or fails the specific requirements.

Standards and Specifications

Quality control follows the establishment of product specification. When new information is available on quality control methods and packing procedures, management, in cooperation with production, sales, and quality assurance, must draft changes in process procedures and specifications or standards. These standards or specifications and process procedures are developed to provide information for production personnel in the packing of desired quality.

A major function of the quality assurance department is to determine the product's deviation from these specifications and, when necessary, make the changes to control the desired level of product quality. The personnel in the production department and the quality assurance department must cooperate to produce a given quality product.

TABLE 14.1A. CANNED OR SIMILARLY PROCESSED FRUITS, VEGETABLES, AND PRODUCTS THEREOF CONTAINING UNITS OF SUCH SIZE AND CHARACTER AS TO BE READILY SEPARABLE

Container Size Group	Lot Size (No. of Containers)[a]							
Group 1, any type of container of a volume not exceeding that of a No. 303 size can	3000[b]	3001 to 12,000	12,001 to 39,000	39,001 to 84,000	81,001 to 145,000	145,001 to 228,000	228,001 to 336,000	336,001 to 480,000
Group 2, any type of container of a volume exceeding that of a No. 303 size can but not exceeding that of a No. 3 cylinder size can	1500[b]	1501 to 6000	6001 to 19,500	19,501 to 42,000	42,001 to 72,500	72,501 to 114,000	114,001 to 168,000	168,001 to 240,000
Group 3, any type of container of a volume exceeding that of a No. 3 cylinder size can, but not exceeding that of a No. 12 size can	750[b]	751 to 3000	3001 to 9750	9751 to 21,000	21,001 to 36,250	36,251 to 57,000	57,001 to 84,000	84,001 to 120,000
Group 4, any type of container of a volume exceeding that of a No. 12 size can	Convert to equivalent No. 10's based on net weights							
Lot inspection								
Sample size (number of sample units)[c]	3	6	13	21	29	38	48	60
Acceptance number	0	1	2	3	4	5	6	7
On-line in-plant inspection								
Sample size (number of sample units)[c]	3	6	6	13	21	29	38	48
Acceptance number	0	1	1	2	3	4	5	6

[a]Under on-line in-plant inspection, a 5% overrun in number of containers may be permitted by the inspector before going to the next larger sample size.
[b]Or less.
[c]The sample units for the various container size groups are as follows: Groups 1, 2, and 3–1 container and its entire contents, Group 4 approximately 2 lb of product. When determined by the inspector that a 2-lb sample unit is inadequate, a larger sample unit may be substituted.

TABLE 14.1B CANNED, FROZEN, OR OTHERWISE PROCESSED FRUITS, VEGETABLES, AND RELATED PRODUCTS THEREOF IN A COMMINUTED, FLUID, OR HOMOGENEOUS STATE

Container Size Group	Lot Size (No. of Containers)[a]							
Group 1, any type of container of 1 lb or less	4500[b]	4501 to 18,000	18,001 to 58,000	58,501 to 126,000	126,001 to 217,000	217,001 to 342,000	342,001 to 504,000	504,001 to 720,000
Group 2, any type of container exceeding 1 lb but not exceeding 3 lb	3000[b]	3001 to 12,000	12,001 to 39,000	39,001 to 84,000	84,001 to 145,000	145,001 to 228,000	228,001 to 336,000	336,001 to 480,000
Group 3, any type of container exceeding 3 lb but not exceeding 10 lb	1500[b]	1501 to 6000	6001 to 19,500	19,501 to 42,000	42,001 to 72,500	72,501 to 114,000	114,001 to 168,000	168,001 to 240,000
Group 4, any type of container exceeding 10 lb	Convert to equivalent number of No. 10's based on net weight for canned products and use Group 3. Convert to equivalent number of 6-lb units on frozen products and use Group 3.							
Lot inspection								
Sample size (number of sample units)[c]	3	6	13	21	29	38	48	60
Acceptance number	0	1	2	3	4	5	6	7
On-line in-plant inspection								
Sample size (number of sample units)[c]	3	6	6	13	21	29	38	48
Acceptance number	0	1	1	2	3	4	5	6

[a]Under on-line in-plant inspection, a 5% overrun in number of containers may be permitted by the inspector before going to the next larger sample size.

[b]Or less.

[c]The sample units for the various container size groups are as follows: Groups 1, 2, and 3–1 container and its entire contents. A smaller sample unit may be substituted in Group 3 at the inspector's discretion. Group 4, approximately 16 oz of product. When determined by the inspector that a 16-oz sample unit is inadequate, a larger sample unit may be substituted.

Measurement

The facilities for a quality control program will vary with the size of the operation, the number of products being packed, and the different grades desired.

The Laboratory

The quality assurance laboratory serves as the center for control and evaluation of quality for all operations. This means that the quality control program starts with the purchase of seed and follows through to the ultimate consumption of the finished product. Some of the functions, duties, and line checks that can be made by the quality assurance staff include

1. Determination of percentage germination and purity of seed soil and tissue analyses,
2. Soil and tissue analyses,
3. Collection and summarization of weather data for use in scheduling of raw products, cans, labor, etc.,
4. Identification of crop diseases and/or insects,
5. Determination of raw product quality,
6. Evaluation and continues checking of processing variables affecting quality,
7. Determination of the efficiency of each unit operation as related to finished product quality,
8. Periodic and continuous checking of water supply, equipment and plant sanitation, and waste disposal,
9. Evaluation of the finished production quality, including the storage life of the product,
10. Development of new products and the improvement of present processing, production, and quality methods.

The ideal laboratory location is close to, but apart from, the preparation lines with North light available. It should be easy to clean, dust-free, soundproof, and well ventilated. The walls and cabinet should be painted with a nongloss or flat white paint.

The size of the laboratory depends primarily on the number of products being packed and the daily volume of business. The minimum recommended size is a floor space of 12 × 18 ft. A typical floor plan for a medium-sized quality control laboratory is shown in Fig. 14.1. This laboratory is 18 x 25 ft and laid out to handle the basic work for a medium-size processing plant. Many other innovations and arrangements are possible.

The grading table should be approximately 8 ft long, 3 ft high, and 30 in. wide. The top of the table should be of nonglossy stainless steel, or painted with a hard finish, nongloss or flat white paint, or covered with a thin sheet of pickled steel. This last top will wear well, will not reflect light, and is

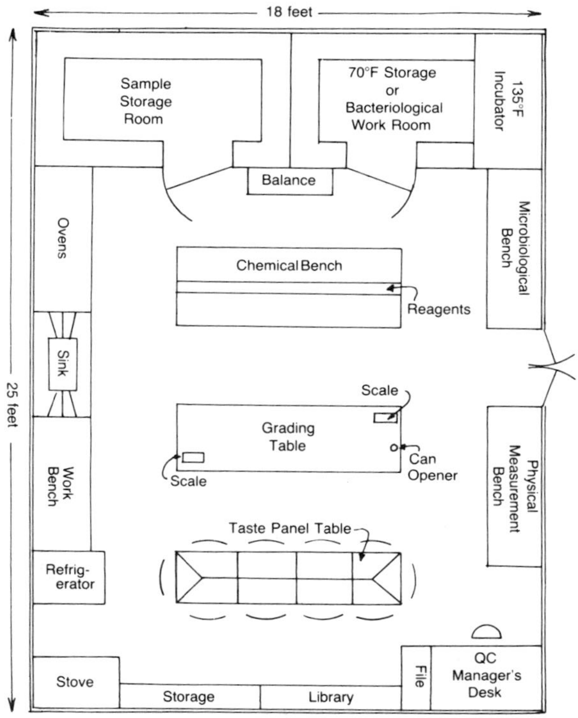

FIGURE 14.1. FLOOR PLAN FOR QUALITY EVALUATION LABORATORY.

exceptionally easy to keep clean. The table should have either North light or a recommended overhead light. For a table of the above size, 4 MacBeth Examolites would be adequate. These lights are the closest match to North daylight available and a good substitution for daylight. The obvious advantage is that the light is uniform day in and day out, and one can grade well at midnight as at midday.

Drawers and closed cupboards under that table should be provided for the storage of grading equipment such as vacuum, pressure, headspace gauges, hydrometers, salometers, trays, pans, and screens. Standard equipment for the grading table also includes a heavy-duty, table-mounted can opener, and one or two grading scales. The scales should read to at least ¼ oz and should have a tare beam for tare weighing empty cans, screens, etc.

The analytical or chemical bench should be approximately 3 ft high and 3 ft wide, with a maximum length of 12 ft. (In Fig. 14.1, the bench is 8 ft long.) The bench should be equipped with water, gas, electric, vacuum and air pressure outlets. The top of the bench should have an acid- and alkali-resistant top with an 4-in. square lead drain sink running down the center. There may or may not be a sink at the end of the bench. Above the bench top and over the center sink, a double-deck shelf (10 in. wide) should be con-

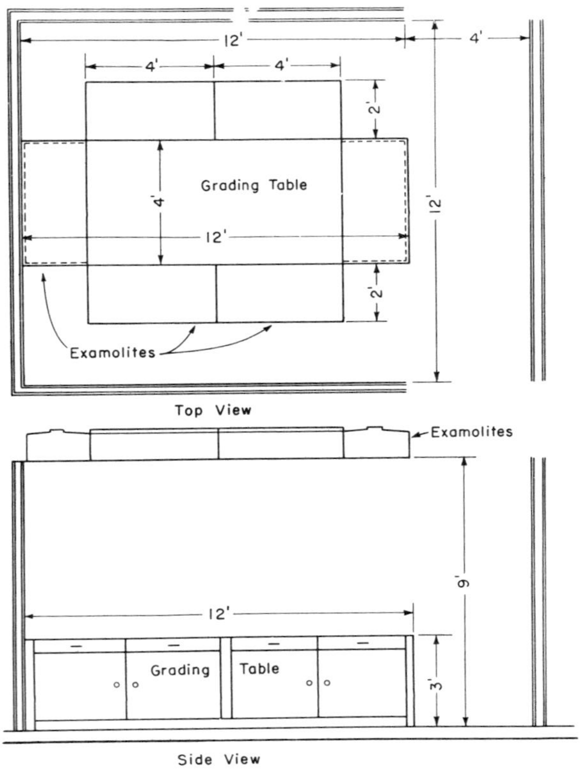

FIGURE 14.2. GRADING TABLE ARRANGEMENT SHOWING LOCATION AND DIMENSIONS TO MACBETH LIGHTS.

structed for the storage of chemicals and reagents. Drawer and closed cupboard space should be provided under the bench for the storage of special analytical equipment.

The taste panel table should be approximately 30 in. high, 2 ft wide, and 8 ft long. Partitions 18 in. high should be constructed for the top of the table to allow panel members to work individually when evaluating the samples. The table should be open underneath so that panel members can sit down when working. Overhead lighting with varying colored lights is necessary to light the samples properly for unbiased taste evaluation. A separate room operated from a separate kitchen is preferable, but the above arrangement is satisfactory for the practical taste evaluation of line samples.

The microbiological and physical testing tables should be approximately 36 in. high and 2 ft wide. Outlet facilities similar to those on the chemical bench are desired, but usually an electric line and a gas line will suffice. The bench top, drawer, and cupboard space should be constructed like the analytical bench. At one end of the microbiological bench, space should be provided for the analyst to sit down to use a microscope. The space should be approximately 26 in. high and 30 in. wide with a 2-in. drawer.

The quality evaluation manager should have one corner of the laboratory set aside for his desk, files, and library. The library and manuals should be available for use by his analysts at any time. Moreover, space should be provided for the immediate and daily filing of reports.

Basic Equipment. Laboratory supply houses and laboratory equipment manufacturing companies can supply most of the suggested basic grading equipment. This equipment includes

1. *Can Opener.* The can opener should be the heavy-duty type. It should cut the top of the can out and not cut the rim off, thus facilitating the measurement of the headspace of the can. For efficient operation and use, it should be oiled periodically and the teeth on the gripping wheel should be cleaned with a stiff wire brush.
2. *Vacuum Gauge.* The vacuum gauge should be approximately 2½ in. in diameter and specially fitted for testing the vacuum in tin or glass containers having metal caps. Preferably, it should be a combination pressure and vacuum gauge. To keep the gauge in good working condition when taking a vacuum reading, the needle should be inserted as near to the outer edge of the can lid as possible to prevent the product from being drawn up into the gauge. The vacuum or pressure reading should not be taken on cans that are swelled and/or badly dented. The needle point of the gauge should be periodically sharpened, and the gauge tube should be kept clean.
3. *Headspace Gauge.* The headspace gauge should be calibrated in 32nds of an inch. The gauge should be kept dry and clean to prevent corrosion. Occasionally the gauge can be wiped with an oiled cloth to prevent rusting and facilitate ease of operation.
4. *Grading Scale.* The grading scale should have a capacity of up to 12 lb, be equipped with a tare beam, and have an accuracy of 1/16 oz. It should be kept on the grading table, preferably close to the can opener. Twice a year, the scale should be dismantled, and the knife edges cleaned lightly with a light emery cloth. After each use of the scale, the weights should be returned to the weight rack. These weights and the scale pan should be kept clean to avoid inaccuracies in weighing.
5. *Grading Screens.* Grading screens are used to determine drained weight. The screens are of two diameters. For No. 3 size can and smaller, an 8-in. diameter screen is used, and for cans larger than No. 3 a 12-in. diameter screen is used. For tomatoes, a screen with 2 meshes to the inch (0.446-in. square opening) is used. In addition, the diameter of the wire for the tomato screen should be 0.054 in. The grading screens should be washed immediately after use to prevent verdigris (green rust on copper). When purchasing screens, it is suggested that the manufacturing companies supply screens of uniform weight. This allows one tare weight to suffice for all screens when determining drained weights on two or more cans at one time.

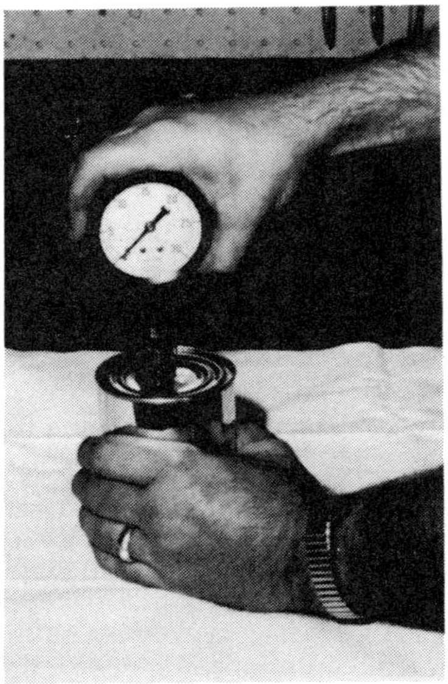

FIGURE 14.3. USING THE VACUUM GAUGE.

FIGURE 14.4. GRADING SCALE TO DETERMINE GROSS AND NET WEIGHTS OF CANNED PRODUCTS.

6. *Grading Trays.* White enameled or plastic trays should be used for grading. The trays should be of two depths, ¾ in. deep for No 3 size can and smaller, and 2 in. deep for cans larger than the No. 3. Trays approximately 10 in. wide by 15 in. long are satisfactory. The trays should be kept clean and stored in a convenient cabinet, preferably under the grading table.

7. *Sizing Gauges.* Sizing gauges are used to size–grade tomatoes. These gauges should be kept clean and stored in a drawer in the grading table. In addition to the standard sizing gauges, a vernier caliper, calibrated to 1/32 in. should be provided for the measurement of defects.

8. *Brine and Sirup Cylinders.* Brine and sirup cylinders are used to determine the color and separation of the juice and to determine the concentration of the brines and the caustic solutions. The cylinders should be of clear, colorless plastic or glass, have a plain rim, and be approximately 12 in. in height and 2 in. in diameter. The cylinders should be kept clean and be stored in a separate compartment of the grading table.

9. *Hydrometers and Salometers.* Hydrometers should be provided for the following different types of solutions and packing media: Salt concentrations (salometers) reading in percent salt, sugar concentrations (Brix hydrometers) reading in percentage sugar in sirup, alcohol hydrometers reading in percentage by volume, and alkali hydrometers reading in percentage by weight. The range of any of the hydrometers is quite narrow; when ordering hydrometers consideration should be given to the solutions being evaluated. After a hydrometer has been used, it should be carefully washed, dried, and put back in the appropriate hydrometer holder.

10. *Thermometers.* Fahrenheit and/or Centigrade thermometers covering a wide temperature range should be provided. Also, piercing ther-

FIGURE 14.5. EVALUATING THE DRAINED WEIGHT OF CANNED TOMATOES.

mometers should be obtained for determining the center can temperature for acid products. The can-testing machine should be obtained for determining the center can temperature for nonacid products like soups. It may be desirable to obtain maximum and minimum thermometers and regular pocket thermometers for routine in-plant quality control work. All thermometers and hydrometers should be kept in a drawer in the grading table.

11. *Miscellaneous Equipment.* Spoons, rubber tray scrapers, cylinder and tray brushes, can marking pencils, rubber stamps, etc., are helpful and essential equipment and supplies for the efficient operation in the quality control laboratory.

12. *Special Equipment.* A colorimeter, pH meter, refractometer, and physical instruments for texture, color, etc., should also be provided. Basic quantitative equipment such as an analytical balance, beakers, burettes, condensers, crucibles, graduated cylinders, desiccators, filtering apparatus, volumetric flasks, funnels, pipettes, and blenders are also necessary.

Microscopic and bacteriological equipment are essential for most products. Basic equipment includes a microscope (preferably binocular type with low power, high power, and oil-immersion lenses plus a mechanical stage), inoculation tubes, Petri dishes, colony counter, dilution bottles, slides, autoclave, incubation oven, etc. A few typical quality control procedures requiring microscopes and bacteriological equipment are mold count, plate count, insect fragment count, water and sewage analysis, and spoilage analysis.

Procedures. When using the quick tests of quality, available procedures must be followed as given, or small deviations may cause large errors in the results. The quality assurance technologist should be thoroughly familiar

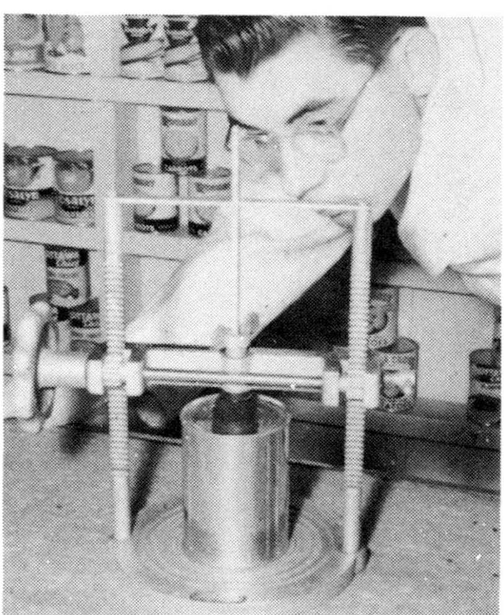

FIGURE 14.6. CENTER TEMPERATURE MEASUREMENT OF CANNED FOODS.

with the Standard Methods of Analysis for food, water, soil, pesticides, sewage, etc., and the published literature on objective tests of quality. If methods are not available, he may have to develop techniques and procedures to meet his own specific set of conditions. All procedures should be written out in complete detail and followed until they are modified.

Reports. A complete write-up of the results is as important as the analysis of the samples. The original should be given to management with copies to the quality assurance files and the applicable departments. The reports should give a complete history of daily quality control activities and recommendations, where applicable. In all reports, the quality assurance technologist should keep in mind why he is writing the report, for whom and to whom, and the type and amount of material to be presented. The details will vary with the type of report. Minimum parts of the report should include

1. Title
2. Purposes or objectives
3. Scope of work
   a. Accomplishments to date, including literature reviewed
   b. Present efforts: what was done and what remains to be done
   c. Statistical presentation of data: tabular and analysis
   d. Newly developed techniques
4. Discussion (if work far enough along)
5. Summary of work by answering objectives
6. Cost to date
7. Future requirements
   a. Laboratory procedures, supplies, apparatus and space
   b. Pilot plant equipment and use
   c. Production runs
   d. Personnel
   e. Estimated costs to complete project
8. Personnel involved in project with project leader signing report

Interpretation

Statistical quality control (SQC) or Statistical Process Control (SPC) is a useful tool for the interpretation of reports. Variation is always present in the measured quality of manufactured products. Variation is from two sources: explained and unexplained. Unexplained variation is inevitable. Explained variation can usually be detected and corrected by appropriate methods.

SQC employs statistical principles and methods to assess the magnitude of unexplained variation and to detect explained variation. It will indicate the limits of variation. These limits are determined by the laws of probability. Statistical quality control is sampling the product, determining the quality variation of the sample, and relating the findings to the entire lot under consideration.

Before discussing statistical quality control charting, terminology should be understood. Statistical texts should be used for more detailed explanations and formulas. To interpret quality control data, the following terms should be understood.

A *histogram* is a vertical or horizontal bar chart in which the adjacent columns have heights or lengths proportional to the number of observations. A *frequency polygon* is a diagram in which a line connects the dots representing the frequency of observations for a given dimension or value.

Measures of central tendency include the arithmetic mean, the median, and the mode. The *arithmetic mean* or *average* in statistics, $\bar{X}$ (read X bar) is the quotient obtained by dividing the sum of a set of reading or observations by the number, n, of observations:

$\bar{X} = \frac{\Sigma x}{n}$ where Σ means sum of. The *median* is the reading or observation above and below which an equal number of observations fall.

The *mode* is the value that occurs most frequently; in the case of a histogram, it is the tallest or longest bar.

Measures of dispersion includes the standard deviation and the range. The *standard deviation*, σ (read sigma), is the square root of the sum of the squared deviation from the sample mean

$$\sigma = \sqrt{\frac{\Sigma(X - \bar{X})^2}{n}}$$

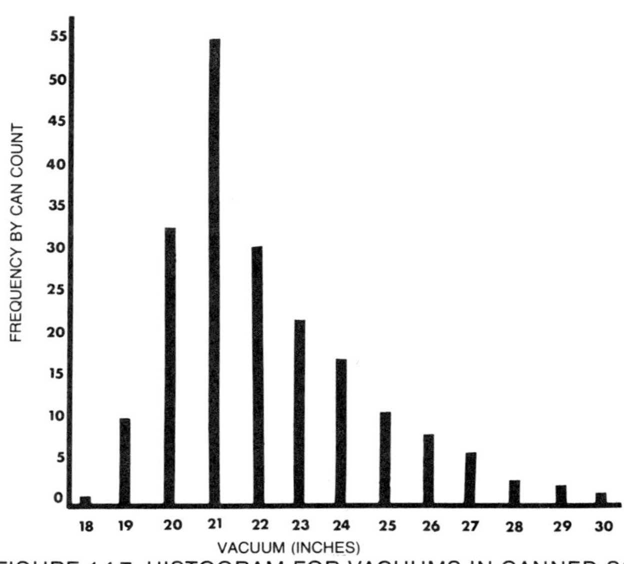

FIGURE 14.7. HISTOGRAM FOR VACUUMS IN CANNED SOUP.

This is used to express the amount of variation in a set of data. The *range*, *R*, is the difference between the largest and the smallest values of a set of observations.

The *normal curve* is a distribution of individual values in which the average, median, and mode are the same. The standard deviation divides the range of a set of data into six equal parts as follows: $\bar{X} \pm 1\sigma = 68.26\%$ of total area; $\bar{X} \pm 2\sigma = 95.46\%$ of total area; and, $\bar{X} \pm 3\sigma = 99.73\%$ of total area.

The standard deviation is useful when determining the degree of variation within a sample: 68% of the values will be within plus or minus 1σ or standard deviation from the average; 95% of the values will be within plus or minus 2σ from the average; and 99% will be between plus or minus 3σ from the average. These values inform management whether the product is within the specified tolerances. For a normal distribution of samples, these values are best illustrated by Fig. 14.8.

Figure 14.9 illustrates the practical application of the expected weight distribution of items for a product or process in control and for a product underweight or out of control.

Table 14.2 shows the SQC Data Form I for the frequency distribution of vacuum in a set of samples. Further, the calculations show these data highly skewed to the left, indicating very high vacuums.

Figure 14.11 indicated the descending cumulative frequency distribution of the data, indicating that 70% of the samples are 21 to 23 in. of vacuum.

Figure 14.12 shows that ascending cumulative frequency data for the same date, indicating that 10% have vacuums over 25 in. As shown, there are several methods of presenting statistical data.

FIGURE 14.8 APPROXIMATE AREAS FOR A TYPICAL NORMAL CURVE.

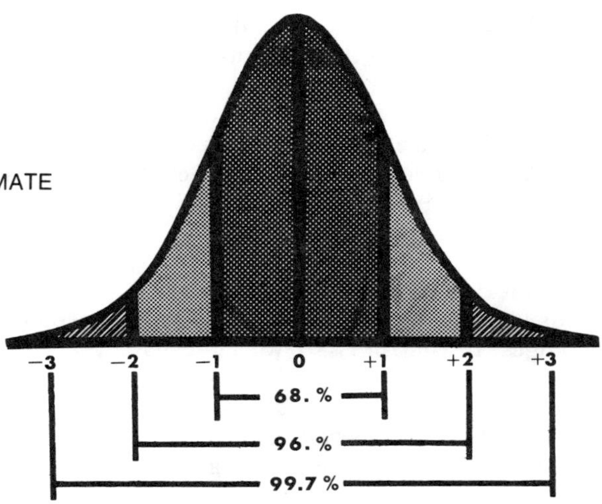

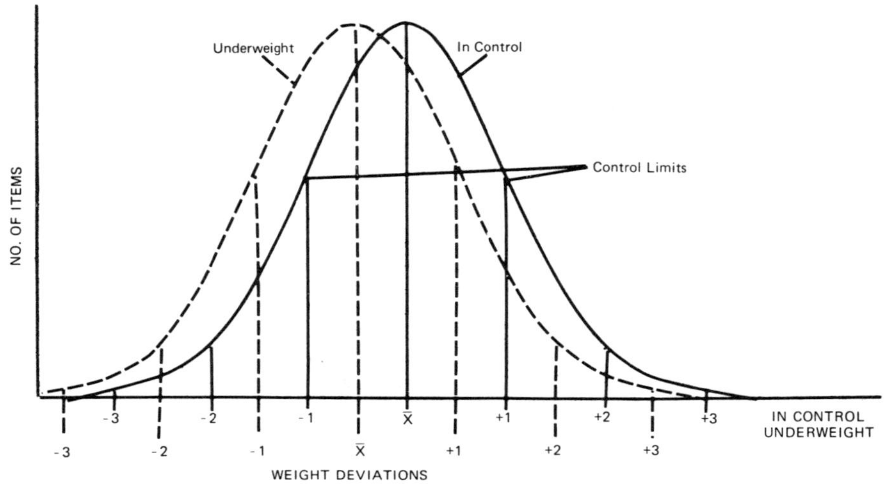

FIGURE 14.9. EXPECTED WEIGHT DISTRIBUTION OF PRODUCTS IN CONTROL AND UNDERWEIGHT (OUT OF CONTROL).

Definitions of the terms used are given in Table 14.3.

$\bar{X}$ *and R chart* is one of the Shewart techniques. It serves very well to estimate the *center* and *spread* of the distribution curve of production run or batch of sampling. $\bar{X}$ and R charts are used primarily to measure quality characteristics called *variables*. Variables are characteristics to which numbers can be affixed, such as dimensions, pounds, and time. $\bar{X}$ and R chart is a graphic means of portraying the variation within samples and/or the variation between samples. Calculated control limits are used to determine whether or not the fluctuations are significant.

The control chart takes time into consideration, whereas the distribution curve does not. If the points do not fall within the control limits, there is an assignable cause present. If the points do fall within the control limits, it means that a constant system of chance is operating. It is possible for a process to be in control and not be at a satisfactory level (statistically in control).

The $\bar{X}$ and R chart is used to study process capability and control, a variable that can be expressed numerically. After a process has been brought under control at a satisfactory level, the control chart indicates significant changes. In control chart work, speed is the essence of effectiveness. If a sample average falls outside the estimated control limits, it can be concluded that the out-of-control average does not come from the same system as the others; in operating terms, it means that an assignable cause is prevalent, must be discovered, and corrected.

TABLE 14.2. SQC DATA FROM I FREQUENCY DISTRIBUTION FOR VACUUM IN CANNED SOUP

Class Interval or Measurement	f (Frequency)	Total (f)	d	fd	fd^2	%	Cumulative Frequency (%) Ascending	Cumulative Frequency (%) Decending
18		1	−4	−4	16	0.5	0.5	100.0
19		10	−3	−30	90	5.0	5.5	99.5
20		33	−2	−66	132	16.5	22.0	94.5
21		55	−1	−55	55	27.5	49.5	78.0
22		31	0	0	0	15.5	65.0	50.0
23		22	1	22	22	11.0	76.0	35.0
24		17	2	34	68	8.5	84.5	24.0
25		11	3	33	99	5.5	90.0	15.5
26		8	4	32	128	4.0	94.0	10.0
27		6	5	30	150	3.0	97.0	6.0
28		3	6	18	108	1.5	98.5	3.0
29		2	7	14	98	1.0	99.5	1.5
30		1	8	8	64	0.5	100.0	0.5

where $n = 200$
$\Sigma fd = 36$
$\Sigma fd^2 = 1030$

$$= c\Sigma fd^2 - (\Sigma fd)^2 = 1\sqrt{\frac{1030}{200} - \frac{(36)^2}{(200)}} = \sqrt{5.15 - 0.0324} = \sqrt{5.1176} = 2.26$$

$2\sigma = 4.52$ $\quad 3\sigma = 6.78$

$$\bar{X} = Z + \frac{(c)\,(fd)}{n} = 22 + \frac{(1)\,(36)}{200} = 22 + 0.18 = 22.18$$

$$CV = \frac{(\sigma)}{\bar{X}}(100) = \frac{(2.26)}{(22.18)}(100) = (0.102)(100) = 10.2\%$$

where n = sample size, f = frequency, d = observation value, c = class interval, Z = value of median, and CV = covariance

If values reflecting the unexplained variation in a process are plotted on a time basis, then statistical limits can be determined within which such values will lie. This forms a statistical control chart. Values falling outside these statistical limits indicate the occurrence of significant variation, usually due to an assignable factor. On the $\bar{X}$ control chart these statistical limits are defined as the upper control limit (UCL) and the lower control limit (LCL). If a given attribute or characteristic exceeds the UCL, then that particular attribute is above its desired value, or higher than normal. Consequently, the processor, in the case of net weight, is giving the consumer the amount exceeding the UCL free of charge. This automatically minimizes his yield and profit. However, if an attribute falls below the LCL, this attribute is of lower quality than the processor seeks to maintain. The $\bar{X}$ is a measurement of the central tendency. Due to unexplained variation, half of the values, approximately, are located above it while the other half are located below it, indicating the average value for unexplained variation.

FIGURE 14.10. FREQUENCY DISTRIBUTION FOR VACUUM IN CANNED SOUP.

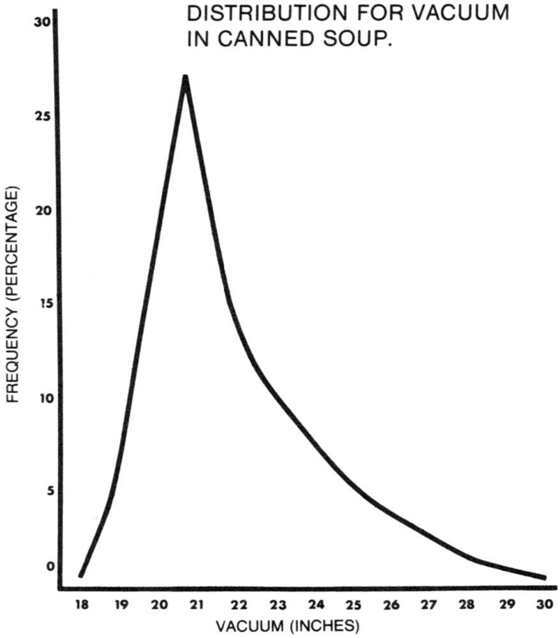

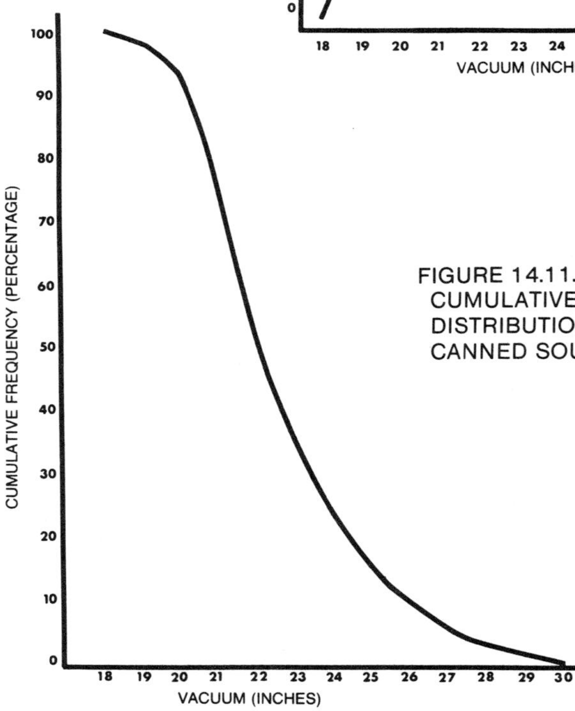

FIGURE 14.11. DESCENDING CUMULATIVE FREQUENCY DISTRIBUTION OF VACUUMS FOR CANNED SOUP.

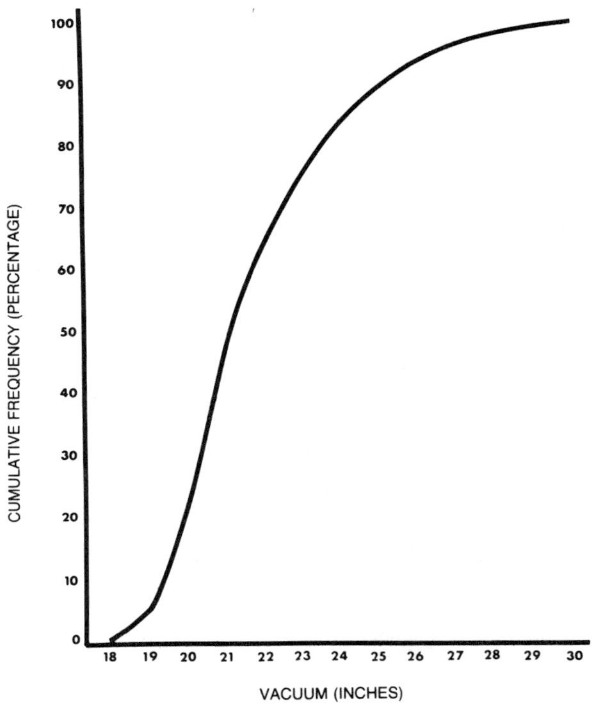

FIGURE 14.12. ASCENDING CUMULATIVE FREQUENCY DISTRIBUTION OF VACUUMS FOR CANNED SOUP.

The range *(R)* chart, also a control chart, indicates the difference between the highest and lowest value, thus showing the variance present in the set of samples. The *R* chart has an upper range limit (URL). The height of the URL is determined by unexplained variation; a range in excess of this value is usually due to an assignable cause.

It is necessary to use both the $\bar{X}$ and R charts together. For example, the $\bar{X}$ chart may indicate a consistent quality, while the range could vary from a small to a large extent, showing the variability among the samples or product.

The data in Table 14.4 may clarify this. Five samples were taken every hour for 10 hr from a given production line. The average, $\bar{X}$, was computed as well as the range. The average of the 5 samples was plotted on an $\bar{X}$ chart

TABLE 14.3. DEFINITION OF TERMS USED IN STATISTICAL QUALITY CONTROL

Symbols	Read	Meaning
m	m	Number of subgroups in a sample
n	n	Total number of sample units or measurements in a sample $n = m\, n_s$
n_s	n sub s	Number of sample units or measurements in a subgroup
X	X	Value of an individual measurement for a variable
Σ	sum of	Sum of a series of numbers (X means the sum of several measurements)
M_i	M_i	Median of all the individual measurements in a subgroup
$\bar{\bar{M}}_i$	M_i double bar	Median of all the individual measurements or subgroup medians (M_i) in a sample
$\bar{X}$	X bar	Average of all individual measurements in a subgroup
$\bar{\bar{X}}$	X double bar	Arithmetic mean of all the individual measurements in a sample. When the average is calculated for each subgroup in a sample for conventional averages, $\bar{\bar{X}}$ is also the average of the subgroup averages.
R	R	A range of measurements, the difference between the highest and lowest measurement within a subgroup
$\bar{R}$	R bar	Average range of all the subgroup ranges
σ	sigma	Standard deviation of the averages. The width of one zone in the normal distribution curve of individual items called a standard deviation. Plus and minus 3σ from the average includes 99% of the normal curve
$s\bar{x}$	s x bar	Standard deviation of the averages
$\bar{X}'_{max}$[a]	X bar prime max	A specified maximum lot average value
$\bar{X}'_{max}$ adjusted	X bar prime max adjusted	$\bar{X}'_{max}$ plus a sampling allowance
$\bar{X}'_{min}$[a]	X bar prime min	A specified minimum lot average value
$\bar{X}'_{min}$ adjusted	X bar prime min adjusted	$\bar{X}'_{min}$ minus a sampling allowance
$\bar{R}'$[a]	R bar prime	A specified average range value
R_{max}[a]	R max	A specified maximum range for a subgroup
$UCL_{\bar{x}}$	Upper control limit for averages	Upper and lower control limits for averages determines the pattern that sample averages should follow if a constant system of a chance is operating. If the averages do not conform to this pattern, there is an assignable cause present
$LCL_{\bar{x}}$	Lower control limit for averages	
UCL_R	Upper control limit for ranges	Upper control limit for range sets the pattern within which sample ranges should fall. If the sample ranges do not conform to this pat, tern, there is an assignable cause present
LRL[a]	Lower reject limit for individual measurements	Lowest value an individual measurement may have without causing the production to be rejected for failure to meet prescribed requirements for individual measurements
LRL_x[a]	Lower reject limit for subgroup averages or medians	Lowest value the average or median of a subgroup may have without causing the production to be rejected for failure to meet perscribed requirements for subgroup averages
LWL[a]	Lower warning limit for individual measurements	This value serves as a warning point that the production may have reached a level where the chances of subsequently finding an individual measurement that will fall below the LRL have increased to a degree that the production may be in danger of rejection

TABLE 14.3. (Continued)

Symbols	Read	Meaning
$LWL_{\bar{x}}$[a]	Lower warning limit for subgroup averages for medians	This value serves as a warning point that the quality of the production may have reached a level where the chances of subsequently finding a subgroup average or median that will fall below $LRL_{\bar{x}}$ have increased to a degree that the production may be in danger of rejection
URL	Upper reject limit for individual measurements	The highest value an individual measurement may have without causing the production to be rejected for failure to meet prescribed requirements for individual measurements
URL_x	Upper reject limit for subgroup averages or medians	The highest value the average or median of a subgroup may have without causing the production to be rejected for failure to meet perscribed requirements for individual measurements
UWL	Upper warning limit for individual measurements	This value serves as a warning point that the quality of the production may have reached a level where the chances of subsequently finding an individual measurement that will exceed the URL have increased to a degree that the production may be in danger of rejection
UWL_x	Upper warning limit for subgroup averages or medians	This value serves as a warning point that the quality of production may have reached a level where the chances of subsequently finding a subgroup average or median that will exceed that URL_x have increased to a degree that the production may be in danger of rejection

[a]These limits and values are to be established and incorporated in the USDA Grade Standards for the various products. These values will be available upon request to: Chief, Processed Products Standardization and Inspection Branch, Fruit and Vegetable Division, AMS, U.S. Department of Agriculture, Washington, DC 20250.

TABLE 14.4. STATISTICAL QUALITY CONTROL RECORD AND DATA FORM

Product *Tomato Soup* Size of Container *300* Code *1234* Plant *WAGCO*

Sample No.	Frequency of Sample Sets by Day Shift for Line 4										$\bar{X}$	$\bar{R}$
	7 AM	8 AM	9 AM	10 AM	11 AM	12 AM	1 PM	2 PM	3 PM	4 PM		
1	18.5	15.2	16.3	19.1	18.7	15.9	16.8	16.0	16.0	16.1		
2	17.0	15.3	14.8	18.4	18.3	15.2	15.8	16.1	16.2	16.0		
3	16.5	18.4	14.6	18.6	17.7	14.8	16.4	16.3	16.5	16.0		
4	16.8	15.0	15.1	16.1	16.2	14.1	15.8	16.0	16.1	16.1		
5	15.0	15.0	15.0	17.5	17.9	15.4	14.9	16.2	16.0	16.2		
Sum of X values	83.8	78.9	75.8	89.7	88.8	75.4	79.7	80.6	80.8	80.4		
$\bar{X}$	16.8	15.8	15.2	17.9	17.8	15.1	15.9	16.1	16.2	16.1	16.29	
R	3.5	3.4	1.7	3.0	2.5	1.8	1.9	0.3	0.5	0.2		1.88

Note 1: $UCL_{\bar{x}}$ (upper control limit for average) = $\bar{\bar{X}} + A_2\bar{R}$
$LCL_{\bar{x}}$ (lower control limit for average) = $\bar{\bar{X}} - A_2\bar{R}$
UCL_R (upper control limit for range) = $D_4\bar{R}$

Note 2: A_2 for five sample numbers in a set is equal to 0.58.
D_4 for five sample numbers in a set is equal to 2.11.
For these calculations see Table 14.5.

and similarly the range plotted on an R chart. The R chart indicated that there was a wide range for the first 2 hr, a narrower one the third hour, widened in the fourth hour, and again narrower the last 3 hr. The narrow range is indicative of greater uniformity. If the quality attribute being evaluated has a narrow range within the UCL and LCL, the operation is running under normal conditions.

A flattening of the distribution attributable to material instability, operator carelessness, machine wear, or other causes is reflected by the appearance of a sample range above the URL. The process control chart thus provides immediate information about the pattern of variation expected from the process, and gives prompt indication of trouble if it is present.

When developing $\bar{X}$ and R charts, the objectives for use of these charts should be determined first. In most cases, the chart is to provide a basis for taking corrective action when the process is out of control. The next decision is the selection of variables to be measured and/or methods to measure them. The size of sample must then be determined. Recommendations on sample size follow:

1. A sample size of 2 should not be used because estimates of limits using an average of 2 may contain large sampling errors,
2. A sample size of 3 can be used satisfactorily but is difficult to calculate,
3. A sample size of 4 is best from a statistical viewpoint because the distribution of $\bar{X}$ is nearly normal, but is harder to calculate than a sample size of 5,
4. A sample size of 5 is the most commonly used, accurate and easy to calculate,

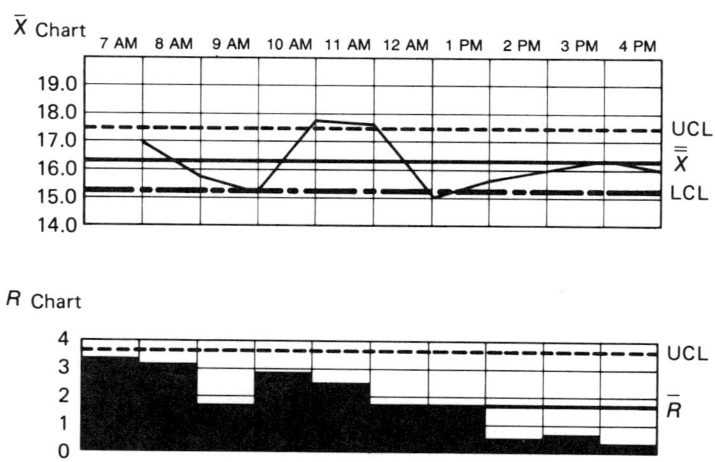

FIGURE 14.13. STATISTICAL QUALITY CONTROL CHART FOR NET WEIGHTS AND AMOUNT OF RANGE.

5. Sample sizes larger than 5 are advantageous when it is desired to make the control chart sensitive to small variations in the process average; the larger the sample size, the narrower the control limits and the easier it is to detect small variations.

The samples selected should be representative of the time interval covered. All methods, measurements, and procedures should be followed strictly; an alteration in any one of these may cause a significant change in the data being collected.

When data are recorded, any conditions that have changed since the last sample was taken should also be recorded. These include such items as changes in operators, product quality, and machine setting.

The chart is originally plotted without control limits. When sufficient data are collected, the control limits computed will be reasonably reliable.

The $\bar{X}$ and R themselves will not correct a situation. They will only tell the operator where to look for the trouble. Once he has found the source of trouble, the operator must take appropriate action to control the particular unit operation within the established limits or specifications. Table 13.5 shows the necessary factors for computation. The formulas for computing these control limits follow.

For Averages	*For Range*
$UCL = \bar{\bar{X}} + A_2\bar{R}$	$UCL_R = D_4R$
$LCL = \bar{\bar{X}} - A_2\bar{R}$	$LCR_R = D_3R$

Standard Deviation or Sigma (σ)

$$\sigma = \frac{\bar{R}}{d_2}$$

TABLE 14.5. FACTORS FOR COMPUTING CONTROL LIMITS

Sample Size (n)	Factors for Averages A_2	Range D_3	Range D_4	Factor for Standard Deviation or Sigma (d_2)	Sample Size (n)
2	1.880	0.0	3.268	1.128	2
3	1.023	0.0	2.574	1.693	3
4	0.729	0.0	2.282	2.059	4
5	0.577	0.0	2.114	2.326	5
6	0.483	0.0	2.004	2.534	6
7	0.419	0.076	1.924	2.704	7
8	0.373	0.136	1.864	2.847	8
9	0.337	0.184	1.816	2.970	9
10	0.308	0.223	1.777	3.078	10
11	0.285	0.256	1.744	3.178	11
12	0.266	0.284	1.717	3.258	12
13	0.249	0.308	1.692	3.336	13
14	0.235	0.329	1.671	3.407	14
15	0.223	0.348	1.652	3.472	15

Where $\bar{\bar{X}}$ = the average for a set of $\bar{X}$ values or the grand average, and $\bar{R}$ = the average for a set of $\bar{R}$ values. All factors in Table 13.5 are based on the normal distribution.

Control limits are placed on charts to provide a basis for action. If a sample average falls outside the estimated control limits, the out-of-control average point probably does not come from the same system as the others. In operating terms, this indicates a major source of extra variability between samples, that is, an explainable cause that an investigation probably will uncover, making corrective action possible.

Some SQC problems require the testing or measuring of the significance of the difference between two sample means. Homes devised a simplified form for carrying through a t^2 test using an F table and a table of reciprocals, and the following formula:

$$F = \frac{(\bar{X}_1 - \bar{X}_2)^2/S^{12}}{{n_1}^{-1} + {n_2}^{-1}}$$

Step 1: Subtract the means and square
Step 2: Form the pooled sum of squares

$$S'^2 = \frac{\Sigma {x_1}^2 + \Sigma x_2^2}{n_1 + n_2 - 2}$$

Step 3: Add the reciprocals of n_1 and n_2 $(1/n_1 + 1/n_2)$
Step 4: Multiply the pooled sum obtained in (2) by the sum of the reciprocals obtained in (3)
Step 5: Divide the squared difference between means obtained in (1) by the product obtained in (4)
Step 6: Compare with a standard F table for 1 and $n_1 + n_2 - 2$ degrees of freedom
Step 7: See example in Table 14.6.

If more sophistication is desired in the interpretation of the SQC data, particularly when a process has two or more factors each affecting the variability, the interpretation of the relative importance can be made using the analysis of variance technique. This statistical technique is based on the principle that the total variance of the process is equal to the sum of the component variances if the factors are acting independently. The advantage of this method is that if several factors are involved in a process or the quality of a product, it is possible to rank their effects on the total variation in order of magnitude. Thus, the information is used to determine where the reduce variability for maximum improvement with a minimum expenditure of time and effort. An example of presenting a two-way analysis of variance is given in Table 14.7.

Statistical interpretations are valuable during the processing of any given product for machine or process control and for quality evaluation.

TABLE 14.6. TESTING THE SIGNIFICANCE BETWEEN TWO SAMPLE MEANS

Method I		Method II	
X_1	X_1^2	X_2	X_2^2
0.020	0.000400	0.010	0.000100
0.020	0.000400	0.022	0.000484
0.015	0.000225	0.029	0.000841
0.003	0.000009	0.009	0.000081
0.015	0.000225	0.036	0.001296
0.014	0.000196	0.023	0.000529
0.010	0.000100	0.010	0.000100
0.028	0.000784	0.025	0.000625
0.020	0.000400	0.021	0.000441
0.008	0.000064	0.034	0.001156
0.010	0.000100	0.010	0.000100
0.031	0.000961	0.010	0.000100
$\Sigma X_1 = 0.194$	$\Sigma X_1^2 = 0.0003864$	$\Sigma X_2 = 0.239$	$\Sigma X_2^2 = 0.005853$
$\bar{X}_1 = 0.0161$		$\bar{X}_2 = 0.0199$	

1. $(\bar{X}_1 - \bar{X}_2)^2 = (0.0161 - 0.0199)^2 = 0.00001444$
2. $\Sigma X'_1{}^2 = \Sigma X_1^2 - (\Sigma X_1)^2/N = 0.003864 - (0.194)^2/12$
   $= 0.003864 - 0.003136 = 0.000728$
   $\Sigma X'^2_2 = \Sigma X_2^2 - (\Sigma X_2)^2/N = 0.005853 - (0.239)^2/12$
   $= 0.005853 - 0.004760 = 0.001093$
3. $(1/N_1 + 1/N_2) = (1/12 + 1/12) = 0.0833 + 0.0833 = 0.1666$
4. $S'^2 = \dfrac{\Sigma X_1^2 + \Sigma X_2^2}{N_1 + N_2 - 2} = \dfrac{0.001821}{22} = 0.0000828$
5. $(S'^2)\,(1/N_1 + 1/N_2) = (0.0000828)\,(0.1666) = 0.00001379448$
6. $F = (\bar{X}_1 - \bar{X}_2)^2/(S'^2)\,(1/N_1 + 1/N_2)) = 0.00001444/0.00001379448 = 0.992$

The F table shows a 5% value of 4.30 for $N_1 + N_2 - 2 = 22$ degrees of freedom which is larger than 0.992. Therefore, no significant difference between these means has been shown.

Machine or process control usually means taking a series of samples from the line to measure given attributes of quality. These measurements are used to determine if such machines as size graders, fillers, and closing machines are within the specified working tolerances. Quality *control* refers to inspection where man is the factor determining the uniformity of the samples. Inspection may apply all along the line, i.e., starting with the receipt of the raw product, or during the process, or with the finished product. Such attributes of quality as color, defects, net and drained weight, and maturity are evaluated.

Several advantages accrue from a statistically controlled process. These include:

1. An increase in personnel efficiency,
2. Supervisors are kept on their toes,
3. Raw materials are economically used,
4. Tooling and machine problems are reduced,
5. Estimates and bids can be made with greater assurance,
6. Sound predictions can be made,
7. Work flow becomes smoother, and
8. Study and analysis of the process lead to improvements.

TABLE 14.7. SQC FORM IV ANALYSIS OF VARIANCE FOR FLAVOR (10 JUDGES AND 10 SAMPLES) OF TOMATO JUICE[a]

	Judge											
Sample	1	2	3	4	5	6	7	8	9	10	Sum	$\bar{X}$
A	1	7	1	4	1	6	8	5	3	7	43	4.3
B	5	4	6	9	10	6	5	6	5	6	62	6.2
C	6	5	8	5	8	7	8	5	6	4	62	6.2
D	9	8	5	3	7	5	9	2	6	5	59	5.9
E	7	2	1	2	6	6	6	7	7	3	47	4.7
F	5	1	1	5	10	8	9	8	5	3	55	5.5
G	6	3	5	1	5	8	5	7	5	3	48	4.8
H	7	6	8	9	4	7	6	8	9	4	68	6.8
I	8	2	1	1	3	2	4	1	2	2	26	2.6
J	9	1	2	1	2	2	6	4	2	3	32	3.2
Sum	63	39	38	40	56	57	66	53	50	40	502	
$\bar{X}$	6.3	3.9	3.8	4.0	5.6	5.7	6.6	5.3	5.0	4.0		5.02

Source of Variance	Sums of Squares	Degrees of Freedom	Mean Square	Calculated "F"	Book "F" 0.01	Book "F" 0.05	LSD @ 0.01
Judges	112	9	12.44	2.75	2.64	1.99	1.55
Samples	162	9	18.00	3.97	2.64	1.99	1.55
Error	367	81	4.53				
Total	641	99					

[a]These data show significant differences among the judges and the samples. It is obvious that judges 2, 3, 4, and 10 are different from judges 1 and 7. However, we need to use the LSD (1.55) to show the difference of judges 5 and 6 to judges 2, 3, and 4. Further, the LSD tells us that judges 1, 5, 6, 7, and 8 are not different. By the same method of interpretation, samples I and J differ from all the other samples except A, etc.

It is the quality assurance technologist's responsibility to see that the measurements are taken in a systematic fashion, and that all the data are recorded and properly interpreted. These functions, if properly carried out, are informative as well as preventive. They inform management of what is happening at all times. If a machine or an inspection shows variation in the samples, they can prevent the product from getting out of control if the adjustment is properly and promptly made.

The obvious advantage of a systematic method of interpreting quality control data is that management can easily see what happens from hour to hour, line to line, year to year, or even season to season. It is a graphic presentation of the statistical interpretation of the observed samples indicating what is taking place on the production line. If properly used, the quality control chart technique can save management money and assure it product uniformity. Moreover, tolerances can be made closer for any particular attribute of quality or machine specification with the ultimate result of improved quality.

The establishment and use of a SQC program is not just another way to keep someone busy, but a tool to force the operator of every unit operation in a food plant to pay strict attention ot the process he is responsible for. It will result in more uniform products at reduced costs. Further, the SQC program has been proved to be an effective method of developing the responsibility of plant personnel for the good of a growing organization.

CHAPTER 15
Quality Control

Quality control has moved from the laboratory and quality assurance personnel to the production floor and each of the unit operators. The main reason being that the control of quality in a food plant must be part of the operators responsibilities. Management with the assistance of quality assurance personnel still establishes the level of quality for the firm and they set forth the specifications for each of the various unit operations in the production process. However, the operator who operates the various unit operations controls the operation to produce uniform products according to established specifications.

Thus, the operator to do his or her job properly needs help and a thorough understanding of the requirements. Further, they need knowledge in how to properly use the tools of quality measurement and control. People do not always perform well because (1) they do not know what to do, (2) they do not know how to do their job, and/or (3) they do not know why they must do the job a certain way. Once the operator is properly trained, he or she must be held responsible for the manufacture of given levels of quality along with producing products efficiently.

All of this implies an immense amount of communication know how. Communication must be up and down the entire production line, it must be between the quality assurance personnel and each of the operators, and it must be between management and the individual operator. Thus, communication, that is, the spoken and written word, the visual examples with actual data and charts, and the many models become two way streets for all to see, observe, and follow. Communication must never stop. There cannot be any barriers as communication is most crucial to the success of the firm. It helps all employees understand the role of quality assurance and quality control programs and what they mean to the future growth and success of the firm. Quality products do make a difference.

To accomplish the above, the individual operator must be helped through proper education. Each operator must be made aware of the significance of quality issues and quality requirements. Each individual operator must change his behavior towards producing quality products all the time. Further, he must make quality mindness a part of his daily routine. He must see his role as part of the team in the production chain. He must understand that the weak link in the chain can and does cause significant problems in producing quality products all the time. Finally, he must see himself and his role in the production chain as the key to making uniform quality products.

All employees want to be involved in the success of their firm and they want to be given an active role in producing quality products efficiently. They will become more motivated and they will contribute more effectively if given a voice as this creates in their mind partial ownership. They will become more loyal and they will demonstrate great enthusiasm for their work and the firm. It is a most simple way of recognizing them and their abilities, talents, and efforts.

It is a proven fact that once management sets forth its plan or policy and follows through with positive actions and plans, employees become highly motivated and produce more effectively as they now see how the plan will benefit them, the firm, the customer, and the whole organization. The first step must be managements willingness to treat employees as part of the team and allow them to make their contribution for the good of the whole.

The results of controlling quality to given levels end up with compliments from customers. The compliments may be in greater repeat business, objective letters, positive phone calls, open discussions, one on one situations, or other positive means of showing appreciation for the company's efforts in providing the customer with what they expect and a little more. The key point to remember is that all quality efforts are in the eyes of the beholder. The beholder is the one who the firm must satisfy all the time.

Most companies today realize that one of the biggest variables in manufacturing is the individual operator. The operator wants and needs to learn how to do it right, what to do, and why it must be done a certain way to produce the kind of product that makes for growth of the firm. For success of the firm, every manager should put people first as people can make a real difference in the production of quality products.

The production employees must become part of the problem solving team and they must understand problem solving techniques. They need to know how to "brainstorm", how to use the "fishbone diagram-CEDAC" technique, and they need to understand the 80-20 rule or PARETO principles. They must learn how to gather and analyze data to solve problems. They need to know how to read and interpret quality control and run charts. They must thoroughly understand that "the close enough" syndrome is not acceptable in a food plant. Their creed must be "to always do it right all the time". They must develop the right attitude and show it along with their enthusiasm for their work and the team. Together as a team, much can be accomplished if given the right environment, the right machinery and proper materials and personnel who are motivated to always do the right job.

As previously indicated, the key to quality control is the actual operator. He or she must be rewarded for his or her efforts and they must be recognized for doing their job. People must be treated as each of us would want to be treated. Money is not everything to a good reliable well trained employee who wants to be with the firm for the long run. If the firm is serious about controlling quality to given standards or specifications, people must come first and they must be treated as associates in the business for long term effects.

Thus, the key to controlling quality of manufactured products is to train the operator, give them the proper tools, hold them accountable, and give them a voice in the operation of the plant. The result will be the manufacture of products which meet the expectations of the customer all the time. The customer does not want to be surprised with the second purchase of the product. They want it to be what they have come to expect. The end result is that management can rest assured that quality is under control and consistent. Further, management can expect its market share to increase and an improvement in the bottom line. The control of quality means you never have to say "I'm sorry". You always know you are doing what you fully intended to do in the manufacture of all of your products all the time.

PROBLEM SOLVING TECHNIQUES

A. BRAINSTORMING PRINCIPLES

(Creation of a 'storm' of ideas)

1. Identify the problem, such as: WHY DO WE ALWAYS HAVE PROBLEMS WITH QUALITY? This should be a clear written statement and written on a flip chart for all to see.
2. Select a LEADER to keep the session moving and select a person to act as RECORDER to record all suggestions on a second flip chart.
3. Use a team approach and get as many people as possible to become part of the team, however, every person must be given a chance to offer his idea(s) one at a time.
4. All ideas are evaluated. None are thrown out. Freewheeling is the name of the session and any unrelated idea is OK and should be encouraged. Further, ideas can come from "hitchhiking" off of someone else's idea.
5. Questions are allowed, but only for clarification of an idea. No one is permitted to criticize, censor or interrupt.
6. Members of the team can pass on their turn, but can submit ideas on later turns.

7. When everyone passes on a complete turn, the brainstorming session is over and the idea list is turned over to a problem solving team, such as a PARETO team.

PROBLEM SOLVING TECHNIQUES

B. PARETO Principles

Pareto principle tells us that 80% of the problems come from 20% of the causes, based upon actual facts or data. The idea behind Pareto is that we can separate the "vital few'' from the "trivial many" (see Figure 1).

1. Again one must clearly identify the problem, that is, WHY DO WE ALWAYS HAVE PROBLEMS WITH QUALITY?
2. Develop a chart with the attributes listed as PROBLEMS charted on the horizontal axis and the frequency of the problems charted on the vertical axis.
3. Each of the PROBLEMS are listed in tabular form and the team is asked to list these in order of importance, that is, the frequency of occurrence in their opinion and calculate the frequency of occurrence and accumulated %, i.e.,

Type of Problem	Number of Complaints	% Frequency	Accumulated % Frequency
Raw Materials	30	32	32
Slicer	20	21	53
Fryer Operator	15	16	79
Seasoning	10	11	90
----	--	--	--
----	--	--	--
----	--	--	--
Package	05	05	100
Total Complaints	95	100	

4. The % frequency is plotted on the vertical axis with contributing attributes to the problem on the horizontal axis as is shown inFigure 1. In addition the cumulative frequency may be plotted on the same chart. Thus, one can observe that 80% of the problems are caused by raw materials, slicer, and fryer operator. The other 5 contributing causes to the problem of quality are the trivial many and only represent some 21% of the causes of the problems.

C. CAUSE AND EFFECT DIAGRAM– CEDAC

Cause and Effect (CEDAC) is a simple technique for dissecting a problem or a process. CEDAC identifies all possible relationships among INPUT and OUTPUT Variables, that is, the six categories on the following skeleton (Fishbone) (Materials, Methods, Measurements, Man, Machinery, and Environment). CEDAC illustrates the relationships of the influencing factors of a particular effect (problem) and its corresponding causes. CEDAC delineates the problem and categorizes the information for use in statistical analysis. The causes can be broadly classified as follows: Materials, Machines, Man (workers), Methods, and Environment (Measurements may be added). The potential contributing causes are then drawn as ribs off of the main branches.

CEDAC organizes the thinking and provides a plan of attack all at the same time. Further, CEDAC brings out all known factors (causes), not just the suspected ones. This is accomplished by using cards to help explain in

FIGURE 1 - PARETO CHART

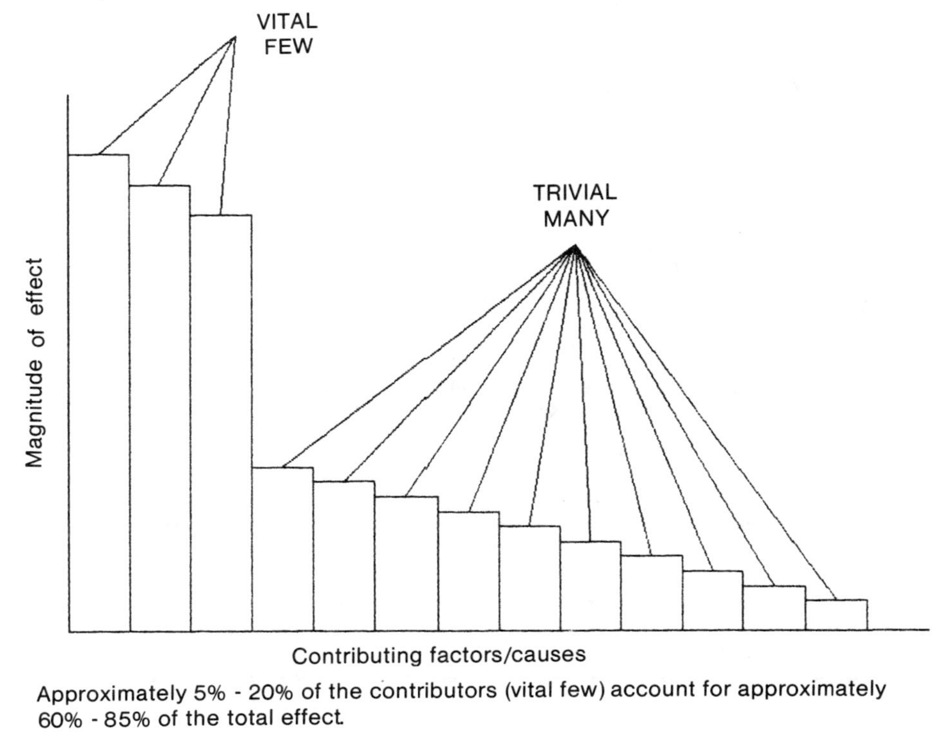

Approximately 5% - 20% of the contributors (vital few) account for approximately 60% - 85% of the total effect

short sentences the details of the cause. In the real world, employees using CEDAC write out their observation, their idea, and their knowhow on a 3 x 5 card and place these cards on the diagram in the appropriate place. This formation is then available to anyone to see, to study, and to add to with another card. Thus, the main purpose of CEDAC is to classify the various causes using the information that is available and thought to have an effect on the cause to help arrive at a quantitative answer.

CEDAC may be used to (1) analyze a process, (2) to solve problems and (3) to make improvements. CEDAC is an effective communication mechanism. One way to use CEDAC is to follow these suggested steps:

1. Form a team of persons concerned about the problem. In our case, the problem WHY DO WE HAVE PROBLEMS WITH QUALITY?
2. List all the various problem areas (causes) on the "Fishbone" diagram under the 4 M's and E.
3. Select the major causes by majority rule.
4. Determine all the capabilities of solving the problem using actual data. (Normally followed up by using statistical methods).
5. Interpret your data and theorize as to the possible causes.
6. Implement suggested solutions and install controls using X Bar and R Charting of results.

FIGURE 2. EXAMPLE OF BASIC STRUCTURE OF CAUSE AND EFFECT DIAGRAM TO SOLVE PROBLEMS.

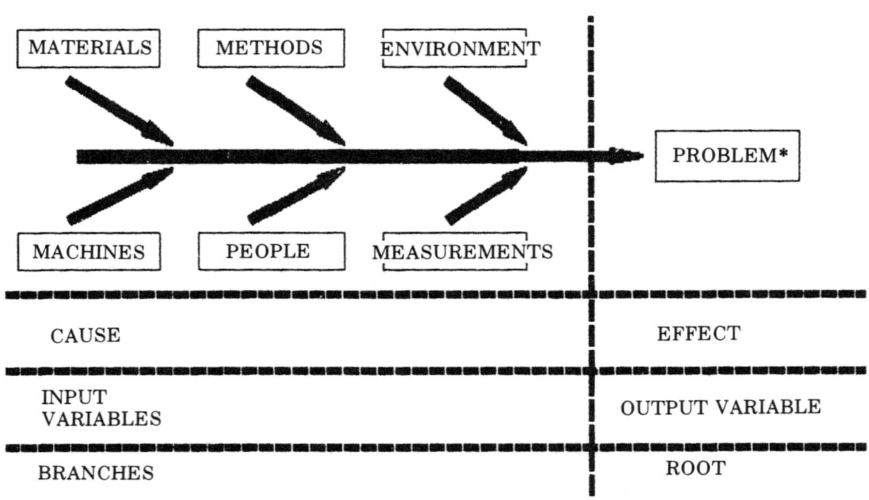

CAUSE AND EFFECT DIAGRAM

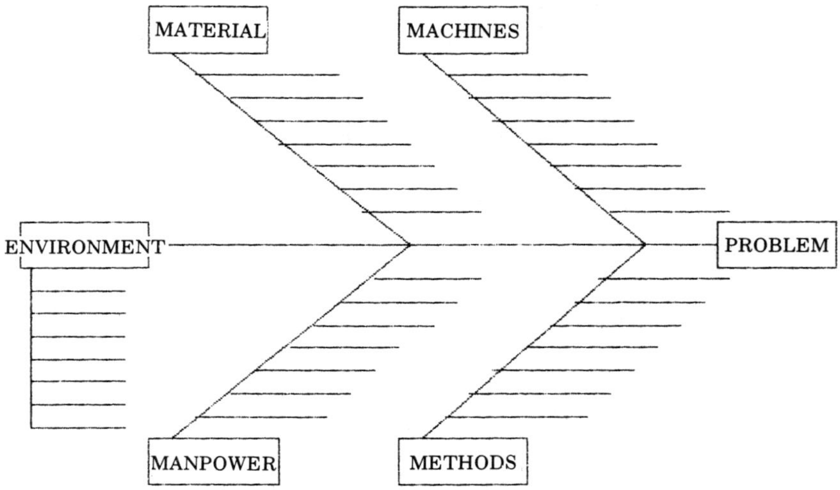

CHAPTER 16
Quality Evaluation of Processed Tomatoes and Tomato Products

If the product is labeled, record the following label information:

1. Name of the product
   (a) Style of pack
   (b) Type of product
2. Net weight
3. Vendor
4. Other information concerning the packing medium including salt, sugar, spices, etc.

The following container information should be recorded:

1. Code mark on container exactly as presented
2. External condition of container
3. Size of container

Next, the following steps should be carried out:

1. Determine and record the gross weight of the container and its contents.
2. Record the vacuum with the aid of a vacuum gauge.
3. Remove the lid by cutting it out from the top.
4. Where applicable, empty the contents on a screen. (Use 8-in. screen for 2½ size can or smaller and a 12-in. screen for cans larger than 2½ size.) Drain for 2 min and weigh the fruit on the screen without disturbing the contents thereof. Record the drained weight.
5. Wash the can or package, dry and determine the net weight by subtracting the weight of the empty container from the gross weight.
6. Note and record the appearance of the interior of the container. Look for corrosion, discoloration or breakage of the enamel, if an enamel-lined container is used.

7. Record the count or size of the commodity, where applicable.

8. Determine the score points for each factor (attribute) of quality in accordance with the applicable United States Department of Agriculture Standards for Grades for the particular commodity. Record the score points for each factor of quality.

9. Ascertain the Grade by totaling the scores for each factor of quality. Keep in mind the limiting rule ("stopper") for any particular factor of quality, where applicable.

TABLE 16.1 USDA ATTRIBUTES OF QUALITY FOR GRADES OF PROCESSED TOMATOES AND TOMATO PRODUCTS

	Score Points					
Attributes	Juice	Catsup	Sauce	Pulp	Paste	Tomatoes
Color	30	25	25	50	50	30
Absence of defects	15	25	25	50	50	30
Consistency	15	25	25	—	—	—
Flavor	40	25	25	—	—	—
Drained weight	—	—	—	—	—	20
Wholeness	—	—	—	—	—	20

[a]The specific score points for each of the attributes of quality and the interpretation of these attributes (factors) of quality is found in the Appendix for the specific U.S. Department of Agriculture Standards for Grades for each of the various canned products.

DETERMINATION OF THE STANDARD OF FILL OF CONTAINER

According to the Food, Drug and Cosmetic Act of 1938, the standard of fill of container prescribes the minimum quantity of contents which must be in the container. The container must be so filled as not to be misleading to the consumer, and be as full as commercially practicable without impairment of the quality of the food product. In order to avoid a charge of slack fill, the can shall be filled to not less than 90% of the capacity of the can, although this figure is official only for canned tomatoes. The measurement of the gross head-space of the can is a reliable check on the fill of container (Procedure A). Other specific standards of fill are based on one of the following standards of measurement: (Procedure B). The drained weight of the food product measured against the water capacity of the can by weight and (Procedure C) a percentage of the total capacity of the can.

Equipment for the determination of the standard of fill includes:

1. Headspace gauge,

2. 2-mesh screen (wire of the screen is of uniform diameter of 0.054 in., woven into square opening of 0.446 in., use 8-in. screen for containers 3 lb. or less and 12-in. screen for containers with contents of 3 lb or over), and

3. Grading scale (accurate to 1/16 oz).

Procedure A

The term "general method for water capacity of containers" means the following method:

1. In the case of a container with lid attached by double seam, cut out the lid without removing or altering the height of the double seam.
2. Wash, dry, and weigh the empty container.
3. Fill the container with distilled water at 68°F to $\frac{3}{16}$-in. vertical distance below the top level of the container, and weigh the container thus filled.
4. Subtract the weight found in (2) from the weight found in (3). The difference shall be considered to be the weight of water required to fill the container.

In the case of a container with lid attached otherwise than by double seam, remove the lid and proceed as directed in clauses (2) to (4) inclusive, except that under clause (3) fill the container to the level of the top thereof.

Procedure B

The term "general method for fill of containers" means the following method:

1. In the case of a container with lid attached by double seam, cut out the lid without removing or altering the height of the double seam.
2. Measure the vertical distance from the top level of the container to the top level of the food.
3. Remove the food from the container; wash, dry, and weigh the container.
4. Fill the container with water to $\frac{3}{16}$-in. vertical distance below the top level of the container. Record the temperature of the water, weigh the container thus filled, and determine the weight of the water by subtracting the weight of the container found in (3).
5. Maintaining the water at the temperature recorded in (4), draw off water from the container as filled in (4) to the level of the food found in (2), weigh the container with remaining water, and determine the weight of the remaining water by subtracting the weight of the container found in (3).
6. Divide the weight of water in (5) by the weight of water found in (4), and multiply by 100. The result shall be considered to be the percent of the total capacity of the container occupied by the food.

In the case of a container with lid attached otherwise than by double seam, remove the lid and proceed as directed in clauses (2) to (6) inclusive, except that under clause (4), fill the container to the level of the top thereof.

Procedure C: Percentage of the Total Capacity of the Can

1. If the container's lid is attached by a double seam, cut out the lid without removing or altering the height of the double seam.
2. Measure the vertical distance from the top level of the container to the top level of the food.
3. Remove the food from the container; wash, dry, and weigh the container.
4. Fill the container with water to $\frac{3}{16}$-in. vertical distance of the water, weigh the container thus filled, and determine the weight of the water by subtracting the weight of the container found in (3).
5. Maintaining the water at the temperature recorded in (4) draw off water from the container as filled in (4) to the level of the food found in (2), weigh the container with remaining water, and determine the weight of the remaining water by subtracting the weight of the container found in (3).
6. Divide the weight of water found in (5) by the weight of water found in (4), and multiply by 100. The result found is the percent of the total capacity of the container occupied by the food.

CHAPTER 17
Color and Color Measurement

Color is one of the most important quality factors associated with the evaluation of all food products. This is particularly true because the consumer notices color first, and this observation often provides preconceived ideas about other quality factors such as flavor or aroma. It is important, therefore, to make a favorable initial impression with a standard and familiar color that the consumer both wants and expects to see.

In the case of tomatoes and tomato products, the degree of color quality practically represents the measure of total quality. The tomato processor, being aware of these facts, must take the necessary action that ensures repetition of high color quality. Precise color measurement is, therefore, the first step in successful color control.

Research on the standardization of tomato products dates back to the work of Gaylord in the 1920s (Gaylord and Cleaver 1927, 1928; Gaylord and MacGillivray 1929). He and his co-workers were the first to demonstrate the importance of raw tomatoes in the production of a high quality, standardized tomato product. Gaylord also showed that yields could be increased, wastes could be reduced, and quality of the overall pack could be improved with the introduction of a grading system (Gaylord and Cleaver 1928). With such advantages, grading systems soon made the "flat-rate" purchase of raw tomatoes obsolete (Spangler 1956). In 1926, United States Standards for Canning Tomatoes were issued, and in 1933 Standards for Quality of Tomatoes for Manufacture of Stained Tomato Products were recommended and soon widely accepted (Francis and Clydesdale 1970). Today, almost all tomatoes for processing are purchased on a graded basis.

When applying grades to tomato products, color is without a doubt the most important single quality factor affecting grower–processor relationships and consumer acceptance. The color grades in use today for tomato juice, canned tomatoes, puree, paste, sauce, and catsup are a result of classic investigations conducted by MacGillivray (1931, 1937) and Marshall (1968) on disk colorimetry during the 1930s. Since other factors, such as flavor, consistency and absence of defects, can be regulated and controlled during processing, color becomes the most important single characteristic of quality. The consumer associates certain color characteristics with fresh and wholesome products. Thus, the extent to which original natural color is present in the processed product is an important criterion of a quality tomato item.

Successful color evaluation of tomatoes and tomato products is dependent to some extent on an understanding of the theory of color and to a greater extent on standardized conditions such as lighting, absence of glare, and the angle of observation.

FACTORS CONTRIBUTING TO TOMATO COLOR

Color in the tomato is a result of the presence of carotenes and carotenol. The carotenoid pigments derive their names from carotene and are, for the most part, polyene compounds of a yellow to red color. Many different types of pigments have been isolated in the tomato fruit, but of major importance to overall color are: (1) alpha-carotene, (2) beta-carotene, (3) gamma-carotene, (4) delta-carotene, (5) lycopene, and (6) 22 xanthophylls (carotenol). The chemical formula of the carotenes is $C_{40}H_{56}$ and of xanthophyll is $C_{40}H_{56}(OH)_2$. The most abundant carotenoid of the tomato is lycopene, which comprises approximately 83% of those pigments present. The chemical structures of lycopene and β-carotene can be shown by the following:

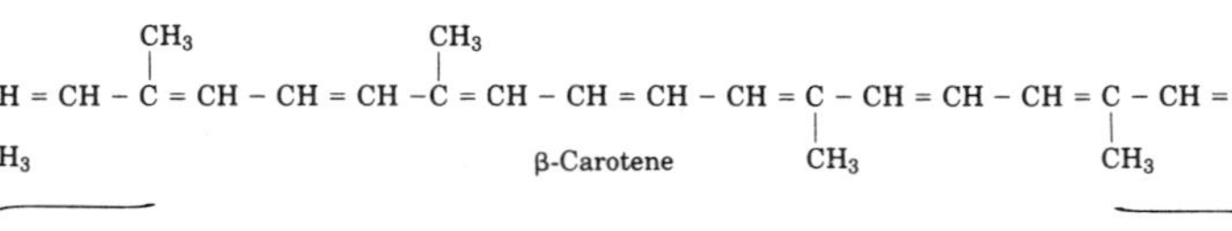

Lycopene

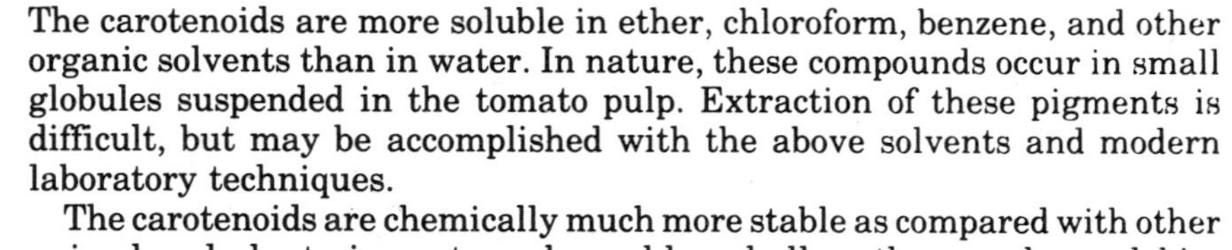

β-Carotene

The carotenoids are more soluble in ether, chloroform, benzene, and other organic solvents than in water. In nature, these compounds occur in small globules suspended in the tomato pulp. Extraction of these pigments is difficult, but may be accomplished with the above solvents and modern laboratory techniques.

The carotenoids are chemically much more stable as compared with other animal and plant pigments such as chlorophyll, anthocyan, hemoglobin, and myoglobin. However, the carotenoids may be partially destroyed under the condition of the low water percentage of tomato products, heating, the presence of metallic ion (Cu^{2+}, Fe^{3+}, etc.), or the presence of oxygen. Because tomato products discolor with reducing lycopene, lycopene destruction should be prevented.

McCallum, in his investigation of the distribution of carotenes and carotenoids in the tomato, found that the outer pericarp was highest in total carotenoids and the locular contents highest in carotene. These two regions of the fruit resulted, respectively, in the best and poorest color. Ellis and Hammer (1943) showed that there is a greater concentration of carotene in the stem than in the blossom end of the fruit, the transverse segments giving intermediate value. These polar variations were found to be associated with illumination and ripeness (McCallum 1955) but not, in most cases, statistically significant (Ellis and Hammer 1943).

Ellis and Hammer (1943) also observed that the carotene content is influenced by various factors. They attempted to correlate the carotene content of the fruit with factors such as size, degree of ripeness, mineral nutrient supply of the product, soil conditions, and variety. Of the several hundred ripe fruits analyzed for carotene:

1. Large fruits were only slightly richer in percentage carotene than small fruit,
2. Wide variations in the supply of the macronutrient elements to the tomato plants growing in some cultivars produced only slight variation in the carotene content of the fruit, even though the variations in the nutrient supply greatly affected growth and fruit fullness of plants,
3. Differences in carotene content were correlated with varietal differences,
4. Ripe fruits produced in the greenhouse, whether in summer or winter, were lower in carotene content than fruit produced outside during the summer, and
5. Fruits picked green and ripened in storage were very much lower in carotene than vine-ripened fruit (Ellis and Hammer 1943).

COLOR PERCEPTION

The human eye consists of focusing elements and sensitive elements. The focusing elements are the:

1. Cornea or white of the eye, which serves primarily to focus the image,
2. Pupil, which is the diaphragm that regulates the amount of light entering the eye,
3. Lens, which makes fine adjustment and brings the image into sharp focus, and
4. Vitreous humor, which maintains the roundness of the eyeball.

Rods and cones are the sensitive elements that transform the optical image into a nerve pattern. The rods are affected by small amounts of light and are responsible for the ability to see by moonlight and starlight. The cones are responsible for day vision and for perception of colors. Cones are distributed over the whole retina but are most concentrated in the center. All cones are directly connected to the brain.

The brain is the site at which the stimulus to the eye is received and interpreted. The eye and the brain are directly connected but no one has yet successfully explained the complete mechanism by which visual objects are perceived. Color perception of tomato products by the human eye has its limitations. Some of these limitations are the eye's inability to distinguish small color differences in a nonhomogeneous surface; the need for a suitable color standard for comparison; the quality of color in the tomato product being graded; and eye fatigue.

Although the human eye has some weaknesses when involved in color evaluation, subjective color determination can provide meaningful results. It is well established that the human eye is capable of distinguishing small color differences when it is able to compare two colors (Geisman *et al.* 1957). It follows then that a standard color chart should be made available for comparison if a uniform color grade is to be maintained by means of a subjective evaluation.

Light and Lighting

Whether the method of evaluation is subjective or objective, the standardization and uniformity of light quantity and quality is a requirement for

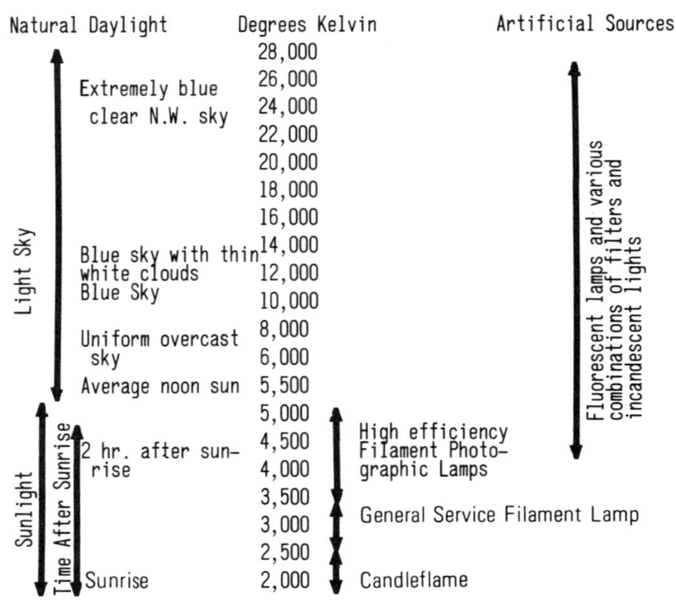

FIGURE 17.1. COLOR TEMPERATURE OF VARIOUS LIGHT SOURCES.

FIGURE 17.2
MACBETH MODEL BBX 324
EXECUTIVE LIGHT
FOR COLOR
GRADING.

proper color evaluation. Investigations on color evaluation show that during daylight hours, natural daylight varies between wide limits: from 12 to over 350 footcandles, from 30,000 to 40,000 degrees Kelvin color temperature, and from yellow through white to deep blue. These wide variations illustrate how very unstable daylight is as a source of light for color grading.

As mentioned in Chapter 6, an artificial lighting source such as the Macbeth Examolite is recommended in inside or outside areas for controlled lighting quality. Color evaluations are thus made at a color temperature equivalent to north sky natural daylight on a moderately overcast sky. The lighting produced from this source is uniform, and illumination is closer to daylight than that obtained with ordinary fluorescent daylight.

The energy distribution curve for a daylight fluorescent tube shows the presence of violet, blue, green, and yellow spectrum lines as well as low energy in red and blue. It is the deficiency in red and blue light (due to inability to produce an efficient phosphor for these parts of the color spectrum) and the presence of spectrum lines that tend to minimize or accentuate color differences depending on the color in question that normally rule out fluorescent daylight as a source of light for accurate color matching. The Examolite light-energy distribution curve has been improved over straight fluorescent tubes, in addition to some suppression of the effect of the spectrum lines.

The color of the surrounding area is also a very important factor in the lighting of a color-grading room. Neutral gray (reflectance of approximately 70%) of a flat finish (Mat 8) is satisfactory in reducing glare and providing a comfortable background free from conflicting color influence.

For more accurate duplication of natural daylight, which is required for color standards and critical color evaluation, the use of a Macbeth-filtered incandescent lamp installation is suggested.

SYSTEMS OF COLOR MEASUREMENT

The Agtron spectrophotometers and Hunter colorimeters described in Chapter 6 are instruments that measure color objectively. It should be understood at this point, however, that these instruments are influential in determining a color grade for tomatoes before processing, i.e., as delivered to the grading station by the grower. For this reason, they were included in Chapter 6, since each may be an integral part of the grading operation. The aim of this section is to discuss systems that may provide a basis for determining color grade or measurement of color quality after processing. As such, the following systems, methods, and instruments are most popular.

Since the publication of the first disk method in 1929 by Newhall for determining the color of agricultural products, the science of subjective color classification has made great progress. In 1931 at Cambridge, England, the International Commission on Illumination adopted certain standards for use in colorimetry which have since been put to practical use throughout the world. As a result, practically all data can be translated to ICI standards, which provide a common language for describing and identifying colors. The following systems are available for color matching and comparison.

Ridgway Charts

The Ridgway charts were first developed in 1886 and enlarged in 1912 by Robert Ridgway of the U.S. Biological Survey. The charts contain 1113 colors and 36 hues that are reduced by regular proportions of white, gray, and black to give systematic groupings.

Maerz and Paul Color Dictionary

The Maerz and Paul Color Dictionary was published in 1930 and contains 7056 colors on 56 charts. The order of hue presentation follows that of the spectrum; the charts are divided into seven main groups. In setting color standards for canned fruits and vegetables, the USDA has made use of these color charts for color-matching purposes, since the dictionary has one of the widest number of colors available. Some inherent disadvantages of using the system are that samples in some places on the charts are very close, so close indeed that they appear to be the same color. In other places there are wide differences between the colors of neighboring samples. Another disadvantage is that the charts are printed and cannot be accurately reproduced in future editions.

Munsell Color System and Charts

The Munsell Color system is based on the use of three visual color attributes, namely, hue, value (lightness), and chroma (saturation). The

Munsell Charts, in 40 hues, contain 982 colors, in comparison with 7056 colors in the Maerz and Paul Color Dictionary. The Munsell Charts, however, have the advantage of being devised on a psychological system of visually equal steps for scales of the three color attributes. Thus, the color chips are spread over the entire color solid, instead of being concentrated in certain areas. By this method, results of color measurements are expressed in terms of color order rather than color mixture, and allow an interpretation of results directly in terms of the visual qualities known in the Munsell System. Data obtained or characteristics of any given color by the Munsell color system can be translated or be given in figures of the ICI System, which is internationally recognized, or they can be translated into many other color systems and notations.

CIE or ICI System

The Commision Internationale de l'Eclavinge system is based on the effect any color may have on an agreed "standard observer." Three beams of fully saturated spectral colors isolated from the spectrum of white light (blue, green, and red) are considered to be the vertices of an equilateral triangle. It is assumed that at the corners of the triangle the amount of the particular color is 100% and that as the light progresses further away from the corner it becomes uniformally weaker, so that as it reaches any point on the opposite side its intensity is zero (Goose and Binsted 1964). Thus, the position of any point in the triangle can be defined mathematically by means of coordinates. This enables values to be quoted which may be converted into actual color (Goose and Binsted 1964).

The following methods and instruments are commonly used to provide a means of color grade determination or a way to measure color quality in tomatoes and tomato products.

Macbeth–Munsell Disk Colorimeter

The Macbeth–Munsell Disk Colorimeter is a method of subjectively measuring color that depends on a visual comparison. The unit was developed especially for the color grading of tomato products such as tomato juice, tomato pulp (puree), tomato catsup, tomato paste, chili sauce, and tomato sauce. The unit has been adopted by the USDA for the grading of tomato products.

For close control or evaluation of color of any sample, disks can be established for specific score points of different grades. Table 17.1 gives Munsell color percentages for scoring tomato juice.

Munsell Color Notation. The definite evaluation of the color of food products according to the Munsell System consists of two parts. The first is the percentage of the different specific colors which, when blended together, give a composite color that exactly matches the sample. The percentage notations of a particular color must always add up to 100%. The second

TABLE 17.1. MUNSELL COLOR PERCENTAGES FOR SCORING TOMATO JUICE

Score Points	Red (5R 2.6/13)	Yellow (2% YR 5/12)	Black[a]	Gray[a]
28[b]	73	16.34	5.33	5.33
26[c]	65	21	7.0	7.0
24.5[b]	59	24.5	8.25	8.25
23[c]	53	28	9.5	9.5

[a]Any combination of neutral black and gray is allowed. For "cold break" extracted and high temperature, short-time sterilized juice a greater percentage of gray may be necessary to match the sample; and for hot-break extracted and conventionally processed juice a greater percentage of black may be necessary to match the sample.
[b]Percentages determined by interpolation and research.
[c]Minimum standards established by USDA.

essential part of each color notation is the description of each of the color cards, i.e., hue, value, and chroma. Each class of agricultural product requires a particular set of color cards. For example, the color of canned tomatoes and tomato products may be matched by varying the proportions of particular colors, namely, definite shades of red, yellow, and combinations of neutral black and neutral gray. Those colors are designated by specific formulas in each of which there is an empirical designation of (1) hue, (2) value, and (3) chroma. The formula of the red color used in matching tomato products is 5R 2.6/13. The 5R indicates the hue, which is the attribute of color that permits colors to be classed as reddish, yellowish, greenish, or bluish. Five principal hues are used in the Munsell system: red, yellow, green, blue, and purple, designated R, Y, G, B, and P. Midway between these are five similar intermediate hues: yellow-red, green-yellow, etc., designated YR, GY, etc. The designation 5R is a pure red free from purple or yellow. The second notation, 2.6, indicates the value or intensity of the black constituent of the color. This is expressed in an arbitrary scale from 0 to 10, where zero is absolute black and 10 is absolute white. A value of 2.6 indicates a considerable proportion of black constituent. The final designation shows the chroma, which expresses the strength or degree of departure of a particular hue from a neutral gray of the same value. The scales of chroma extend from 0 for a neutral gray to 10, 12, 14, or farther, depending on the strength or saturation of the individual color.

TABLE 17.2. TOMATO COLOR DISK SPECIFICATION

Color Notation	Disk No. 1	Disk No. 2	Disk No. 3
	U.S. Grade A (U.S. Fancy Color)		
Red (5r 2.6/13	65%	65%	65%
Yellow (2.5YR 5/12)	21%	21%	21%
Black (N1)	7%	14%	0%
Gray (N4)	7%	0%	14%
	U.S. Grade C (U.S. Standard Color)		
Red (5R 2.6/13)	53%	53%	53%
Yellow (2.5YR 5/12)	28%	28%	28%
Black (N1)	9.5%	19%	0%
Gray (N4)	9.5%	0%	19%

The required color cards, which can be obtained only from the Munsell Color Co., Newburgh, NY, are cut in the form of Maxwell Disks, which are uniform circles of each of the cards, with a small hole at the center and a single radial slit from the center to the edge so that the disks can be placed upon one another and slipped together by means of the radial slit. This permits any desired proportion of one or more disks to be exposed as a segment of the single circle that is visible. When these disks are held together with a suitable binding post at the center and spun at a speed great enough to eliminate flicker, the color seen by the eye is the sum of the different segments exposed. The disks should be slipped together in such a direction that the spinning will premit them to lie flat rather than to fly apart. The amount of each color card exposed is changed until the combined effect exactly matches the color of the sample. The amount of each color exposed is measured by the percent of the circumference occupied by each segment.

The requirement of color for Grade A tomato pulp, according to the USDA, is that it shall be "equal or better than that produced by spinning a combination of the following Munsell disks: R65% (5R 2.6/13) glossy finish, Y21% (2, 5YR 5/12) finish, glossy; black and gray 14% total or any combination of the two (Black N1 glossy finish and/or Gray N4 mat finish). For minimum Grade C, 53% Red, 28% Yellow, and 19% of the Black and Gray in any combination are used according to the USDA and the FDA."

Similar disks could be established for other tomato products enabling the packer to score samples within a specific range, if desired, in addition to conforming to the minimum USDA standards. The Agtron Corp. of Sparks, NV have developed plastic disks conforming to the USDA Munsell grade specification for use with the colorimeter. These offer the advantage of resistance to wear and moisture.

Description of Macbeth–Munsell Colorimeter. The Macbeth–Munsell Colorimeter consists of an arrangement of two spinning disks mounted directly beneath a color-corrected light source with controlled viewing conditions. The light source is composed of two R40 300-watt reflector flood lamps used with two 7¼-in. Macbeth daylight filters. The light source and filter combination is mounted on a deck in the upper portion of the unit and is enclosed with a special nonselective diffusing glass. The light source-filter combination produces the closest duplication of north sky daylight (7500° Kelvin) that is commercially available. The filters used are carefully graded and selected for identification. All Type No. 1 disk colorimeters have a temperature of 7500° Kelvin unless otherwise specified.

The center section of the unit, enclosed by two hinged doors, constitutes the viewing mask support (horizontal bar), attached to a positioning guide for a constant angle of viewing. The interior of the viewing area is painted a light neutral gray (Munsell N8) in order to standardize the surrounding conditions when color judgments are made. The sample holder consists of two parts, which include a tray and a 3½-in. diameter by ⅞-in. deep sample

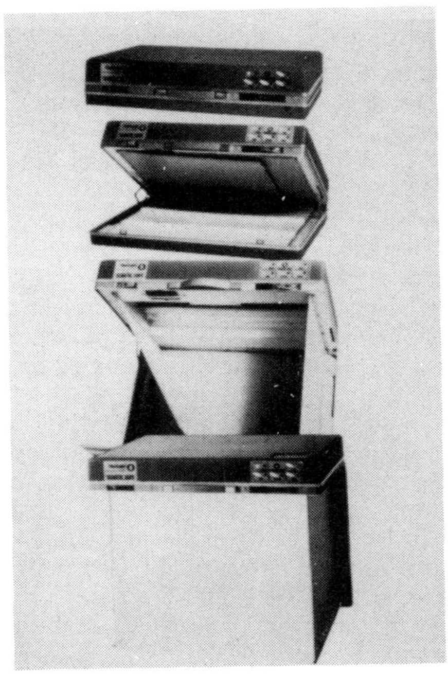

FIGURE 17.3. MACBETH "THE JUDGE" PORTABLE BRIEF CASE SIZE COLOR MATCHING BOOTH.
Unfolds from a brief case to a light booth in several easy steps.

cup. The cup fits against a guide on the tray, which in turn, slides onto the viewing surface by means of ways or tracks. The sample holder is constructed in this manner so that samples that spill over from the cup will be caught by the tray, which is easily removed for cleaning. On both sides of the sample holder are mounted the spinning disk motors. The standard size of the Munsell standard color disk is 3¾-in. in diameter. The disks are punched in the center with a ¼-in. hole, which fits over the spindles of the spinning disk motors. The color disks are held in place on the spindle by a knob that is a friction fit for the spindle. The plastic disks are drilled and mounted similarly.

Operation of the Colorimeter. Suggested rules for operation of the colorimeter for grading tomato products are as follows.

1. Place the colorimeter in a position so that the operator does not face into a strongly contrasting light but not necessarily in a darkened room.
2. Fill the sample cup level with the tomato product and place between the two spindles of the colorimeter.
3. By spinning Grade A or Grade C disks on each side of the sample,

determine which of the three combination of black and gray most nearly looks like the sample. Use disks with this combination for both A and C for the grading comparison.

4. After determining the proper black and gray combination for the products place the Grade A disk on the left spindle and the Grade C disk on the right spindle.

5. Turn the lamp and spinning motors on and adjust the viewing mask at the 45° position and in such a manner that when the eye is about 18 in. from the mask, a section of the sample will be seen through one aperture of the mask and a section of the spinning disk will be seen beside it in the adjacent aperture in the mask. Make all grading comparisons in this manner with the optional use of a hand mask. The hand mask is used to provide what is known as "aperture colors." This means that with the use of such a mask the outside factors which distract color judgments are removed.

6. If there is a perfect color match, assign score points corresponding to the color disk matched, in accordance with the standard. Other score points are assigned by interpolation using the two disks as reference points, or disks can be made up for score points.

7. Do not adjust scores assigned, even if they appear to be wrong when the sample is viewed in natural daylight. The color temperature and light-energy distribution of the artificial light source in the colorimeter remains constant, while that of natural daylight varies over a wide range, depending on the time of day, time of year, latitude, atmospheric conditions, surrounding contrast, and other factors. Extensive experimentation has proved that grading under these conditions with a constant light source of the proper color and intensity is comparable to that under the most ideal natural daylight conditions available, and is much more uniform than when care is not used to wait for the natural daylight conditions.

Hunterlab Color and Color Difference Meter

The Hunterlab Color and Color Difference Meter is a tristimulus colorimeter that measures color on three scales by the use of three filters that approximate the X, Y, and Z functions of the CIE system. Three values are obtained for each color measured: Rd (45 degrees–0 degree luminous reflectance) or L (visual lightness on a scale of 0–100; 0, perfect black; 100, perfect white) depending on the type of measuring circuit selected; a, on a scale whereby plus is red, zero is gray, and minus is green; and b, on a scale whereby plus is yellow, zero is gray, and minus is blue. The unit of color measurement for these three scales is the National Bureau of Standards unit of color difference devised by Judd. Hunter values may be converted to CIE values by the following equations:

(1) $Rd = 100Y$

(2) $a = 175fr(1.02 - Y)$

(3) $b = 70fr(Y - 0.8477)$

(4) $fr = 0.51\ (21 + 20Y)/(1 + 20Y)$

or if using the L measuring circuit:

(1) $L = 100Y^{1/2}$
(2) $a_L = 175(1.02X - Y)/Y^{1/2}$
(3) $b_L = 70(Y - 0.847Z)/Y^{1/2}$

In using the Hunter instrument, color standards are necessary for accurate color measurements, as is the case in tristimulus colorimetry. The standards are tile plates which are manufactured in a number of different colors and have assigned to them standard values for L, a, and b. The standard selected for color measurement of any given substance should approximate as closely as possible that particular substance. This is commonly known as a "hitching post" standard for color measurement.

In operation, the Hunter instrument is standardized according to known values as assigned to each standard. The color of the sample is determined by reading the three tristimulus values from the dials. The readings may be plotted in Munsell Chromaticity Charts as shown in Fig. 15.6. Thus, hue or chroma values can be determined from Hunter a and b values for any given L value. More recent models incorporate digital automatic readouts and eliminate the necessity for standardizing the instrument. Further, conversions to other color systems are part of the newer models.

It is essential that the sample to be measured be a homogeneous mass and free from air bubbles. The instrument should be located where there is

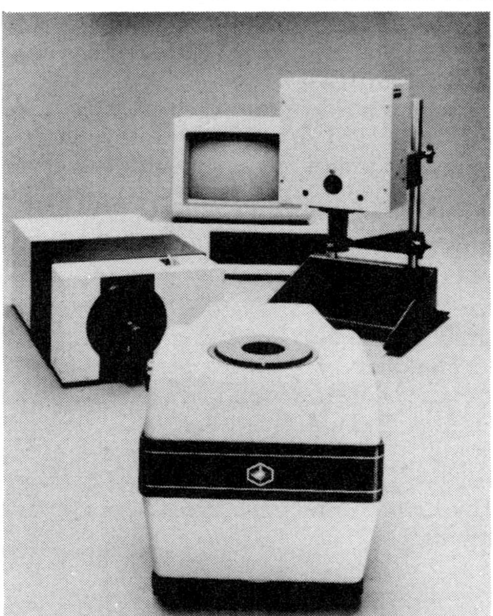

FIGURE 17.4. HUNTERLAB'S COLOR EVALUATING SYSTEM.

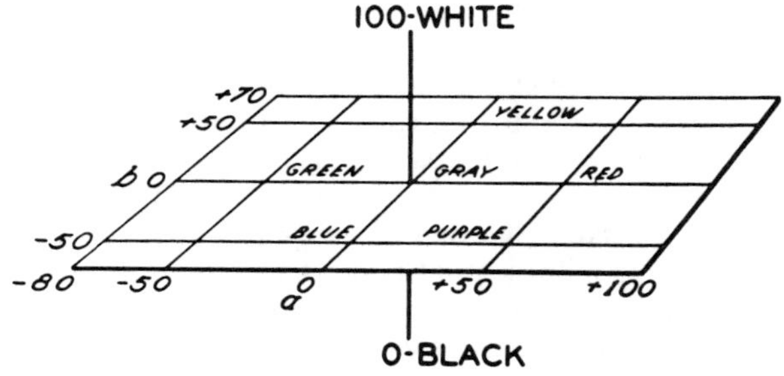

FIGURE 17.5. DIAGRAM SHOWING DIMENSIONS OF THE HUNTER L, a, b, COLOR SOLID.

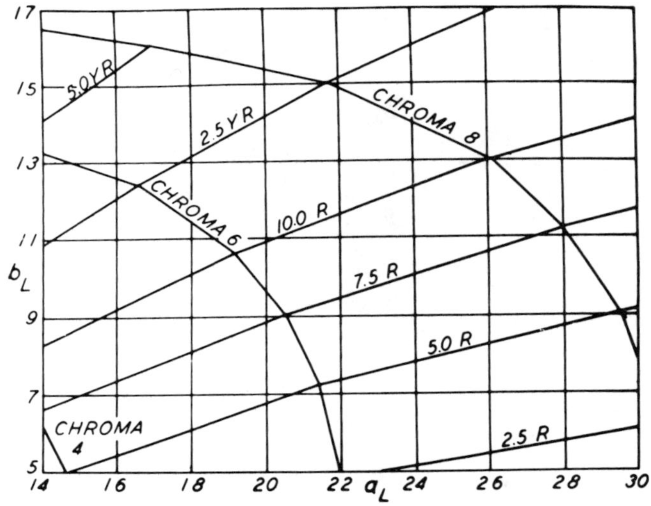

FIGURE 17.6. MUNSELL HUE AND CHROMA COORDINATES FOR VALUE 3, IN TERMS OF HUNTER a_L b_L.

medium or subdued illumination, no drafts, and relatively dry air of constant temperature. The procedure for operation varies with each particular model but the basic principle is either manual or automatic measurement of the three values.

Figure 17.7. (Hunter +a vs. +b values for tomato juice) illustrates how the instrument might be used to control color quality. The circled numbers show the correlation between subjective and objective color measurement of tomato juice. Samples of juice were subjectively evaluated and scored in conjunction with the USDA Quality Grade Standards (Grade A: 26–30 points; Grade C: 23–25 points; and Grade D, 0–22 points). The color of the samples was then objectively measured by means of the Hunter Color and Color Difference Meter and the respective $+a$ and $+b$ values plotted. As such, the instrument could, in this instance, be used to evaluate the efficiency of any number of persons making a subjective evaluation of tomato juice.

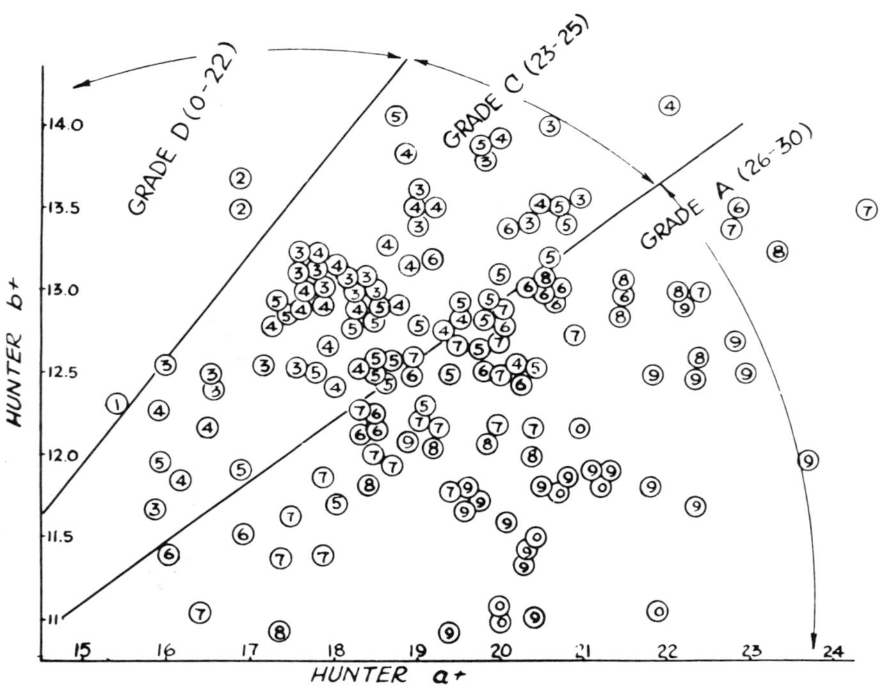

FIGURE 17.7. ACTUAL TOMATO JUICE SCORE ON USDA 20-30 COLOR BASIS (FIRST DIGIT ELIMINATED ON CHART) AS RELATED TO HUNTER a AND b COORDINATES WITH GRADES A AND C LINES SUPERIMPOSED.

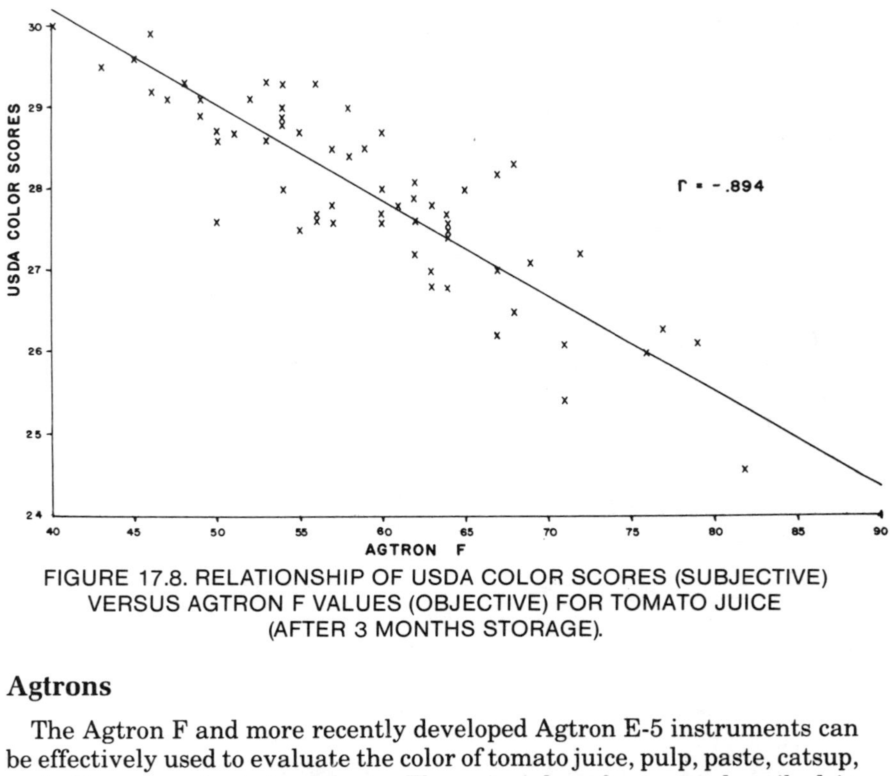

FIGURE 17.8. RELATIONSHIP OF USDA COLOR SCORES (SUBJECTIVE) VERSUS AGTRON F VALUES (OBJECTIVE) FOR TOMATO JUICE (AFTER 3 MONTHS STORAGE).

Agtrons

The Agtron F and more recently developed Agtron E-5 instruments can be effectively used to evaluate the color of tomato juice, pulp, paste, catsup, and other comminuted products. The principles of use are described in Chapter 17. Figure 17.8 illustrates the relationship of Agtron F versus USDA color scores for the USDA Standards of Grade for Tomato Juice. A correlation of 0.894 is highly significant, and this quick objective method is most worthy of substitution for the subjective color evaluation of tomato products color.

REFERENCES

EASTMOND, J.E., PETERSON, J.E. and STUMPF, R.R. 1951. Observation of color changes in some processed and stored foods. Food Technol. *5* (3) 121–128.

ELLIS, G.H. and HAMMER, K.C. 1943. The carotene content of tomatoes as influenced by various factors. J. Nutr. *25*, 539–553.

EVANS, R.M. 1948. An Introduction to Color. John Wiley & Sons, New York.

FRANCIS, F.J. and CLYDESDALE, F.M. 1970. Color measurement of foods: XVII. Tomatoes and tomato products. Food Prod. Dev. *4* (1) 88–102.

GAYLORD, F.C. and CLEAVER, H.M. 1927. Grading tomatoes for quality. Purdue Univ. Exp. Stn. Bull. *317*.

GAYLORD, F.C. and CLEAVER, H.M. 1928. Buying tomatoes on grade. Purdue Univ. Argic. Exp. Stn. Bull. *328*.

GAYLORD, F.C. and MacGILLIVRAY, J.H. 1929. Buying tomatoes on grade. Purdue Univ. Agric. Stn. Bull. *336*.

GEISMAN, J.R. *et al.* 1957. Color evaluation of food and agricultural products. Unpublished. Dep. Hortic., Ohio State Univ., Columbus.

GOOSE, P.G. and BINSTED, R. 1964. Tomato Paste, Puree, Juice, and Powder. Food Trade Press, London.

GOULD, W.A. 1952. Artificial light for visual color evaluation of fruits, vegetables. Food Packer *33* (11) 33–35.

GOULD, W.A. 1953. Here's where we stand on color grading of fruits and vegetables. Food Packer *34* (2) 42, 44, 96, 99.

GOULD, W.A. 1977. Food Quality Assurance. AVI Publishing Co., Westport, CT.

GRAVES, M. 1952. Color Fundamentals. McGraw-Hill Book Co., New York.

JPN. NATL. FOOD RES. INST. 1974. Tomato processing technology. Ser. Promoting Spread of Food Technol. *9*. Japan National Food Research Institute, Tokyo.

JUDD, D.B. 1952. Color in Business, Science, and Industry. John Wiley & Sons, New York.

KASKEY, G.T. 1971. Agtron Instruments in Color Quality Control. Magnuson Engineers., San Jose, CA. Apr. 21.

LA COSTE, R. 1952. Processors to Control Color of Tomato Juice. Canner/Packer (Feb.).

MacGILLIVRAY, J.H. 1928. Color of different regions of a tomato and a method for color determination. Proc. Am. Soc. Hortic. Sci. *25*, 37–40.

MacGILLIVRAY, J.H. 1931. Tomato color as related to quality in the tomato canning industry. Purdue Agric. Exp. Stn. Bull. *350*.

MacGILLIVRAY, J.H. 1937. Spectrophotometric and colorimetric analyses of tomato pulp. Proc. Am. Soc. Hortic. Sci. *35*, 630.

MARSHALL, M. 1968. LIH. Color technology. Chem. Eng. *75* (17) 146–156.

McCALLUM, J.P. 1955. Distribution of carotenoids in the tomato. Food Res. *20* (1) 55–59.

MURRAY, H.D. *et al.* 1952. Colour in Theory and Practice. Chapman and Hall, London.

SPANGLER, R.L. 1956. Standardization and inspection of fresh fruits and vegetables. U.S. Dep. Agric. Agric. Misc. Publ. *604*.

WHIPPLE, S.R. 1952. Grading tomatoes for color. Canner/Packer *114* (9) 14–20.

CHAPTER 18
Tomato Solids

In recent years much emphasis, particularly by the products and concentrate processors, has been given to the use of tomato solids or portions thereof to predict finished product yields, consistency, and overall quality. The data in Table 18.1 shows how the type and range of solids is found in the tomato. This variation is due to variety/cultivar, maturity of the tomato at harvest, areas of production, and/or cultural conditions and practices during production such as, irrigation and/or water stress.

Workers in California state that there is an inverse relationship between soluble solids content and yields. Varieties with high yields tend to have lower soluble solids content. Conversely varieties with high solids content tend to have lower yields. Garvey and Hewitt have uncovered a small wild tomato native only to the remote Galapagos Islands off the coast of Ecuador that has very high soluble solids. This small orange colored fruit belongs to the species *Lycopersicon cheesmanni* and has soluble solids which range between 12-14% (degrees Brix). In extensive gene mapping and breeding work they have developed tomatoes with solids over 9.0% (degree Brix). Lines have now been released and some authorities believe that tomatoes with solids as high as 15% (degree Brix) are possible or on the horizon.

COMPOSITION OF THE TOMATO

It should be pointed out that total solids and water make up the composition of the tomato. Of course portions of the water content are removed in processing with the total solids, thus, being more concentrated. Many concentrated products are manufactured with solids in excess of 40% (40 °Brix). These same solids are diluted back to given concentrations depending on the types of finished products being manufactured.

TOTAL SOLIDS

Due to the present time consuming methods of measurement of the total solids content, other components of the total solids are measured, such as the degree Brix. The total solids is the better method of indicating tomato quality, but the technique is not fast enough for production line operations.

FIGURE 18.1. VACUUM OVEN EQUIPMENT FOR TOTAL SOLIDS MEASUREMENT.

The official method is to dry a sample of the tomato in a flat bottom dish in *vacuo* at 70 ℃ for 2 hours, or by drying at atmospheric pressure at 100 ℃. for 6 hours, or by drying in a microwave to constant weight. All of these methods are too time consuming and a direct reading of the tomato juices is the most common method of predicting the solids content using a refractometer.

DEGREE BRIX/SOLUBLE SOLIDS

The refractometer reading is not a true sugar reading but is the measure of soluble solids in the tomato. Some processors use this Brix value and multiply this by the tons from a given acre and arrive at an estimate of tons of tomato solids from a given field. This has been used as a method of predicting yield per acre when making concentrated products.

The refractometer measures the refractive index. This is a measure of the speed of light passing through a substance compared to the speed of light passing thru air. Water has a refractive index of 1.330 which means that light travels through water 1.3 times slower than through air. To say that tomato juice has an index of 1.3446 means that light is slowed by a factor of 1.3446 when light passes from air to tomato juice. Thus there is a direct relationship between the soluble solids content of the tomato and the refractive index of a solution. By referring to Table 18.2 one can see the relationship of the refractive index to sucrose solutions or degree of Brix.

Some refractometers read in both Refractive Index and degree Brix. Many processors and buyers use the degree Brix to indicate the soluble solids content of the tomato product. For Ketchup, 1% should be added to the Brix reading to obtain the true Total Solids content of the product. Since the refractive index is a measure of the rate of passage of light through a substance, it is affected by temperature. If a material expands when heated it will become less dense. Light will then be slowed down less and the refractive index will decrease. For a sugar solution, the change amounts to 0.5% sugar for every 5.6 ℃; that is, if the refractive index of a solution measures 20% at 20 ℃. the same sample would measure only 19.5% when heated up to 26 ℃. Thus, the temperature of the sample must be known exactly and the results interpreted accordingly.

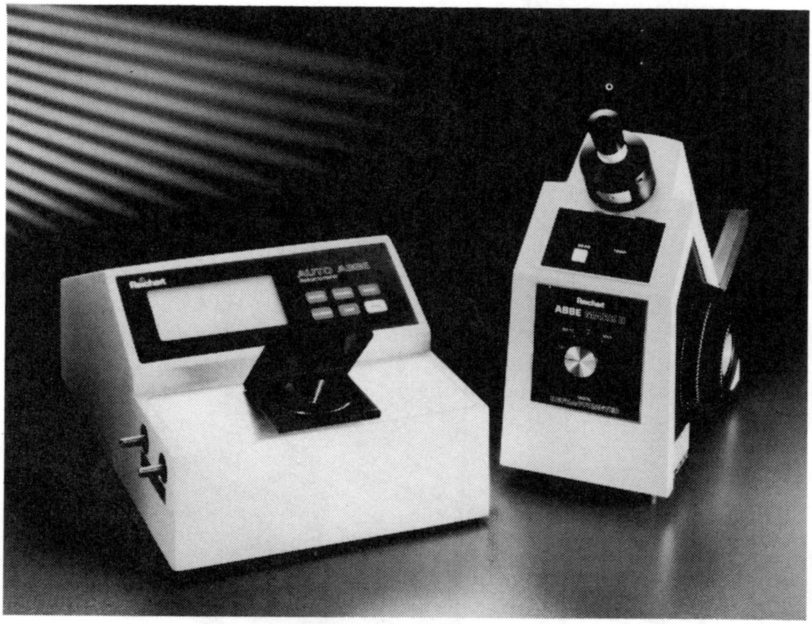

FIGURE 18.2. BENCH MODEL REFRACTOMETER WITH FITTINGS FOR TEMPERATURE CONTROL.
Courtesy of Leica Inc.

Suggested procedure when using the Abbe' Refractometer:

1. After standardizing the refractometer with distilled water (refractive index of 1.330), place a few drops of the serum on the carefully cleaned dried fixed prism of the refractometer and slowly bring the two prisms together and clamp them shut.
2. Turn on the in-line switch and adjust the light arm for proper illumination.
3. Bring the borderline into view with the coarse adjustment. (The color of the borderline may have to be compensated by adjusting the compensation dial so that the borderline is faintly red on each side.)
4. Observe the crosshairs sharply by focusing the eyepiece if necessary and bringing the borderline up to the intersection by means of the coarse or fine hand control.
5. Read the index by depressing the momentary contact switch to the fourth decimal place.
6. Open the prisms and clean very carefully with distilled water and a piece of lens paper. Do not scratch the prisms. When not in use, keep lens paper between the prisms and keep them closed.

When working with tomato materials with a high concentration of insoluble solids, such as pulp and paste, it is advisable to filter out the insoluble solids from the serum (soluble solids). Removal of the insoluble solids does not affect the reading as the refractive index is only going to measure the soluble solids. If the product is difficult to filter, use Klerzyme (Walenstein Co.) or Pectinol R (Rohm and Haas Co.) as a filter aid. Either of these materials can be mixed with the tomato materials and filtering is made much easier. Use only a 1% of the filter aid and correct the refractive index reading accordingly.

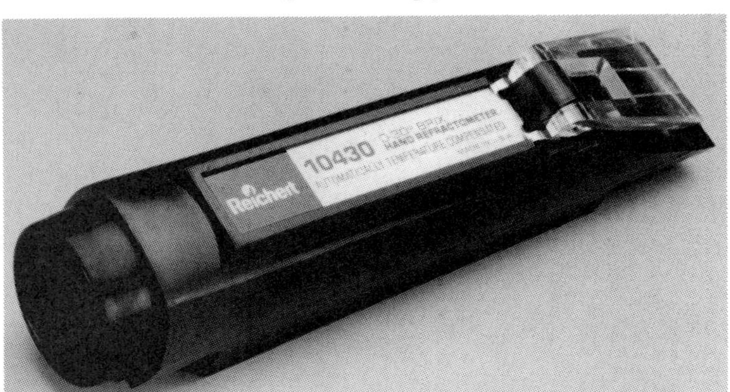

FIGURE 18.3. HAND REFRACTOMETER.
Courtesy of Leica Inc.

WATER INSOLUBLE SOLIDS

Another portion of the total solids content of the tomato that is as important as the Brix value, is the water insoluble solids content. This measurement is the difference between the Total Solids content and the Brix value. This water insoluble portion contains the protopectin substances. This value has been used to calculate the Total Solids/Water Insoluble Solids ratio and has been used to predict consistency and or Bostwick values of the concentrated products. Most workers today do not use this value as it does not tell the whole story.

ALCOHOL INSOLUBLE SOLIDS

Another aspect of the tomato total solids is the often overlooked alcohol insoluble solids content. This fraction contains the pectic substance and

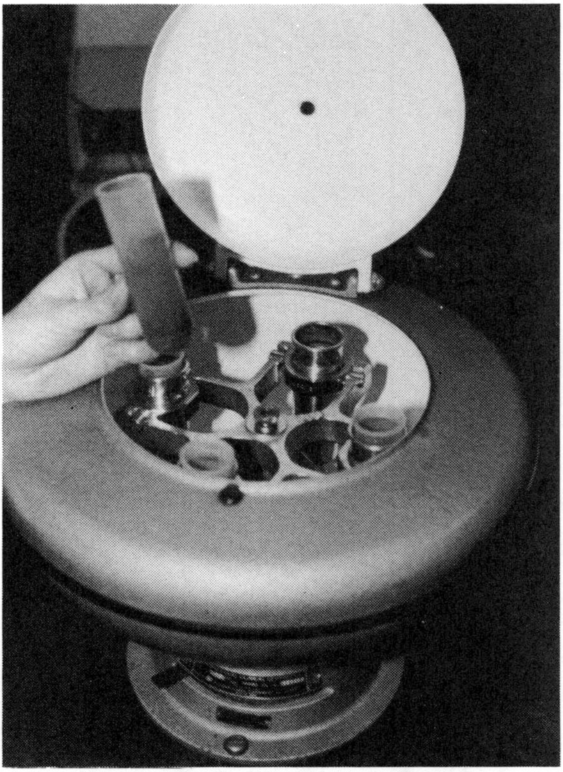

FIGURE 18.4. CENTRIFUGE FOR DETERMINING PRECIPITATE TOMATO SOLIDS.

other higher polysaccharides (cellulose, araban-galactan mixture and the xylans). Some people feel that this fraction has a lot to do with the consistency of the tomato products.

BLOTTER TEST

Faster quicker methods are desired and some workers have come up with novel techniques that are quite helpful, Nelson and co-workers proposed a serum separation evaluation procedure called the "blotter test". Several workers have established possible standards when using this technique, but the real value of the test is to find out if the serum separates from the solids content. Ideally high quality tomato products should not "weep" or separate on standing. The theory being that when one uses ketchup or other tomato solids they should not separate. For example, when one pours Ketchup on a bun, does the bun become soaked or wet from the ketchup? Good ketchup will not "wet" the bun.

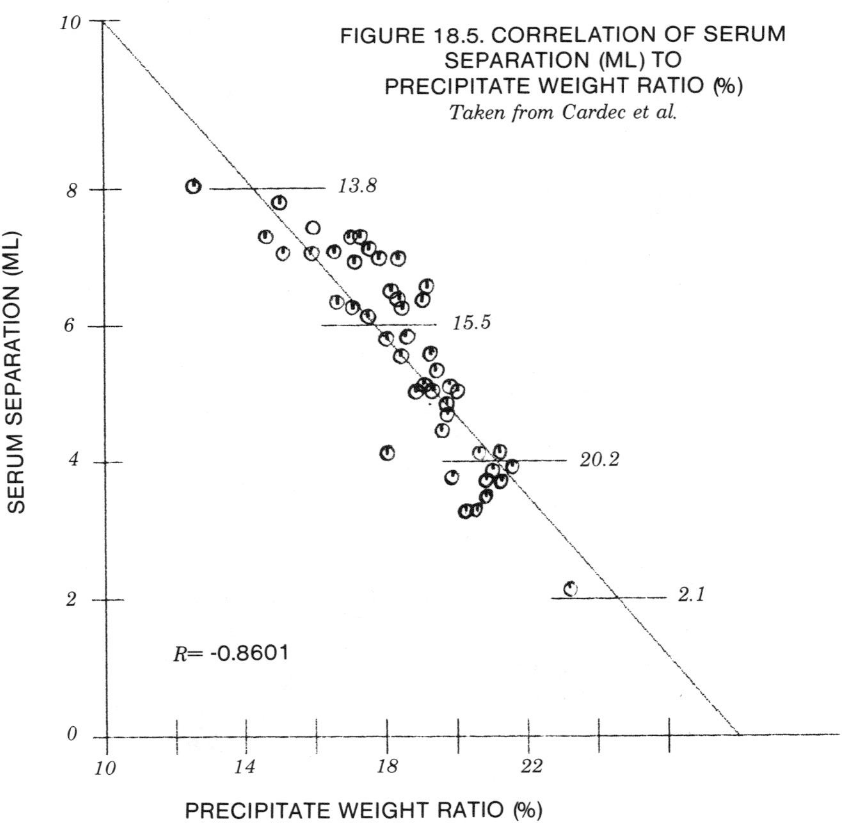

FIGURE 18.5. CORRELATION OF SERUM SEPARATION (ML) TO PRECIPITATE WEIGHT RATIO (%)
Taken from Cardec et al.

PRECIPITATE WEIGHT RATIO

Another technique is to determine the precipitate weight ratio as reported by Caradec et al. In this test one weighs approximately 40 grams of a tomato sample into a 50 ml pre-weighed centrifuge tube and centrifuges the sample at 12,800 g for 30 minutes at 4 ℃. The supernatant is removed by inverting for 3 minutes and the precipitate weight is calculated.

SERUM SEPARATION

Another technique is to measure the serum separation. Caradec suggests the construction of a cone from 42 mesh stainless steel on a 60° angle and approximately 10 cm in length. 15 ml of tomato juice is placed in the cone and allowed to drain for 5 minutes. The amount of serum separation in ml is recorded and used as a measure of consistency.

FIGURE 18.6
RELATIONSHIP OF SPECIFIC GRAVITY TO BRIX

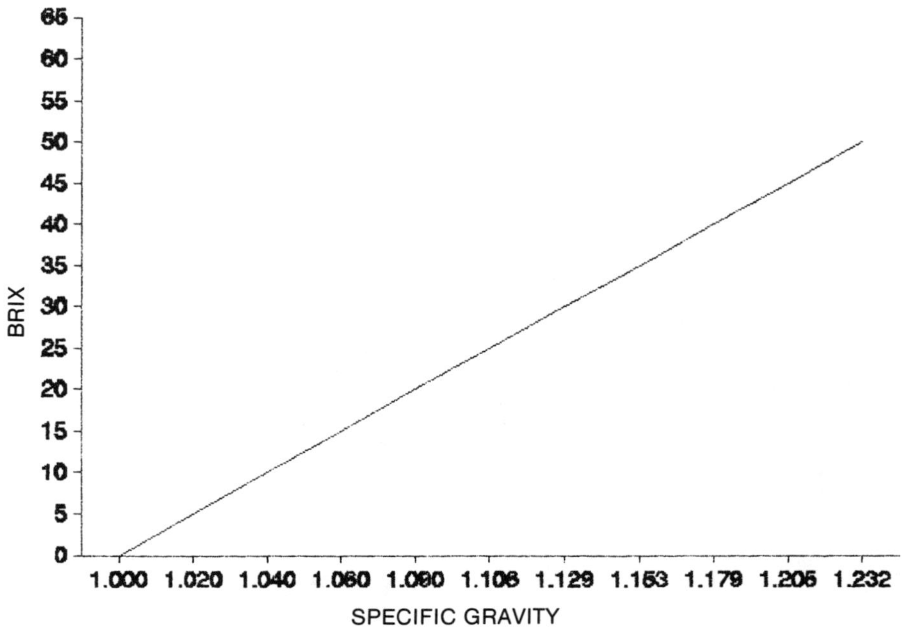

TABLE 18.1 — RANGE OF WATER AND SOLIDS IN THE TOMATO

TOMATO FRACTION	CONTENT IN %
Water content	91.9 to 95.5
Total solids	4.5 to 8.1
Soluble solids	4.0 to 7.0
Water insoluble solids	0.55 to 1.60
Alcohol insoluble solids	0.50 to 1.50

SPECIFIC GRAVITY (AOAC 32.026)

Determine specific gravity at 20/20 ℃ using Gay-Lussac or similar small neck bottle without cap.

1. Clean and calibrate the bottle at 20°C. using water to fill bottle to level full. Wipe bottle dry and weight immediately.

2. Cool the sample to 16 to 18 ℃. and fill calibrated bottle with the tomato pulp and centrifuge for 1 minute at approximately 1000 rpm. Add enough pulp to fill bottle to the top and centrifuge again. Remove bottle from centrifuge and take the pulp temperature inserting the thermometer into the pulp so that air is not introduced. When the temperature is just at 20°C., remove the thermometer and add enough pulp at the same temperature to have the bottle slightly overfilled. Strike off even with a straight edge and clean outside of bottle and weigh at once to nearest 0.01 gram.

Calculate the specific gravity as follows:
Sp. Gr. = Weight of pulp in bottle/weight of water that bottle held.

REFERENCES

CARADEC, P.L. and NELSON, P.E. 1985. Effect of Temperature on the Serum Viscosity of Tomato Juice. J. of Food Sci. 50: 1497-1498.

CARADEC, P.L., P.E. NELSON and N. TAKADA. 1985. Tomato Products: A New Serum Separation Measurement. J. of Food Sc. 50: 1493-1494.

GARVEY, T. CASEY and J.D. HEWITT. 1988. Increasing Soluble Solids in Processing Tomatoes by Using Genes from a Wild Tomato Relative. California Processing Tomatoes News and Views (11) No. 4: 1-2.

GETCHELL, R.N. and SCHLIMME, D.V. 1985. Particle Size of Water Insoluble Tomato Solids Measured by Laser Instrumentation. J. of Food Sc. 50: 1495-1496.

KALLAS, N.N. 1981. The Effect of Tomato Processing Extraction and Size Reduction Operations on the Mold and Rot Fragment Counts. Ph.D. Dissertation. The Ohio State University.

MARSH, GEORGE L. 1986. Tomato Solids: Their Meaning and Use. California Processing Tomatoes News and Views Vol. 9 (2): 2.

TABLE 18.2 — RELATIONSHIP OF REFRACTIVE INDEX TO SUCROSE (DEGREE BRIX) AT 20 °C.
(Taken from International Scale of Refractive Indices of Sucrose)

Refractive Index	% Sucrose	Refractive Index	% Sucrose	Refractive Index	% Sucrose
1.3385	3.8	1.3660	21.3	1.3935	36.8
1.3390	4.1	1.3665	21.6	1.3940	37.0
1.3395	4.5	1.3670	21.9	1.3945	37.3
1.3400	4.8	1.3675	22.2	1.3950	37.6
1.3405	5.2	1.3680	22.5	1.3955	37.8
1.3410	5.5	1.3685	22.8	1.3960	38.1
1.3415	5.8	1.3690	23.1	1.3965	38.3
1.3420	6.2	1.3695	23.4	1.3970	38.6
1.3425	6.5	1.3700	23.7	1.3875	38.8
1.3430	6.8	1.3705	23.9	1.3980	39.1
1.3435	7.2	1.3710	24.2	1.3985	39.3
1.3440	7.5	1.3715	24.5	1.3990	39.6
1.3445	7.8	1.3720	24.8	1.3995	39.9
1.3450	8.2	1.3725	25.1	1.4000	40.1
1.3455	8.5	1.3730	25.4	1.4005	40.4
1.3460	8.8	1.3735	25.7	1.4010	40.7
1.3465	9.1	1.3740	26.0	1.4015	40.9
1.3470	9.5	1.3745	26.3	1.4020	41.2
1.3475	9.8	1.3750	26.6	1.4025	41.4
1.3480	10.1	1.3755	26.9	1.4030	41.7
1.3485	10.4	1.3760	27.1	1.4035	41.9
1.3490	10.8	1.3765	27.4	1.4040	42.2
1.3495	11.1	1.3770	27.7	1.4045	42.4
1.3500	11.4	1.3775	28.0	1.4050	42.7
1.3505	11.7	1.3780	28.3	1.4055	42.9
1.3510	12.0	1.3785	28.6	1.4060	43.2
1.3515	12.4	1.3790	28.8	1.4065	43.4
1.3520	12.7	1.3795	29.1	1.4070	43.7
1.3525	13.0	1.3800	29.4	1.4075	43.9
1.3530	13.3	1.3805	29.7	1.4080	44.2
1.3535	13.6	1.3810	30.0	1.4085	44.4
1.3540	13.9	1.3815	30.2	1.4090	44.7
1.3545	14.3	1.3820	30.5	1.4095	44.9
1.3550	14.6	1.3825	30.8	1.4100	45.2
1.3555	14.9	1.3830	31.1	1.4105	45.4
1.3560	15.2	1.3835	31.3	1.4110	45.6
1.3565	15.5	1.3840	31.6	1.4115	45.9
1.3570	15.8	1.3845	31.9	1.4120	46.1
1.3575	16.1	1.3850	32.2	1.4125	46.4
1.3580	16.4	1.3855	32.4	1.4100	46.6
1.3585	16.7	1.3860	32.7	1.4135	46.9
1.3590	17.0	1.3865	33.0	1.4140	47.1
1.3595	17.4	1.3870	33.3	1.4145	47.4
1.3600	17.7	1.3875	33.5	1.4150	47.6
1.3605	18.0	1.3880	33.8	1.4155	47.8
1.3610	18.3	1.3885	34.1	1.4160	48.1
1.3615	18.6	1.3890	34.3	1.4165	48.3
1.3620	18.9	1.3895	34.6	1.4170	48.6
1.3625	19.2	1.3900	34.9	1.4175	48.8
1.3630	19.5	1.3905	35.1	1.4180	49.0
1.3635	19.8	1.3910	35.4	1.4185	49.3
1.3640	20.1	1.3915	35.7	1.4190	49.5
1.3645	20.4	1.3920	36.0	1.4195	49.7
1.3650	20.7	1.3925	36.2	1.4200	50.0
1.3655	21.0	1.3030	36.5		

CHAPTER 19

Consistency (Viscosity) of Tomato Products

Viscosity or consistency is one of the most important factors to be considered in determining the overall quality and acceptability of many tomato products. For many tomato products, viscosity is paramount as a quality attribute in determining the acceptability of the product by the consumer and is an integral part of the quality grade standard requirements (E.E. Judge & Sons). Webster defines viscosity and consistency in a synonymous manner, that is, as the degree of solidity, degree of density. The food technologist may define viscosity as the measure of a fluid's internal friction, or the measurable resistance when one layer of fluid is made to move in relation to another (Minard, Undated). To the consumer, the term may imply cohesiveness, smoothness, or the degree of gelling and firmness. As such, the consumer probably evaluates consistency second only to initial color. For this reason, the measurement, control, and maintenance of viscosity is mandatory for high-quality products.

The food processor involved in the canning of such items as tomato juice, tomato catsup, tomato paste, tomato soup and other tomato sauces is well aware of the importance of this quality attribute. He utilizes several methods and instruments to objectively evaluate and control the consistency of his finished product. The instrument best suited to measure the viscosity or consistency of a given product usually depends on the rheological properties of that product. Some consistometers and viscometers are designed specifically for a given product; others are more sensitive and give better readings in conjunction with the classification and type of material to be analyzed.

CLASSIFICATION

A classification and definition of viscosity according to its rheological properties are as follows:

Viscosity is a measure of the resistance offered by a fluid to relative motion of its parts. More precisely, it is the ratio of resistance to shear to rate of shear. The fundamental unit of viscosity measurement is the poise. A material requiring a shear stress of 1 dyne per cm to produce a rate of shear of one inverse second has a viscosity of 1 poise or 100 centipoises. The viscosity of water at room temperature is approximately 1 centipoise (Anon. 1954).

Apparent viscosity is a measure of the viscosity that must be related to a particular rate of shear for non-Newtonian substances (Anon. 1954). A Newtonian liquid is one in which resistance to shear is directly proportional to rate of shear. An example is water. A non-Newtonian liquid is one in which resistance to shear is not linearly related to rate of shear. Examples are chocolate, catsup, and cream style corn. Figure 19.1 shows the relationship of shear rate to shear stress with different types of fluids.

A *Bingham plastic solid* is a material that requires an initial value of shear stress to start the flow and then has a linear shear relationship. General plastic solids are substances that also require an initial force to start flow, but the shear relationship thereafter is nonlinear. A thixotropic material is any material having a viscosity dependent on its previous shearing history. A pseudoplastic product has a consistency that decreases with an increase in the shear rate.

The viscosity or consistency of any given product therefore depends on its resistance to shear as it is subjected to varying rates of shear. Exactly how it behaves when the above conditions are met results in its ultimate classifi-

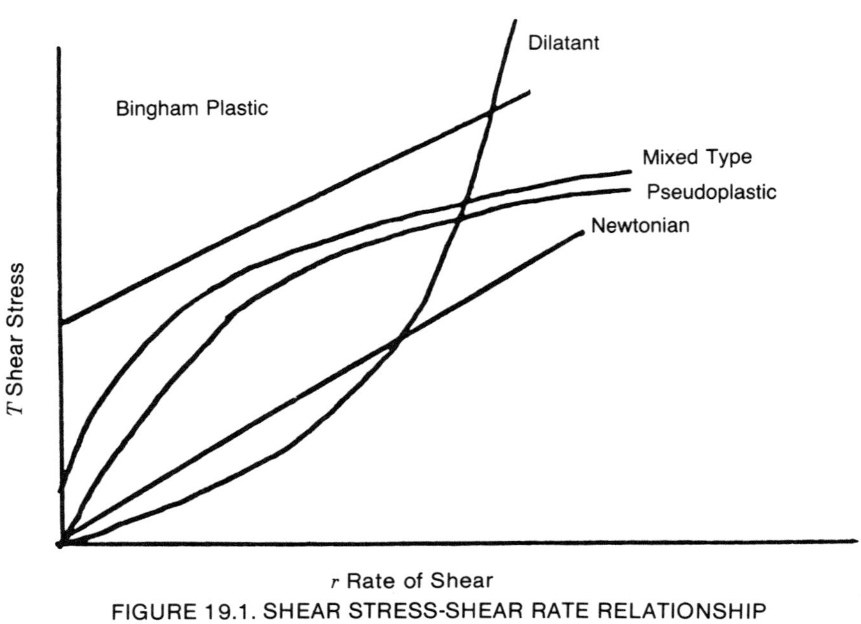

FIGURE 19.1. SHEAR STRESS-SHEAR RATE RELATIONSHIP OF DIFFERENT TYPES OF FLUIDS.

cation. Most food products exhibit non-Newtonian characteristics, the majority falling into the pseudoplastic category.

Tomato products do not follow the simple Newtonian *(R)* model in which a linear relationship between shear stress *(t)* and shear rate *(dr/dt)* is obeyed:

$$R = u\ dr/dt$$

where u = a simple viscosity constant. Instead of being able to characterise the flow behavior with a simple viscosity constant (u) for tomato products one requires at least two parameters and a more complex model.

Tomato products are pseudoplastic fluids in which the apparent viscosity (the ratio of shear stress to shear rate) decreases with increasing shear rate. So the flow behavior of tomato products can be written as the following:

$$T = kr^n$$

where T = shear stress; r = shear rate; k = consistency index; n = flow behavior index.

The apparent viscosity (u^a) of tomato products is usually present in the following form:

$$u^a = T/r = kr^n/r = kr^{n-1}$$

MEASUREMENT

Measurement of viscosity can be accomplished best by utilizing the following instruments in conjunction with the specified tomato products.

Tomato Juice

The consistency of tomato juice is such that Grade A flows readily and has a normal amount of insoluble tomato solids in suspension with little tendency to settle out; while Grade C flows readily and has a normal amount of insoluble tomato solids in suspension with no marked tendency to settle out (Anon. 1971). It is readily seen that the above statement is a broad generalization offering no specific value or numerical range in juice consistency measurement. The following methods have been used to varying extents and provide quite satisfactory to marginal results.

Modified Efflux-Tube Viscometer and GOSUC Consistometer. The modified efflux-tube viscometer was developed at the Ohio State University Food Processing and Technology Laboratory (Gould 1971). It consists of a blown glass reservoir sloped into a 2-mm orifice on the efflux and a ¾-in. orifice at the top, which allows pouring the sample into the reservoir through a funnel or filling the reservoir by vacuum. A piece of rubber tubing is attached to a metal plug with a 5/64-inch (2-mm) precision-bore orifice. This allows the flow to be stopped by a pinch clamp and the instrument to be accurately standardized by adjusting the length of the rubber tubing. The instrument is standardized by filling it with water at room temperature and adjusting the metal plug to efflux 200 ml in 32 sec.

Kluter reports that the GOSUC consistometer is a valuable tool in tomato juice consistency measurement in that it gives satisfactorily reproducible results when compared to more expensive instruments (Kluter 1959). Viscosity measurement utilizing the GOSUC consistometer is easily accomplished by timing with a stopwatch the efflux of tomato juice between two graduate marks. Results are recorded in seconds per 200 ml.

USDA Viscometer. The USDA Viscometer is similar in operating principle to the GOSUC consistometer in that it also measures flow through an orifice in seconds. It consists of a funnel with a 400- to 500-ml capacity having a stem cut to a 3/4-in. length. To this funnel is attached a 1/4-in. rubber hose having a vertical length of 3 in. A pinchcock is attached to the hose to control flow. Inserted at the bottom of the rubber hose is a metal precision tube 1-in. long having a 5/64-in. (2-mm) bore. This four-part assembly is attached to a ring stand. A 200-ml volumetric flask is stationed underneath this assembly to receive the material under test (U.S. Dep. Agric. 1958).

Tomato juice viscosity may be evaluated with the USDA Viscometer by pouring 250 ml into the funnel after adequate agitation of the sample. The temperature is measured and recorded. The pinchcock is opened, and the efflux of 200 ml of juice is measured by the volumetric flask and timed with a stopwatch. Standardization is accomplished by adjustment of the rubber tubing or orifice.

Kluter indicates that low correlations of the USDA viscometer with other types of gravity-flow orifice-constriction instruments are evident. This is due to lack of shearing rate produced by a decreasing hydrostatic head, flowing in a conical slant, and frequent plugging of the orifice (Kluter 1959).

Capillary Viscometer. The Capillary Viscometer was designed and developed by the California Packing Company Laboratories. The cup or reservoir of the instrument is a 1½-in. diameter Lucite tube with a wall thickness of ¼ in. The length of the reservoir is 7 in. The bottom of the reservoir contains a 1-in. Lucite plug which is sloped at an angle to the opening. Inserted into the opening is a 12-in. precisionbore Pyrex tube. The glass tube is vertically enclosed in a chromeplated brass tube having a sleeve of Tygon between the glass and outer brass covering. The instrument is standardized by filling the reservoir with water at 75 ±1°F and adjusting to an efflux of 13 sec. The distance to this point is marked on the outside of the glass reservoir (Kluter 1959).

Tomato juice viscosity measurement may be accomplished by filling the reservoir with a thoroughly mixed sample, taking care to properly prime the entire length of the viscometer. After complete filling and leveling is accomplished, the flow is released by removing a finger held underneath the

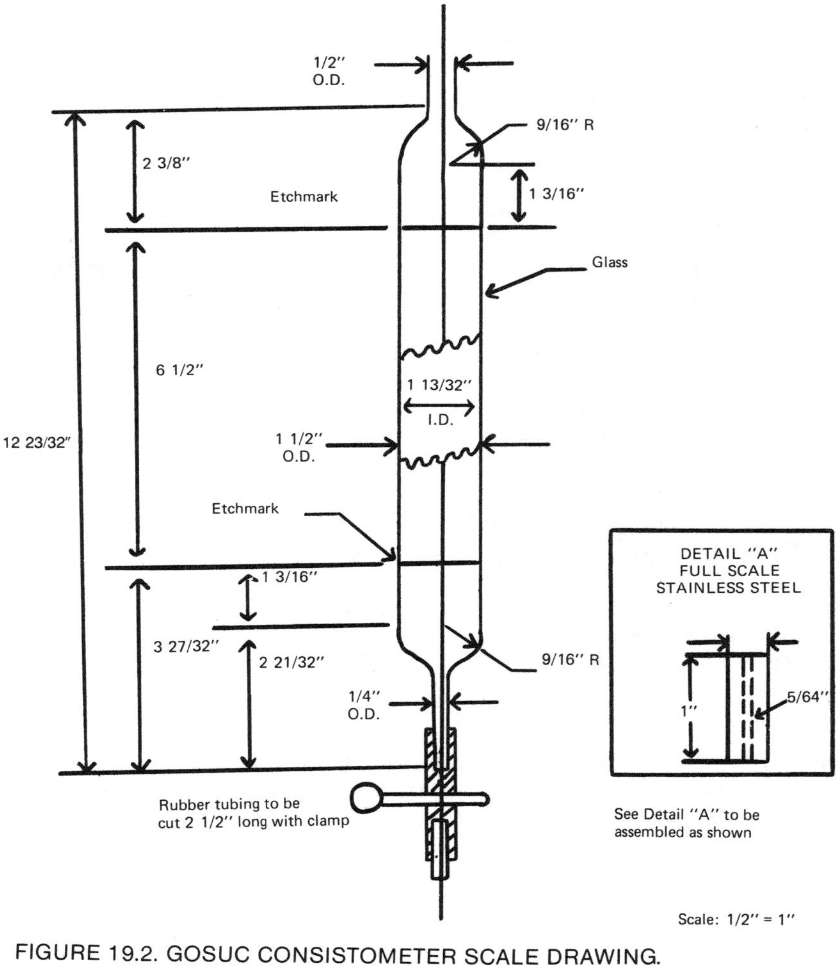

FIGURE 19.2. GOSUC CONSISTOMETER SCALE DRAWING.

bottom of the tube. A stopwatch is used to time the efflux of sample accurately to the previously standardized mark. Results are recorded in seconds.

Stormer Viscometer. The Stormer Viscometer consists of six basic parts: (1) test cup, (2) rotor, (3) driving weight, (4) winding drum, (5) brake mechanism, and (6) revolution counter. The gears and pension mechanisms are mounted inside a dust-proof cap placed on top of the upright supports (Gould 1953A).

The viscosity is determined by measuring the time required for a definite number of revolutions of a rotating cylinder immersed in a sample placed in test cup. The temperature of the sample is maintained at a desired level by means of a water or oil bath. The rotating cylinder is driven by a falling weight through a series of gears. A revolution counter is attached to the spindle of the rotating cylinder or other rotor (Anon. 1971).

Viscosity measurement is accomplished by first placing the instrument on a horizontal shelf or table so that the driving weight is vertically mobile through a distance of approximately 40 in. The rotor is then attached to the chuck and secured with the screw on the chuck. The driving weight should next be adjusted with gram weights in unison to the viscosity of the test material. The sample is placed in the sample cup, or, if desired, the product's container; in either instance, the sample should be uniform and temperature-controlled. The platform is raised to immerse the rotor in the sample in a centered position until the top of the rotor is covered to a depth of ½ in. With the brake on, the driving weight is raised by turning the handle of the rewinding drum counterclockwise until the weight nearly touches the pulley. The dial is set by releasing the brake by a quarter turn of the brake-control knob. The rotor is allowed to revolve until the pointer on the dial is located between 80 and 90; then set the brake. The purpose of locating the pointer 10 to 20 graduations to the right of zero is to permit the rotor to turn a sufficient number of revolutions to get a running start before checking with the stopwatch. With stopwatch in hand, the brake is released and the time is measured in seconds required for 100 revolutions of the rotor as indicated by the revolution counter. The brake is set and the number of seconds recorded (Gould 1953A).

Kluter found certian drawbacks in using the Stormer to measure tomato-juice viscosity. Since a minimum amount of weight is needed to drive the rotor, the spindle may have a tendency to stick (Kluter 1959).

Other Methods. Other viscosity-measuring instruments such as the Brookfield Viscometer and Gardner Mobilometer could be used to measure tomato juice viscosity but are better suited to more viscous materials than tomato juice.

Catsup

Catsup consists of two parts: a thick syrup and tomato fiber. The proportion of syrup to tomato fiber and the characteristics of the syrup are the principle factors that determine the consistency of the catsup; the sugars, acid, protein, mineral salts, and tomato flavor from the tomatoes have little effect on it. The amount of pectin in the catsup does affect the finished product. The dissolved pectin makes the liquid portion somewhat viscous. The thickness or body of the catsup is largely determined by the viscosity of the liquid and the proportion of insoluble tomato fiber present (Smith, Undated).

Other factors affecting the consistency of catsup are:

1. The maturity and variety of the tomatoes,
2. Excessive cyclone pressure (Smith),
3. Variation in size of finisher screen,

4. The method of preparing the pulp (hot-break pulp gives a better body than cold-break pulp),

5. Milling, and

6. The final pH of the finished product (adjustment of pH can aid in control of juice consistency).

The relationship between the consistency of the tomato catsup and these factors can be illustrated as the following.

The USDA Grade Standard for Catsup refers to the viscosity of the product and the tendency to hold its liquid portion in suspension. Grades A and B show not more than a slight separation of free liquid when poured on a flat grading tray, are not excessively stiff, and flow not more than 9 cm in 30 sec at 68°F (20°C) in the Bostwick Consistometer (Judge & Sons 1971). The temperature of the catsup must be 68°F (20°C), as above these values catsup will flow much faster in the Bostwick Consistometer, as illustrated in Fig. 19.4. Grade C catsup may show a noticeable, but not excessive separation of free liquid when poured on a flat grading tray, is not excessively stiff, and flows not more than 14 cm in 30 sec at 68°F (20°C) in the Bostwick Consistometer.

The following are methods utilized in the measurement of catsup viscosity.

Bostwick Consistometer. The Bostwick Consistometer (CSC Sc Co. Inc.) is used to determine the consistency of a viscous material by determining how far the material flows under its own weight along a level surface in a given period of time. The Consistometer consists of a metal trough closed off near one end by a gate that can be opened almost instantaneously. The shorter end of the trough, the reservoir, has walls that are carefully measured and leveled along the top. The longer end is graduated in 0.5-cm steps starting 1 cm from the gate. The graduations are numbered at each centimeter. It is equipped with a two-way level and two leveling screws on the reservoir end. The gate slides vertically between two grooved posts. It is pushed up by two springs in the posts, and is held down in a closed position by one end of an L-shaped trigger, which hooks over the top of the gate. When the free end of the trigger is pushed down, it rotates around its switch and the gate flies up. Necessary accessories are a stopwatch or other timing device, straightedge such as a spatula, and a thermometer (Gould 1971).

Catsup consistency is measured by first adjusting the leg screws to precisely level the instrument. The gate is then closed and held in this position by engaging the trigger release mechanism. Catsup should be prepared by mixing with distilled water to produce a 12% concentration prior to the consistency reading. Catsup is then poured into the holding compartment and leveled off even with the sides of the compartment. The gate is then released, allowing the catsup to flow over the centimeter scale for 30 sec. The farthest point of flow on the scale at the termination of this time period is recorded as the index of consistency for the sample (Gould 1953B). Read the Bostwick to the nearest 0.1 cm.

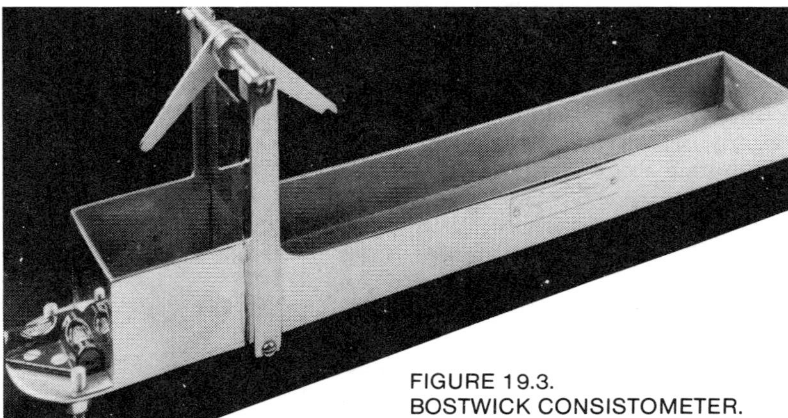

FIGURE 19.3. BOSTWICK CONSISTOMETER. *Courtesy of USDA.*

The Bostwick Consistometer is the consistometer that measures the shear stress under the fixed condition of the shear rate, although efflux-tube viscometers are the consistometers that measure shear rates under fixed conditions of shear stress.

Recently, a new type of the Bostwick Consistometer was developed in Japan. Advantages of this consistometer can be pointed out as the following.

1. Because the metal plate of the bottom is the shape of a "v," the flow of samples regularly makes a single point. In case of the earlier model, the flow of samples is an irregular shape. Thus, one can read the point of the flow exactly.
2. The consistometer is put on a horizontal plastic plate adjusted to level position with two water levels being at right angles to each other. Thus, it is not necessary to adjust the water levels between measurements.
3. In Japan this new consistometer is also used to dertermine the tendency to hold the liquid portion of catsup in suspension. They determine the distance between the point of colorless liquid and the point of red body in a given period of time.

Brookfield Viscometer. Another instrument that may be used to measure catsup consistency is the Brookfield Viscometer. This viscometer may be used to control final catsup consistency in that measurements may be taken on catsup that is both hot and cold. Since catsup is classified as plastic material when hot and a thixotropic material when cold, substantially different viscosities can be satisfactorily measured by varying the shear rate and shear force.

The Brookfield Viscometer consists of a synchronous motor, a spindle that is rotated in the substance to be examined, a dial, and pointer. The

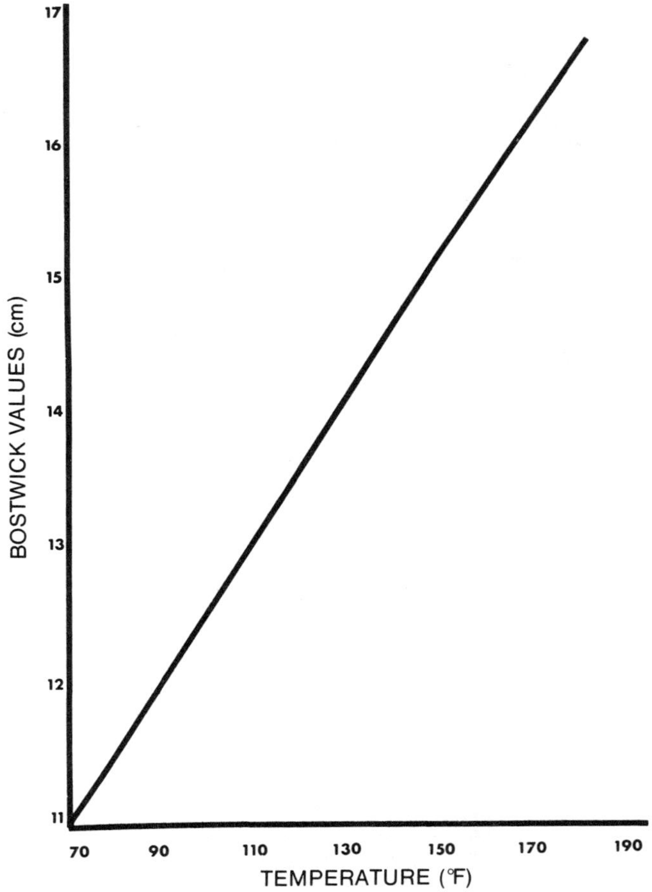

FIGURE 19.4. EFFECT OF TEMPERATURE ON FLOW OF TOMATO CATSUP.

pointer is connected to the power unit through a calibrated spring. The torsion on the spring is a measure of the shearing force exerted on the sample by the rotating spindle. The viscometer works on the principle that a measurable drag is imposed when its rotating spindle is immersed in a material under observation. The values obtained can be expressed as centipoise (Trusler 1953; Oliver 1948).

The viscometer is equipped with a number of different-size spindles, a spindle guard, which fastens into position around the spindle by means of threaded logs, a speed selector to regulate spindle speed, a clutch, which protects the beryllium–copper spring from severe initial torque and which also locks the pointer on the dial to permit readings, a float-level to level the instrument, and an optional stand to secure the instrument.

To measure catsup viscosity, the guard and spindle are immersed in a beaker or suitable container to a depth indicated by a groove on the spindle. The clutch lever is depressed and the motor started by snapping on the switch located below the dial on the right side of the instrument. The clutch lever is then released, and rotation of the spindle is allowed to continue

TABLE 19.1. SCORE POINT GUIDE FOR CONSISTENCY OF TOMATO CATSUP AND SAUCE

Grade	Points	Consistometer Readings	Free Liquor
		Tomato Catsup	
A	25	6.1–7.5	None or slight amount (1/8 in.)
	24	7.6–8.0 or 5.1–6.0	
	23	8.1–8.5 or 4.6–5.0	
	22	8.6–9.0 or 4.1–4.5	
C	21	9.1–10.0 or 3.6–4.0	Noticeable but not excessive (1/4 in.)
	20	10.1–11.0 or 3.1–3.5	
	19	11.1–12.5 or 2.6–3.0	
	18	12.6–14.0 or 2.0–2.5	
SStd.		More than 14.0 or less than 2.0	Excessive
		Tomato Sauce	
A	25[a]	9.10	0
	24[a]	12,11	1/16
	23	13	2/16
	22	8 or 14	3/16
C	21	15 or 7	Slightly more than 3/16 in.
	20[a]	16 or 6	1/4 in.
	19[a]	17 or 5	5/16 in.
	18	18 or 4	3/8 in.
SStd.		More than 18.0 or less than 4.0	More than 3/8 in.

[a] Many other combinations would merit these scores, for example, a Consistometer reading of 9 cm and 2/16 in. of free liquor would score 24 points. A consistometer reading of 10 and 3/16 in. free liquor would score 23 or 24 points, depending on the effect of the free liquor.

FIGURE 19.5. CONCENTRATED TOMATO PRODUCTS DILUTION CHART - GRAMS OF WATER TO BE ADDED to 100 GRAMS OF PRODUCT TO OBTAIN 12 PERCENT N.T.S.S. FOR CONSISTENCY DETERMINATION

INSPECTION AID NO. 93

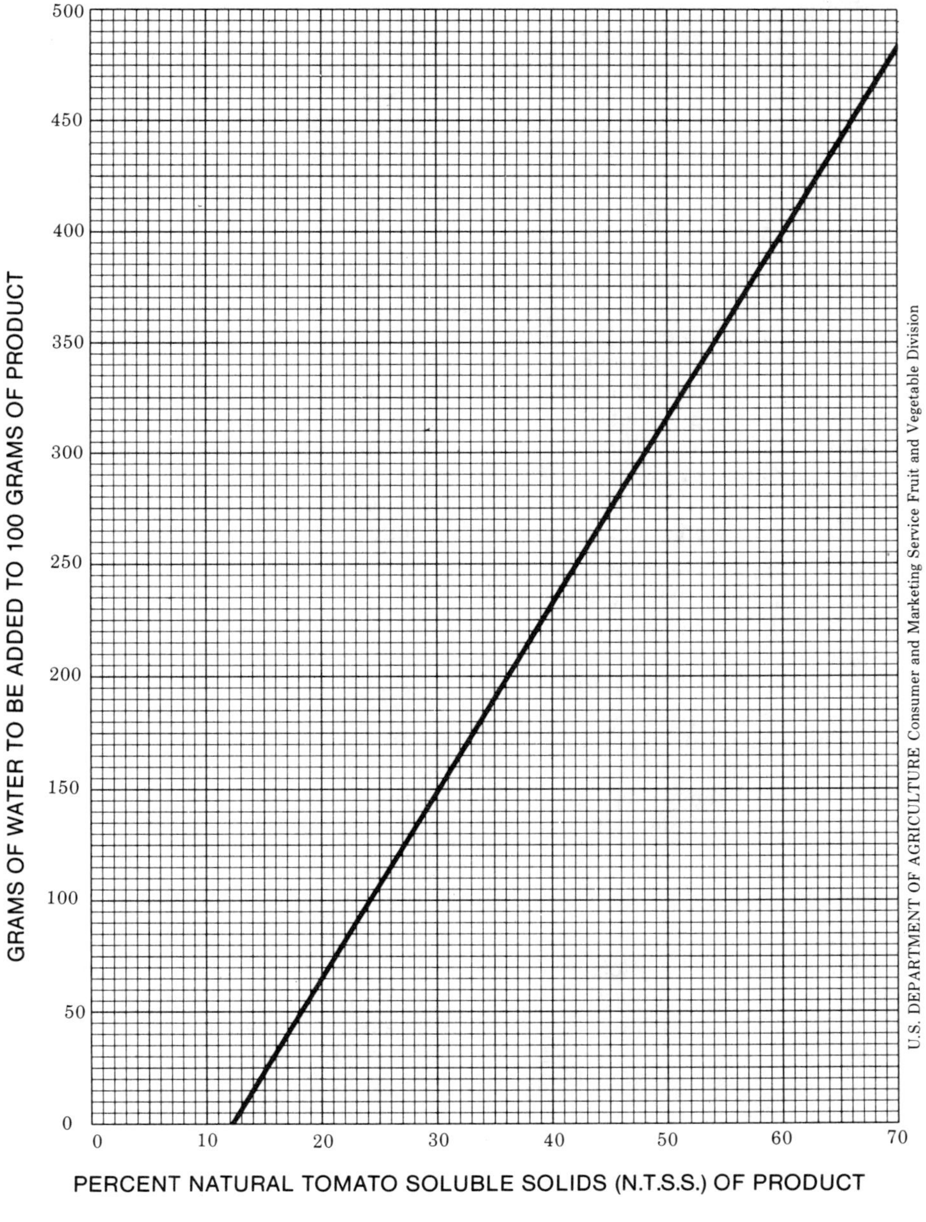

until 8 or 10 revolutions have been made, or until a constant reading without appreciable variation from each revolution has been obtained. The clutch lever may then be depressed and the motor turned off. A dial reading is maintained for recording or observation as long as the clutch is depressed. The dial reading may then be converted into centipoise units by means of a graph or factor-finder supplied by the manufacturer.

The temperature of the sample must be controlled very closely. Higher temperatures usually produce a lower viscosity, and vice versa. This situation is magnified when using a rotational type viscometer. The temperature of the sample should always be checked before checking viscosity and standardized accordingly. Materials have been encountered that changed viscosities 50% per °C (Anon. 1954).

Adams Consistometer. The Adams Consistometer and Bostwick Consistometer are similar in principle in that both measure the flow produced by a sample contained in a respective reservoir when released. Satisfactory correlation coefficients have been realized between the two, which indicates that a fairly accurate comparison can be made from the results of each.

The essential parts of the Adams consistometer are (1) the leveling screws; (2) a hollow, truncated cone with an inside bottom diameter of 3 in., inside top diameter of 2 in., and a height of 4 27/32 in.; and (3) a polished circular measuring plate with a minimum diameter of 12 in. (Gould 1953B).

Catsup consistency is measured in the following manner:

1. The consistometer is leveled by adjusting the leg screws in order to obtain a uniform flow over the disk,
2. The hollow truncated cone is held down on the center of the disk,
3. The sample is filled into the cone and leveled off with a spatula, taking care to avoid air pockets,
4. The cone is raised allowing the sample to flow over the measuring disk for 30 sec, and
5. The consistency of the product is determined by recording the extent of flow of the sample at four equidistant points on the disk.

The average of these readings is then taken and recorded as the consistency according to the number of radial lines encompassed (Gould 1953B).

Blotter Test. The Blotter Test is a quick visual check often used to demonstrate claims for high concentration of tomato solids in their particular brands of catsup (Smith, Undated). A spoonful of catsup is dropped on a blotter and allowed to stand for a given period of time, usually 3 min. Results can be standardized and controlled by using standard grade white desk blotters, accurate sample measurement, and precise timing, as described by Davis *et al.* (1954). If the catsup soaks into the blotter and forms a

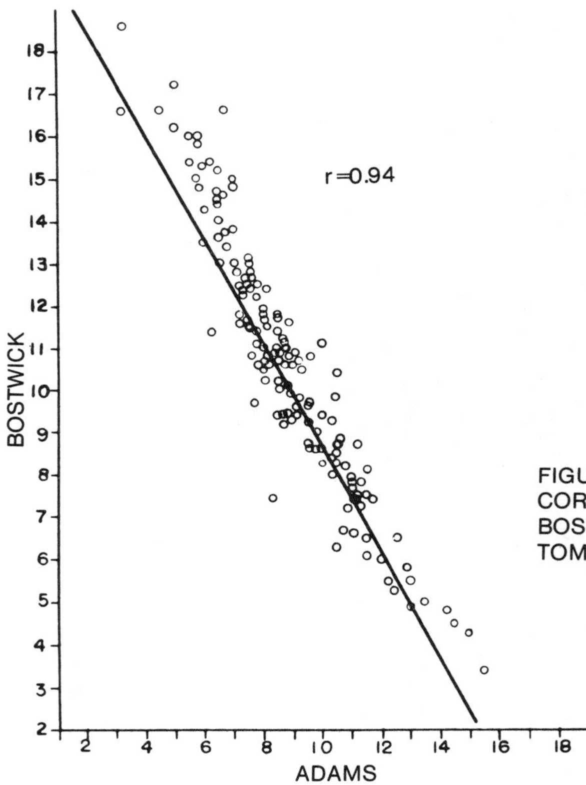

FIGURE 19.6. CORRELATION OF BOSTWICK TO ADAMS FOR TOMATO PULP.

wide ring of colorless liquid around the red center it is called a poor test, and this is said to indicate a lack of the proper amount of tomato solids in the catsup. If the catsup stands in a ball and only a narrow ring of liquid, less than ¼ in., appears in half an hour, the test is said to be satisfactory.

Smith (Undated) makes the following points concerning the blotter test:

1. Thin catsups give a poor blotter test.
2. Many thick catsups having a high concentration of tomato solids also give a poor blotter test.
3. A sugar acid syrup containing as much as 50% sugar and 2% acid gives a poor blotter test.
4. The insoluble solids of tomato catsup have nothing to do with the blotter test. The syrup part of catsup gives just as good a blotter test as does the heavy fiber potion . . . from this it is evident that the blotter test is a function of the soluble solids in the catsup, aided by the fine insoluble particles of cell protoplasmic material in the syrup.

FIGURE 19.7. ADAMS CONSISTOMETER.

Continuous Measurement of Hot Catsup. Catsup viscosity is measured directly and continuously by a rotational type viscosity recorder, as described by Richardson (1954). The torque imposed on a constant speed, rotating spindle varies with viscosity, and is converted into measurements of capacitance which are directly proportional to the viscosity and are transmitted as such to the recorder. A temperature bulb mounted in the sampling line provides a continuous, recorded measurement of temperature with the viscosity record.

Within the transmitter a synchronous motor turns the measuring assembly at a constant rate of 50 rpm. A shaft that drives the torque spindle is connected through a beryllium–copper spring to capacitor plates in the housing. Variations in viscosity produce variations in torque on the spindle, which varies the tension of the beryllium–copper spring and alters the relative position of the capacitor plates. The changes in capacitance reflect changes in catsup viscosity (Richardson 1954).

Tomato Paste

Tomato paste, being a mixture of liquid, insoluble solids, and coagulated flocs of pectin, does not lend itself to examination with the usual viscometer (Underwood and Keller 1948). As such, no guidelines for paste consistency are listed in the Quality Grade Standards. Perhaps the best data available that are indirectly related to paste consistency are the tomato solids concentration; for extra-heavy concentration, 39.3% or more; for heavy concentration, 32.0% or more, but less than 39.3%; for medium concentration, 28.0% or more, but less than 32.0%; for light concentration, 24.0% or more, but less than 28.0%. Since a low-consistency paste may still read high in solids content as measured by the refractometer, it is desirable, if possible, to measure paste consistency with an objective instrument.

Penetrometer. The Penetrometer measures the degree of penetration by a selected instrument into the material being tested, as produced by a given force applied over a given area for a measured length of time at a specified temperature. Penetration cones and plunger rods of different weights are furnished with the instrument to cover a wide range of consistencies (Underwood and Keller 1948).

A constant penetration time interval must be used in order to obtain accurate and reproducible results. A time interval of 5 sec is found to be adequate. Also, the degree of penetration increases with a rise in temperature. For every 10°C rise in temperature, there is an increase in penetration of 6 scale divisions over a temperature range of 0° to 80°C. Therefore, all consistency values must be obtained at the same temperature. A temperature of 25° C is a convenient temperature on which to standardize.

Tomato paste consistency is measured in the following manner:

1. Place the Penetrometer on a firm, flat surface, and level it accurately by means of the leg screws,
2. Set the dial pointer to zero,
3. Mix a quantity of paste to be measured just enough to insure homogeneity,
4. Fill a container more than level full with paste at the desired temperature (a No. 2 can that has been cut down or other suitable container, having a capacity of around 300 ml is satisfactory),
5. With one sweep of a straightedge, level off the top of the sample even with the edge of the container,
6. Pass the straightedge once again across the paste to produce a smooth, homogeneous surface,
7. Immediately place the sample on the centering ring on the base of the instrument directly under the penetration cone,
8. Lower the cone so that its tip just touches the surface of the paste, using the mirror provided for this purpose,
9. Simultaneously press the penetration cone release and the starter button on the interval timer, and
10. Allow a penetration time of 5 sec.

After completion of the 5 sec, lock the cone in place and depress the dial pointer lever as far down as it will go. Read the depth of penetration or *P* reading from the scale on the dial and record as 1/10 mm. In taking duplicate readings, remove most of the paste from the sample container and follow the same procedure for refilling, leveling, and adjusting the temperature before making the second measurement. Always be sure that no air is worked into the paste or that the paste is packed by pressing (Underwood and Keller 1948).

Potentiometric Viscometer. The Potentiometric Viscometer is designed to measure consistencies in food products such as tomato purees, catsups, and pastes. The instrument, as described by McColloch and Beavens, consists of two rotors, one spiral-shaped and the other shaped in a manner similar to a tuning fork. The spiral-shaped rotor is best suited to measure tomato paste consistency. The rotor consists of a stainless steel tapered wire helix. Its action is such that any path formed by a given turn of the spiral is continuously closed by the action of other turns of the wire. The test medium is submitted to a force tending to pump it up the turns of the spiral, and it is mainly the force with which the medium resists this pumping action that forms the basis of the consistency measurement in tomato pastes. The total range of viscosities covered by two rotors is 300 to 10,000 centipoises (Gould 1953B; McCulloch and Beavens 1954).

Resistance of the test medium to movement of the rotor causes an angular displacement of the rotor relative to the driving motor and against the force of the torque spring. When the force of the torque spring just balances the resisting force of the medium, the rotor turns at a synchronous velocity with a fixed angular displacement relative to its no-load position in relation to the motor. This displacement is reflected in the position of the slider on the potentiometer winding. When a small voltage is applied across the potentiometer winding, the displacement of the slider may be continuously indicated in terms of a small current or voltage drop by a milliammeter or a voltmeter. This reading is proportional to the angular displacement of the rotor, and therefore to the viscosity or consistency of the medium under observation (McCulloch and Beavens 1954).

The viscosity of tomato paste is measured in the following manner:

1. Bring the sample to constant and uniform temperature of 25°C,
2. Adapt the spiral rotor to the instrument,
3. Adjust the voltmeter to read 0.00 to 2.50 volts,
4. Lower the rotor into the sample so that the top of the working portion of the rotor is about 0.25 in. below the upper surface of the sample (the reading can be taken without removing the paste from its original container),
5. Start the rotor, and
6. Read and record the voltmeter scale reading after the rotor has run for 1 min.

The scale reading obtained is converted to the equivalent centipoise value by means of a calibrator curve, and the consistency of the sample is then expressed in terms of centipoises of apparent viscosity (McCulloch and Beavens 1954).

A calibration curve can be prepared for obtaining standards of reference by measuring viscosities of materials approximating paste and plotting volt readings. The use of another viscometer is suitable for preparing a set of secondary standards for comparison. Lastly, calibration curves can be prepared by using standard solution of known viscosity.

McCulloch and Beavens report a range of 1650 to 5100 centipoises apparent viscosity obtained from the sampling of 20 samples of tomato paste from different manufactures; 50% had consistencies in the range of 3000 to 4000 centipoises apparent viscosity (McCulloch and Beavens 1954).

Tomato Pulp

Tomato pulp or puree differs from tomato paste in that the soluble solids content is less than 24%. The factors for grading and scoring each is the same as indicated in the U.S. Standards for Grades. The segregation of concentration lists extra-heavy concentration as having a tomato solids reading of 15.0% or more, but less than 24.0%; a heavy concentration as having 11.3% or more, but less than 15%; a medium concentration as having 10.2% or more, but less than 11.3%; and a light concentration as having 8.0% or more, but less than 10.2% (Judge & Sons 1971).

Viscosity measurement in tomato pulp does not present the same problem as does tomato paste. However, the heavier consistencies approach that of paste, and, as such, require similar means to measure viscosity. The apparent viscosity of pulp can be measured by means of the Penetrometer, as described for paste, or by using the Potentiometric Viscometer. When using the latter, the fork rotor would probably be best suited for pulp, especially with high concentrations.

The viscosity of pulp can be measured and controlled when hot by utilization of a rotary-type viscometer, such as the Brookfield, as described for catsup viscosity. The continuous kettle-type monitoring system employed for catsups could be installed as another means of control.

Lighter concentration of pulp could probably be measured with efflux-type viscometers or capillary viscometers. Heavy to extraheavy consistencies can successfully be measured by using falling-ball or falling-plate devices. Success in measuring pulp viscosity with selected measurement devices probably depends on which concentration is to be measured, as light concentrations may be watery and extra-heavy concentrations may be fairly well set up.

Tomato Sauce

The factors of quality for tomato sauce and the scoring factors are the same as for tomato catsup, except that there is no Grade B tomato sauce; hence all the references to Grades A and B in tomato catsup are to Grade A in tomato sauce, and in tomato sauce, there is no reference to discoloration in the neck of the bottle (E.E. Judge & Sons).

The consistency of tomato sauce is such that Grade A flows not more than 14 cm (9 cm for catsup) in 30 sec at 20°C in the Bostwick Consistometer. Grade C tomato sauce flows not more than 18 cm (14 cm for catsup) in 30 sec at 20°C in the Bostwick Consistometer (E. E. Judge & Sons).

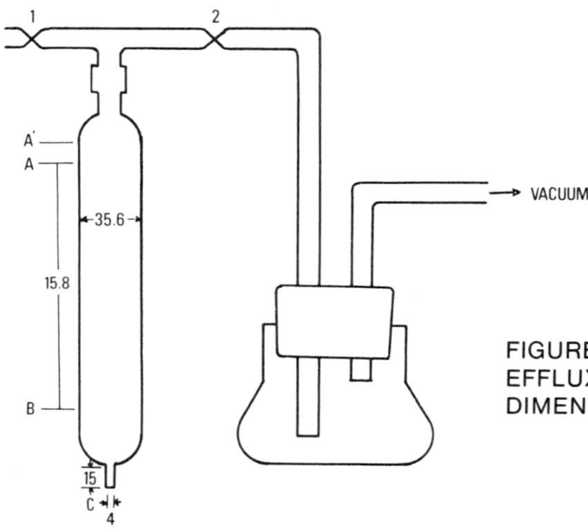

FIGURE 19.8.
EFFLUX TUBE VISCOMETER.
DIMENSIONS IN MILLIMETERS.

Tomato Soup

The viscosity measurement and control of tomato soup is very important, since variances in viscosity will alter the heat penetration during the processing operation. For this reason, it is desirable to measure the viscosity before and after the application of a heat process to achieve commercial sterility. Most attention is paid to the condensed form, since the ready-to-serve variety of tomato soup is watery and viscosity measurement is not as critical or difficult.

An inherent problem exists in the viscosity measurement of soup where starches, flours, and other thickening agents are added. Before filling, tomato soup may exhibit some dilatant properties, i.e., as the product sets and agitation is applied, thickeners continue to expand and, as such, the viscosity increases. Therefore, timing, amount of agitation, and temperature must be closely controlled when measuring viscosity before the filling operation. After processing and cooling, the viscosity continues to increase significantly up to a period of two weeks; depending on formulation, it may continue to increase slowly for an indefinite period.

When the viscosity is measured in tomato soup after a given time, it may exhibit pseudoplastic properties. The consistency decreases as any amount of shear is applied due to thickener breakdown. This may reach a certain point before stability is established and before any meaningful reading can be taken. This problem is especially apparent in rotational-type viscometers.

It is important, then, to measure viscosity in tomato soup both before and after processing, in addition to keeping accurate control of the time–temperature variables. To measure viscosity before processing, a rotary-type viscometer is probably the most accurate method. The Brookfield Viscometer, FMC Consistometer, or the Fisher Electroviscometer can be successfully employed. Regardless of which method is used, the sample should always be taken after a specified period, after addition of the thickener, volume standardization, and uniform agitation.

Brookfield Viscometer. The description and use of the Brookfield Viscometer is detailed under viscosity measurement of catsups. For its use in soup viscosity measurement, reference should be made to that section.

FMC Consistometer. The FMC Consistometer is comprised of two basic parts: (1) the turntable, which rotates at a constant speed, and (2) a measuring head calibrated in arbitrary units from 0 to 100 (Gould 1953C).

Viscosity measurement of tomato soup before processing is accomplished by adhering to the following procedure (Gould 1953C):

1. Obtain sample after batch completion; a 3-min figure versus a standard time interval between batch completion and sampling is sufficient
2. Adjust temperature to desired level after pouring sample into a 303 × 406 can
3. Lower the measuring head into the sample
4. Start the turntable
5. Make a reading after 30 sec (referred to as the dynamic reading)
6. Stop the turntable and make another reading after 60 sec (referred to as the static reading)

Fisher Electroviscometer. The Fisher Electroviscometer is another type of rotational viscometer that measures the viscosity of the sample by means of a patented torque–magnetic–electrical system. Viscosity is measured in absolute units or centipoises.

The operating principle of the Electroviscometer is that a sample under test is rotated about a stationary bobbin. The sample cup rotates at a constant speed. The rotating sample exerts a torque on the bobbin and the viscosity is determined. Since the bobbin is attached directly to a patented coil, its torque tends to turn the coil. The coil, however, is in a magnetic field and resists the tendency to turn when current is passed through it. The restoring torque from a regulated voltage power supply tends to swing the coil back to its original position. The force necessary to keep the coil from turning registers on a meter calibrated directly in centipoises (Gould 1953C).

Tomato soup viscosity is evaluated in the following manner:

1. Obtain a sample 3 min after completion of batch,
2. Fill sample cup with soup, adjust temperature to desired level and put cup into the holder,
3. Start the motor,
4. Adjust the control, and
5. Take the reading directly in centipoises.

Viscosity measurement after processing is best achieved by using falling-body or limit-of-flow techniques. These may help reduce variances and drift encountered by thickener breakdown when using rotary viscometers.

Methods that are suitable for determination of processed soup viscosity are the Bostwick Consistometer, the Adams Consistometer, or the Gardner Mobilometer. Again, however, the instrument that might be ideal for soup viscosity measurement depends to a great extent on formulation. Some brands of condensed soup may be too viscous to flow from a Bostwick Consistometer, while others may be too thin.

Use of the Bostwick Consistometer and Adams Consistometer is described in the section for catsup viscosity measurement, and reference to that should be made for anticipated use.

Gardner Mobilometer. The Gardner Mobilometer measures the consistency of products by determining the time in seconds for the plunger to fall in the food product through the distance between two reference points (usually 10 cm) on the plunger rod (Gould 1953A). It consists of 5 basic parts: (1) base plate with bubble leveling device; (2) sample cylinder; (3) plunger and weight pan; (4) a bracket to support the weight pan; and (5) a water-cooling jacket (Gould 1953A).

Tomato soup viscosity is measured by the following procedure:

1. Place the instrument on a horizontal shelf and level it by adjusting the leveling screws,
2. Remove the cylinder, fill to a uniform height with tomato soup, and replace the cylinder in its position inside the water jacket,
3. Attach the desired disk to the plunger,
4. Adjust the plunger in the cylinder until the disk is just immersed (approximately 0.5 in.) in the sample and add the derived amount of weights to the weights pan, and
5. Release the plunger and record with a stopwatch the time for the plunger to pass through the desired distance.

The number of seconds required for the plunger with uniform driving weight to travel through 10 cm is used as a measure of the consistency of the food sample.

If all food substances requiring viscosity measurement were Newtonian in nature, the problem associated with such measurement would be markedly decreased. Then again, there would not be much need for any detailed instrument descriptions, since possibly one or two instruments could handle those materials where control is desired. Non-Newtonian fluids require complex and often expensive equipment to produce useful and acceptable results. Visual observation and acceptance may always be the final basis of determining whether a product is too thick, too thin, or just right.

FACTORS AFFECTING CONSISTENCY IN TOMATO PRODUCTS

One of the most important factors affecting viscosity of unconcentrated tomato juice or pulp is the cultivar used (Gould 1978).

The viscosity of tomato juice, also, depends on the break temperature. Bel-Haj (1981) showed that a break temperature of 200°F (93°C) gave juice with higher viscosity than juice made at a break temperature of 150°F (66°C) or 175°F (79°C). Prolonged heating at high temperatures caused a reduction in viscosity due to denaturation of pectin (Doesburg 1965).

A screen size of 0.033 gives a viscosity that is higher than when a 0.023 screen is used. In addition, faster pulper blade speeds tend to increase consistency significantly. Milling has been used to increase viscosity and works by shear action (see Milling). These factors are important in order to obtain a product with a good consistency.

REFERENCES

ANON. 1954. Viscosity measurement. Foxborro Co., Foxborro, MA, Tech. Inform. Bull. *1-A-70a*.

ANON. 1971. Stormer Viscometer. A.H. Thomas Co., Philadelphia.

BEL-HAJ, H.M. 1981. Effect of cultivars, break temperature, pulping and extraction methods on the viscosity of tomato juice. Ph.D. Dissertation. Ohio State Univ., Columbus.

DAVIS, R.B., DEWEESE, D. and GOULD, W.A. 1954. Consistency measurements of tomato puree. Food Technol. *8* (7) 330.

DOESBURG, J.J. 1965. Pectic substances in fresh and preserved fruits and vegetables. I.B.V.T. Commun. *25*. Inst. of Research on Storage and Processing of Hortic. Produce, Wageningen, Netherlands.

GOULD, W.A. 1953A. Consistency in processed foods, Part II. Canner/Packer *34* (4) 42, 58–60.

GOULD, W.A. 1953B. Measuring consistency in foods. Canner/Packer *34* (3) 44, 66, 70.

GOULD, W.A. 1953C. Consistency in processed foods. Canner/Packer *34* (5) 44, 66, 70.

GOULD, W.A. 1971. Tomato Processor Quality Control Handbook. Dep. Hortic., Ohio State Univ., Columbus.

GOULD, W.A. 1978. Quality evaluation of processed tomato juice. J. Agric. Food Chem. *26*, 1006.

JUDGE, E.E. & SONS. The Almanac of Canning, Freezing, Preserving Industries, Edward E. Judge & Sons, Westminster, MD.

KLUTER, R.A. 1959. The objective measurement of tomato juice consistency. M.Sc. Thesis. Ohio State Univ., Columbus.

McCOLLOCH, R.J. and BEAVENS, E.A. 1954. Measurement of food characteristics. Agric. Food Chem. *2* (19) 986.

MINARD, R.A. (Undated.) An industrial rotational viscometer and its use with materials of varying complexity. Pap. *54-26-2*. Stoughton, MA.

OLIVER, E.M. 1948. Measurement of enamel-ship consistency by means of the Brookfield viscometer. J. Am. Ceram. Soc. (May).

RICHARDSON, L.M. 1954. Continuous measurement of ketchup viscosity in the finishing kettle. Pap. *54-16-2*. Presented at 1st Int. Instr. Soc. Am., Philadelphia, Sept.

SMITH, H.R. (Undated.) The consistency of tomato catsup. Natl. Canners Assoc., Washington, DC, Mimeo Rep. *310415*.

TRUSLER, A.B. 1953. Comparative viscosities of coconut oil liquid soaps. J. Am. Oil Chem. Soc. *30* (3) 100.

UNDERWOOD, C.J. and KELLER, G.J. 1948. A method for measuring the consistency of tomato paste. Fruit Prod. J. and Am. Food manuf. *28* (9) 103.

U.S. DEP. AGRIC. 1958. United States standards for grades of canned tomato juice. Fed. Reg. Jan. 16.

CHAPTER 20
Total Acidity and pH

Practically all foods contain an acid or a mixture of acids. These acids may occur naturally, may be produced by action of microorganisms, or may be added in such products as catsup or chili sauce during their manufacture. In all cases, the acids present are largely responsible for the tart or sour flavor. Total acidity determinations are useful as a measure of this tartness. Total acid is usually determined by titrating an aliquot of sample with a base of known strength using a suitable indicator to determine the end point. In the case of highly colored foods, such as tomatoes, accurate determination of the end point is very difficult when using an indicator; thus, it is easier and more accurate to use electrometric methods. The titration is usually calculated and reported in terms of the predominant acid: in the case of tomatoes, citric acid.

Total acid may be expressed on three different bases, which are outlined below:

1. As ml of 0.1 *N* NaOH per 100 ml of sample:

a. Equation: $A = \dfrac{B \times C}{D}$

where: A = ml of 0.1 *N* NaOH are 100 ml of juice
B = 100 (used because the calculation is based on a 100-ml sample)
C = volume of 0.1 *N* NaOH used
D = volume of sample used

b. Example: If 12.26 ml of 0.1 *N* NaOH are required to titrate a 10-ml sample, it would take 12.26 × 10 or 122.6 ml of 0.1 *N* NaOH per 100 ml of sample.

$$A = \frac{100\text{ ml} \times 12.26\text{ ml}}{10\text{ ml}} = 122.6\text{ ml}$$

2. As grams of acid in a sample aliquot:

a. In general, the predominant acid in the juice or product is used in expressing grams of acid present. It is customary to express acids in tomato products as citric acid.

b. Equation: $W = V \times N \times \text{mEq wt}$

where: W = g of acid in aliquot
V = volume in ml of NaOH titrated
N = normality of NaOH (0.1 N)
mEq wt = milliequivalents of acid, 0.064 for citric acid

c. Example: If 12.26 ml of 0.1 N NaOH are required to titrate a 10-ml sample of a tomato product, the acid content of the sample aliquot would be 0.0785 g.

$$W = 1.26 \times 0.1 \times 0.064 = 0.0785 \text{ g/10 ml or } 0.00785 \text{ g/ml}$$

3. As percent of acid in a sample aliquot:

a. Equation: $$Z = \frac{V \times N \times \text{mEq wt}}{Y} \times 100$$

where: Z = % of acid in sample
V = volume in ml of NaOH titrated
N = normality of NaOH (0.1 N)
mEq wt = milliequivalents of acid. 0.064 for citric acid
Y = volume in ml or weight in g of sample

b. Example: If 12.26 ml of 0.1 N NaOH are required to titrate a 10-ml sample of a tomato product, the % acid in the sample would be 0.785.

$$Z = \frac{12.26 \times 0.10 \times 0.064}{10} \times 100 = 0.785\%$$

TABLE 20.1. COMMON ACIDS FOUND AND USED IN FOODS AND THEIR WEIGHTS AND FACTORS

Acid	Formula Wt	Equivalent Wt	Acid mEq Factor
Acetic	60.05	60.05	0.0060
Butyric	88.10	88.10	0.0088
Citric	192.12	64.04	0.0064
Lactic	90.08	90.08	0.0090
Malic	134.09	67.05	0.0067
Oleic	282.46	282.46	0.0282
Oxalic	90.04	45.02	0.0045
Succinic	118.09	59.05	0.0059
Stearic	284.47	284.47	0.0284
Tartaric	150.08	75.04	0.0075

pH DETERMINATION

The term pH is the symbol for hydrogen-ion concentration. The hydrogen-ion concentration of a food is a controlling factor in regulating many chemical and microbiological reactions. Because the hydrogen-ion concentration expressed in moles is cumbersome, the pH scale as developed by Sørensen is commonly used.

The pH scale ranges from 0 to 14. A neutral solution has a pH of 7.0. A lower scale reading indicates an acid solution and a value above 7.0 indicates an alkaline solution. The pH scale is logarithmic rather than linear in character. Therefore, a pH of 5.0 is 10 times as acid as a pH of 6.0

There are two principal methods used to measure pH. One is the colorimetric method, which depends on the use of an indicator solution that produces a characteristic color at a given pH. Because there is no one indicator that produces characteristic colors for the the entire pH range, it is necessary to use several indicators to cover the entire range from pH 0 to pH 14.

TABLE 20.2. RELATIONSHIP OF pH VALUE TO CONCENTRATION OF ACID (H+)

pH Value	Concentration	
0	10,000,000	
1	1,000,000	
2	100,000	
3	10,000	Acidity
4	1,000	
5	100	
6	10	
7	0	Neutral
8	10	
9	100	
10	1,000	
11	10,000	Alkalinity
12	100,000	
13	1,000,000	
14	10,000,000	

Most of these indicators are weak organic acids that are capable of existing in two tautomeric forms in equilibrium with each other. When placed in a solution on the acid side of their neutral point, the equilibrium will shift to the acid form, and the acid color of the indicator will appear. The reverse is true when the indicator is placed in a solution on the alkaline side of the neutral point. Table 20.3 lists some of the indicators used for pH measurement.

To measure the pH of a solution, the appropriate indicator is added to an aliquot of the solution. The color of this solution is then compared with the color of the indicator at known pH's. The pH of the solution will be the same as that of the indicator standard that has the same color. A serious objection

TABLE 20.3. INDICATORS SHOWING BOTH pH RANGES AND COLOR CHANGES

Indicator	pH Range	Color
Acid cresol red	0.2– 1.8	Red-yellow
Acid *meta*cresol purple	1.2– 2.8	Red-yellow
Benzo yellow	2.4– 4.0	Red-yellow
Bromphenol blue	3.0– 4.6	Yellow-blue
Bromcresol green	3.8– 5.4	Yellow-blue
Methyl red	4.4– 6.0	Red-yellow
Chlorphenol red	5.2– 6.8	Yellow-red
Bromthymol blue	6.0– 7.6	Yellow-blue
Phenol red	6.8– 8.4	Yellow-red
Cresol red	7.2– 8.8	Yellow-red
*meta*Cresol purple	7.6– 9.2	Yellow-purple
Thymol blue	8.0– 9.6	Yellow-blue
Phthalein red	8.6–10.2	Yellow-red
Tolyl red	10.0–11.6	Red-yellow
Acyl red	10.0–11.6	Red-yellow
Parazo orange	11.0–12.6	Yellow-orange
Acyl blue	12.0–13.6	Red-blue

to colorimetric pH determinations is that they are not reliable when used with highly colored samples, as the natural color will obscure the color of the indicator.

The other method used to measure pH is to measure the potential developed between two electrodes when immersed in a solution. Several types of electrodes have been developed for this purpose. The most useful type used at this time is the glass-calomel system and an associated potential-measuring device. This device measures the voltage development between the glass electrode and the calomel electrode. The voltage is a measure of the hydrogen-ion activity and each change of 1 pH unit corresponds to a change of 0.059 volt. This developed voltage is measured by balancing the potential of the electrode system against a known potential provided by the interval battery or line voltage and a calibrated slide-wire potentiometer. The scale of the electric meter is calibrated in terms of pH rather than millivolts.

While the glass-calomel system is the one most widely acceptable, it should be remembered that other electrode systems are useful in certain operations. The choice of electrode system will depend on the use intended and the character of the material being tested.

In order to measure the pH of an unknown it is necessary to reduce the sample to liquid form. In the case of tomatoes, extraction of the liquid by means of a hand-operated juicer will provide sufficient liquid to make the pH determination. This can be made directly on the product where an electric pH meter is used. When using a colorimeteric determination, it is necessary to first filter the puree to obtain a solution free of solid particles.

The actual mechanics of making a pH measurement are relatively simple. Using an electrode system, a sample of the unknown liquid is placed in a small beaker, the electrode inserted, and the pH measurement made. The

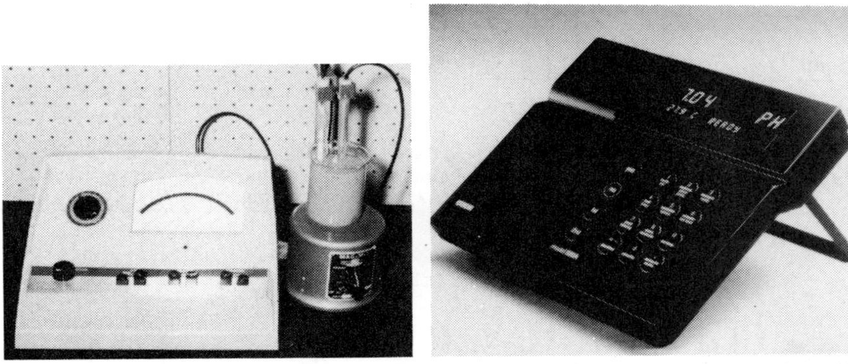

FIGURE 20.1. Glass Electrode pH Meter set up with magnetic stirrer for pH determination of tomato juice and pulp (left). Digital Electronic Model pH Meter courtesy of Wilkens-Anderson Company.

adjustments necessary and the operating instructions will vary with the apparatus used. Usually concise operating instructions are furnished with the instrument.

When using a colorimetric procedure, a definite volume of liquid is placed in a calibrated tube and then a definite amount of indicator is added. To another tube just the unknown is added. This second tube is aligned in back of the tubes containing the standard indicator solution, in order to compensate for any turbidity or slight amount of color in the sample which would cause a change in the observed color. There are several types of colorimetric pH apparatus available at moderate cost. Any chemical supply house catalog will list several.

The pH of a solution is a measurement of the free hydrogen ion H+ concentration. In pure solutions of an acid or base, it is proportional to the normal concentration of the acid or base. However, in solutions of fruits or vegetables, this is not the case. Any such solution will contain colloids and buffer salts, which will influence the pH reading. For this reason, it is possible to have two solutions with the same pH but with widely different total acid concentrations.

For example, in testing tomato varieties it was found that two varieties had a pH of 4.25, but one had a total acid of 0.50% and the other of 0.35%. Because of the presence of buffers in food products, the total acid content does not have as much significance as pH measurements in determining the types of reaction that will take place.

In the canning of foods, one of the most important factors affecting the sterilization times and temperatures is the actual pH value of the food. The lower the pH value, that is, the higher the amount of acidity in the food, the lower the degree of heat required for sterilization. Examples of typical pH values for some canned foods classified with respect to their pH values is shown in Table 20.4.

TABLE 20.4 FRUITS AND VEGETABLES CLASSED ACCORDING TO ACIDITY

Group No.	Group Description	pH	Examples of Food Products
I	Nonacid	7.0–5.3	Corn, lima beans, and peas
II	Semiacid or medium acid	5.3–4.6	Beets and pumpkin
III	Acid I	4.6–3.7	Apricots, pears, and tomatoes
IV	Acid II	3.7 and below	Applesauce, grapefruit, and pickles

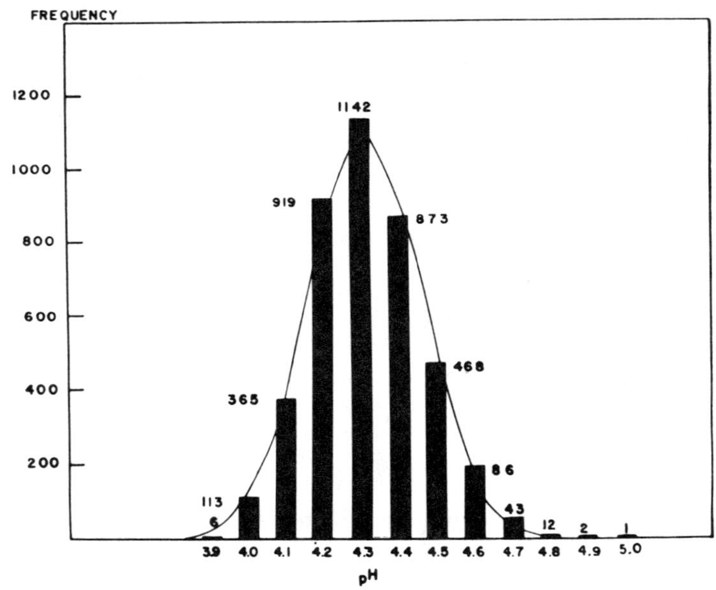

FIGURE 20.2. FREQUENCY DIAGRAM OF pH VALUES OF TOMATO (1961) SURVEY.

It is usually considered that a pH of 4.6 is the dividing line between acid and nonacid foods. This usually means, with a product having a pH of 4.6 or less, that the germination of bacterial spores from organisms such as *Clostridium botulinum* will be inhibited after proper sterilization.

Even with any given product the pH may vary considerably. Some of the most important factors believed to have an effect on the actual pH values of a product are (1) cultivar, (2) maturity, (3) seasonal variations due to growing conditions, etc., (4) geographical areas, (5) handling and holding practices prior to processing, (6) processing variables, and (7) salt.

In Table 20.5, 10 varieties of tomatoes are listed with their pH values. Further, for each variety there are four classifications of maturity of the tomatoes: High U.S. No. 1's, Low U.S. No. 1's, High U.S. No. 2's, and Low U.S. No. 2's.

The classifications of these tomatoes were based entirely on the color of the individual fruits. All tomatoes were graded by a USDA federal-state inspector. He selected 10 tomatoes from each of 10 cultivars at each of 3 or more harvests for each of the above 4 color classifications.

TABLE 20.5. pH RELATIONSHIPS BETWEEN MATURITIES AND CULTIVARS OF TOMATOES

Cultivar	High 1	Low 1	High 2	Low 2	Average
Purdue 1361	4.70	4.71	4.26	4.66	4.58
Rutgers	4.57	4.61	4.64	4.44	4.56
Red Top	4.73	4.40	4.45	4.34	4.48
Tipton	4.53	4.54	4.48	4.50	4.51
Kokomo	4.73	4.62	4.75	4.45	4.62
Ace	4.86	4.64	4.66	4.63	4.69
Red Jacket	4.45	4.43	4.36	4.32	4.39
Improved Garden State	4.45	4.51	4.44	4.54	4.48
Long Red	4.63	4.74	4.56	4.50	4.63
Early Red	4.48	4.29	4.31	4.32	4.35
Average	4.61	4.55	4.49	4.48	4.53
L.S.D. 0.01 level: quality, 0.078; Cultivar, 0.102					

The juice was extracted from these tomatoes and the pH determined. One can see from these data significant variations between several of the varieties. Ace and Long Red have the highest pH, while Early Red and Red Jacket have the lowest pH. Within a variety, it should be noted that the Low U.S. No. 2's are significantly lower in pH value than are the High U.S. No. 1's. These data indicate two important points:

1. A processor should have knowledge of the cultivar of tomato he is processing. It may be necessary to adjust the process times and temperatures to provide adequate sterilization. Using careful control procedures, he could blend cultivars and perhaps rest assured that his sterilization conditions are safe, thus preventing spoilage, or use FDA-permitted organic acid additions to tomatoes to adjust the pH of the product to a level below 4.6.

2. Quality of the raw material may seriously affect the sterilization conditions unless known acid is carefully controlled.

In conclusion, pH is one of the quality control checks on which many food processors have not fully relied. It is a simple measurement requiring little time to accomplish. Further, little cost is required to provide the adequate equipment. It is a quality control check that will assist in the prevention of spoilage. It is one of the more important factors accounting for flavor

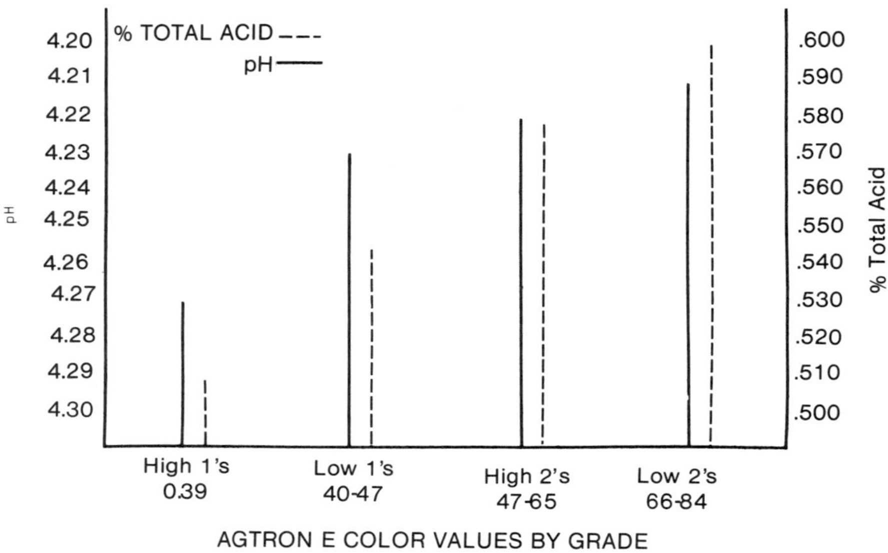

FIGURE 20.3. RELATIONSHIP OF pH AND PERCENTAGE TOTAL ACID VALUES ACCORDING TO MATURITY. AVERAGE VALUES FOR ONE SEASON WITH SAMPLES REPRESENTING ONE CULTIVAR OF TOMATOES.

changes in many products. Further, it is most important from the standpoint of water treatment (Anon. 1966), waste disposal (Anon. 1970), and detergent efficiency (Gould 1957). As an example of the latter, the corrosiveness and sanitizing power of chlorine solutions increase as the pH is lowered. Thus, the quality control technologist should know the pH of the water before establishing the chlorine level of his detergent.

REFERENCES

ANON. 1966. Acidification of whole tomatoes. Canning Trade (Sept. 19) 10.

ANON. 1970. The practice of adding acid to tomatoes. Canner/Packer (May) 18–19.

GOULD, W.A. 1957. Changes in pH values after time, temperature of cook. Canner/Packer *38* (9) 22;*38* (10) 16–17, 20.

CHAPTER 21

Defects and Material Other Than Tomatoes

This subject is not one normally found or discussed when operating under perfect conditions; however, it is obvious to most in the industry that Material Other than Tomatoes (MOT), Extraneous Vegetable Matter (EVM), filth, and other unwanted materials do occur. Some of these materials have been brought on by changes in culture, harvesting, and handling systems during recent years. Further with the emphasis on quantity production practices without full emphasis on quality, conditions have changed drastically. Some solutions are coming to the forefront with electronic color and dirt sorters on harvesters, the development of tomatoes more suited to existing systems, and the modern emphasis on better control of mechanical harvesting and handling operations.

Much of the defect and extraneous material problems are brought into the factory from the field operations. Further, these problems become accentuated during specific harvest periods within any given season. Rains, 'glut periods', holding loads longer than necessary under detrimental conditions, careless operators of mechanical harvesters and handling equipment, certain cultivars that are not truely adaptable to mechanical harvesting and handling, and lack of adequate help compound the defect and MOT, EVM, filth and other problems.

The following are some specific defects and some acceptable methods for measuring same. Some of the ways of determining the amount of defects in a load of tomatoes or in a processed tomato product are quite obvious and will not be discussed. These include such things as insects, rodents, stems, sticks, stones, etc.

MOT AND OTHER MATERIAL

The first method is to adequately sample the incoming load of tomatoes and remove a given weight of the sample. This sub sample should then be washed in a simulator to the conventional washing system to remove the mud, dirt, sand, trash, vines, etc. The washed sample is then reweighed and the percent MOT etc. is calculated by dividing the original weight by the washed weight and multiplying by 100 to convert the number to percentage. The result is the percent unusable tomatoes in a given load. Some firms have tolerances as low as 2% while other firms may allow up to 6 or more percent. (See Chapter 6).

The following material is essentially taken from the USDA Grading Manuals and USDA Standards for Grades for Tomatoes and Tomato Products (see references).

SAND AND INORGANIC RESIDUES

Sand and Inorganic Residue is one of the most common problems on extracted and comminuted and concentrated products. It can be a problem with even canned tomatoes. Here its a matter of pouring the contents of the can over a screen and washing the tomato with a fine spray and catching the washings in a suitable container. After settling for 2 minutes and decanting off the serum and juice, the sand and inorganic residue will be in the bottom of the container.

For juice and concentrated products the following modified procedure is taken from the USDA Inspector's Handbook and is as follows:

1. Pour 600 ml of a sample into a 1 liter beaker.
2. Fill beaker with water, stir well with a spoon, remove spoon and agitate with a rotary motion for 1 minute.
3. Allow to settle for 2 minutes or longer, if necessary, and decant ¾th of the liquid.
4. Add, at least, 500 ml of water and repeat the process in 2 and 3 above until liquid is fairly clear.
5. Continue adding the juice or tomato product in such portions and repeating the process until the entire contents of the container are used. Thoroughly rinse the container and add the rinse water to the beaker. (Obviously, if the sample came from containers larger than a #10 can it would not be necessary to use more than a 5 lb sample from a given container).
6. Agitate, allow to settle, and decant all but about 150 ml of the liquid.
7. Swirl the remainder in the beaker, causing the sand or grit to collect in a circle in the center of the bottom of the beaker and examine for sand or grit. (I prefer to pour the remainder in the beaker onto a #10 Wattman filter paper in a 4 inch Buchner funnel and with a low vacuum pull the liquid through the paper and leave the sand and inorganic residue on the filter paper for actual enumeration of sand and inorganic particles).

DARK SPECS, SEEDS, PIECES OF SEEDS, PEEL, HARD CORE MATERIAL

In most normal operations one would not expect to find dark specks, pieces of seeds, whole seeds, peel, or core material present in juice or concentrated products; however, broken screens, unclean equipment, improper processing causing burn-on, over extraction, and/or errors in judgement can cause these defects to appear in processed products. In Grade A Tomato Juice the combination of defects should only slightly affect the appearance or drinking quality of the product. Whereas, in Grade B Tomato Juice, the presence of defects should not seriously affect the appearance or drinking quality of the product.

The USDA has established guidelines to interpret the above two statements for various defects based on color differences of the defects, size of the defect particles, and the types of the defects. Light brown or black specks less than 1/32nd of an inch are classed as minor defects, while dark brown or black specks between 1/32 and 1/16 of an inch are major defects, and dark brown or black specks larger than 1/16th of an inch are classed as severe defects in tomato juice. Light brown specks, peel or seed particles 1/16 to 1/8th of an inch are considered as major defects and light brown specks, peel or seed particles larger than 1/8th of an inch are severe defects in tomato juice. The tomato juice is simply poured onto clean white trays or glass bottom trays in a very thin layer (maximum of 3/8ths inch thick) and examined by looking at the sample using sizing gauges to measure the defects. (I prefer the glass tray and hold this over a white light for observing the defects should they be present). The data in Tables 21.1 and 21.2 show the criteria for evaluating tomato juice for defects:

TABLE 21.1 — ACCEPTABLE (AC) / REJECTION (RE) NUMBERS FOR VARIOUS DEFECT CLASSIFICATIONS FOR TOMATO JUICE

Grade	Defect Classification						
	Severe		Major		Total		Whole Seed Tolerance
	AC	RE	AC	RE	AC	RE	
A	1	2	4	5	8	9	Not more than 1 whole seed/500 ml
B	2	3	8	9	16	17	Not more than 3 wholeseeds/500 ml

TABLE 21.2 — SCORING OF DEFECTS IN TOMATO JUICE

Grade	Points	Severe	Major	Minor/Total	No. of Seeds /500 ml
A	15	0	0	2	0
	14	0	2	6	0
	13	1	4	8	1
B	12	1	4	10	1
	11	1	6	12	2
	10	2	8	16	3

DEFECTS IN CATSUP

The defects in catsup are quite similar to those described for tomato juice and are caused by similar circumstances. Again the USDA in the U.S. Standards for Grades for Tomato Catsup has defined the types and kinds of defects along with objective scoring practices as shown in Tables 21.3, 21.4, and 21.5.

TABLE 21.3 — CLASSIFICATION OF TYPE AND KIND OF CATSUP DEFECTS

Color, Kind, Type, or Size of Defect	Classification		
	Minor	Major	Severe
LIGHT BROWN specks or seed particles 1/31 to 1/16 inch	X		
More than 1/16 inch but not more than 1/8 inch		X	
More than 1/8 inch			X
DARK BROWN or BLACK SPECKS			
1/32 inch or less	X		
More than 1/31 inch but not more than 1/16 inch		X	
More than 1/16 inch			X
PEEL			
More than 1/8th inch but not more than 3/16th inch	X		
Over 3/16 inch		X	

TABLE 21.4 — ACCEPTANCE (AC) AND REJECTION (RE) NUMBERS FOR VARIOUS DEFECTS BY CLASSIFICATION FOR TOMATO CATSUP

Grade	Defect Classification						
	Severe		Major		Total		Seeds
	AC	RE	AC	RE	AC	RE	
A	2	3	6	7	12	13	Not more than 1 seed/100 oz.
C	5	6	15	16	30	31	Not more than 5 seeds per 100 oz.

TABLE 21.5 — SCORING GUIDE FOR DEFECTS IN TOMATO CATSUP

Grade	Points	Severe	Major	Minor/Total
A	25	0	0	0 - 2
	24	0	1 - 2	3 - 5
	23	0	3 - 4	6 - 8
	22	1	5	9 - 10
	21	2	6	11 - 12
C	20	3	7 - 9	13 - 18
	19	4	10 - 12	19 - 24
	18	5	13 - 15	25 - 30

The scoring of the defects in canned tomatoes is probably the most difficult of all the tomato products because of the types of defects that could be in the product. These include blemished tomatoes, discolored tomatoes, extraneous vegetable materials, objectionable core material, and peel (attached or loose in the juice or container).

Peel is considered a defect whether attached to the tomato unit or loose in the sample unit except when scoring unpeeled tomatoes. The peel should be spread out and the surface area measured and scored according to the scoring system as outlined in Table 21.5.

Blemished tomatoes are tomatoes that have the appearance of scarred raised scabby tissue, dark tough areas around the core, dark tissue around the blossom ends, and any other well-defined unsightly objectionable areas. Defects or imperfections are discolored portions such as sunburn areas, cloudy spots, and internal browning. Cloudy spots that are large, whitish and unsightly or pale yellow are considered a discolored portion. Internal browning (tomato mosaic) is considered discolored if it slightly affects the appearance of the product. Internal browning that materially affects the appearance is scored as blemished.

Core material is considered objectionable if the stem scar is over ¼th inch in diameter and if there is tough or fibrous material associated with the core.

Table II in the Appendix for Canned Tomatoes summarizes the tolerances permitted for the above defects for each of the tomato Grades, that is, Grade A, Grade B, and Grade C.

Extraneous vegetable material (EVM) includes harmless material such as stems, calyx bracts, tomato leaves, internal sprouted seeds, and other material not necessarily related to the tomato plant. One EVM point is allowed in each 100 ounce increment of product for Grade A; 2 EVM points are allowed in Grade B; and 4 EVM points in Grade C. The factors are weighted as follows:

Each 1/16th square inch of flat material (bract, leaf, etc. equals one (1) EVM point.
Each 1/2 inch in length of tomato vine or other round tomato vegetable material equals four (4) EVM points.
Each stem (with/without) bract equals four (4) EVM points.
Each tomato containing sprouted internal seeds over 5/16th inch equals one (1) EVM point.

To calculate the number of EVM points allowed in a sample, divide the total net weight of the sample representing the lot by 100. This is the number of EVM points allowed for Grade A in the sample. This number multiplied by two (2) is the number of EVM points allowed in Grade B. The number of EVM points allowed in Grade A for the sample multiplied by four (4) is the number of EVM points allowed in Grade A.

The USDA cites the following example for interpretation as follows:

> "21 sample units comprise a sample of #10 cans of whole tomatoes with a total net weight of 2,142 ounces. The total ounces, 2,142 divided by 100 equals 21.42 EVM points allowed in Grade A for stems, sepal bracts, and internal sprouted seeds. The grader found 3 stems (12 points), 6/16 sq. in. sepals (6 points), 3 tomatoes with internal sprouted seeds (3 points) or a total of 21 points. The sum of this example is less than 21.42 points allowed for Grade A, therefore, it meets the EVM lot tolerance for U. S. Grade A. for EVM'.

REFERENCES

ANON. U.S. Department of Agriculture Canned Tomatoes Grading Manual, April 1990.

ANON. U.S. Department of Agriculture Tomato Catsup Grading Manual, September 1990.

ANON. U.S. Department of Agriculture Tomato Juice Grading Manual, April 1986.

ANON. U.S. Department of Agriculture Method of Analysis for Tomato Products-Sand and Inorganic Residue. February 1968.

CHAPTER 22
Flavor and Flavor Evaluation

The flavor factor is important from the pesticide and chemical additive standpoint and from the standpoint of determining suitable varieties of tomatoes for processing, maturity levels, product formulation, and processing methods, as well as the actual storage periods and conditions of the processed products. Flavor is also important from the standpoint of acceptability of food products for repeat sales.

There are many techniques used for the evaluation of flavor in processed foods. However, the subjective (human evaluation) approach is still one of the best methods to determine the actual acceptance of the product. There are two general types of flavor evaluation methods in present-day use, namely, the consumer acceptance (preference) method and the panel difference method.

The consumer acceptance method is usually employed when a new product is being developed, a change in the manufacturing procedure of the standard product is deemed advisable, or for constant quality checking of the manufactured product versus competitors' products. This type of flavor evaluation is ideally conducted with a true cross section of the market in which the manufacturer wishes to promote his product. The number of members on the panel varies. Such factors as income, food habits, age, and sex may have considerable influence on consumer acceptance of the product.

The consumer acceptance testing method is being used by several large companies. However, it is very expensive for routine quality control of manufactured fruits and vegetables. The panel difference method is without doubt the best approach to use here.

In practice, the panel difference method should be used prior to the consumer acceptance test. If the panel finds the difference between the samples significant, then to find the actual consumer acceptance or rejection, the product should be submitted to the consumer acceptance panel.

At the present time flavor-difference panels can be classified into six general types: (1) paired comparison tests, (2) triangle test, (3) dilution test, (4) ranking test, (5) numerical scoring test, (6) descriptive terms.

Before discussing these general types of difference testing methods, consideration should first be given to the following factors:

1. The actual selection of the judges (their availability, sensory acuity, health, etc.).
2. The size of the panel (a small well-selected and well-trained panel is believed to be better than a large unselected and untrained panel).
3. The preparation of the sample (temperatures of the product, cooking and serving procedure, size of the sample, and quality differences other than flavor).
4. The conditions of the judging room (use of booths, control of odors, type of lighting, temperature, humidity, etc.), and
5. The number of samples to serve at any one time (if the samples vary widely in their flavors, more samples can be evaluated at one time than if they vary less).

In the paired comparison method, two unknown samples are submitted to the judges, and the judges are asked which sample has the better flavor. In practice, the judges may be given a preliminary sample (warm-up sample) and then they are asked to tell which of the two unknowns is the same as the warm-up sample. This method is used by some panels.

In the triangle method, the judges are given three samples, two of which are duplicates. The judges are asked, "Is there a difference among the samples?" This method is very good when evaluating a small number of samples and when attempting to determine small differences in flavor between the two samples.

The dilution testing method is valuable when determining differences with homogeneous materials. This method is successful when used to find actual quantities of insecticides present in processed products and product formula variation. The success in using this type of panel is determined to a great extent by the threshold acuity (level) of the panel members, at least, when working in the lower concentrations of the flavoring materials.

The ranking test finds excellent application for routine quality control of ingredient mixing, production changes, etc. Here the judges are asked to rank the samples in decreasing or increasing order of flavor intensity. With this method, samples from a day's run can be ranked in order of flavor differences. Then, to find out if the differences between the best and the poorest samples are significant, a triangle test can be conducted.

In commercial practice, however, the numerical scoring method has probably found the greatest application. In most USDA standards for juice, pulp, catsup, and other products where flavor is a factor, the technologist actually scores the product according to its worth on a numerical scale. In some standards, flavor may hold a prominent place, while in others it may be entirely neglected. In the routine quality control of plant production, the technologist can score the product; then the extremes can be further evaluated by a panel using the triangle method. Probably the greatest problem with the numerical scoring method is the lack of a consistent and constant reference point. Therefore, the use of a trained panel to score the product continuously is the only satisfactory method known. The technologist should set up a panel of flavor judges and constantly refer samples to this panel throughout the campaign to make certain that the quality meets product specifications.

The last difference test is the descriptive terms method. Here, little application has been made in the processing industry. However, the method offers great possibilities. The Arthur D. Little Laboratory's approach, the flavor profile system, would seem to offer much to flavor evaluations from a control angle, at least. Judges attempt to describe the flavor or give their impressions of it.

Regardless of the method used, the results are of little value unless they are taken carefully and unless the panel can repeat a test on the same samples at any given time. This means that in each test check samples should be used. In this way, one can detect good and poor judges because they must score the check samples the same each time. Further, judges should be selected on their ability to evaluate flavor differences rather than on their relative position in the company.

Interpretation of data obtained from any of the above methods is without doubt the most difficult phase of present-day flavor evaluation. Dr. John W. Tukey's quick and easy statistical procedure is highly recommended for the interpretation of the flavor data. It is reported to be simple and yet accurate. This method of analysis permits an evaluation of results by simply adding the appropriate totals, the ranges (differences between highest and lowest levels), and multiplying the sum of the ranges by a factor obtained from a reference table.

The summary tabulation form (Table 22.1) was the one developed specifically for this type of flavor evaluation.

Four steps are presented here. Table 22.1, the summary sheet made from the score sheets of a 15-member panel, shows the results of tests on tomato juice.

A. The first step eliminates from the summary sheet the judges who failed to find much difference between the three "treatments."

B. The scores of all remaining judges are tested to see if there was a measurable difference among the three juices. This is done by finding the overall significant difference.

C. The extent of the difference between any two of the juices is then determined by finding the least significant difference.

D. Finally, the importance of the samples rejected is assessed. This is done by testing the percentage of replicate samples judged "not acceptable."

In the example, which is followed step by step, the terminology conventional among food technologists is used. For those not familiar with these

TABLE 22.1. SUMMARY SHEET MADE FROM SCORES OF 15 JUDGES TESTING TOMATO JUICE FOR OFF-FLAVORS

Treatment	Replicate	Flavor Difference Scores for Indicated Judge																R of N	
		1	2	3	4	5	6[a]	7	8[a]	9	10	11	12[a]	13	14	15[a]	Total	Sum	NA[b]
A	I	2	5*[c]	2	3	7*	7*	3	1	1	4	5*	3	6*	5*	3			
Code	II	3	1	2	2	6*	3	3	3	3	3	3	1	9*	3	1			
19225	III	4	7	2	3	4*	5	7*	1	2	3	3*	2	7*	2	1			
	IV	3	6*	2	3	4*	7*	5*	1	2	2	1	1	1	2	3			
	Sum	12	19	8	11	21	22	18	6	8	12	12	7	23	12	8	136	15	13
	Range (R)	2	6	0	1	3	4	4	2	2	2	4	2	8	3	2	% NA = 36.1		
B	I	2	5*	4	3	5*	5*	7*	1	6*	3	7*	3	6*	7*	1			
Code	II	5*	5*	5*	3	5*	3	7*	1	6*	7*	5*	3	6*	5*	3			
29225	III	6*	9*	1	5	6*	5*	5*	3	5*	5*	3*	1	7*	1	1			
	IV	5	9*	4*	3	6*	5*	7*	1	5*	2	3*	2	3*	2	5*			
	Sum	18	28	14	14	22	18	26	6	22	17	18	9	26	15	10	191	14	28
	Range	4	4	4	2	1	2	2	2	1	5	4	2	1	6	4	% NA = 77.8		
C	I	2	1	3	3	1	3	1	1	5	1	1	1	1	3	1			
Control	II	1	1	1	2	1	6*	1	1	3	1	1	1	1	3	1			
09225	III	2	1	2	1	1	3	1	3	2	1	1	1	1	3	1			
	IV	1	1	4*	1	1	5*	1	4	2	1	1	1	1	1	1			
	Sum	6	4	10	7	4	17	4	9	12	4	4	4	4	10	4	49	8	0
	Range	1	0	3	2	0	3	0	3	3	0	0	0	0	2	0	% NA = 0		
Grand sum of ranges		7	10	7	5	4	9	6	7	6	7	8	4	9	11	6			
Grand range of sums		12	24	6	7	18	5	22	3	14	13	14	5	22	5	6	142 Grand R of Totals		
O.S.D.	(1.25)	8.8	12.5	8.8	6.3	5.0	11.3	7.5	8.8	7.5	8.8	10.0	5.0	11.3	13.8	7.5			

[a] These judges (3, 6, 8, 12, 14, 15) were unable to distinguish flavor differences and their data were eliminated from consideration.
[b] NA = not acceptable.
[c] An asterisk was placed next to sample scores when sample flavor was judged not acceptable.

terms, "treatment" refers to the material under test. In the example two tomato juice samples with off-flavors, Treatments A and B are used with a sample of juice with no off-flavors, Treatment C (Control).

"Replicates" could be interpreted as "rounds." Each judge gets three small cups containing a sample of each "treatment" (coded of course) in each replicate. The test panel shown on the summary sheet had the procedure repeated four times, giving four replicates.

The detailed steps below should be followed to determine whether or not there is a statistical difference between any two treatments.

For Each Judge

1. The sum of the scores in all the replicates for each treatment. For example, for Judge 1, the sum of the scores for Treatment A is 12, for Treatment B, 18, etc.

2. The range (difference between the highest and lowest score) within each treatment (i.e., Judge 1, Treatment A, the range is 2, for Treatment B, 4, etc.).

3. The grand sum of the ranges computed in 2 above, i.e., for Judge 1 it is 2 + 4 + 1 = 7.

4. The grand range of the treatment sums computed in (1) above, i.e., the highest treatment sum for Judge 1 occurred in Treatment B (18), the lowest sum in Treatment C. The difference between these two is the grand range of 12 for Judge 1.

5. Calculate the overall significant difference (O.S.D.) between treatment sums as follow: Enter Table 22.1 and obtain the appropriate factor in the 5% column for the number of replicates and treatments used. For example, in the illustration (Table 22.1. the number of replicates being 4 and the number of treatments 3, the factor at the 5% level of significance is 1.25. Enter this factor on the summary form and multiply the grand sum of the treatment ranges for each judge by this factor. The product which is the O.S.D. is then entered at the bottom of the column for each judge, i.e., for Judge 1 it is 1.25 × 7 = 8.8. A comparison of this product with the grand range of the treatment sums will indicate whether or not the judge rated the various treatments significantly different. The range between the highest and lowest treatment score should be greater than the O.S.D. value, i.e., for Judge 1 one grand range of treatments sums is 12, which is higher than the O.S.D. value of 8.8. Whenever the range between the highest and lowest treatment sum for a given judge is equal to or less than the O.S.D. value, it indicates lack of ability to distinguish between any of the treatments.

6. Evaluation of judge performance: It will be noted in Table 22.1, after completing the calculations for O.S.D. values for the 15 judges, that judges 3, 6, 8, 12, 14, and 15 had grand range of treatment sums which were equal to or lower than their individual O.S.D. values at the 5% level. These judges, therefore, were unable to distinguish flavor differences, and in the illustration (Table 22.1) their data were eliminated from further consideration.

For Each Treatment

1. Add the sums of replicate scores for each of the judges not eliminated. For example, the total of the sums for Treatment A for the 9 remaining judges is 136, for B, 191, and for C, 49. These figures should be entered in the total column to the right side of the summary tabulation form.

2. Compute the range of the sums for the remaining judges in each treatment and enter this value in the column on the right side of the summary tabulation form headed Range of

TABLE 22.2. MULTIPLIERS OF THE RANGE FOR COMPUTING OVERALL SIGNIFICANT DIFFERENCE IN EVALUATING JUDGES' PERFORMANCES

No. of Judges or Replicates	No. of Treatments and Significance Level															
	2		3		4		5		6		7		8		9	
	5%	1%	5%	1%	5%	1%	5%	1%	5%	1%	5%	1%	5%	1%	5%	1%
	Multipliers															
2	3.43	7.92	2.37	4.42	1.78	2.96	1.40	2.06	1.16	1.69	1.00	1.39	0.87	1.20	0.78	1.03
3	1.91	3.14	1.44	2.14	1.13	1.57	0.94	1.25	0.80	1.04	0.70	0.89	0.62	0.78	0.56	0.69
4	1.63	2.47	1.25	1.74	1.01	1.33	0.84	1.08	0.72	0.91	0.63	0.78	0.57	0.69	0.51	0.62
5	1.53	2.24	1.19	1.60	0.96	1.24	0.81	1.02	0.70	0.86	0.61	0.75	0.55	0.66	0.50	0.59
6	1.50	2.14	1.18	1.55	0.95	1.21	0.80	0.99	0.69	0.85	0.61	0.74	0.55	0.65	0.49	0.59
7	1.49	2.10	1.17	1.53	0.95	1.21	0.80	0.99	0.69	0.84	0.61	0.74	0.55	0.65	0.50	0.59
8	1.49	2.08	1.17	1.52	0.96	1.21	0.81	0.99	0.70	0.85	0.62	0.74	0.55	0.66	0.50	0.59
9	1.50	2.09	1.18	1.53	0.97	1.22	0.82	1.00	0.71	0.85	0.62	0.75	0.56	0.66	0.51	0.60
10	1.52	2.10	1.20	1.55	0.98	1.23	0.83	1.01	0.72	0.86	0.63	0.75	0.57	0.67	0.52	0.61
11	1.54	2.11	1.21	1.56	0.99	1.24	0.84	1.02	0.73	0.88	0.64	0.77	0.58	0.68	0.52	0.61
12	1.56	2.13	1.23	1.58	1.00	1.25	0.85	1.03	0.74	0.89	0.65	0.78	0.59	0.69	0.53	0.62
13	1.58	2.15	1.25	1.60	1.02	1.27	0.86	1.05	0.75	0.90	0.66	0.79	0.59	0.70	0.54	0.63
14	1.60	2.18	1.26	1.62	1.03	1.28	0.87	1.06	0.76	0.91	0.67	0.80	0.60	0.71	0.55	0.64
15	1.62	2.20	1.28	1.64	1.05	1.30	0.89	1.08	0.77	0.92	0.68	0.81	0.61	0.72	0.56	0.65
16	1.64	2.22	1.30	1.65	1.06	1.31	0.90	1.09	0.78	0.93	0.69	0.82	0.62	0.73	0.56	0.66
17	1.66	2.24	1.31	1.67	1.08	1.33	0.91	1.11	0.79	0.95	0.70	0.83	0.63	0.74	0.57	0.67
18	1.68	2.27	1.33	1.69	1.09	1.34	0.92	1.12	0.80	0.96	0.71	0.84	0.64	0.75	0.58	0.68
19	1.70	2.30	1.34	1.71	1.10	1.36	0.93	1.14	0.81	0.97	0.72	0.85	0.65	0.76	0.59	0.68
20	1.72	2.32	1.36	1.73	1.11	1.38	0.93	1.15	0.82	0.98	0.73	0.86	0.65	0.77	0.59	0.69

Source: Condensed from tables prepared by Thomas E. Kurtz, Richard F. Link, John W. Tukey, and David L. Wallace.

Judge Sums. In this example, for Treatment A the highest sum, 23, was given by Judge 13 and the lowest sum of 8 by Judge 9. The range is therefore 15, and is entered on the next column on the same line as the 136.

3. Count the number of asterisks in all of the replicates for a single treatment for the 9 remaining judges and record the number in the column at the extreme right of the sheet headed Number Not Acceptable. In the example given in Table 22.1, number of asterisks for Treatment A for 9 judges is 13. For later evaluation of the significances, this number should be converted to a percentage. In the example cited, 13 is 36.1% of the total 36 evaluations made for Treatment A.

All Treatments, All Judges

1. Determine the range of the total scores for each treatment that were computed for B-1 above. In Table 22.1 this value is 142, which is obtained by subtracting 49 (sum for Treatment C) from 191, the sum for Treatment B. Enter this figure on the summary tabulation form on the lower right hand at the base of the Total Column.

2. Obtain the judge total by adding the range of judge sums for each treatment. In the example cited, the 3 ranges are 15 for A, 14 for B, and 8 for C, making a total of 37. This figure should be entered at the base of the column headed Range of Judge Sums.

3. The next step is to determine the overall significant difference (O.S.D.) values to determine whether any significant differences exist among treatments. To do this, obtain the appropriate factors from Table 22.2. These factors are found in Table 22.2 in the column for the number of treatments that were used and on the line for the number of judges. In the example used, the appropriate factors are found in column 3 "for number of treatments" and on line 9, "for number of judges." For significance at the 5% level, the figure is 1.18 and at the 1% level, 1.53. Multiply this factor by the judge range total, 37. This will give 43.7 at the 5% level and 56.6 at the 1% level. In the example, Table 22.1, the grand range of totals is 142, which greatly exceeds the O.S.D. value of 56.6, and, therefore, a highly, significant difference among treatments exists.

4. The next step is to determine the least significant difference (L.S.D.) values which will permit an evaluation of differences between any two treatments. It is necessary, however, that the O.S.D. value be significant before L.S.D. values are calculated and specific comparisons between treatments are made. (To be significant the O.S.D. value at the 5% level of significance must be less than the grand range of totals.) The procedure for calculating L.S.D. values is essentially identical to that used for O.S.D. In the example, the judge range total, 37, is multiplied by the L.S.D. factor found in Table 18.3 in the column headed 3 (number of treatments) and on the line 9 (number of judges). The L.S.D. value at the 5% level is $0.98 \times 37 = 36.3$; at the 1% level it is $1.34 \times 37 = 49.6$. Therefore, the differences between any two treatment totals (A — C = 87, B — C = 142, and B — A = 55) are significant at the 1% level in this illustration.

Table 22.4 contains the minimum percentage not acceptable that is necessary for significance at the 1% level for the indicated number of flavor difference evaluations. For example, with the 36 flavor difference evaluations used in this illustration, a minimum percentage of 20.5 is needed for significance. Hence, the percentages rated not acceptable in both Treatments A and B (36.1 and 77.8) are highly significant.

TABLE 22.3. MULTIPLIERS FOR COMPUTING THE LEAST SIGNIFICANT DIFFERENCES WHEN USING THE SIMPLIFIED FLAVOR DIFFERENCE PROCEDURE

No. of Judges or Replicates	No. of Treatments and Significance Level															
	2		3		4		5		6		7		8		9	
	5%	1%	5%	1%	5%	1%	5%	1%	5%	1%	5%	1%	5%	1%	5%	1%
							Multipliers									
2	3.43	7.92	1.76	3.25	1.18	1.96	0.88	0.39	0.70	1.07	0.58	0.87	0.50	0.74	0.44	0.63
3	1.63	3.14	1.14	1.73	0.81	1.19	0.63	0.91	0.52	0.73	0.44	0.61	0.38	0.53	0.33	0.46
4	1.91	2.47	1.02	1.47	0.74	1.04	0.58	0.80	0.48	0.68	0.40	0.55	0.35	0.48	0.31	0.44
5	1.53	2.24	0.98	1.37	0.72	0.98	0.56	0.77	0.47	0.63	0.40	0.54	0.34	0.47	0.30	0.43
6	1.50	2.14	0.96	1.32	0.71	0.96	0.56	0.76	0.46	0.62	0.40	0.53	0.34	0.46	0.30	0.42
7	1.49	2.10	0.96	1.33	0.71	0.96	0.56	0.76	0.47	0.63	0.40	0.53	0.35	0.46	0.31	0.42
8	1.49	2.08	0.97	1.33	0.72	0.97	0.57	0.77	0.47	0.63	0.41	0.54	0.35	0.47	0.31	0.42
9	1.50	2.09	0.98	1.34	0.73	0.98	0.58	0.77	0.48	0.64	0.41	0.55	0.36	0.48	0.31	0.43
10	1.52	2.10	0.99	1.35	0.74	0.99	0.59	0.78	0.49	0.65	0.42	0.55	0.37	0.48	0.32	0.43
11	1.54	2.11	1.00	1.35	0.74	0.99	0.59	0.79	0.49	0.65	0.42	0.56	0.37	0.49	0.32	0.43
12	1.56	2.13	1.01	1.36	0.75	1.00	0.60	0.80	0.50	0.67	0.43	0.57	0.38	0.50	0.33	0.44
13	1.58	2.15	1.03	1.38	0.76	1.01	0.61	0.81	0.51	0.68	0.43	0.57	0.38	0.50	0.34	0.45
14	1.60	2.18	1.04	1.39	0.77	1.03	0.62	0.82	0.52	0.69	0.44	0.58	0.38	0.50	0.34	0.45
15	1.62	2.20	1.06	1.42	0.79	1.05	0.63	0.84	0.52	0.69	0.45	0.60	0.39	0.52	0.35	0.46
16	1.64	2.22	1.07	1.43	0.80	1.07	0.64	0.85	0.53	0.70	0.45	0.60	0.40	0.53	0.35	0.46
17	1.66	2.24	1.08	1.44	0.81	1.08	0.65	0.86	0.54	0.72	0.46	0.61	0.40	0.53	0.36	0.48
18	1.68	2.27	1.10	1.46	0.82	1.09	0.65	0.86	0.54	0.72	0.47	0.62	0.41	0.54	0.36	0.48
19	1.79	2.30	1.11	1.48	0.83	1.10	0.66	0.88	0.55	0.73	0.47	0.62	0.42	0.56	0.37	0.49
20	1.72	2.32	1.13	1.51	0.83	1.10	0.67	0.89	0.56	0.74	0.48	0.64	0.42	0.56	0.37	0.49

TABLE 22.4. MINIMUM PERCENTAGE OF "NOT ACCEPTABLE" JUDGEMENTS REQUIRED FOR SIGNIFICANCE AT THE 1% LEVEL

No. judgments	20	25	30	31	32	33	34	35	40
Min %	31.8	26.5	23.5	23.0	22.0	21.5	21.0	20.5	18.5
No. judgments	45	50	55	60	65	70	75	80	
Min %	17.0	15.5	14.5	13.5	12.5	11.5	11.0	10.5	

REFERENCE

GOULD, W.A. 1958. Flavor: Can it be measured? Canner/Packer *38* (13) 10–14.

CHAPTER 23
Drosophila and Insect Control

Drosophila is the generic name of a group of small flies that feed on and breed in plant material (Wilson 1952). Although the fly is a constant menace to all fruit canners, the emphasis here is on their detrimental effects on tomatoes and tomato by-products.

The drosophila fly may be referred to as a vinegar gnat, fruit fly, sour fly, or pomace fly. Common to the species are populations of *Drosophila pseudoobscura*, *D. immigrans*, *D. hydie*, *D. busckii*, and the species causing the most problem, *D. melanogaster*.

The fly, as such, is not the subject of as much concern as the eggs it deposits and the subsequent stages of development of the fly in the life cycle. For many years it was erroneously assumed that the drosophila fly laid eggs only on fermenting and rotting plant products where the adults were feeding (Wilson 1952). Field observations have revealed that eggs are deposited on fresh growth cracks in sound tomatoes on the vine and in cracks resulting from the picking operation. The eggs are deposited by means of an adhesive substance common to the fly, and they are not easily or readily removed. Because of this, the egg stubbornly resists removal by all known washing methods. As such, drosophila represents a potential threat of ultimate seizure of the finished product in conjunction with Section 402(a) (3) of the Federal Food, Drug and Cosmetic Act, which defines a food as adulterated if it consists in whole or in part of any filthy, putrid, or decomposed substance.

LIFE CYCLE, HABITS, AND OTHER FACTORS

The life cycle of *D. melanogaster* varies from 5 to 8 days, with an average summer cycle of 7 days. Other cycles for species at 85°F (30°C) are *D. immigrans*, 9 days; *D. busckii*, 11 days; and at 70°F (21°C), *D. pseudoobscura*, 14 days. Starting with the adult, the life cycle progresses as follows:

Adult

The adult of the *D. melanogaster* species is a transparent-winged fly, about 1/8 in. long. The body is yellowish, and the abdomen is crossed by dark bands (Anon. 1962A). Females live longer than males; higher temperatures markedly decrease the life-span of each. The female fly may begin laying eggs during the second day of adult life and continues at the rate of about 26 per day. A single fly may lay as many as 2000 eggs (Wilson 1952).

Eggs

The eggs are pearly white, elongate, and about 1/50 in. long. Eggs of *D. melanogaster* have two appendages or filaments attached near the head end. Eggs of other species may be slightly longer and have two to four appendages.

Individual eggs are too small to be seen easily by the unaided eye. Eggs usually are laid with their appendages or filaments above the surface of the medium in which they are placed (Anon. 1962A). These serve as respiratory organs (Pepper et al. 1953). The eggs may then hatch into larvae in about 24 hr.

This time may be drastically reduced, however, because if flies do not find suitable media on which to lay, the eggs will be retained in the body. The fly is capable of holding eggs and depositing them shortly before hatching or actually depositing live larvae which hatch within the fly itself (Wilson 1952).

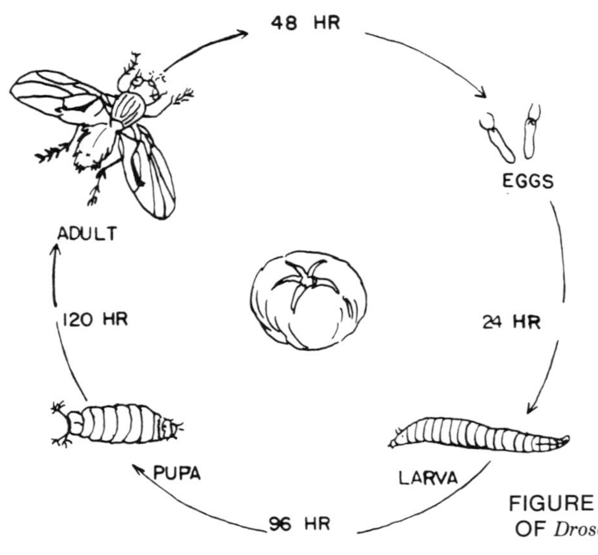

FIGURE 23.1. LIFE CYCLE OF *Drosophila Melanogaster.*

Larvae

The larvae, commonly referred to as Maggots, may be quite transparent, or they may appear to be cream colored or some other shade caused by food in the gut (Anon. 1962A). When first hatched, larvae measure about 1/30 in. (Wilson 1952). They shed their skin three times and may mature into the pupa stage in about 4 days, at which time they are about 1/4 in. long. Larvae may confine their feeding to the cracks in fruit and free juice. Upon reaching maturity, they seek a dry place in which to pupate.

Pupae

The pupae are about 1/8 in. long. At first they are yellowish-white, then turn amber, and finally turn brown within a few hours. The anterior end of a pupa is broader and flatter than the posterior end, and it has two stalklike structures that bear the respiratory organs (Anon. 1962A). The pupal period may require 5 or more days during transformation into an adult fly (Pepper et al. 1953).

Five factors affect the activity of the drosophila fly: (1) temperature, (2) moisture, (3) food, (4) light intensity, and (5) air movement (Michelbacker 1958).

A temperature of approximately 55°F (13°C) is close to the lower threshold of activity, an optimum temperature is 75 to 80°F (24 to 27°C) and a temperature over 90°F (32°C) eliminates most activity. It has been found that fruit exposed to the sun on a hot day may reach 125°F (52°C) or higher; this is sufficient to kill the larvae. The flies usually are more active in the morning and late afternoon. During the heat of the day, they seek shelter in the shade of rank tomato vines and in grass and weeds (Anon. 1962A).

Moisture favors adult feeding and maintains moist surfaces for egg deposition. Further, it prevents the rapid drying out of breeding sources (Michelbacker 1958). Rainy periods at any time during the season are favorable to activity by directly lowering the temperature and light intensity.

Drosophila, like most other insects, are attracted from considerable distances by the odors of plant materials, especially after fermenting begins. Adults feed on fermenting plant materials; apparently, yeast constitutes an important part of their diet (Wilson 1952). There is a marked difference in the food habits of the adults and larvae. Ideal breeding media for the larvae are sound, ripe fruits with fresh moist cracks or breaks in the skin (Michelbacker 1958).

Adult flies are most active when the light intensity approaches that of about sundown. As such, most activity occurs between the hours of 6 to 8 AM and 4 to 8 PM when the light and temperature are lower. The rate of egg deposition during these periods has been found to be 25 to 35 times the rate of egg deposition during the rest of the day (Anon. 1960).

Most flight occurs when it is calm, and the direction of flight is into the drift. Strong winds tend to pin down the flies (Michelbacker 1958). Wind velocities above 15 mph curtail flight, but flies may remain active, feeding, mating, and egg laying in the mass of tomato vines, provided other conditions are favorable (Wilson 1952).

DROSOPHILA CONTROL BEFORE AND DURING HARVESTING

The control of fly eggs and larvae in the finished product must begin at the onset of the tomato season. Adherence to the following points has proved successful in drosophila control before and after harvesting.

1. Do not locate tomato fields near peach orchards or melon fields. Do not locate tomato fields near fruit or sweet potato storage houses, or near places where organic refuse is deposited (Anon. 1962A).
2. Grow a good cultivar of tomatoes that is crack resistant. Ripe tomatoes that have fresh cracks or varieties that crack easily upon harvesting attract egg-laying flies.
3. Plant in rows far enough apart to permit cultivation and spraying to be carried out with minimum damage to plants and fruit. Wide row spacings will also minimize picker damage to plants and fruit during harvesting. Where high per acre plant populations are desired, space plants closer together (Anon. 1960).
4. Keep fields free from weeds and grass. Such growth provides shade and lower temperatures for the flies during hot weather and facilitates egg laying.
5. Provide clean roadways, preferably running across the rows at convenient distances, for stacking the field containers and for use by the pickup trucks. Fruits and plants crushed during harvest operations by trucks, containers, or pickers provide an excellent breeding ground for drosophila (Anon. 1960).
6. Do not delay field spraying so as to allow large populations of drosophila to develop. Insecticide applications should start as soon as the flies are present in the field. Their presence may be detected by the slit-tomato test.

To make the test, obtain several ripe tomatoes. Make two vertical slits about 1½ in. long on opposite sides of each. Do not cut through the wall of the tomato. Squeeze the tomato enough to slightly open the slits and free some juice. At about 4 PM, place one of the slit tomatoes at each of several random locations in the tomato field. Collect them at about 9 AM the following morning and examine the slits under a magnifying glass. If eggs are present, start insecticide applications (Anon. 1962A).

Once started, insecticide applications should be repeated at 4- or 5-day intervals as long as the presence of drosophila flies or eggs indicate control is needed (Anon. 1962A).

For field application to the tomato plants, one of the following materials can be used (Anon. 1962A):

a. Diazinon: ¾ lb of active ingredient per acre in a spray or dust, or 1 lb per acre in granules, or

b. Malathion: 2 lb of active ingredient per acre is a spray or in granules,

c. Aldrin: ½ lb of active ingredient per acre in a spray, dust or in granules.

Any of these formulations can be applied with power equipment or band seeders. Sprays can be applied at a rate of 50 to 200 gal. per acre (Anon. 1962A).

The fly and egg control of harvested fruit is best accomplished with the aid of pyrethrum dust, as described in later paragraphs.

7. Do not delay the first picking. Harvest frequently thereafter, to prevent the accumulation of overripe cracked tomatoes.

8. Remove tomatoes from the field as soon as possible after harvesting. If at all possible, do not let harvested fruit remain overnight in the field or at the factory.

9. Do not be careless or rough in handling tomatoes during or after harvesting (Anon. 1962A). Studies have revealed that if adult flies are abundant and active, up to 75% of the eggs found on the fruit at the canning plant will have been laid in these fresh cracks after the fruit is picked (Anon. 1960).

10. Do not overfill baskets or boxes or containers or crush the fruit when stacking or loading filled containers.

11. Dust the harvested fruit during and after loading. Extensive experiments and industry experience have revealed that pyrethrum dust containing 0.1% stabilized pyrethrin oxide applied as dust to tomato fruit will repel drosophila flies and largely prevent egg deposition for 24 hr. The dust should be applied to lug boxes, hampers, and other containers of tomatoes from the top and through the sides of the stack (Anon. 1960). Other fogs, mists, and sprays are available for use; pyrethrum dust, however, has proved to be the most successful (Anon. 1960).

12. Use clean, dry containers for harvesting the fruit. Bickley and Ditman (1953) indicate that treating containers in an insecticide dip can effectively reduce egg deposition.

DROSOPHILA CONTROL AT THE PLANT AND DURING PROCESSING

Processing

There are a number of plant areas and unit operations that must receive close scrutiny in the event of a quality tomato pack. Specifically, areas that initiate the tomato process, such as grading stations, receiving stations, and storage yards around the cannery, must be given a proper amount of attention.

Harvested loads arriving at the processing plant should be inspected in some manner. Loads that fail to meet minimum quality standards as set forth by the processor should be rejected. Also, a system of scheduling to prevent long lines and backups should be maintained. Accepted loads should be dusted immediately with pyrethrum dust.

Areas around the receiving station and cannery should be kept clean and free from refuse. Frequent inspections of all involved areas should be made. Surfaces in contact with transported tomatoes should be hard and level so that they can be kept clean. The importance of sanitation in the canning operation in and around the plant cannot be overemphasized. Waste material carelessly overlooked can support thousands of drosophila.

The success of each unit operation will determine the acceptability of the tomato pack. Since no one unit operation is successful in satisfactorily decontaminating tomatoes, a number of operations must perform with utmost efficiency.

Common among these unit operations are fluming, soaking and washing, sorting, and trimming. Each of these operations is discussed in detail under "Preparation of Tomatoes for Processing," Chapter 7. As stated in that chapter, drosophila eggs and larvae removal are best accomplished when the flume is sufficiently long and air or steam agitated, when the soak tank is heated and supplied with sufficient amounts of detergent and chlorine, and when tomatoes are spray-washed with adequate amounts of water at a proper pressure. The results of adding detergents to flumes and soak tanks are also covered in Chapter 7. The consensus is that tomatoes soaked in a low-foaming detergent with an alkaline solution (0.2% concentration of a lye solution) at a temperature of 130°F (55°C) for 3 min will significantly reduce egg and larva loads.

Egg reductions up to 94% using warm 140°F (60°C) lye solutions have been reported (Anon. 1961). The addition of alkali to initially low pH, low foaming detergents can effect an increase in egg removal of from 35 to 90%. Thus, proper soaking in a detergent–alkaline solution not only drastically reduces drosophila eggs and larvae, but can also reduce the amount of water otherwise required to achieve similar results. Lastly, the addition of chlorine to the soak tank and frequent water changes help inhibit the thermophile bacteria buildup facilitated by the high temperature water.

After a proper amount of soaking, the fruit is sprayed to complete the cleansing operation. The spray nozzle, location, and pressure (as described in Chapter 7) are so constructed as to remove caustic and detergent residues along with loosened eggs and larvae. It is advantageous to chlorinate the spray water. This may be done in the same manner as soak tank chlorination to achieve a concentration of 6 to 8 ppm. Even though the water in the spray rinse is chlorinated, it should be discarded after the rinsing operation. The eggs and larvae contamination along with the mold and rot make this water unsuitable for use (Gould 1971).

The tomatoes are then conveyed over inspection tables where they must be meticulously sorted and trimmed. In many instances, this is the last opportunity to remove undesirable fruit, i.e., culls, rot, mold, and any other extraneous material. The efficiency of the trimming, sorting, and inspection may be the most important factor in determining the acceptability of the pack.

Certain criteria must be met to ensure the success of this operation. First, sorters and trimmers must be properly trained and supervised. A person properly trained but lacking adequate supervision does not perform at peak efficiency. Likewise, a person adequately supervised but poorly trained in specifically what to remove or trim cannot perform satisfactorily. In addition, the inspector should only be expected to work at top efficiency for a relatively short length of time without a break or change of jobs. Sorters and trimmers should be segregated. An initial sort of culls and undesirable fruit should be performed before the trimming operation. Sorters must not have the responsibility of trimming.

The conveyor should be of the roller type at the sorter station. Conveyors over 30 in. wide should have sorters and trimmers stationed on both sides. Loads over one layer deep on the belt must not be permitted. Ideal belt speed should be 20 to 25 ft per min and be adjustable as the situation demands. Other suggestions for the sort and trim operation are listed in Chapter 7. Lighting above the belts should be adequate and at an angle which eliminates glare. Light sources depicting north sky daylight (7500° Kelvin) are recommended. Lastly, all sorted and trimmed material should be disposed of in such manner as not to contaminate clean, washed fruit.

A system of random tomato inspections at regular intervals at the end of the sort belts is one method of determining the efficiency of the sort and trim operation. A trained inspector can grade the fruit and make adjustments on the spot before the product is canned. Some canners use a statistical quality control method in which limits based on tomato defects are established. Inspections that fall outside established limits are considered "out of control" and appropriate action is immediately taken.

The final check the processor must make during the canning operation is a laboratory analysis to determine the amount, if any, of drosophila eggs or larvae present in the finished product. This gives day-to-day assurance that a quality pack is being maintained.

METHODS OF DROSOPHILA DETECTION

The purpose of performing a fly egg and larvae count is to accurately determine their presence and, as such, to ascertain exactly how many are present. The ideal method for making this determination should be simple, fast, and accurate; i.e., the method should show a definite number of eggs or larvae, if indeed, such exist.

There are a number of methods available for the detection of drosophila eggs and larvae. The tolerance for eggs and larvae as set forth by the Food and Drug Administration are listed in Table 23.1.

TABLE 23.1. FDA DEFECT ACTION LEVEL FOR DROSOPHILA EGGS AND LARVAE

Product	Sample Size (g)	No. of Eggs	No. of Larvae
Tomatoes	500	10 or	
		5 and	1 or 2
Tomato juice	100	10 or	
		5 and	1 or 2
Tomato purée	100	20 or	
		10 and	1 or 2
Tomato paste, pizza, and other sauces	100	30 or	
		15 and	1 or 2

Different methods for the detection of drosophila eggs and larvae are described.

GOSUL Method

This is a simple and fast method of detection that can be applied either to the processed or unprocessed product. The method consists of the application of a 100-watt long-wave ultraviolet light source to a series of filters that have had the sample placed upon them. Moisture is removed from the sample by means of a Büchner Funnel with the aid of applied vacuum. As mentioned, advantages include simplicity and speed; the entire test can be conducted in 20 to 30 min, along with a very high percentage of accuracy. Expenditure for equipment is nominal (Geisman and Gould 1963).

Equipment

1. 100-watt long-wave (3660) ultraviolet light source
2. #5 Büchner funnel
3. Several 18.5 or 20-cm "sharkskin" filter papers
4. 2000-ml filtering flask
5. Two 500-ml beakers
6. 9 in. × 9 in. glass or plastic plate
7. Aspirator pump or other vacuum source
8. Binocular wide-field microscope (10×) or 10× hand lens
9. 100-ml graduate cylinder
10. Teasing needle
11. One pair tweezers

The reagents required are vinegar (40 or 50 grain) or 4 or 5% acetic acid.

Procedures

1. Thoroughly mix sample. This can be accomplished by vigorously shaking the sample in a container with an up-and-down motion 200 times.

2. Open container and pour a 100-ml aliquot of sample into a graduated cylinder.

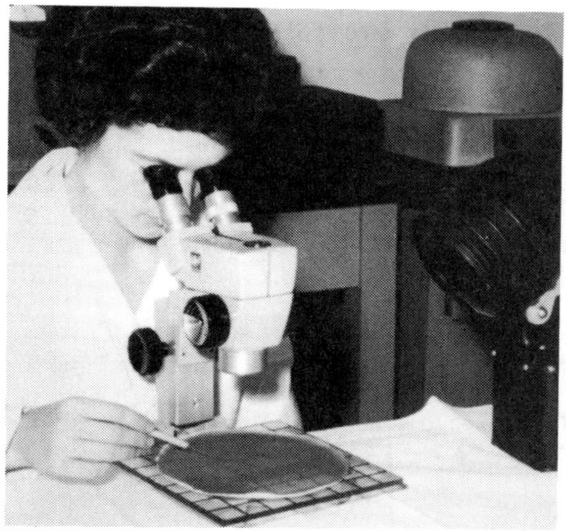

FIGURE 23.2. BLACK LIGHT EXAMINATION FOR FLY EGGS.

Tomato Juice

1. Seat #5 Büchner funnel securely into a 2000-ml filtering flask and attach flask to aspirator pump or other vacuum source. Moisten an 18.5 or 20-cm "sharkskin" filter paper with water spread evenly in funnel.
2. Pour approximately 50-ml aliquot onto the seated filter paper. This can be accomplished by carefully pouring sample into center of paper and rotating flask and filter to spread juice to sides. Apply vacuum to filtering flask to remove moisture. Paper should be nearly dry before it is removed from funnel. Repeat this operation with the rest of the samples and apply rinse water from graduated cylinder to last paper.
3. Remove dry filter paper from flask with aid of tweezers and place on a previously lined glass or plastic plate. Plate may be lined into 1-in. square to aid in examination.
4. Using a 100-watt long-wave ultraviolet light source and a binocular wide-field microscope, examine the paper for eggs and larvae. For best results, the light source should be positioned so that beams strike the paper at approximately a 45° angle. Eggs and larvae appear blue-white and may be easily detected.
5. Record number of eggs and larvae separately for each sample and report as number per 100 g.

Tomato Pulp and Paste

1. A 100-ml aliquot of tomato pulp should be diluted with 200 ml warm

(100°F) water and 100 ml tomato paste diluted with 200 ml of warm water. This aids in filtering the sample.

Catsup

1. Dilute catsup with 250 ml vinegar (40 or 50 grain) or 4 or 5% acetic acid solution (100°F water may be used if acid is not available). Stir mixture thoroughly. The acid aids in the removal of the sugar present and thus prevents a haze which would otherwise form.
2. Follow steps (1) to (5) as for juice (requires 6 to 8 filter papers).

Unprocessed Tomato Juice

1. Pour only 34 ml of raw juice onto filter paper at one time. Spread juice evenly, as before (requires 3 filter papers).
2. Follow steps (1) to (5) as for juice, except use 34-ml portions of aliquot instead of 50-ml.

Recommendations

1. Sample should be thoroughly mixed before analysis.
2. Aliquot must be spread evenly and thinly over surface of the filter paper in order to prevent some eggs and larvae from being covered with tomato fibers.
3. Best counting results are obtained in a darkened room.
4. After an analyst has become familiar with the counting technique it is possible to use a 10× hand lens instead of the microscope without loss of accuracy.

AOAC Method

This method relies on the principle that fly eggs and larvae will settle out of a two-medium solution, whereas most insects or hair float. Being relatively dense, the eggs and larvae will settle in a water–gasoline mixture. A separatory funnel is used instead of trap flasks that are utilized to recover insects or insect parts. The gasoline–water mixture is vigorously stirred and allowed to settle, permitting most of the plant tissue to rise and thereby releasing the eggs and larvae. At regular intervals, a 15 to 20 ml portion is drained off. After an adequate amount of draining, the water phase plus any eggs and larvae is filtered through a 10×× bolting cloth. The cloth is used instead of paper because the openings are fine enough to catch even small larvae but not large enough to trap tomato tissue and fiber that would obscure the microscopic examination (Wilson 1952).

The equipment and procedure are as follows (Gould 1971):

Equipment

1. 2000-ml separatory funnel
2. Ringstand
3. No. 10 sieve
4. 600 ml beaker

5. 10×× bolting cloth
6. Hirsch funnel
7. 10× binocular wide-field microscope
8. Teasing needle

Reagents

1. Gasoline

Procedure

Comminuted Products

1. Mix sample thoroughly and transfer 100 g to a 2-liter separatory funnel.
2. Add 20 to 30 ml gasoline and shake thoroughly, releasing pressure as necessary.
3. Fill separator with water in such a manner as to produce maximum agitation.
4. Place separatory funnel in a ringstand and let settle. At 15-min intervals during 1 hr, drain 15 to 20 ml from funnel, and gently shake funnel with a rotary motion to facilitate settling out of the fly eggs and larvae.
5. If drained liquid contains seeds, pass it through a No. 10 sieve, and rinse seeds and sieve thoroughly, recovering both liquid portion and rinse water in beaker.
6. Filter through 10×× bolting cloth in Hirsch funnel.
7. Examine for eggs and larvae at about 10×. If fly eggs or larvae are found in this examination, continue separation, and draining as above for additional hour.

Canned Tomatoes

1. Pulp entire contents of can in such a way that a minimum number of eggs and larvae are crushed or broken. This may be done by passing the material through a No. 6 or No. 8 sieve and adding seeds and residue remaining on the sieve to the pulp.
2. Place 500 g of the well-mixed pulped tomatoes in a 6-liter separatory funnel.
3. Add 125 to 150 ml gasoline and about 1 liter of water.
4. Shake vigorously, releasing pressure as necessary.
5. Fill funnel with water.
6. Place funnel in ringstand and let layers separate.
7. At 15-min intervals during 1 hr drain 25 to 30 ml from bottom of funnel and gently shake funnel with rotary motion to facilitate settling of fly eggs and larvae. Each portion may be examined at once or combined with subsequent portions.
8. Pass drained portions through a No. 10 sieve and rinse seeds and sieve thoroughly, recovering both liquid portion and rinse water in a beaker.
9. Filter through 10×× bolting cloth in Hirsch funnel.
10. Examine cloth for eggs and larvae at about 10×.

11. If eggs or larvae are found in this examination, continue separation and draining as above for an additional hour.

Staining Method

This is another relatively simple method that is based on the staining of tomato solids with crystal violet dye in preference to fruit fly eggs and maggots. They appear pearly white in comparison with the blue-colored juice.

Equipment

1. Beaker, 600-ml or 500-ml
2. Glass funnel with rubber tubing and pinch clamp
3. Glass plate with under surface painted flat black
4. Flood lamp or strong light source
5. 10× magnifying lens
6. Thin spatula

Reagents

1. 1% crystal violet dye; dissolve 1 g of dye in 99 g alcohol.

Procedure

Tomato Juice

1. Weigh a 100-g sample of juice in a beaker and dilute with an equal volume of water.
2. Add 4 ml of 1% crystal violet dye to the sample. Mix and allow to stand for 1 min.
3. Pour the dyed juice into the glass funnel with rubber tubing and pinch clamp.
4. Allow the sample to slowly flow downward over the glass pane with light source on. Position the light to eliminate reflected light and glare.
5. Accurately count eggs and larvae present. The unstained eggs and larvae appear pearly white and are obvious in contrast with the dark blue tomato material and black surface upon which they are viewed.
6. Remove any eggs or larvae with a thin spatula in question and observe under a 10× magnifying lens.
7. Wash the beaker and funnel thoroughly after being emptied and inspect in same manner.
8. If there is a question relating to the sample upon completion of the assay, it is repeated by pouring the sample back over the viewing surface.

DETERMINATION OF INSECT FRAGMENTS IN TOMATO PRODUCTS

Almost all emphasis has been placed up to now on the drosophila problem plaguing the tomato processor. There is, however, another similar problem facing the processor: the removal of insects in or on the products delivered to the plant for processing. Failure to do so would result in as much cause for seizure as infestation with drosophila eggs or larvae.

Successful control of insects and insect fragments on raw product is fortunately synonymous with that applied to drosophila. If adequate measures have been taken to control egg and larvae infestation, fragment contamination of the finished product should also be minimal.

Proper preharvest spraying in the fields, dusting after harvest, and in-plant washing, sorting, and trimming are of course essential. Two other factors, touched on briefly in preceding pages, are also essential in insect control, namely, insect control around all areas of the plant plus insect control inside the plant.

Control of insects outside the plant depends on basic and sound sanitation practices. Any insanitary area or place containing organic refuse will draw insects in abundant numbers. Immediate cleanup and continuous vigilance concerning any insanitary condition are essential.

Weeds and lawns should be kept under control. Trees and shrubs should not be located near the plant. Removal or consistent spraying of such greenery is a must.

Lastly, the routine use of one of a number of insecticides outside the plant cannot be overlooked. The first application should begin before the start of the canning season (Anon. 1962A). Repeat applications at about 10-day intervals during the season. Spray the outside walls of the plant and inside walls and ceilings of sheds used for holding tomatoes or containers. Spray waste hoppers to the point of runoff both top and underneath.

One of the following insecticides may be used to form a spray (Anon. 1962A):

1. Diazinon: 13 oz of active ingredient per 10 gal. water
2. Naled: 13 oz of active ingredient per 10 gal. water
3. Ronnel: 13 oz of active ingredient per 10 gal. water

These formulations should not be allowed to drift on unprocessed fruit. They cannot be applied while the plant is in operation or inside the plant. Automatic atomizers or misters can be used to apply pyrethrum spray mist outside the plant while in operation.

Control of insects inside the plant can be maintained by instituting similar measures. Of course, the first concern should be to keep insects from entering the plant. Reduction of outside populations by the above-described methods decreases the number that may seek entry. Standard practices to exclude insects at ports of entry include screening of all windows and doors with fine mesh screen, use of fly fans around employee entrances, and the closing off of all gaps and holes around walls, ceilings, and piping. Some processors use insect electrocuters to control insects during periods of darkness. Lastly, pyrethrum may be used inside the canning plant during periods when the plant is temporarily shut down and before cleanup. Atomizers can be used to apply this insecticide in enclosed areas. The objective is to fill the area with a mist or fog. When used in the plant, a thorough wash-down of all equipment should follow (Anon. 1960).

A laboratory analysis of the finished product must then follow to determine if any insects are present. The method most commonly used is oil flotation in a Wildman trap flask. Unlike eggs and larvae, insect material and insect fragments float, chiefly because insect cuticle is wet by oil and not wet by water (Harris and Reynolds 1960).

Insect Fragment Determination

The Wildman trap flask method utilizes the principle that oil will float on water as insects rise with the oil. The amount of lift exerted on an insect fragment depends partly on the specific gravity of the adhering oil compared to that of the aqueous medium (Harris and Reynolds 1960). The oil phase can then be examined under a microscope after filtering with the aid of vacuum. The equipment and procedure is as follows (Gould 1971):

Comminuted Tomato Products

Equipment

1. 2-liter Wildman trap flask
2. Büchner or Hirsch funnel
3. 1 box of rapid suction filter paper with hard finish and 1 to 2 cm wider than the interval diameter of the funnel
4. Petri dishes
5. Wash bottle

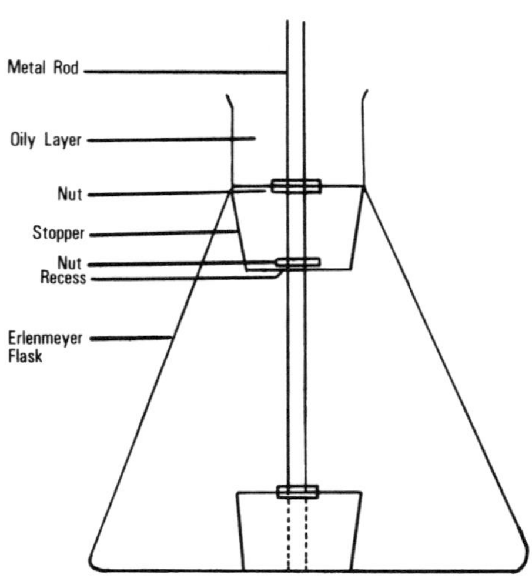

FIGURE 23.3. WILDMAN TRAP FLASK.

6. Castor oil
7. 95% alcohol
8. 20 to 30× wide-field microscope
9. 500-ml beaker

Procedure

1. Place 200 g of any tomato product except paste (where 100 g is used) in Wildman trap flask.
2. Add 20 ml castor oil and mix well.
3. Add enough warm tap water (approximately 100°F) to fill flask. At

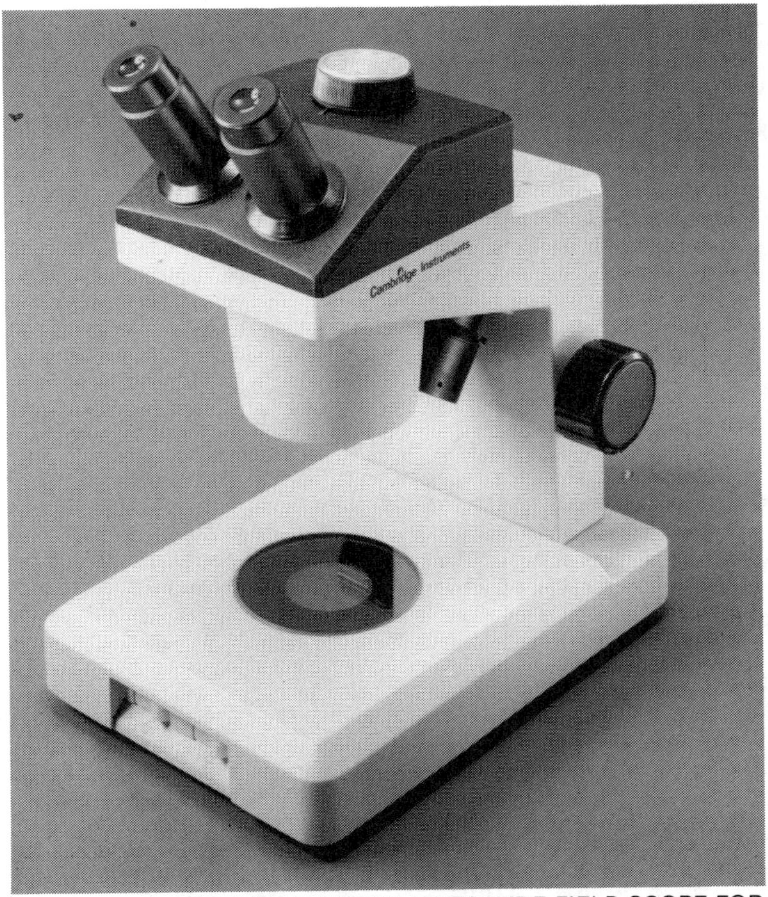

FIGURE 23.4. CAMBRIDGE INSTRUMENTS WIDE FIELD SCOPE FOR INSECT FRAGMENT EVALUATION.
Courtesy of Leica Inc.

first the bubbles may tend to bring up tomato tissue, but gentle rotary stirring with the metal rod will cause material to settle.

4. Let stand, with occasional stirring, for 30 min; then trap off oil layer into beaker.

5. Wash out the neck of the flask with alcohol to remove all castor oil.

6. Add additional warm water to fill Wildman flask, stir and let stand 10 min. Then repeat trapping operation.

7. Pour oil layer from beaker on filter paper which has been fitted into filter funnel and apply vacuum to filter flask until paper is dry. Be sure to wash beaker and filter paper with alcohol to remove oil.

8. Examine filter paper under 20 to 30× wide-field microscope and record any insect fragments.

SUMMARY

The positive control of drosophila and insect fragments in the finished products is possible if certain precautions are taken and other procedures followed. Proper sanitation in and around the plant, proper pre- and post-harvest practices, and proper soaking, washing, trimming, and sorting at the plant can combine to produce a quality pack.

Almost all practices and procedures cited here relate to the average problem of the tomato processor. There are, however, certain variations encountered with drosophila activity according to regional locations. The Eastern and Midwestern areas may encounter drosophila activity at any time of day during the latter part of the harvest season. The Western regions, however, encounter most activity in the early morning and early evening or on overcast days (Anon. 1960).

Due to the fact that most drosophila egg deposition occurs on fresh cracks caused by rough handling or the picking operation, the mechanical harvesting of the fruit may eventually reduce the problem if varieties are grown that resist cracking or injury. Nevertheless, the elimination of cracking and injury to harvested fruit would undoubtedly reduce the drosophila problem.

REFERENCES

ANON. 1960. *Drosophila* Control. Natl. Food Processors Assoc., April, Washington.DC.

ANON. 1961. Research Information. Natl. Food Processors Assoc., Berkeley, CA, Bull. *49,* Jan.

ANON. 1962A. Controlling *Drosophila* flies on tomatoes. U.S. Dept. Agric. Farmer's Bull. *2189.*

ANON. 1962B. Post harvest *Drosophila* control programs for tomato processors. Canning Trade *85* (2) 12, 13, 17.

BICKLEY, W.E. and DITMAN, L.P. 1953. *Drosophila* Problem in the Canning of Tomatoes. Md. Agric. Exp. Stn., College Park, Publ. *190*, Contrib. *2513*.

GEISMAN, J.R. and GOULD, W.A. 1963. Detection of *Drosophila* eggs and larvae in tomato products. The Dart *2-763*. Ultra-Violet Products, San Gabriel, CA.

GOULD, W.A. 1971. Suggestions for *Drosophila* Control in Canning Plants. Tomato Processor Quality Control Handbook. Dep. Hortic., Ohio State Univ., Columbus.

HARRIS, K.L. and REYNOLDS, H.L. 1960. Microscopic-Analytical Methods in Food and Drug Control. Food Drug Control, Food Drug Tech. Bull. *1*. U.S. Dep. Health, Educ., Welfare. Food Drug Admin., Washington, DC.

JUDGE, E.E. & Sons. The Almanac of the Canning, Freezing and Preserving Industries, Edward E. Judge & Sons, Westminster, MD.

MICHELBACKER, A.E., 1958. Biology and ecology of *Drosophila*. *Drosophila* Conf., Dec. 8–9, West. Reg. Res. Lab. Albany, CA.

PEPPER, B.B., REED, J.P. and STARNES, O. 1953. *Drosophila* as a pest of processing tomatoes. Ext. Serv. Coll. Agric. Rutgers Univ., New Brunswick, NJ. Ext. Bull. *266*, July.

SIEGEL, M. 1958. The tomato menace. Canning Trade *80* (Dec. 29).

WILSON, J.J. 1952. Chairman of Special Committee on Tomato Products Sanitation. Suggestions for Improving Sanitary Conditions in the Harvesting, Transportation and Processing of Tomatoes. Natl. Food Processors Assoc., Washington, DC. July 15.

CHAPTER 24

Mold – Counting Methods and Principles

INTRODUCTION

Mold counts are only one of several criteria for determining the acceptability of the processed tomatoes and tomato products. The Food and Drug Administration states that "(a) Although mold count is conclusive evidence of inclusion of substantial amounts of rot, mold count is not the only way of establishing that comminuted tomato product contain decomposed tomato material. (b) Where factory observations or other evidence reveals that comminuted tomato products contain rot not caused by mold, such rot, as well as that caused by mold, will be taken into account in applying the provisions of the Federal Food, Drug and Cosmetic Act against adulteration. (c) The *blending* of tomato products adulterated with tomato rot, of whatever kind, with tomato products made from sound tomatoes, or with other sound food, renders the blend adulterated."

THE MICROSCOPE

Proficiency in the microscopic examination of tomato products for mold requires skill in the use and care of the microscope (Anon. 1968). For mold-counting work, the microscope should be a basic compound microscope, either monocular or binocular. The latter is most desirable (U.S. Dep. Agric. 1967).

It should be equipped with 10× Huygenian ocular (eyepiece) containing Howard micrometer disk. Field diameter for the 10× objective−10× ocular combination must be 1.382 mm. (In the purchase of a new microscope, this diameter should be specified; a used or existing microscope should be modified by a microscope service specialist to give the correct field diameter. Some models of older microscopes have adjustable draw tubes. These can be manipulated to give the proper field diameter; frequent checking of the diameter is necessary to ensure that the body tube length has not been inadvertently changed.) It should also be equipped with an Abbé condenser and achromatic objectives of 10× and 20×. The microscope should contain a mechanical stage and substage lamp with daylight glass.

Construction of the Microscope

The basic compound microscope (Fig. 24.1) consists of two sets of optics mounted in a holder called the body tube. The set nearer the specimen, called the objective, magnifies the specimen a definite amount. The second lens system, the ocular, further magnifies the image formed by the objective, so that the image seen by the eye has a magnification equal to the product of the magnifications of the two systems. Each objective and ocular has its magnification factor engraved on it. Microscopes are generally supplied with several objectives, each having a different power. For mold-counting, two objectives are required: the 10× (16 mm) for counting, and the 20× (8 mm) for confirming identity of questionable filaments. The Howard micrometer disk is located within the ocular.

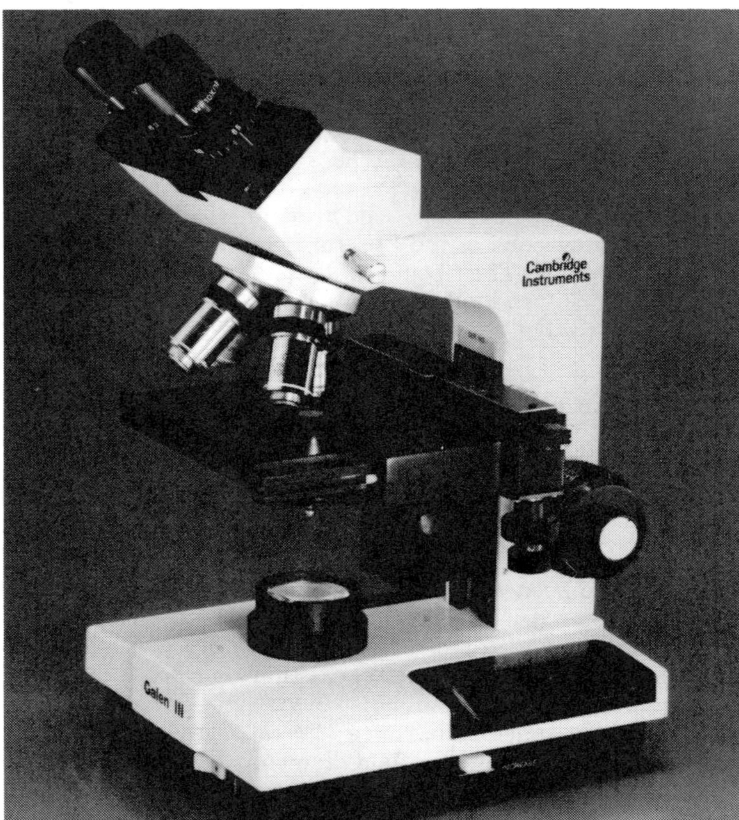

FIGURE 24.1. CAMBRIDGE INSTRUMENTS COMPOUND MICROSCOPE EQUIPPED WITH MECHANICAL STAGE FOR MOLD COUNTING.
Courtesy of Leica Inc.

The body tube (containing the ocular and objective) and a stage to support the specimen are mounted on a stand. Depending on the model of microscope, either the tube or the stage is equipped with a rack-and-pinion focusing mechanism. A mechanical stage is provided to enable the analyst to move the Howard slide with controlled longitudinal and lateral movements. With regard to moving the slide, an apparent movement of the slide to the left in the microscope field is actually a movement to the right on the stage, since the image formed by a compound microscope is seen inverted and reversed.

A condenser and mirror are located beneath the stage. The function of the mirror is to focus light rays into the condenser. The mirror has two surfaces: one plane and one concave. Choice of surface depends, for the most part, on the type of illumination and the objective being used. With the low-powered objectives and with the microscope lamps in common use, the concave mirror generally gives more uniform illumination. Experience with the equipment will dictate whether the concave or plane mirror is preferable.

The condenser contains a series of lenses, which assist in illuminating the field by providing a cone of light of sufficient angle to fill the aperture or opening of the objective. Some microscopes are equipped with condensers that are divisible. For low-power observations, such as at 100× magnification, the top element of the condenser can be removed to achieve more uniform illumination. A rack-and-pinion device, permitting vertical travel of the condenser, facilitates proper light adjustment. The condenser is provided with an iris diaphragm, which controls the amount of light entering the condenser and the angle of the emitted cone. Proper adjustment of the condenser is of extreme importance because it enables the analyst to control the light to achieve maximum visibility of microscopic particles in the specimen. If insufficient illumination reaches the specimen, visibility will be impaired; excess illumination will render invisible the smallest objects in the specimen (Anon. 1968).

Proper Use of the Microscope

1. Place the slide containing the specimen onto the mechanical stage and move it to a point where the specimen is approximately centered over the opening in the stage.
2. By means of the coarse adjustment, bring the objective to a point where it *almost* touches the cover glass. Do this *only* while observing the movement from the side of the microscope with your eyes near stage level. *Do not look into the eyepiece while making this adjustment*, because by doing so, you might damage the objective or break the cover slip.
3. While looking into the eyepiece, adjust the mirror to direct the light from the substage lamp, through the condenser and specimen, into the

objective. Then adjust the condenser height and the iris diaphragm opening as necessary to give even but not brilliant illumination over the entire field.

4. Look through the eyepiece and slowly move the objective away from the slide, by means of the coarse adjustment, until the image is fairly distinct. This movement must be slow so that the specimen image will not be missed. If there is too much light, so that the specimen cannot be seen easily against the glare, turn the disk diaphragm under the microscope stage until a smaller opening (aperture) is under the specimen.

5. Focus with the fine adjustment until the specimen is clearly defined. The specimen under examination may be moved as desired by means of the mechanical stage. During examination of each field of a Howard slide, the fine adjustment must be in almost continuous use in order to examine the total volume of sample within that field.

Use of binocular microscope differs from that of monocular microscope in that two additional adjustments must be made: (1) an adjustment of the separation of the two eyepieces to adapt it to the interpupillary distance of the analyst's eyes; and (2) an adjustment of the focusable eyepiece. The latter is done by focusing on the specimen through the fixed eyepiece (use fine adjustment); then adjust the focusable eyepiece until the specimen is clearly defined. A single sharp image will now be seen through both eyepieces (Anon. 1968).

Daylight may be used for illuminating the microscope by turning the mirror so that light from a window is reflected into the microscope. If the concave side of the mirror is used, the light is more intense, because the concave mirror acts like a lens to condense the light from the whole area of the mirror onto that covered by the objective.

Daylight is of variable quality and not always available. Consequently, artificial light is generally used for microscopy. Sunlight, except for special applications, should never be allowed to fall on the mirror or into the lenses of the microscope. It is not good for the microscope and is likely to injure the eye of the observer.

Two kinds of microscope illuminators are in common use. One is a simple lamp which has a uniformly illuminated surface of ground or opal glass. The other type has a focusable condensing lens to project an image of the lamp filament into the microscope. A small light source of the illuminated surface type may be placed in front of the microscope, or the mirror may be removed and the lamp set underneath the microscope condenser.

The lamp should be moved toward or away from the mirror until the field examined is evenly filled with light. When a concave mirror is used, this distance may easily be determined by placing a pencil on the lamp and then moving the lamp until the pencil point is in focus with the specimen. To make sure that the illumination is good, remove the eyepiece from the microscope and look at the back lens of the objective. This lens should be evenly and completely filled with light. If not so filled, the lamp and mirror should be readjusted until the best illumination is obtained (Anon. 1968).

Care of the Microscope

Successful microscopy requires skill and the proper care of the instrument. The microscope is a precision instrument made from valuable materials by expert workmen; with reasonable care, it will give many years of reliable service.

The microscope should be carried by its arm in an upright position and, when not in use, should be placed in its case or properly covered to protect it from dust. When the microscope is brought from a cold to a warm room it should be allowed to warm up gradually before being used (Richards 1958).

The lenses must be kept meticulously clean. Dust should be loosened and brushed off with a camel's hair brush or blown off with an aspirator, such as an all-rubber ear syringe (available from a drug store) (Richards 1958;

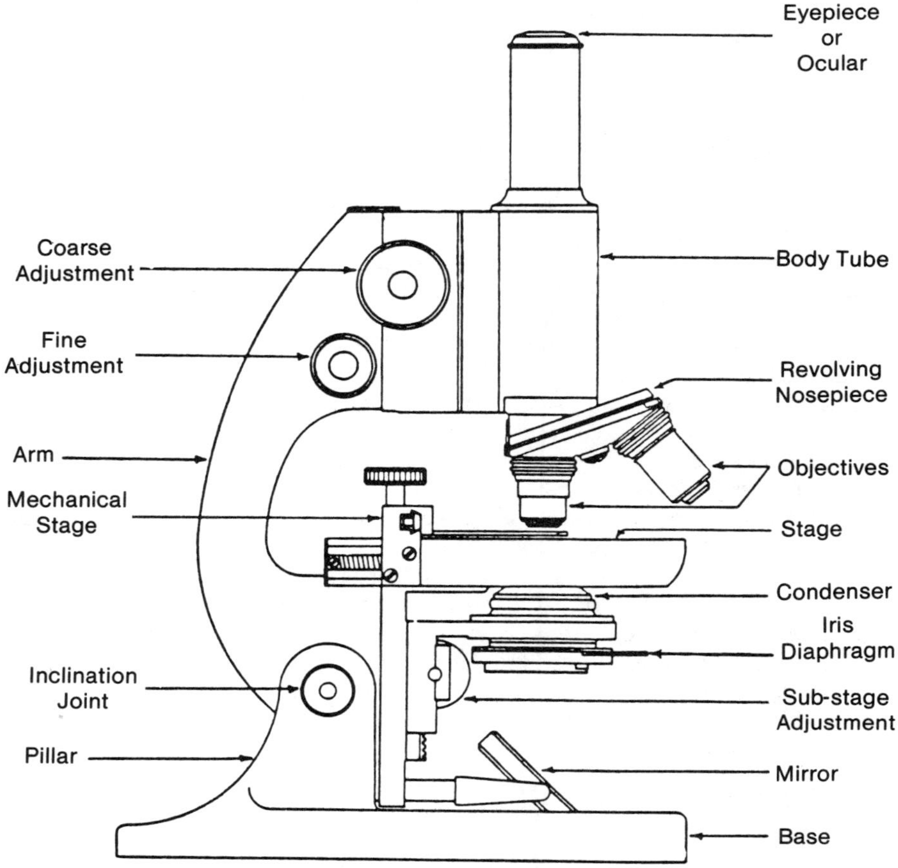

FIGURE 24.2. THE MECHANICAL PARTS AND ARRANGEMENT OF A TYPICAL MICROSCOPE

Anon. 1968). Lens paper (no substitute) may be used to remove smudges, such as those caused by eyelashes touching the upper lens of the eyepiece (Anon. 1968). Alternatively, a cotton swab wetted with acetone can also be safely used to clean the lens and will not leave lint associated with lens paper.

Dust on the eyepiece lenses is seen as specks which rotate when the eyepiece is turned while looking through it. Dirt on the objective prevents clear vision and the object appears as if it were in a fog. If a wet preparation touches the objective lens, the lens will have to be cleaned before one can see clearly through it. An eyepiece should always be kept in the tube to prevent dust from collecting on the back lens of the objective or on the prisms.

Should dust settle on the prisms or on the glass protection plates of the binocular body, blow it off with air from an aspirator. Compressed air from laboratory pipes may contain traces of moisture or of oil from the compressor and should not be used unless an absorbent cotton filter is placed on the discharge tube. If the field does not appear clear, it is well to examine the lower surface of the objective with a magnifying glass. Any dirt or damage to the lens may then be seen easily.

Objective lenses are carefully adjusted at the factory, and should not be taken apart except where they have been made to separate. If they must be taken apart, this should be done at the factory, where facilities are available for testing the reassembly.

The dry objectives, the condenser, and the eyepiece may be cleaned with distilled water when a liquid is necessary and the lens should be wiped dry with fresh lens paper immediately after cleaning.

In tropical regions, mold may grow on dirty lens surfaces when the relative humidity and the temperature exceed 80°F (27°C). When the optical surfaces are kept clean, mold growth can be avoided, or minimized, by keeping the optics in a desiccator, or the microscope in a warmed cabinet.

The surface of the microscope is finished with enamel or metal plating and requires little more care than keeping it clean and free from dirt. These finishes resist most laboratory chemicals; ordinarily a little mild soap and water are all that is necessary for cleaning.

Careless handling or dropping may disturb the adjustment of the optical parts of the microscope. If the instrument does not seem to perform properly and there is no dirt on the objective or the eyepiece, it may mean that some of the prisms have become shifted. Do not attempt to adjust any of the prism systems, but send the instrument to the factory, where tools and tests are available for adjustment and for making certain that the adjustment has been done properly (Richards 1958).

HISTOLOGY OF THE TOMATO

In addition to training in the use and care of the microscope, the analyst should also study the microscopic appearance of each part of a sound tomato. To accomplish this, he should cut very thin slices of each of the various parts of the tomato, place them on a clean glass slide in a drop or two of water, cover with a thin glass cover and examine under 100× magnification (Anon. 1968). Before taking sections of the tomato, it may be advisable to place the tomato in alcohol in order to harden the fruit (Eisenberg 1952).

With the razor blade cut a small section of the epicarp (skin) from the tomato. If this section contains flesh cells, they may be removed from the epicarp by soaking in a chloral hydrate solution (Eisenberg 1952). Epicarp cells are greenish-yellow in color and polygonal in shape. Although separated by the middle lamella, the individual cells fit together closely and show definite cell wall outlines. The cell walls are rather thick and amber in color. When examined at 100× magnification, a piece of tomato skin closely resembles the surface of hammered aluminum.

Mount, in a similar manner, a thin cross section of mesocarp, flesh cells, and examine under low power. Flesh cells are clear, thin walled, and considerably larger than skin cells. They vary in shape from oval to circular and because of their appearance are often referred to as "cellophane footballs." The cells lying just beneath the skin and next to the edge of the seed cavity are smaller and narrower. The cell contents are sparsely granular and appear to be made up of tapering lines which, in reality, are folds in the cell wall.

Fibrovascular bundles are the white, threadlike veins in the tomato flesh that carry moisture and nutrient throughout the fruit. Under the microscope they appear dark in color, and the vascular elements of the bundles resemble a series of tightly coiled springs. Rectangular, brick-shaped companion cells may frequently be seen attached to the coiled bundles. In comminuted tomato products, the bundles are broken up into many small

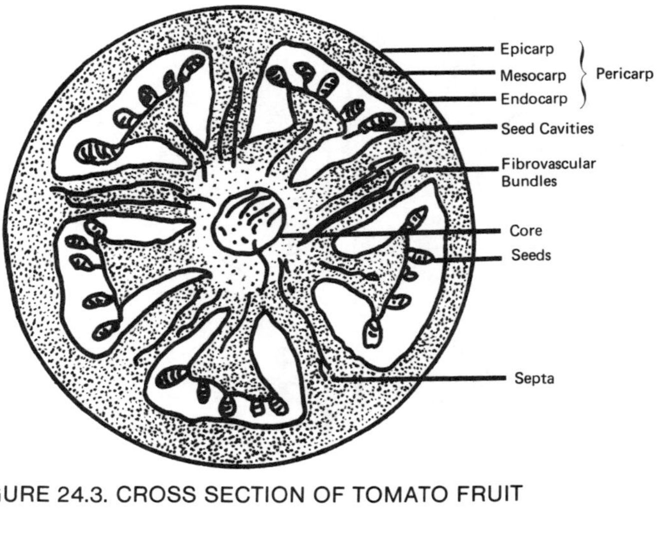

FIGURE 24.3. CROSS SECTION OF TOMATO FRUIT

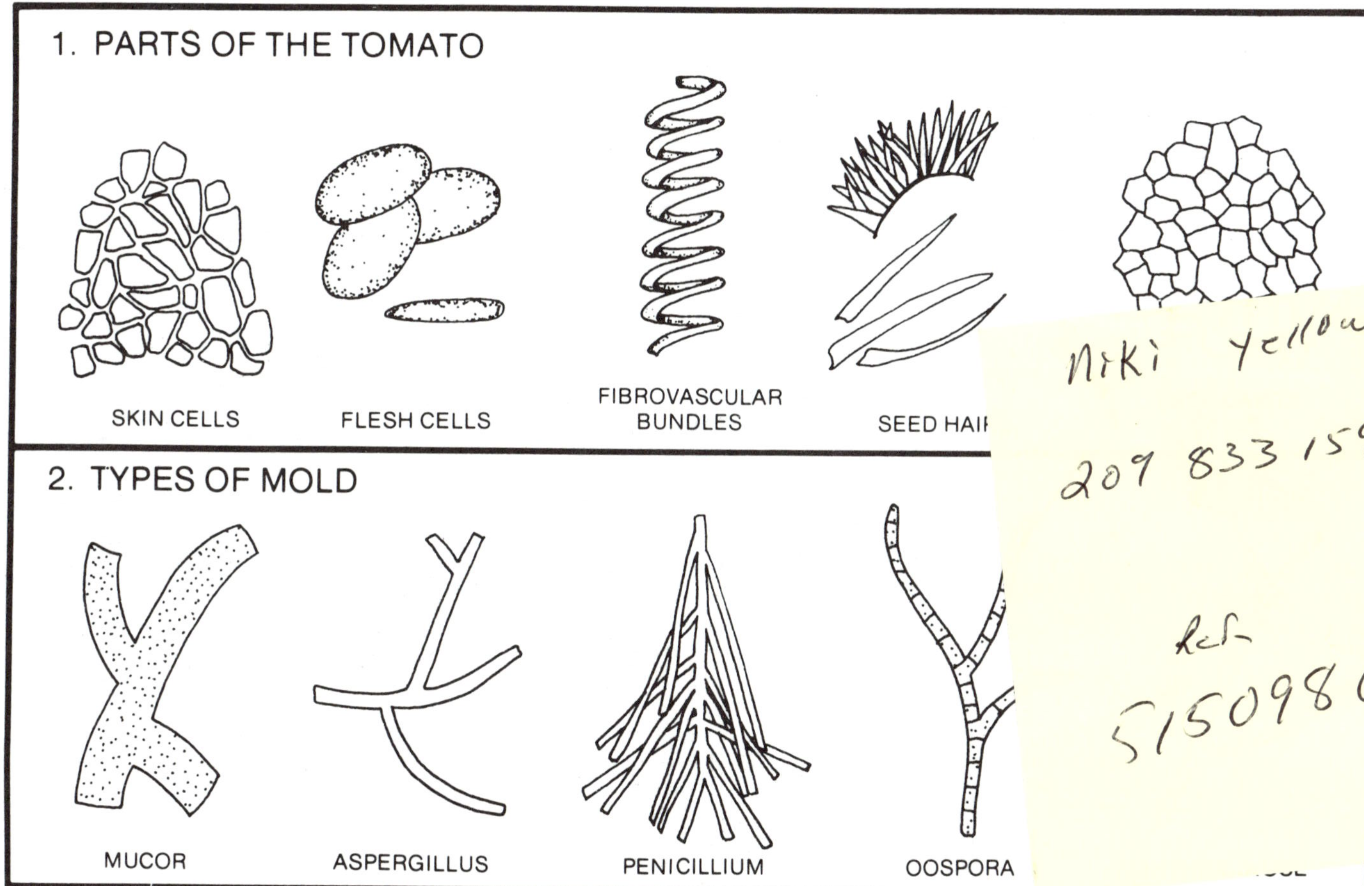
1. PARTS OF THE TOMATO
SKIN CELLS
FLESH CELLS
FIBROVASCULAR BUNDLES
2. TYPES OF MOLD
MUCOR
ASPERGILLUS
PENICILLIUM
OOSPORA

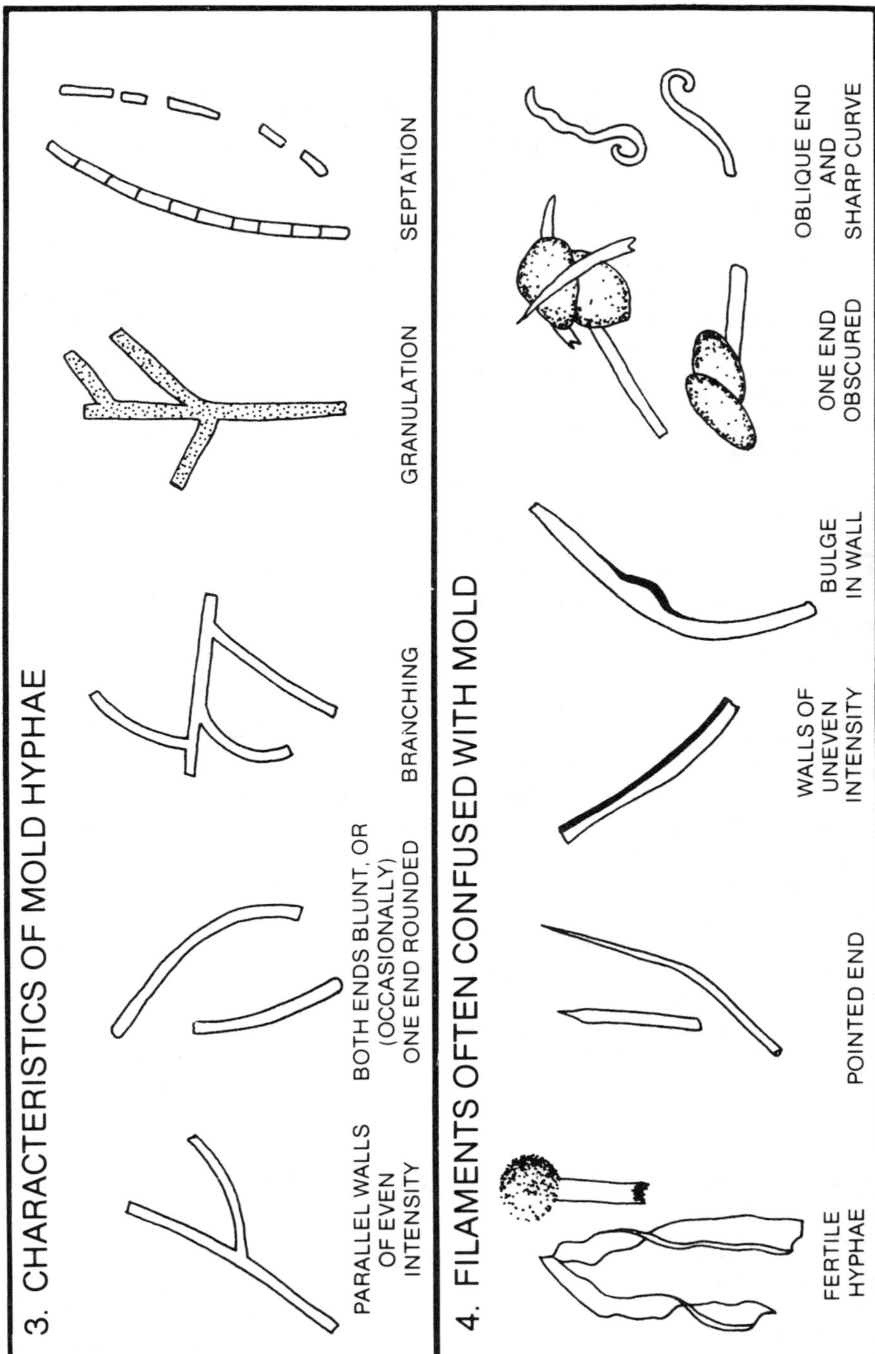

FIGURE 24.4. FILAMENTS FOUND IN TOMATO PRODUCTS.

pieces and occasionally a coil-like spring (resembling the letter "s" in shape) will be found in the finished product. These broken vascular cells can usually be distinguished from mold be a small loop or enlargement on the end, or an end that is pointed or slightly frayed.

Cut a section of the seed-cavity lining cells with a razor blade and place in the chloral hydrate solution. Soaking the section will remove some of the gelatinous material which interferes with the examination of these cells (Eisenberg 1952). The cells lining the seed cavity are smaller than flesh cells and are thin walled. They are irregular in shape and might be compared to the pieces of jigsaw puzzle with interlocking edges. The cell contents are almost clear, except for occasional flecks of cellular material (Anon. 1968).

Remove some of the core cells and place a section on a slide, add water and a cover glass, and examine under low power (U.S. Dep. Agric. 1967). Core cells are small and round and have rather thick walls. The contents of the larger cells are rather clear, while the smaller cells have a light amber color due to the more dense cellular materials. The large cells tend to occur in clumps surrounded by many small cells (Anon. 1968).

Lift out a portion of the gelatinous mass containing seeds and covering cells and place in the choral hydrate solution. Soak and remove a seed from the solution. Hold the seed with the tweezers and cut the seed in half with the razor blade. Place sections on a slide, add a cover glass, and press gently to separate the cells. Examine under low power (Eisenberg 1952). The internal seed cells vary in shape from rectangular bricklike cells through cubelike to almost ovoid or egg-shaped cells. Also, considerable color variation exists, depending on the compactness of the cell structure and amount and size of the cell contents. The darker cells are more compact and appear to contain more cytoplasm.

Sometimes portions of stem or sepals are encountered in tomato products. Make thin tangential sections using a razor blade, clean in chloral hydrate solution, place on a slide, and study under the microscope. Stem cells vary in shape and size depending upon their location in the stem; however, they are usually rectangular in appearance. Compared to the flesh cells of the fruit, the stem cells are rather small. They are characterized mainly by the presence of chlorophyll, giving a green coloration to the cell. The epidermal hairs of the stem are long and tapering, and are usually divided into three segments.

HOWARD MOLD COUNT METHOD OF TOMATO PRODUCTS

Mold growth in significant quantities is not found on sound tomatoes. The growth of mold on tomatoes breaks down the tomato tissue producing a rot (Anon. 1968). Therefore, the presence of large numbers of mold filaments in tomato products indicates the use of unfit moldy raw material or contamination of the pulp by unclean, moldy equipment (Cruess 1958). Federal and state food law enforcement authorities consider the amount of mold in canned tomato products to be an index of the care used by the packer, particularly in the sorting and trimming operations, to keep rot out of canned products (Anon. 1968).

As increasing proportions of visible rot are allowed to remain on the tomatoes going to the extractor, the mold count on the finished product increases, but the relationship is not an exact one due to the nature of the rot that is present. The character of the rotten portion varies greatly. Rots caused by some species of molds are soft and will be broken up in the extraction procedure. Other types of rot are hard and tough, and will be largely discharged with the skins and cores without contributing much mold to the finished product (Anon. 1968).

In order to determine the degree of mold contamination and thus check the efficiency of the operations, the Howard Mold Count Method is used by food technologists (Eisenberg 1952).

The Howard Mold Count Method was developed in 1910 by B.J. Howard of the Bureau of Chemistry and Soils, U.S. Department of Agriculture, the federal agency then charged with the enforcement of the Food and Drug Act of 1906. Since the decay of tomatoes is largely caused by molds, Howard conceived the idea of using the occurrence of mold filaments in the comminuted finished product as a means of ascertaining the presence of decayed tomato materials. Investigational studies were started in the fall of 1908 as a result of a conference between Howard and Dr. W.D. Bigelow, the chief of the Food Division of the Bureau of Chemistry. Studies were conducted by Howard in canning factories, and the official procedure was first described in U.S. Circular 68 (Feb. 13, 1911) entitled, "Tomato Ketchup Under the Microscope, with Practical Suggestions to Insure a Clean Product."

The Howard Method was designed for two purposes: (1) to give the manufacturers a method for checking their product by determining if their sorters and trimmers were doing good work; and (2) to enable the federal food law enforcement agencies to prevent shipment in interstate commerce of tomato pulp or other strained tomato products made from moldy or decomposed material.

Howard stated that the only equipment necessary for mold counting was a good compound microscope giving magnifications of approximately 90× and ordinary slides (Howard 1911). His instructions were to place a drop of the product to be examined on a microscope slide, place a cover glass over it, and press down until a film of the product about 0.1 mm thick was obtained. A film thicker than this was too dense to be examined, and a thinner one gave a very uneven preparation.

When a satisfactory slide was obtained, it was to be examined at about 90× magnification. Each field of view was to cover approximately 1.5 mm^2, and approximately 50 fields were to be examined. The presence or absence of mold filaments in each field was noted, and then the slide was moved to a completely new field. If a field had more than one filament or clump of mold, it was still to be counted as one positive field. The percentage of the 50 fields showing mold present was then calculated. Results with this method showed that homemade catsup had practically no positive fields, some manufactured catsup had from 2 to 5%, and in some factories every field counted showed the presence of mold. Howard stated that with reasonable care, a mold count of 25% positive fields was possible (Howard 1911).

Methods for estimation of yeast, spores, and bacteria were also published at this time, but counts on these organisms are infrequently done now.

Howard and Stephenson (1917) published slight modifications of this method, which had been developed during further study. By this time, a Howard mold-counting cell, which was a modification of the Thoma-Zeiss blood-counting cell, had been developed. The cell had an unruled center disk about 19 mm in diameter, and was constructed to give a sample thickness of 0.1 mm when properly prepared. Further instructions given were to clean the Howard cell so that Newton's rings were produced between the slide and

FIGURE 24.5.
HOWARD MOLD COUNTING CHAMBER.

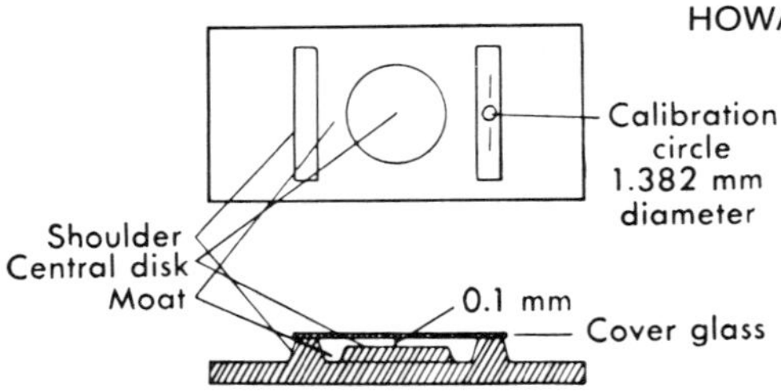

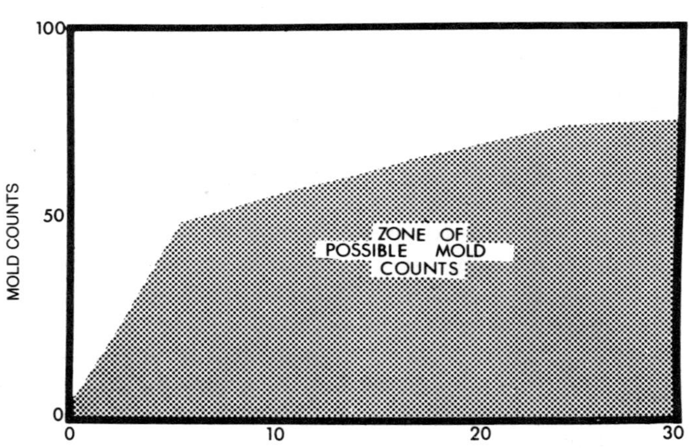

FIGURE 24.6. RELATION BETWEEN PERCENTAGE BY WEIGHT OF ROT AND MOLD COUNT (LABORATORY SAMPLES).
From Goldblith et al. (1961).

cover glass. There were precise directions for preparing the slide. Using a knife blade or scalpel, a drop of thoroughly mixed sample was to be spread evenly over the central disk, so that molds and tomato fiber would not be concentrated in the center. No excess liquid was allowed to be drawn into the moat of the slide.

It was again reported that each field of view was to be 1.5 mm^2, but it was added that this size could be obtained by adjusting the diameter of the microscope field to 1.382 mm. Again 50 fields prepared from two or more slides were to be examined, and it was further stated that no field could be considered positive unless the aggregate length of the mold filaments exceeded one-sixth the diameter of the field. These changes tended to make mold counting more defined and less variable when done by different counters. This method was adopted as a tentative method of the AOAC for the microanalysis of tomato products.

A modified method for slide preparation was presented in 1956 by Kopetz *et al.* (1957). This modification was basically the inclusion of careful manipulation of the cover glass instead of the official prespreading technique. The cover glass could either be lowered at a slant from one edge or rapidly lowered from a parallel position so that the sample spread evenly over the entire surface. Kopetz stated that there were several advantages to this modification, e.g., elimination of scratching the glass, and that there were no significant differences in distribution of mold filaments. However, the modified technique was not accepted in the official final action.

A few techniques to make mold counting easier without changing the official method have been developed. Beach reported that a 25-square grid drawn on the microscope stage and used with a rectangular Howard cell was

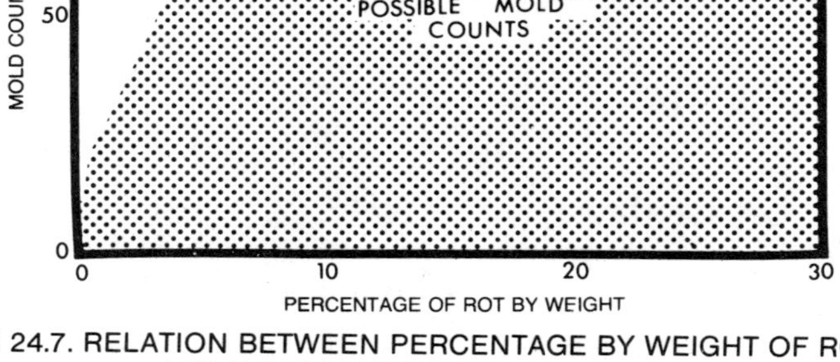

FIGURE 24.7. RELATION BETWEEN PERCENTAGE BY WEIGHT OF ROT AND MOLD COUNT (FACTORY SAMPLES).
From Goldblith et al. (1961).

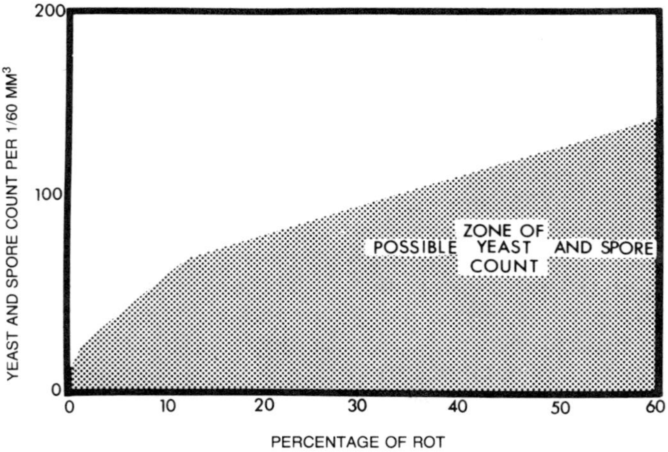

FIGURE 24.8. PERCENTAGE BY WEIGHT OF ROT, YEAST, AND SPORE COUNT.
From Goldblith et al. (1961).

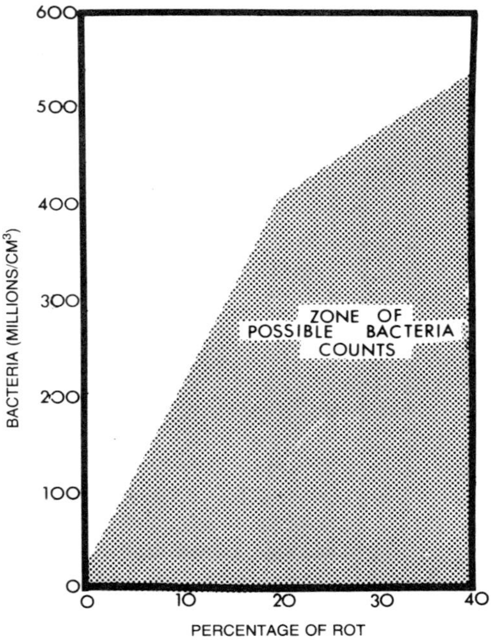

FIGURE 24.9. PERCENTAGE BY WEIGHT OF ROT AND BACTERIAL COUNT.
From Goldblith et al. (1961).

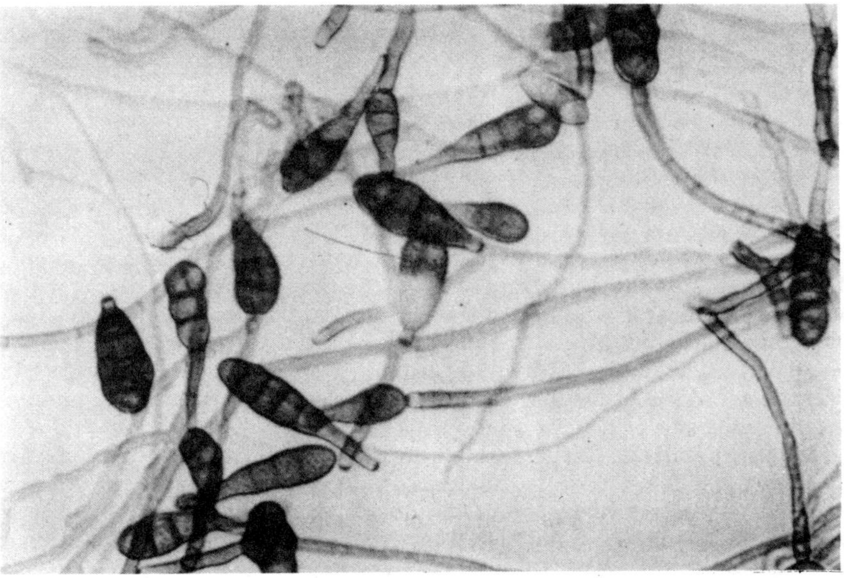

FIGURE 24.10. *ALTERNARIA*, A CAUSE OF ROT IN TOMATOES, SHOWS INDIAN CLUB-SHAPED SPORES OR CONIDIA. FILAMENTS ARE SEPARATE AND OF MEDIUM THICKNESS (x450).
From Beuchat (1978).

beneficial in training mold counters faster (Beach 1951). Pitman reported a method to lessen the eye fatique of the counter (Pitman 1943). The lower end of a camera cable release was attached to the microscope, and a piece of paper that moved with the slide was attached to the mechanical stage. Each time a positive field was counted, the cable was used to punch a hole in the paper, and thus the counter did not have to look up from the microscope.

The Howard method has been criticized as not being truly scientific, and as admitting wide variations in results among competent counters. Although discrepancies have been encountered at times among analysts, the method has been used by the tomato industry for over 70 years, during which time no superior one has been advocated. It is now judicially recognized as the official method of determining the presence of mold in tomato products, and, on the basis of the counts obtained, as an index of whether decayed tomatoes were included in the product.

CHARACTERISTICS OF MOLD

To the naked eye, the vegetative form of a mold resembles a bit of cotton and is called the mycelium. Each thread or filament of this mycelium is called a mycelial thread or hypha. As with most fruits and vegetables, such growth on tomatoes results in and/or occurs with partial to entire rot and decomposition of the fruit. Hence, the Howard mold count to determine the amount of mold hyphae present is actually an index of the extent of decomposition of the tomatoes used in the product.

Mold filaments have been classified in a number of ways. Perhaps the classification that is of most significance to the person making a Howard mold count is the one which divides them on the basis of function into two types: vegetative and fertile. The vegetative filaments (hyphae) are those that grow underneath the surface of the tomato or other substances, and the fertile filaments (hyphae) are those that bear the fruiting bodies or reproductive spores and grow above the surface. The combination of these two form the entire mold plant, the submerged vegetative hyphae serve to anchor the mold more solidly like the roots of a plant, and the fertile aerial hyphae act as the branches, on the ends of which the spores (corresponding to the seeds of a plant) are produced.

Practically all the fertile aerial hyphae will be removed as the tomatoes pass through the soak tank, followed by a vigorous spray wash. The submerged vegetative hyphae, on the other hand, can be removed only by careful trimming and sorting, and if not removed they go into the finished product. It is, therefore, these vegetative hyphae that the counter must recognize and distinguish from other filaments with which they are closely associated. Rot itself consists usually of fruit tissue and mold.

Mold hyphae in all cases are tubular although they may appear to be flat under the microscope. Although different molds show great differences in diameter, in most instances, the diameters of the tubes of any one filament are uniform and the cell walls appear under the microscope as parallel lines. Two notable exceptions are molds of the *Mucor* type and *Oospora*, which are often tapering. Because of these two important general characteristics, parallel walls of even intensity must be emphasized.

Mold filaments are also somewhat brittle and will break instead of forming sharp bends, like the sharp curves and oblique ends characteristic of broken spiral tubes from the fibrovascular bundles. Molds also break to form blunt ends and, except for growing ends which are slightly rounded, the ends of such filaments will always be blunt and not sharp or jagged.

Growing molds have living protoplasm within the tubular structure and may present a granular or stippled appearance. This characteristic may persist after the mold is killed in processing and is prominent in the *Mucor* and *Rhizopus* molds.

Many molds found in food products contain cross walls, which separate the mold filament into short sections; such molds are referred to as septate. The presence of cross walls assists in identifying positively many otherwise doubtful filaments. However, cross walls are generally absent in *Mucor* mold.

Most molds show an abundance of branching, and branches are frequently an aid in the positive identification of mold. Here again, however, many fragments do not show branching in a comminuted food products.

One condition frequently observed in mold filaments found in canned tomato products is the breaks that appear to be present in a given filament. Such a filament is usually referred to as "broken" mold. Sometimes close observation of the filament under proper lighting and proper magnification will reveal fine parallel lines connecting the granulated or other visible parts of the filament. This condition apparently develops within the transparent cell wall or sheath where the protoplasm of the mold separates, possibly at certain of the original cell walls, to form small-to-large spaces between the granular area of the filament.

To overcome uncertainty in the identification of questionable filaments, as well as to check the official count of government agencies, all counters must be familiar with the following set of rules published by the National Food Processors Association (Pitman 1943).

Only filaments which have at least one of the following characteristics shall be classed as mold hyphae:

1. Parallel walls of even intensity with both ends definitely blunt
2. Parallel walls of even intensity with characteristic branching
3. Parallel walls of even intensity with characteristic granulation
4. Parallel walls of even intensity with definite septation
5. Occasionally encountered, parallel walls of even intensity with one end blunt and the other end rounded
6. Occasionally encountered, slowly tapering walls of even intensity with characteristic granulation or septation.

As mentioned before, fertile or aerial hyphae are almost never seen in commercial products. Also, since they would be confused with other filaments, they are not counted as mold except when there are spores attached making identification positive.

All these characteristics are used in counting mold; however, if any one of these can be proved to be correct or true of molds, one need not look for the other five. This method is empirical and must be followed exactly as given in order to obtain satisfactory results.

GENERA OF MOLDS FREQUENTLY ENCOUNTERED

The Howard mold count procedure does not require the identification of mild filaments, as this is immaterial in scoring a given field. Also, the identification of specific molds is based largely on the type of fruiting body produced, which generally is impossible to determine definitely from the examination of the hyphae remaining in commercially sterilized food products. However, some knowledge of the molds commonly encountered on tomatoes may be of interest in control of mold-count level. The following discussion includes several genera of molds found on tomatoes and resulting in the presence of mold filaments.

Alternaria

Alternaria species are among the commonest saprophytes on decaying vegetation of all kinds. A few alternaria, however, can attack living plants. Infection with the mold *Alternaria tomato* produces a disease known as "nailhead spot" on the tomato fruit. This disease is now controlled largely by the use of varieties of tomatoes resistant to the infection. Another species, *Alternaria solani* causes "early blight" and is one of the more common diseases of tomatoes in the Central States. Fruits may be infected at any stage of their growth and first symptoms are minute tan spots. As the spots become older and larger, the centers are more definitely sunken and become grayish-brown. Warm rainy weather is most conducive to infection and rapid development of the disease on the foliage and fruits. Molds of this group may be recognized by the Indian-club shaped multicelled spores of rather large size. The filaments are septate.

Aspergillus

The word aspergillus is derived from a special type of brush (aspergillum) used in the ceremony (the "Asperges") for sprinkling holy water, and the genus can be recognized by this characteristic appearance of the spore-bearing structure, the conidiophore. The unbranched conidiophore arises from an enlarged cell of the vegetative mycelium and terminates in a small

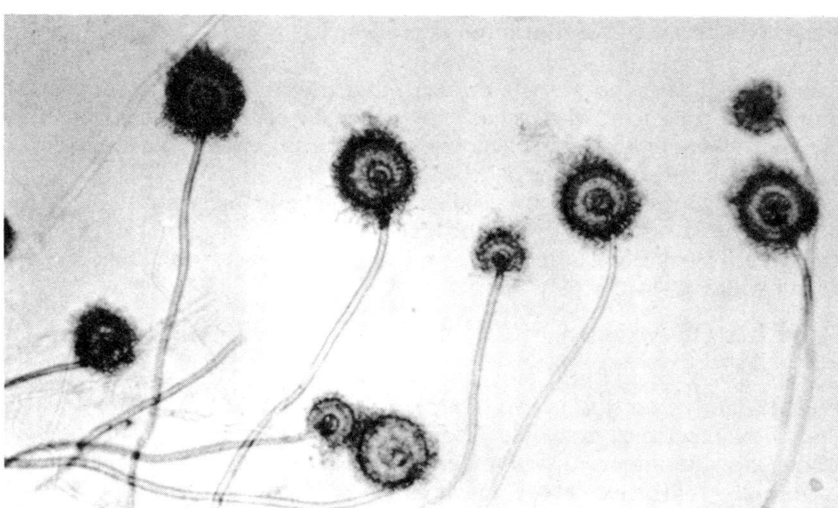

FIGURE 24.11. *ASPERGILLUS OCHRACEOUS* SHOWING MYCELIA AND CONIDIAL HEADS (x100).
Courtesy of Continental Can Company

club-shaped portion. From this, there arise a number of little stalks or projections of characteristic bottle shape, at the tips of which are found chains of conidia or spores. Growth of this mold is not common on tomatoes but it may occur in cracks and other injuries in both the field and in the boxes after harvesting.

The mycelium of the aspergilli is usually septate and members of the species are numerous in soil, and particularly on dried vegetable matter such as hay and grains. The aerial growth of some species also appears as black-colored heads and may be confused with *Mucor*.

Colletotrichum

Colletotrichum phomoides, responsible for anthracnose rot, penetrates apparently healthy, uninjured fruits and develops under the skin of the tomato, first producing small circular, sunken, water-soaked spots. These spots soon become darker and more depressed, similar in appearance to cavities produced by smallpox. The spores themselves are pinkish and are splashed about by rains to infect surrounding fruit. The rot usually occurs in the latter part of the tomato season, as the fruit is most susceptible to

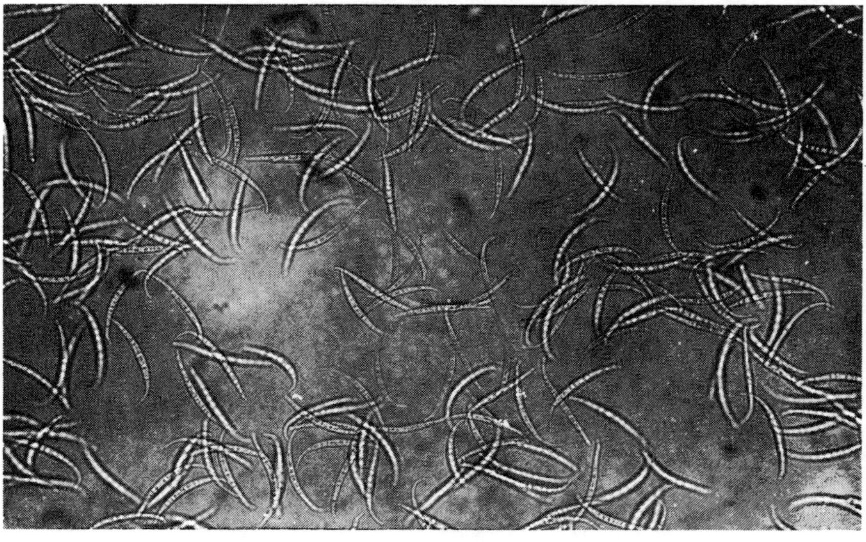

FIGURE 24.12. *FUSARIUM* SHOWING SICKLE-SHAPED SPORES (x200).
Courtesy of Continental Can Company

infection just as it begins to ripen. Usually the infection advances so rapidly and develops so far underneath the surface that an infected tomato must be destroyed. The disease is one of the most widespread and destructive tomato fruit rots in the Central States. The filaments of this mold are very fine and septate, and the spores are small, elliptical structures, produced abundantly underneath the tomato skin.

Fusarium

A great many plant diseases are caused by some form of the *Fusarium* mold which is a large, widespread genus. *Fusarium oxysporum lycopersici* may attack tomatoes at any stage of the plant's development. The fungus enters the roots, passes up the stem, and causes wilting of the foliage and eventual death. Fusarium is most active at temperatures between 80° and 90°F (27° and 32°C) and if conditions are favorable to its growth, it can live almost indefinitely in the soil. The several-celled spores are sickle- or canoe-shaped in appearance and the mycelium is septate.

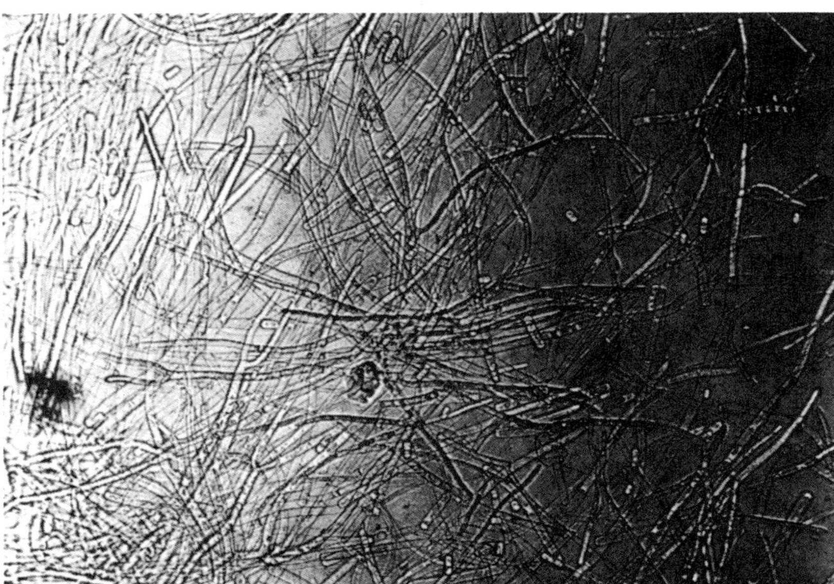

FIGURE 24.13. *OOSPORA* ON TOMATO SHOWING GRANULATION AND BRANCHING (x150).
From Weiser et al. (1971).

FIGURE 24.14. *PENICILLIUM* WITH THE BUSHLIKE CONIDIAL HEADS (x200). *Courtesy of Continental Can Company.*

Mucor and *Rhizopus*

Mucor and *Rhizopus* are often classed together because of their resemblance in appearance and effect on tomatoes. Development of these molds occurs largely in growth cracks and other injuries, both in the field and in the boxes after harvesting. Members of the family to which they belong, frequently referred to as the bread molds, are found abundantly in soil, in manure, on fruits, and especially on starchy foodstuffs. The growth of the mold through the tomato breaks down the tissue completely to give a soft, mushy type of decay commonly referred to as "leakers." These molds are easily differentiated from other groups of molds by the coarse, aseptate mycelium, by the abundant and loosely meshed aerial mycelium, and by the black or brown color of the spores on long stalks, the mycelium being white or gray.

Oospora (Oidium)

Oospora lactis is best known in connection with dairy products, but is found in the canning factory and under certain conditions on tomatoes in the field. It forms a white, slimy mass that gives off a characteristic odor; because of its resistance to heat and antiseptic it can be eradicated only by keeping the equipment scrupulously clean.

This genera of mold is set apart by the lack of well-defined fruiting bodies and propagates by the separation of cells from any part of the mycelium to form individual "oidia," a unit comparable to the spore or conidium. The oidia are usually rectangular with rounded corners. The filaments are septate.

Penicillium

The penicillium or green mold, as it is known, is another type that will develop both in the field and in the plant in growth cracks and other injuries. The name itself comes from the Latin meaning "little brush," and the genus is characterized by fan-shaped, fingerlike fruiting hyphae with numerous spores produced in chains along the ends of the fingerlike extensions. As a matter of interest, several different cheeses, e.g., Roquefort, Camembert, Gorgonzola, etc., are ripened by the action *Penicillium* spp. Also, some of the *Penicillium* species, particularly *Penicillium notatum*, are capable of producing the antibiotic, penicillin. Growth of this mold usually shows the presence of blue-green spots. The mycelium is septate.

Phytophthora

This fungus *Phytophthora infestans* causes "late blight" disease in tomatoes. The severity of this disease fluctuates from year to year depending on weather conditions. The most important conditions favoring the sudden and destructive development of late blight are continued humid weather and cool temperatures. These conditions occur when cool, dewy nights, 60 °F, or lower, are followed by moderate cloudy days. High summer temeratures and dry weather check the development of the disease. The fruit usually start rotting at or near the stem end and soon spread over the entire fruit. However, it can occur anywhere on the tomato fruit. Rotted areas are greenish-black with a rather firm but slightly wrinkled surface. The mold hypha is non-septate, irregular in diameter and contains highly refractive cytoplasm.

AOAC MOLD COUNT PROCEDURE

Preparation of Sample

For canned tomatoes and on products containing whole tomatoes, mold counts will be made on the drained liquid only, unless examination, visual observations, or other history indicate that mold counts should be made on pulped tomatoes also. If necessary to make mold counts on drained tomatoes, pass such tomatoes through laboratory cyclone pulper.

Tomato Juice and Tomato Sauce. Use the juice or sauce as it comes from the container without dilution.

Tomato Paste, Tomato Puree, Concentrated Tomato Juice, and Dehydrated Tomato Powders. Add sufficient water to the puree, paste, or powder to make a mixture and have a total solids content of 7.9 to 8.8% (refractive index 1.3446 to 1.3460, corrected to 20℃).

Chili Sauce, Pizza Sauce, and Similar Products. Pass the contents of the container through the laboratory pulper to remove seeds and large particles. If the container is a No. 10 can size (or larger) of chili or pizza sauce, pulp a 20-oz aliquot of a well-mixed sample from individual containers. If tomatoes are pulped, use entire contents of individual containers if container is of No. 10 can size and smaller. For larger container, use a well-mixed 5-lb aliquot.

Tomato Catsup. The current official method for counting mold in tomato catsup requires a 1:1 dilution, by volume, of the catsup with one of the following stabilizing solutions:

(1) 0.5% Sodium Carboxy Methyl Cellulose (CMC) (Hercules Powder Co., Wilmington, DE). To prepare the solution place 500 ml of boiling water in a high speed blender. With blender running, slowly add 2.5 gms. of the CMC and blend for one minute. Treat the solution with heat or vacuum to remove air bubbles. After solution has cooled, add 2 ml of formaldehyde per 100 ml of solution as a preservative, mix well and store in closed container.

(2) 3% to 5% Pectin Solution. This is prepared by following the same procedure as for the CMC solution except using 15 to 25 grams of pectin.

(3) 1% Algin Solution. This is prepared by following the same procedure as for the CMC solution except using 5 grams of Algin.

The official method for mixing the stabilizing solution with the catsup is to place 50 ml of stablilizing solution in a 100 ml graduate cylinder and adding 50 ml of a well mixed catsup sample to the cylinder by displacement and mix thoroughly while in the cylinder. Drop the catsup directly into the solution. This can best be done by using a spoon instead of pouring from a container into the cylinder. If catsup is allowed to run down side of the cylinder, the graduations will be difficult to read. The mixed tomato catsup may then be transferred to a suitable container and the mold determined directly. An alternate method is to place tared beaker on triple beam balance and add 50 ml of stabilizer (weight of 50 ml of stabilizer can be predetermined so volume need not be taken each time) and next add 50 ml of catsup (eight of 50 ml of catsup predetermined) and mix thoroughly and determine mold count.

Equipment, Materials, and Reagents

1. Howard Mold Counting Chamber and the 33 × 33 cover glass.
2. Microscope: The microscope should be the basic compound microscope

Slide No. __________ Positive Fields __________ Analyst ______________ Date ________

Place the slide on the stage with number at the right.

The circles below correspond to the 25 fields on a slide. In each circle, sketch all of the mold filaments seen in the corresponding field, indicating the relative size and position of each filament. Below each circle put (+) or (-) according to your findings.

1 2 3 4 5

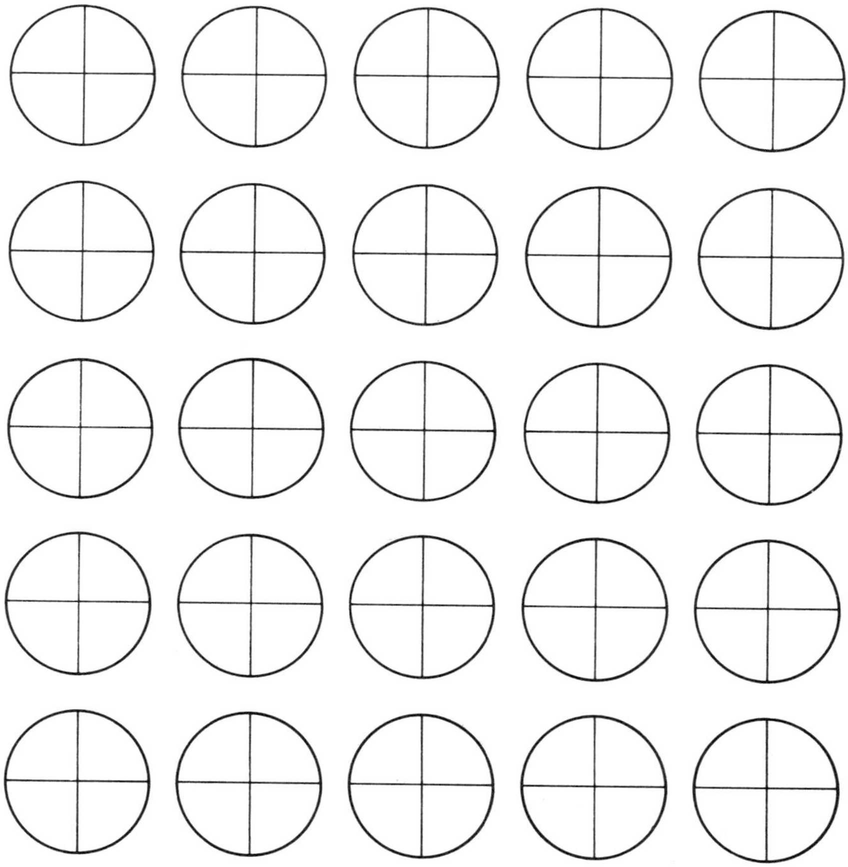

FIGURE 24.15. MOLD COUNT RECORD FORM.

and can be either a monocular or binocular. The latter is most desirable. It should be equipped with an Abbé condenser and achromatic objectives of 10× and 20×. The 10×−16 mm objective must be calibrated with the eyepiece, 10× Huygenian ocular with drop in micrometer disk cross ruled in sixths, to give a field diameter of 1.382 mm. This diameter is checked by comparing the field with the field with the circle etched on the right-hand rail of the chamber. If the diameter is not 1.382 mm, the microscope must be standardized or erroneous results will be obtained. The microscope should contain a mechanical stage and a substage lamp with daylight glass.

Procedure

The following is quoted from the AOAC 10th Edition, 1965:

Clean Howard cell, so the Newton's rings are produced between slide and coverglass. Remove cover and with scalpel (or knife blade) place portion of the well-mixed sample upon central disk; with same instrument, spread evenly over disk, and cover the glass so as to give uniform distribution. Use only enough sample to bring material to edge of disk. (It is of utmost importance that portion be taken from thoroughly mixed sample and spread evenly over slide disk. Otherwise, when cover slip is put in place, insoluble material, and existing molds, may be more abundant at center of mount.) Discard any mount showing uneven distribution or absence of Newton's rings, or liquid that has been drawn across moat and between cover glass and shoulder.

Place slide under microscope and examine with such adjustment that each field of view covers 1.5 mm. (This area, which is essential, may frequently be obtained by so adjusting draw-tube of monocular scopes only—binocular scopes must be adjusted at the factory—that diameter to field becomes 1.382 mm. When such adjustment is not possible, make accessory drop in ocular diaphragm with aperture accurately cut to necessary size. Diameter of area of field of view can be measured by use of stage micrometer, or the calibration generally etched on shoulders of Howard moldcounting cell. When instrument is properly adjusted, quantity of liquid examined per field is 0.15 cu mm.) Use magnification of 90× to 125×. In those instances where identifying characteristics of mold filaments are not clearly discernible in standard field, use magnification of about 200× (8 mm objective) to confirm identity of mold filaments previously observed in standard field.

The latest Howard Mold Counting Chamber has a rectangular chamber (15 × 20 mm) rather than a disk, as just described.

Modifications in Slide Preparation

Two alternate techniques are also acceptable. They are to overcome the uneven distribution of the insoluble solids and entrapment of air bubbles between the central disk and the cover glass. These techniques substitute careful manipulation of the cover glass for the prespreading technique. Both methods utilize the uniform spreading, equal distribution action resulting from lowering the cover glass into place.

Inclined Cover Glass Technique. Using a spatulate instrument, transfer a portion of a well-mixed sample to an area on the central disk halfway between the center of the disk and the edge opposite the analyst (use of a dissecting knife is helpful). One edge of the cover glass is rested in a slanting position on the edges of the slide shoulders nearest the portion of test materials. The cover glass is lowered slightly until it is almost touching the material; then it is lowered rapidly but gently into place. This causes the material to spread evenly over the entire surface of the disk.

Parallel Cover Glass (or "Drop") Technique. Using a spatulate instrument place a portion of a well-mixed sample on the approximate center of the central disk. The cover glass is then held in a position parallel to the surface of the central disk and is lowered slowly until slight contact with the sample and cover glass is seen. While maintaining contact with the sample, the cover glass is alternately lowered and raised very slightly two or three times then, without stopping, is lowered rapidly but gently until contact is made with the shoulders of the slide. This causes the sample to spread evenly over the entire surface of the disk.

In these techniques, as in the prespreading method, the cover glass should not be lowered too rapidly or a part of the sample may splash onto one or both shoulders, thereby ruining the count. Also, it should not be lowered too slowly, as the insoluble material will not spread uniformly. With practice, one can soon learn to control this step in solids preparation and consistently prepare slides showing evenly distributed insoluble material.

Any mount showing uneven distribution of insoluble material, absence of Newton's rings, or liquid that has been drawn or splashed across the moat onto the shoulders must be discarded. A slide should not be counted unless it is properly prepared.

Counting Procedures

After the slide, microscope, and light are properly prepared the sample is ready for counting. If the results are to be expressed as "percentage of mold," proceed as follows:

Single Plan. From each of two or more mounts examine at least 25 fields taken in such manner as to be representative of all sections of the mount. Observe each field carefully, noting presence or absence of mold filaments and recording results as positive when aggregate length of not more than three filaments present exceeds one-sixth of the diameter of field. Calculate proportion of positive fields from results of examination of all observed fields and report as percent of fields containing mold filaments.

Although the "percentage of positive fields of mold" is calculable from a minimum of 50 examined fields of the product, an examination of at least 100 fields is more desirable and reliable.

If the acceptability of the sample is to be determined by the multiple mold count procedure, the following criterion is to be used.

Multiple Plan. Under this procedure, increments of 10 fields are counted according to the standard Howard Mold Count technique; the results are then applied to the appropriate sampling plan (see Table 20.1) to determine if the sample unit meets specifications, fails specifications, or whether additional fields must be counted.

These tables are designed to give the same producer-consumer protection as a single plan consisting of 100 fields. The advantages of a multiple plan are

1. A decision can be made on very good sample units, or very, bad sample units by counting considerably less than 100 fields.
2. An objective guide is provided as to the number of fields to count in order to properly classify the sample units.
3. It may be used as a systematic screening procedure for quality control during processing.

Application

1. Referring to the attached table, select the plan that corresponds to the appropriate mold count acceptance level for the product being tested.
2. Prepare the Howard mold count slide according to the official method and make counts in increments of 10 fields.
3. After counting the first 10 fields:
   a. Accept the sample unit if the number of positive fields does not exceed the first value under the column *c*.
   b. Fail the number unit if the number of positive fields equals or exceeds the first value under the column *r*.
   c. Continue counting an additional 10 fields if a decision to accept or fail cannot be reached on the first 10 fields, i.e., if the number of positive fields falls between *c* and *r*.
4. If additional counting is required, continue in increments of 10 fields, comparing the cumulative results with the appropriate steps in Table 20.1 until a decision to accept or fail is reached. In some instances, it may be necessary to count 120 fields.
5. The number of fields tested (N_c) and the number of positive fields found are cumulative. The results are recorded on a work sheet (Fig. 20.15) as number of positive fields per number of fields counted and not as percent positive fields.

The plan is designed to count a maximum of four slides, 30 fields per slide. The rectangular type cell is best suited for 30 fields per slide; however, the same plan is applicable to the round cell except that it may be necessary to prepare an extra slide if necessary to count 120 fields.

Lost Acceptance Criteria

1. Sampling Plan Table for Lot Acceptance is as follows:

No. of mold counts	2	4	9	13
Ac[1]	0	1	2	3

(For example, the Ac for one through 3 mold counts is 0, the Ac for 4 through 8 is 1 and so forth.) Whenever possible the number of mold counts made on lot in-plant inspection should correspond to the sampling plans above.

2. By the use of multiple sampling plans in Table 24.1 classify each sample unit in one of the following categories:
   a. Acceptable: meets the specific requirement.
   b. A Deviant: fails the specific requirement but meets the next higher level.
   c. Worse than a Deviant: fails the next higher level.
3. Accept the lot as meeting the specific requirement for mold if:
   a. None of the sample units is "worse than a deviant";
   b. The number of "deviant" sample units in the sample does not exceed the acceptance number (Ac) in the sampling plan for Lot Acceptance.

The procedure for classifying a sample unit as a "deviant" or "worse than a deviant" is illustrated as follows, using the following portions of the multiple sampling plan from Table 24.1 at the 40% and 50% levels:

40% Level			50% Level		
N_c	c	r	N_c	c	r
10	0	8	10	1	9
20	4	13	20	6	15
30	7	17	30	11	19
40	11	21	40	16	25
50	15	25	50	21	30
60	20	29	60	26	34
70	24	33	70	31	40
80	28	36	80	26	45

The following counts were made first testing at the 40% level. Example:

N_c	Cumulative Number of Positive Fields Found
10	4
20	8
30	11
40	16
50	21
60	26
70	30
30	36

After counting 80 fields, 36 positive fields were found. This fails the 40% level. These same results are now taken, 10 fields at a time, and compared with the 50% level. No decision can be reached until the third group of 10. The cumulative number of positive fields found after counting 30 fields is 11. The sample unit meets the 50% level, to the 40% requirement rather than "worse than a deviant."

[1]Provided these deviants are acceptable at the next highest level and further provided rot counts are well within limits.

TABLE 24.1. MULTIPLE SAMPLING PLAN FOR MOLD COUNTS[a]

N_c	10%		20%		30%		40%		50%		60%		70%		80%	
	$-c$ = =	r	c	r	c	r	c	r	c	r	c	r	c	r	c	r
10	[1]	4	[1]	6	0	7	0	8	1	9	2	10	4	10	5	[2]
20	[1]	5	0	8	2	10	4	13	6	15	8	17	10	19	13	20
30	1	6	3	10	5	13	7	17	11	19	14	23	17	26	20	28
40	1	7	5	12	8	16	11	21	16	25	20	29	24	32	28	36
50	2	8	7	14	11	19	15	25	21	30	26	35	31	40	36	44
60	3	9	9	16	14	22	20	29	26	34	32	41	38	48	44	52
70	4	10	11	18	17	25	24	33	31	40	38	46	45	54	52	60
80	5	11	13	20	20	28	28	36	36	45	44	52	52	62	61	68
90	6	11	15	22	23	31	32	40	42	49	50	58	60	68	70	76
100	8	12	17	23	26	35	37	43	47	54	57	64	66	75	78	83
110	10	13	20	23	30	37	42	47	53	58	63	69	74	80	86	90
120	12	13	24	25	36	37	48	49	60	61	72	73	84	85	96	97

[a] N_c, the cumulative number of fields to count; c, the maximum cumulative number of positive fields permitted to accept the sample unit for the appropriate percent mold level; r, the minimum cumulative number of positive fields necessary to fail the sample unit for the appropriate percent mold level.
[b] The sample unit cannot be accepted at this level.
[c] The sample unit cannot be rejected at this level.

TABLE 24.2 — FDA REGULATORY ACTION GUIDANCE (SEIZURE) FOR VARIOUS TOMATO PRODUCTS*

Tomato Article	6 Subsamples Examined
Tomato juice	Average 24% or above & all subs above 20%.
Tomato paste or puree	Average 45% or above, & all subs above 40%.
Tomato sauce, undiluted	Average 45% or above, & all subs above 40%.
Canned tomatoes with or without added tomato juice	Average above 15 % and all subs above 12%. Based on examination of packing medium drained from canned tomatoes.
Canned tomatoes packed in tomato puree juice.	Average above 29% and all subs above 25%. Based on examination of packing medium drained from canned tomatoes.
Pizza sauce (based on 6% total solids after pulping).	Average above 34% and all subs above 30%.
Tomato soup and tomato products.	Average of 45% or above and all subs above 40%.
Tomato Catsup	Average 55% or above
Tomato powder (Except spray dried)	Average of 45% or above and all subs above 40%.
Tomato powder (spray dried)	Average of 67% or above.

*Taken from FDA Compliance Policy Guides 7114.30

Food and Drug Administration action levels for mold count for the various processed tomato products are shown as tolerances in Table 24.2. FDA recognizes the fact that special processing equipment such as mills and homogenizers increase the mold count. Under such conditions, allowances over the published tolerances may be made. In recognition of this fact, it is advisable to make mold counts both before and after the product is processed through any mill or comminuting equipment. Such information can be helpful in verifying the increase in mold counts resulting from the use of this equipment.

REFERENCES

ANON. 1954. The Howard Mold Count Method as Applied to Tomato Products. American Can Co., Maywood, IL.

ANON. 1968. Mold Counting of Tomato Products. Continental Can Co. Tech. Center, Chicago.

BEACH, P.L. 1951. Speed your mold count with this handy guide. Food Eng. *23* (10) 151.

BEUCHAT, L.R. 1978. Food and Beverage Mycology. AVI Publishing Co., Westport, CT.

CRUESS, W.V. 1958. Commercial Fruit and Vegetable Products, 4th Edition. McGraw-Hill Book Co., New York.

EISENBERG, M.V. 1952. Observations and suggestions of factory control of rot and extraneous matter in tomato products. National Food Processors Association Conv. Issue *1371*.

GOLDBLITH, S.A., JOSLYN, M.A. and NICKERSON, J.T.R. 1961. An Introduction to the Thermal Processing of Foods, AVI Publishing Co., Westport, CT.

HOWARD, B.J. 1911. Tomato ketchup under the microscope, with practical suggestions to insure a clean product. U.S. Dep. Agric. Bur. Chem. Circ. *68* (Feb. 13).

HOWARD, B.J. and STEPHENSON, C.H. 1917. Microscopical studies on tomato products. U.S. Dep. Agric. Bull. *581* (Oct. 6).

JARVIS, B. 1977. A chemical method for the estimation of mould in tomato products. J. Food Technol. *12*, 581–591.

KOPETZ, A.A., TROY, V.S. and McCALLUM, M.R. 1957. Modified slide preparation for the official Howard mold count method. J. Assoc. Off. Agric. Chem. *40*, 905.

OLSON, N.A. 1980. The effects of milling on mold counts in tomato products. Food Technol. *34* (6) 50–56.

PITMAN, G.A. 1943. Mold counting recording device. J. Assoc. Off. Agric. Chem. *26* (4) 511.

RICHARDS, D.W. 1958. The effective Use and Proper Care of the Microscope. American Optical Co., Buffalo, NY.

RIDE, J.P. and DRYSDALE, R.B. 1972. A rapid method for the chemical estimation of filamentous fungi in plant tissue. Physiol. Plant Pathol. *2*, 7–15.

SMITH, R.H. 1954. Instructions in microanalytical methods. J. Assoc. Off. Agric. Chem. (Feb.).

U.S. DEP. AGRIC. 1967. Methods of analysis for tomato products, mold count. U.S. Govt. Printing Office. Washington, DC, *1967*.

WEISER, H.H., MOUNTNEY, G.J. and GOULD, W.A. 1971. Practical Food Microbiology and Technology, 2nd Edition. AVI Publishing Co., Westport, CT.

CHAPTER 25
Spoilage of Canned Tomatoes and Tomato Products

Spoilage may occasionally occur in canned foods filled hot or thermally processed after closing (Troy *et al.* 1963). There are two reasons why bacterial spoilage may occur in canned foods after they have been processed:

1. Failure to destory all bacteria capable of subsequent growth in the product during the heat process, or
2. Contamination of the product after an adequate heat process

The first is commonly referred to as underprocessing and the second as leakage. Canned tomatoes offer no exception to these two generalizations (Bohrer and Reed 1948).

Canned food products are divided at pH 4.6 into two broad classifications: *low-acid* products, such as corn, peas, and green beans, and acid products, such as fruits, tomatoes, and tomato products. The low-acid products must be processed under pressure at temperatures of 250°F(121°C) or equivalent to destroy the heat-resistant spores of certain bacteria which, if not destroyed, will grow in the product and cause spoilage. Acid products ordinarily do not require such a severe process, since they have a lower pH than low-acid products. Because of this, the spore form of the organism is usually destroyed quite readily, and, even if present, it may not germinate to cause spoilage in the product. Processing of these products at temperatures of 212°F (100°C) or even less is usually sufficient to prevent growth of the organism (Anon. 1948)

The organisms usually found in tomatoes are non-spore-forming bacteria of relatively low heat resistance which are called lacto bacilli. Processing to a center temperature of 185°F (85°C) for a water cool or 170°F (76°C) for an air cool has been commercially successful and is sufficient to destroy the usual nonsporing types (Bohrer and Reed 1948).

Tomato juice, although an acid product, has heat-process requirements that are somewhat more exacting than those of other acid foods such as fruits. During the 1937 season in California, heat-processing for 25 min at

212°F failed to prevent spoilage in tomato juice, as a result of the high pH of the tomatoes and the presence of heat-resistant bacteria (Troy *et al.* 1963).

Spoilage of tomato juice has been caused by spore-forming organisms having greater resistance to heat than the common aciduric types. One of these resistant types is a gas-forming anaerobe, *Clostridium pasturianum*; but the flat-sour organism *Bacillus thermoacidurans* is the resistant type that dominates the subject of adequate processing of tomato juice (Tressler and Joslyn 1961).

FLAT-SOUR SPOILAGE

Flat-sour spoilage in commercially canned tomato juice was first encountered in 1931. Since that time, outbreaks of this type of spoilage have occurred almost yearly, many times reaching serious proportions and causing severe economic losses to the canner. Flat-sour tomato juice has an off-odor and -flavor. The flavor has been described as "medicinal," slightly acid, sour, and highly disagreeable, depending on the stage of spoilage development (Troy *et al.* 1963).

Berry investigated the first reported outbreak of flat-sour spoilage in commercially canned tomato juice (Berry 1933). He found the causative organism to be a spore-forming bacterium of soil origin. Based on its thermophilic nature, its tolerance to heat and its ability to grow in acid products, Berry named the organism *Bacillus thermoacidurans*. Smith *et al.* (1946) regarded *Bacillus thermoacidurans* as the same as *Bacillus coagulans*. In confirmation, after a careful study of the two species, Becker and Pederson stated that there is no justification for considering *Bacillus thermoacidurans* as a species distinct from *Bacillus coagulans*, and the latter name has priority (Becker and Pederson 1950).

Pederson and Becker found that although the vegetative cells of some strains could grow in tomato juices of pH 4.15 to 4.25, the heated spores could not germinate and grow in tomato juice that had been adjusted to a pH lower than 4.32 (Pederson and Becker 1949).

Characteristics of Flat-Sour Spoilage of Tomato Juice

As indicated by the term "flat-sour," the cans do not swell and there is no way of detecting the spoilage until the can is opened. In fact, gas production by the organisms is absent or so slight as to result in no measurable loss in vacuum in any cases thus far investigated. This has precluded the possibility of separating the spoiled cans by the flip vacuum or heating methods that are sometimes successful in segregating flat-sour in nonacid vegetables. In most instances observed, a decided off-flavor and odor is the first noted effect of the growth of organisms, followed by development of acidity and sourness in flavor as the spoilage progresses. The obnoxious off-flavor or very sharp sour taste characteristic of the spoilage, once encountered, is easily recognized (Pearce 1940). As the organism grows in the tomato juice, the pH of the juice drops from 4.5 to 3.5 (Weiser *et al.* 1971).

Heat Resistance of Spores

The bacteria causing flat-sour spoilage are facultative anaerobes of soil origin. In tomato juice, they grow best at temperatures of about 130° to 140°F (54° to 60°C) (Rice and Pederson 1954). The heat resistance of flat-sour spores varies greatly with individual strains of bacteria. Thermal resistance studies reported by Berry (1933), using a concentration of 10 million spores per 1 ml heated in tomato juice at pH 4.5 showed an extrapolated "F_0 value" (destruction time in minutes at 250°F) of 0.33 min. These results are based on subculture data and not on product incubation. Wessel and Benjamin (1941) reported that in tests employing "flash" processes on fresh tomato juice inoculated with *B. coagulans* (using 20,000 spores per 1 ml), destruction was nearly complete in flash-processed juice having a calculated sterilizing value equivalent to 0.73 min at 250°F, and was definitely completed at 1.1 min. Sognefest and Jackson (1947) indicated that practically all strains were destroyed at a sterilizing value equivalent to 0.7 min at 250°F, and the choice of a sterilizing value equivalent to 0.7 min seemed to be a reasonable one for the commercial application of the flash sterilization method.

Recent studies of outbreaks of flat-sour spoilage of tomato juice showed the causative organisms to have a heat resistance in juice of 5 to 37 min at 200°F and 1 to 10 min at 212°F, based on product incubation and using a spore concentration of 10,000 spores per ml. Their maximum thermal resistance when heated and incubated in juice was below 0.7 min at 250°F. It is obvious from these data that merely heating the juice to 200° to 205°F and closing and holding the canned juice at these temperatures for a short time will not destroy the more heat-resistant flat-sour organisms. Also, the holding period of 3 min suggested for presterilized juice after closing and before cooling is not adequate for destruction of heat-resistant flat-sour spores if they are present in the juice.

Causes of Flat-Sour Spoilage

While the organisms are of soil-borne types and commonly enter the plant with the raw product, they are not normally present in sufficient numbers to constitute a spoilage hazard unless given opportunity to multiply in that plant.

Factors that may contribute to flat-sour spoilage of tomato juice are

1. Use of unsound, poor quality raw stock,
2. Rough handling of raw stock during hauling and before and during washing operations,

3. Poor fruit washing operations,
4. High pH of the raw juice,
5. High level of contamination,
6. Insufficient heat treatment,
7. Contaminated plant equipment, and
8. Poor plant sanitation.

Controlling Flat-Sour Spoilage

Prevention of flat-sour spoilage in canned tomato juice starts with the raw product in the field and involves each successive step in the preparation and processing operations. Generally, there are two methods of controlling flat-sour spoilage in canned tomato juice.

1. Reduce the number of spores of *Bacillus thermoacidurans* organisms in both product and equipment by proper sanitary procedure.
2. Process the product sufficiently to destroy any spores of this organism if they are present.

Quality of Raw Stock. Only sound tomatoes of high quality should be allowed to enter the cannery. Use of a "dry sorting belt" before the tomatoes are flumed into the cannery is an effective means for discarding soft, decayed, or moldy tomatoes which are conducive to the growth of flat-sour organisms. Bash (1964) found higher spore counts on the less firm tomatoes when handled in the large dry bulk boxes.

Fruit Handling. Tomatoes should be handled rapidly after picking and transporting to the cannery. Hampers and lug boxes must not be overfilled, thus preventing crushing of the bottom layer of fruit during the hauling operation. Hampers, lug boxes, and baskets should be washed and cleaned before returning to the truck for reuse in the field.

Care must also be taken to avoid crushing or breaking of the tomatoes in bulk handling, particularly during the washing operations, as contamination taken into tomato tissue may not be adequately removed in ordinary washing operations and thus becomes a potential spoilage hazard.

Tomato Washing Operations. Failure to employ efficient fruit washing operations is one of the chief factors contributing to flat-sour spoilage of tomato juice. Flat-sour bacteria are of soil origin and may be carried into the factory unless removed by thorough washing operations. The primary purpose of the tomato washing operation, therefore, is to ensure that only clean tomatoes enter the chopper or extractor.

Field soil is the major source of flat-sour bacteria spores that spoil canned tomato juice. If proper sanitation is maintained in the plant, then field soil is the only significant source of these heat-resistant spoilage organisms. The amount of soil on tomatoes received at the cannery is surprisingly large, particularly where the soil in the growing areas is heavy and rain or irrigation cause its adherence to the tomato. This is more pronounced in machine harvesting under wet conditions.

One of the most important and difficult requirements of tomato washing is the prevention of soil build-up in soak tanks and flumes. This is especially true when the water is recirculated. Figure 21.1 shows the relationship between numbers of flat-sour type spores and the concentration of soil in tomato wash water. In most instances, tomatoes leaving a flume or soak tank are more heavily contaminated with bacteria than when sampled from the field box or basket. One solution to this problem is the use of chlorinated water during the final rinse. Chlorine concentration of 15 to 20 ppm is recommended.

pH of the Juice. The pH or effective acidity of the juice is another major factor affecting flat-sour spoilage, as the acid tolerance of different strains of flat-sour organisms may vary greatly. A general increase in the pH of unprocessed tomatoes has been observed in most of the tomato-growing areas in the United States and Canada during the past few years, in some instances exceeding pH 4.5. The pH is a very definite factor of influence on the thermal resistance of bacteria and on their ability to germinate. The higher the pH, the lower is their heat resistance. Pederson and Becker (1949) studied in detail the critical pH for germination of *B. coagulans* spores in tomato juice and showed that a progressively higher pH was

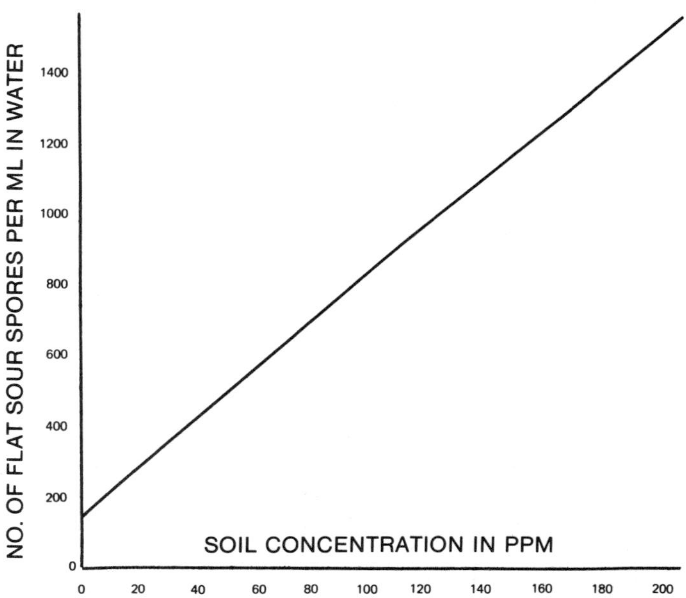

FIGURE 25.1. RELATIONSHIP BETWEEN NUMBER OF FLAT-SOUR-TYPE SPORES AND THE CONCENTRATION OF SOIL IN TOMATO WASH WATER.

necessary for germination as the period of heating was extended. This increased pH of unprocessed tomatoes has aggravated the flat-sour spoilage hazard and has necessitated using more severe processes. Flat-sour spoilage is most prevalent in tomato juice having a pH of 4.35 and higher, although the spores of some strains are reported to have germinated and grown and produced spoilage in juice having a pH as low as 4.1. Knock *et al.* (1959) reported that unheated strains of their South African strain of *B. coagulans* would germinate in tomato juice at pH 4.28.

Level of Contamination. Rice and Pederson (1954) investigated the effect of spore concentration in the inoculum of *B. coagulans* in tomato juice and reported that the minimum pH at which growth was able to occur depended on the concentration of spores in the inoculum, as shown in Table 25.1. These data show that with a high level of contamination, a considerably lower pH is required to inhibit spore germination; and conversely, even though the contamination is low, if the pH of the juice is high, spoilage may occur. Several investigators have reported that under certain conditions, low concentrations of flat-sour spores may cause spoilage. White (1951) reported flat-sour spoilage with juice of initial pH of 4.40 with a spore count of 1 to 3 spores per 1 ml.

TABLE 25.1. RELATIONSHIP BETWEEN SPORE CONCENTRATION AND LOWEST pH AT WHICH GROWTH OCCURS

Culture No.	Spore Concentration per ml	Lowest pH at Which Growth Occurred
710	65,000,000	4.24
	650,000	4.31
	6,500	4.37
	650	4.41
43P	22,000,000	4.19
	22,000	4.24
	220	4.31
	22	4.37

Water Activity

The growth of microorganisms depends on many factors. One of the most important of these factors is water activity (a_w). Most bacteria in tomatoes are inhibited below an a_w of 0.91, but some yeast and fungi may grow in foods with an a_w as low as 0.60. It is important to consider how microorganisms are inhibited with increasing concentration (Fig. 25.1) (Birnbaum *et al.* 1977).

Another reason for inhibition of spoilage in concentrates is the presence of hydroxymethyl furfural which depends on storage conditions and enzyme inhibition. This can be easily quantitated by a rapid HPLC method.

Insufficient Heat Treatment. An important factor in the prevention of flat-sour spoilage of tomato juice is the use of a thermal process adequate to destroy the heat-resistant flat-sour spores present in the juice. Unless an adequate process is employed, the surviving spores may germinate and grow in the product and cause spoilage.

It is also important that presterilized juice and juice packed by the hot fill–hold–cool procedure be filled sufficiently hot to destroy low heat-resistant spoilage organisms in the container and on the cover, during a short holding period between closing and cooling. A minimum closing temperature of 200°F and holding the filled cans of juice at this temperature for 3 min prior to water cooling is adequate under most conditions to destroy low heat-resistant organisms.

Contaminated Plant Equipment. Flat-sour organisms are not normally present in canning plant equipment in sufficient numbers to constitute a spoilage problem. However, if these bacteria are carried into the plant as a result of poor washing of the raw stock, they may "seed" certain pieces of equipment and, through multiplication, become a serious source of contamination, that is, dead ends in pipelines where the juice remains static for extended periods of time.

Wooden and other porous materials provide ideal conditions for the growth and multiplication of these organisms. The use of wooden equipment in contact with the tomatoes or juice should be avoided, as the juice is absorbed in the pores of the wood and cannot be entirely removed. Such equipment has largely been replaced by metal equipment. However, it should be emphasized that the use of metal equipment per se does not ensure complete protection against flat-sour spoilage of tomato juice: the design and accessibility for cleaning of metal equipment are of primary importance.

SPOILAGE OF CANNED TOMATOES

There are several types of spoilage associated with this product, although the loss through spoilage is not great in tomatoes. These are as follows:

1. Hydrogen swell is a physiochemical reaction between the metal of the can and the acids in the fruits, and usually no microorganisms are involved.

2. Bacterial swell is usually caused by one or more members of the lactobacillus group. These organisms are gram-positive, non-spore-forming, nonmotile rods; they produce acid and carbon dioxide which cause the can to swell. Tomatoes and their products are an ideal medium for the growth of the organisms of this group, of which *Lactobacillus lycopersici* is the principal one. This organism can tolerate the high temperatures used in the processing of the product, or it may gain entrance into the can through a slight defect (Weiser *et al.* 1971). In tomato juice with a pH 4.4, the organism is killed at 149°F (65°C) in 3 to 5 min or in 1 min at 168.8°F (76°C) (Pederson 1929).

Other members of the lactobacillus group have been reported as troublesome bacteria in canned tomatoes and tomato pulp, causing a gaseous spoilage. Among these organisms, *Lactobacillus gayoni* has been reported to be more resistant to heating than *Lactobacillus lycopersici*, requiring from 15 to 18 min at 168.8°F (76°C) to kill all the organisms. The most rapid growth of this organism takes place at about 88°F (30°C), although temperatures of from 64.5° to 113°F (18° to 45°C) are favorable. No growth could be obtained at 50°F (10°C) or at 131°F (55°C).

Leuconostoc pleofructi was found to be the least resistant to heat of this group. The organism is destroyed at 149°F (65°C) in 3 min.

Other *Lactobacillus* organisms have been isolated from canned tomatoes, tomato pulp, and low-acid catsup which cause spoilage in tomato products. These organisms are *L. pentoaceticus*, *L. mannitopeum*, and *L. plantasum* (Pederson 1929).

SPOILAGE OF CATSUP

Molds and yeasts find a favorable environment in catsup and reflect the general sanitary practices used in the processing of this product. In fact, mold counts are a routine laboratory procedure followed during processing. The presence of yeasts gives a yeasty flavor along with gas formation, certainly an undesirable defect in catsup (Weiser *et al.* 1971).

Detailed procedures for detection and isolation of spoilage in canned foods are given in the Laboratory Manual for Food Canners and Processors (Natl. Food Proc. Assoc. 1968). Tables 25.2 and 25.3 serve as a key to probable causes of spoilage in canned foods in the pH range of 4.0 to 4.6 for unconcentrated (natural) and concentrated products, respectively. Procedures for examination of foods, either routinely or when implicated in disease outbreaks, are presented in a detailed laboratory manual of methods by Weiser *et al.* (1971). Therefore, for quality control assurance and analysis for detection of spoilage, the reader is referred to the above cited texts.

It is most important in spoilage work to determine first the evaluation of the container. This involves the external examination of the container for any leakage due to lack of integrity of the seams, rough handling defects, or excessive straining of the container (buckling) during processing. The evaluation of the container includes the visual external examination of the can for any defects and the evaluation of the closures as to body hook, cover hook, overlap, width of seam, and tightness ratings. Closing machine records should be examined for the codes in question to ascertain if any abnormalities exist. After the product has been removed from the container for bacteriological evaluation, the internal condition of the can should be studied for defects, including detinning, if any, side-seam characteristics and any other possible abnormalities. Further, the processing records should be evaluated as to the effectiveness of the process to conform to the requirements for the specific commodity.

TABLE 25.2. KEY TO PROBABLE CAUSE OF SPOILAGE IN CANNED FOODS: GROUP 3. ACID FOODS - pH RANGE 4.0 TO 4.6 (EXCEPT CONCENTRATED FOODS)

	Characteristics of Material in Cans						
Condition of Cans	Odor	Appearance	Gas (Co_2 and H_2)	pH	Smear	Cultures	Diagnosis
Swells	Normal to "metallic"	Normal to frothy (cans usually etched or corroded)	More than 20% H_2	Normal	Negative to occasional organisms	Negative	Hydrogen swells
	Sour	Frothy; possibly ropy brine	Mostly CO_2	Below normal	Pure or mixed cultures of rods, coccoids, cocci or yeasts	Growth, aerobically and/or anaerobically at 30°C and possibly at 50°C	Leakage or gross underprocessing
	Sour	Frothy; possibly ropy brine; food particles firm	Mostly CO_2	Below Normal	Pure or mixed cultures of rods, coccoids, cocci or yeasts	Growth, aerobically and/or anaerobically at 30°C and possibly at 50°C (if product received high exhaust only spore formers may be recovered)	No process given
	Normal to sour cheesy	Frothy	H_2 and CO_2	Normal to slightly below normal	Rods—med. short to med. long, usually granular; spores seldom seen	Gas anaerobically at 50°C and possibly slowly at 30°C	Underprocessing—thermophilic anaerobes
	Butyric acid	Frothy; large volume gas	H_2 and CO_2	Below normal	Rods—bipolar staining; possibly spores	Gas anaerobically at 30°C. Butyric acid odor	Underprocessing—butyric acid anaerobes
	Sour	Frothy	Mostly CO_2	Below normal	Short to long rods	Gas anaerobically; acid and possibly gas aerobically in broth tubes at 30°C. Possible growth at 50°C	Gross underprocessing—lactobacilli
No vacuum and/or cans buckled	Normal	Normal	No H_2	Normal to slightly below normal	Negative to moderate number of organisms	Negative	Insufficient vacuum caused by: 1. Incipient spoilage 2. Insufficient exhaust

(continued)

TABLE 25.2. (Continued)

Condition of Cans	Characteristics of Material in Cans: Odor	Appearance	Gas (Co_2 and H_2)	pH	Smear	Cultures	Diagnosis
Flat cans (0 to normal vacuum)	Sour to "medicinal"	Normal	—	Slightly to definitely below normal	Rods—possibly granular in appearance	Growth without gas at 50°C and possibly at 30°C growth on thermoacidurans agar	3. Insufficient blanch 4. Improper retort cooling procedure 5. Overfill Underprocessing—*B. coagulans* (spoilage of this type usually limited to tomato juice)
	Normal	Normal to cloudy brine; possibly moldy	—	Slightly to definitely below normal	Pure or mixed cultures of rods, coccoids, cocci, or mold	Growth, anaerobically and/or aerobically at 30°C and possibly at 50°C	Leakage, or no process given

Source: National Food Processors Association (1968).

TABLE 25.3. KEY TO PROBABLE CAUSE OF SPOILAGE IN CANNED FOODS: GROUP 4. CONCENTATED ACID FOODS - pH BELOW 4.6; AND HIGH ACID FOODS - pH BELOW 4.0.

	Characteristics of Material in Cans						
Condition of Cans	Odor	Appearance	Gas (CO_2 and H_2)	pH	Smear	Cultures	Diagnosis
Swells	Normal to "metallic"	Normal to frothy (cans usually etched or corroded)	More than 20%	Normal	Negative to occasional organisms	Negative	Hydrogen swells
	Normal	Normal to frothy; scorched	All CO_2	Normal	Negative to occasional organisms	Negative	"Frothy fermentation" (this spoilage is limited to concentrated products)
	Normal	Normal to frothy; possibly surface	Mostly CO_2	Below normal	1. Short to long rods 2. Pure or mixed cultures of short to long rods, cocci, coccoids or yeasts	1 and 2: growth aerobically and/or anaerobically with gas production at 30°C and possibly at 50°C	1. Underprocessing or leakage 2. Gross underprocessing or leakage
No vacuum and/or cans buckled	Normal	Normal	No H_2	Normal to slightly below normal	Negative to moderate number of organisms	Negative	Insufficient vacuum caused by: 1. Incipient spoilage 2. Insufficient exhaust 3. Insufficient blanch 4. Inproper retort cooling procedure 5. Overfill
Flat cans (0 to normal vacuum)	Normal to sour	Normal to cloudy brine; possibly moldy	—	Normal to definitely below normal	Pure or mixed cultures of rods, coccoids, cocci or mold	Growth, aerobically and/or anaerobically at 30°C and possibly at 50°C	Leakage or underprocessing (if mold present, leakage

Source: National Food Processors Association (1968).

Two other aspects in evaluation of spoilage problems are the efficacy of the cooling in terms of product temperature, and residual chlorine in the cooling water. Finally, spoilage problems may result from management changes in equipment without adequate evaluation of the new equipment in terms of product preservation. It behooves all concerned with production, processing, and quality assurance to evaluate any change from previous successful practices before the "go ahead" is given to operate new equipment or to make changes in the parameters of a specific unit operation. Disastrous results of processors going bankrupt only serve to emphasize the latter point. Evaluation of spoilage is only an afterthought. Prevention of all spoilage to optimize efficiency and improve quality should be the ultimate goal of all food processors.

REFERENCES

ANON. 1948. Flat-sour spoilage of tomato juice. Continental Can Co. Res. Dep. Bull. *16*.

BASH, W.D. 1964. Effects of handling and holding practices on the aerobic heat resistant bacterial spore population of mechanically harvested tomatoes. Ph.D. Dissertation. Ohio State Univ., Columbus.

BECKER, M.E. and PEDERSON. C.S. 1950. The physiological characters of *Bacillus coagulans (Bacillus thermoacidurans)*. J. Bacteriol. *59*, 717.

BERRY, R.N. 1933. Some new heat-resistant acid-tolerant organisms causing spoilage in tomato juice. J. Bacteriol. *25*, 72.

BIRNBAUM, D.G., LEONARD, S., HEIL, J.R., BUHLERT, J.E., WOLCOTT, T.K. and ANSAR, A. 1977. Microbial activity in heated and unheated tomato serum concentrates. J. Food Process. Preserv. *1* (2) 103–118.

BOHRER, C.W. and REED, J.M. 1948. Spoilage control for canned tomatoes. Natl. Food Proc. Assoc. Res. Lab. Presented at Meet. of Tidewater Canners Assoc. of Va., Irvington, VA., May.

KNOCK, G.C., LAMBRECHTS, M.S., HUNTER, R.C. and RILEY, F.R. 1959. Souring of South African tomato juice by *Bacillus coagulans*. J. Sci. Food Agric. *10* (6) 337.

NATL. FOOD PROCESSORS ASSOC. 1968. Laboratory Manual for Food Canners and Processors, Vol. 1, 3rd Edition. AVI Publishing Co., Westport, CT.

PEARCE, W.E. 1940. Sources of contamination in the manufacture of tomato juice. Am. Can Co. Presented at Canning Prob. Conf. of Natl. Food Proc. Assoc., Chicago, Jan.

PEDERSON, C.S. 1929. The types of organisms found in spoiled tomato products. N.Y. State Agric. Exp. Stn. Tech. Bull. *150*.

PEDERSON, C.S. and BECKER, M.E. 1949. Flat sour spoilage of tomato juice. N.Y. State Agric. Exp. Stn. Tech. Bull. *287*.

RICE, A.C. and PEDERSON, C.S. 1954. Factors influencing growth of *Bacillus coagulans* in canned tomato juice. I. Size inoculum and oxygen concentration. Food Res. *19*, 115.

SMITH, N.R., GORDON, R.E. and CLARK, F.E. 1946. Aerobic Mesophilic Sporeforming Bacteria. U.S. Dep. Agric. Misc. Publ. *559*.

SOGNEFEST, R. and JACKSON, J.M. 1947. Presterilization of canned tomato juice. Food Technol. *1* (1) 78.

TRESSLER, D. K. and JOSLYN, M.A. 1961. Fruit and Vegetable Processing Technology. AVI Publishing Co., Westport, CT.

TROY, V.S., BOYD, J.M. and FOLINAZZO, J.F. 1963. Spoilage of Canned Foods Due to Leakage. Continental Can Co., Metal Div. Res. Dev. Dep., Chicago, May.

WEISER, H.H., MOUNTNEY, G.J. and GOULD, W.A. 1971. Practical Food Microbiology and Technology, 2nd Edition. AVI Publishing Co., Westport, CT.

WESSEL, D.J. and BENJAMIN, H.A. 1941. Process control of heat resistant spoilage organisms. Fruit Prod. J. *20* (6) 178.

WHITE, L.S. 1951. Spoilage bacteria in tomato products. Food Res. *16*, 422.

CHAPTER 26
Composition of Tomatoes

The yield and quality of tomato products depend in great measure upon the composition of the raw material. Nevertheless, tomatoes are purchased by canners and tomato product manufacturers without regard to this fact (Saywell and Cruess 1932).

SOLIDS

Tomatoes contain usually from 7 to 8.5% of total solids, of which about 1% is skins and seeds. The composition of strained tomatoes obtained by cycloning is shown in Table 26.1 (Baker and Wright 1935; Bolcato 1936). The percentage of solids in tomatoes varies through wide limits for a number of reasons, such as variety of tomatoes, character of soil, and especially the amount of rainfall during the growing and harvesting season. Brooks and MacGillivray (1928) have shown that, at the end of prolonged dry spell during the harvesting season, ripe tomatoes are more firm and contain more tomato solids than during a wet season. Saywell and Cruess (1932) found a variation of from 4.56 to 9.55% total solids in tomatoes grown commercially in California in a single season.

Inasmuch as tomato products, such as pulp and catsup, are evaporated to a definite specific gravity, consistency or percentage of solids, their yield per ton of tomatoes varies with the composition of the raw tomatoes used in their manufacture.

CARBOHYDRATES

The free sugars of commercial varieties of tomatoes are predominantly reducing sugars (Miladi *et al.* 1969). The quantity of sucrose found in tomatoes is so negligible that it may be ignored for all practical purposes (Goose and Binsted 1964). Sucrose rarely exceeds 0.1% on a fresh weight basis.

The polysaccharides in tomatoes comprise about 0.7% of tomato juice. Pectins and arabinogalactans constitute about 50% of this; xylans and arabinoxylans about 28%; and cellulose about 25% (Miladi *et al.* 1969).

TABLE 26.1. COMPOSITION OF TOMATOES

Constituent	%
Total solids	7.0 −8.5
Insoluble solids	1.0
Soluble solids	4.0 −6.0
Sugar	2.0 −3.0
Acid	0.3 −0.5
Soluble protein and amino acid	0.8 −1.2
Mineral constituents	0.3 −0.6
Salt (sodium chloride)	0.05−0.1

The reducing sugars, which usually make up from 50 to 65% of tomato solids, are mainly glucose and fructose (Goose and Binsted 1964). The total sugar content in the fresh tomato is found to vary from 2.19 to 3.55% (Smith, Undated). Davies (1968) reported that in general, more fructose than glucose was present. Miladi *et al.* (1969) indicated that total fructose was found to constitute 1.70% of the fresh juice of the variety Heinz 1370 grown in Ohio, as compared to glucose, which was found to constitute 1.51% of the juice. They also detected and measured alpha and beta isomers of the sugars. Alpha sugar was found to be higher in concentration than the beta forms. Alpha-fructose was also highest in concentration as compared to the other free sugars, with a content of 1.46 g in 100 g of fresh tomato juice. Beta-fructose was found to be the lowest in concentration with 0.24 g in 100 g of juice. Beta-glucose was found to be higher than alpha-glucose. The significance of these isomers in tomato juice is not yet known.

PROTEINS AND AMINO ACIDS

There are 19 soluble amino acids in fresh tomato juice. Miladi *et al.* (1969) reported that glutamic acid comprises up to 48.45% of the total weight of amino acids in fresh tomato juice. Second highest in concentration is aspartic acid. The amino acid with the lowest measurable concentration is proline. Processing of fresh tomato juice at 220°F for 20 min results in a substantial increase in the free amino acids as a result of denaturation and partial hydrolysis of protein. The greatest increase occurs in glutamic and aspartic acids, alanine, and threonine. Asparagine and glutamine disappear during processing due to the loss of amide ammonia (NH_3) to form glutamic and aspartic acids, which partially account for the increase in ammonia in canned juice. It could also be due to glutamine and asparagine deamination and formation of pyrollidone carboxylic acid. The amino acids of fresh and processed tomato juice are presented in Table 26.2.

ACIDS

The acid in tomatoes is generally considered to be almost entirely citric, and free acids are almost always determined as citric monohydrate. Some workers have reported the presence of malic acid in quantities often ex-

TABLE 26.2. AMINO ACID CONTENT OF FRESH AND PROCESSED TOMATO JUICE

Amino Acid	mg Amino Acids in 100 g Tomato Juice	
	Fresh	Processed
Aspartic acid	5.5	51.6
Threonine	1.0	9.0
Serine	2.3	12.7
Unknown	—	0.6
Asparagine and glutamine	7.8	—
Glutamic acid	21.9	212.5
Proline	0.1	0.4
Glycine	0.3	1.2
Alanine	1.0	9.0
Valine	0.4	1.7
Methionine	0.2	0.9
Isoleucine	0.6	3.8
Leucine	0.6	3.0
Tyrosine	0.5	3.4
Phenylalanine	1.4	10.8
Lysine	0.9	5.1
Histidine	0.9	7.5
Arginine	0.7	4.4
Total	45.1	337.6

TABLE 26.3. ORGANIC ACIDS IN FRESH AND PROCESSED TOMATO JUICE

Acid	mEq/Liter	
	Fresh	Processed
Acetic	1.06	1.56
Lactic	1.37	1.46
Succinic	0.60	0.49
Alpha-ketoglutaric	1.10	0.53
Pyrrolidone-carboxylic	0.81	8.10
Unknown	0.17	0.28
Malic	3.72	5.39
Citric	60.92	66.92

ceeding citric acid, while traces of tartaric, succininc, acetic, and oxalic acids have also been reported. Recent chromatographic analysis reported by Miladi *et al.* (1969) have separated eight organic acids from tomato juice. Malic acid was found to be the second major organic acid in fresh juice, whereas pyrrolidone carboxylic acid was found to be the second major organic acid in the processed juice. Processing of tomato juice results in an increase in total acid. It was found that acetic acid is increased by 32.1%, apparently due to oxidation of aldehydes and alcohols during processing, and deamination of amino acids, such as alanine to pyruvic. Also an increase in citric and malic acids after processing was noted. Crean (1969) indicates that sugars can decompose on heating in the presence of acids to give acetic, lactic, fumaric, and glycollic acids. The organic acids in fresh and processed tomato juice are presented in Table 26.3.

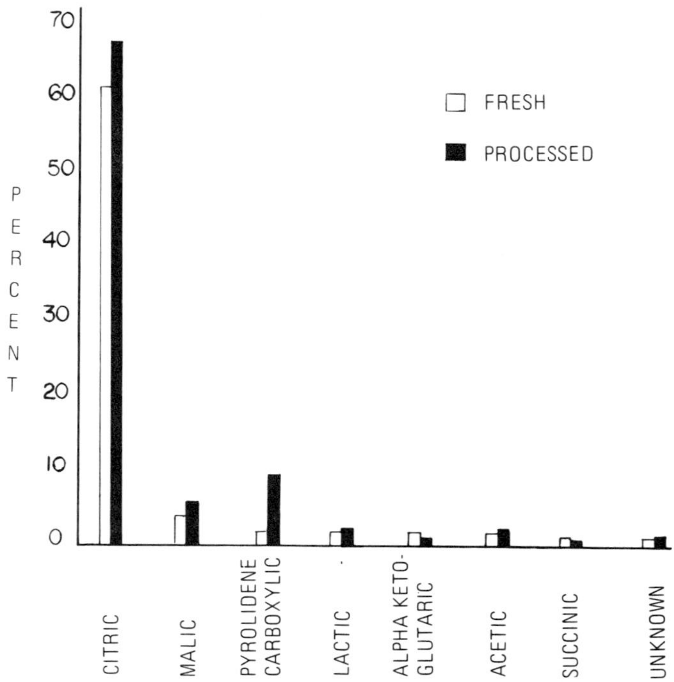

FIGURE 26.1. PERCENTAGE OF ACIDS IN FRESH AND PROCESSED TOMATO JUICE.

MINERALS

The quantity of minerals varies between 0.3 and 0.6%. Also salt (sodium chloride) varies between 0.05 and 0.1%. The minerals play a secondary role in the quality of the finished product. The composition of the tomato and tomato products is shown in Table 26.4 (U.S. Dept. Agric. 1963). It is interesting to note that the levels of minerals, such as calcium and phosphorus, in tomato products vary greatly in the literature as can be seen by comparing USDA values to Lopez and Williams (1981).

PECTIN IN TOMATO

Pectin is a natural constituent of ripe tomatoes. It is formed between the microscopic cells which make up the fleshy red tissues, cementing these together (Smith, Undated). Pectins are polymers of alpha-delta-galacturonic acid linked 1-4. Like the majority of polysaccharides, pectins vary in chain length and hence molecular weight. They are also esterified to varying degrees with methyl groups (Crean 1969). The low-ester pectins, those in which fewer than 50% of the groups are esterified, are known as pectic acids; the higher-ester pectins are called pectinic acids.

TABLE 26.4. COMPOSITION OF TOMATO AND TOMATO PRODUCTS, 100 GRAMS

	Tomato		Tomato Juice				Tomato Purée			
	Fresh	Canned	Regular	Concentrated	Dehydrated	Cocktail	(Pulp)	Catsup	Chili Sauce	Tomato Paste
Water (%)	93.5	93.7	93.6	75.0	1.0	93.0	87.0	68.6	68.0	75.0
Food energy (calories)	22	21	19	76	303	21	39	106	104	82
Protein, g	1.1	1.0	0.9	3.4	11.6	0.7	1.7	2.0	2.5	3.4
Fat, g	0.2	0.2	0.1	0.4	2.2	0.1	0.2	0.4	0.3	0.4
Carbohydrate:										
total, g	4.7	4.3	4.3	17.1	68.2	5.0	8.9	25.4	24.8	18.6
fiber, g	0.5	0.4	0.2	0.9	3.1	0.2	0.4	0.5	0.7	0.9
Ash, g	0.5	0.8	1.1	4.1	17.0	1.2	2.2	3.6	4.4	2.6
Calcium, mg	13	6	7	27	85	10	13	22	20	27
Phosphorus, mg	27	19	18	70	279	18	34	50	52	70
Iron, mg	0.5	0.5	0.9	3.5	7.8	0.9	1.7	0.8	0.8	3.5
Sodium, mg	3	130	200	790	3934	200	399	1042	1338	38
Potassium, mg	244	217	227	888	3518	221	426	363	370	888
Vitamin A (I.U.)	900	900	800	3300	13100	800	1600	1400	1400	3300
Thiamin, mg	0.06	0.05	0.05	0.20	0.52	0.05	0.09	0.09	0.09	0.20
Riboflavin, mg	0.04	0.03	0.03	0.12	0.40	0.02	0.05	0.07	0.07	0.12
Niacin, mg	0.7	0.7	0.8	3.1	13.5	0.06	1.4	1.6	1.6	3.1
Ascorbic acid, mg	23	17	16	49	239	16	33	15	16	49

The growing plant forms first an insoluble compound called "protopectin," which binds the cells firmly together. As the fruit ripens to full maturity this protopectin is changed into pectin, which still holds the cells in place but less rigidly, so that the fruit is no longer hard. Further growth of the tomato allows the pectin itself to be broken down into soluble compounds which have little binding power, so that overripe fruit is soft and mushy.

These transformations of the pectinous materials within the tomato are brought about by the action of enzymes formed within the cells of the plant as it grows. Protopectinase has the specific power of transforming protopectin into pectin. Another enzyme known as pectinase or polygalacturonase can further break down the long pectin chains into shorter ones. A third enzyme, pectan or pectinesterase, can remove the methyl ester groups from the molecule, thus transforming pectinic to pectic acids. The tomato fruit is particularly rich in these enzymes. These enzymes can be classified more minutely: endopolygalacturonase, exopolygalacturonase, endopolymethylgalacturonase, and exopolymethylgalactruonase by their actions. The actions of these enzymes can be shown as follows:

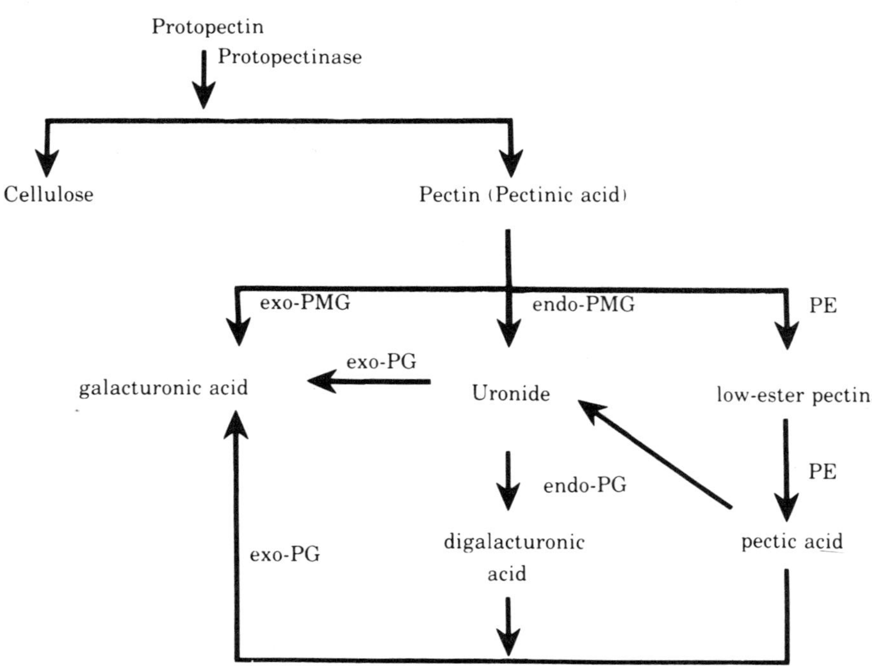

PE: Pectinesterase
PG: Polygalacturonase
PMG: Polymethylgalacturonase

While they are formed only in the growing fruit, their activity does not cease when the fruit is harvested or when it is crushed and screened, and they can play a significant role in determining the texture of processed tomato products. Another enzyme that can also act on tomato solids is cellulase which may be more important in affecting texture (Hall 1963; Foda and McCollum 1970).

Commercial canning varieties of tomatoes show quite uniform amounts of total pectinous material in the raw tomatoes, the amount varying from 0.17 to 0.23%. The transformation of protopectin to pectin in tomatoes which takes place during the last stages of ripening is of great importance for the consistency of the finished product. From the pink stage of ripening to the full red-ripe state there is a large increase in available pectin. Thus, careful selection of full red-ripe tomatoes not only makes for a good tomato color, but also provides more natural pectin for the finished product (Kohman *et al.* 1930).

NUTRIENT COMPOSITION OF TOMATOES AND TOMATO PRODUCTS

Fresh tomatoes, tomato juice, and other processed tomato products make a significant contribution to human nutrition due to the concentration and availability of several nutrients in these products and to their widespread consumption.

Vitamin C, the antiscorbutic vitamin necessary for normal metabolism, wound healing, and collagen synthesis, is correctly associated by consumers with tomato juice and other tomato products. Whole red-ripe tomatoes contain nearly all the vitamin C activity in the reduced ascorbic acid form. Dehydroascorbic acid has been reported to be from 1 to 5% of the total ascorbic acid in tomatoes (Davey *et al.* 1956; Bauerfeind and Pinkert 1970). The ascorbic acid concentration in fresh ripe tomatoes is about 25 mg per 100 g. Thus, a small tomato supplies about 40% of the adult United States Recommended Daily Allowance (RDA) of 60 mg and about two-thirds of the Recommended Daily Allowance (RDA) of 40 mg for children. An 8-oz serving of tomato juice supplies approximately 35 mg of ascorbic acid or about 60% of the adult RDA and 85% of the RDA for children. Thus it is entirely possible to meet ascorbic acid needs in the diet from tomatoes and their products alone.

Tomatoes are also a good source of vitamin A, present in the form of carotene. Fresh ripe tomatoes and tomato juice contain 1000 International Units (IU) of vitamin A per 100 g. Booker *et al.* (1940) gave a figure of 1150 IU per 100 g. On the basis of these figures, a small tomato or glass of juice should supply 20% or more of the adult recommended daily allowance (RDA) of 5,000 IU. It is clear, therefore, that the tomato makes a very important contribution to the vitamin A requirement of the human diet.

In relation to the average consumption of tomatoes and the RDA for vitamins, tomatoes provide significant amounts of vitamins A and C. In addition, they provide small amounts of the B complex vitamins: thiamin, niacin, and riboflavin (see Table 26.4.) Baker and Wright (1935) reported that tomato pulp contains 120 mg of thiamin per 100 g. Munsell (1940) found 69 mg per 100 g of ripe fruits and juice. On this basis a small tomato contains only about one-tenth of the RDA for an adult man.

Early studies indicated the riboflavin content of tomatoes to be rather low. Using the florometric method, Hudson reported a value of 52 mg riboflavin per 100 g tomato. Lanford *et al.* (1941), employing the rat assay method, found 37.3 mg per 100 g which was a higher value than they obtained for either apples or pears. On the basis of these studies it is evident that tomatoes make a very small contribution to the 1.7 mg of RDA of riboflavin required for adults.

Early work by Goldberger and Wheeler (1927) revealed that tomatoes have a definite but feeble pellegra-preventive action. According to a review by Bacharach (1941), tomatoes contain less than 0.015 mg of niacin per 100 g, a small amount compared to the RDA of 20 mg for an adult. Tomatoes may be a major source of these B complex vitamins only to persons consuming very large quantities of tomatoes in their diet.

Of the minerals present in tomatoes (see Table 26.4) iron is the most important in terms of providing adequate nutrition. An 8-oz serving of tomato juice provides approximately 2.0 mg iron in the reduced ferrous state. This concentration is important both because it is 10 to 20% of the RDA of iron and because it is consumed in a product that also provides ascorbic acid, which helps retain the iron in its reduced state and is necessary for iron absorption (Natl. Acad. Sci.-Natl. Res. Counc. 1968). The contribution of tomato products to iron nutrition in the future may be seen as increasingly important.

Factors Affecting the Nutrient Composition of Fresh Tomatoes

The nutritive value of processed tomato products depends both on the initial nutrient concentration in the fresh tomato and on the effects of processing and storage of the finished product. Factors affecting nutrient concentrations in fresh tomatoes include heredity, soil and plant nutrition, cultivation and handling practices, and maturity.

Ascorbic acid has been shown to vary with varieties of tomatoes (Anderson *et al.* 1954; Hamner *et al.* 1945; Scott and Kramer 1949). Maclinn and Fellers (1938) found a variation from 15 to 22 mg per 100 g of different varieties grown side by side on the same soil. On the other hand, Currence (1939) reported that statistically differences between varieties seldom occur in field experiments. He emphasized the relationship between a particular variety and the environmental conditions in determining ascorbic acid content.

Soil conditions and plant nutrition have been indicated as altering the ascorbic acid content of tomatoes. Hester and Kohman (1940) claimed a relationship between the ascorbic acid content of tomatoes and the soil type in different locations. However, part of their differences may have been due to factors other than soil. Hester found that application of potassium fertilizer to certain soils resulted in increased yield and ascorbic acid content of tomato fruits and that in soils deficient in manganese the application of manganese produced a similar effect (Hester 1940).

Cultivation practices also contributed to ascorbic acid content variation. Light intensity previous to maturity as well as treatment with growth regulators and insecticides have been shown to alter ascorbic acid contents (Hamner *et al.* 1945; Oza and Rangneda 1972; Umale 1969; Poe *et al.* 1939). Gamma irradiation of tomato fruit increased their ascorbic acid content in experiments by Satler *et al.* (1970).

The maturity of the tomato fruit also affects the ascorbic acid content (Hamner and Maynard 1942). Some reports have indicated that ascorbic acid content of mature green fruits is essentially the same as for fully ripened ones (Maclinn and Fellers 1938; Pope 1972; Scott and Kramer 1949). Others have indicated that ascorbic acid content increases with red-ripeness (Baker and Wright 1935; House *et al.* 1929). House *et al.* (1929) reported that ethylene-ripened tomatoes are lower in ascorbic acid than vine-ripened fruit.

Certainly environmental conditions have a marked influence upon ascorbic acid content. Currence (1939) reported weekly variations associated with environmental conditions. Variations in the values for canned juices from year to year reported by Hanning (1936) more closely correlated with seasons than with canning procedures.

The vitamin A content of fresh tomatoes, like that of vitamin C, is subject to wide variation. It is apparent, however, that many of these variations are associated with the methods of determining the value, whether chemical or biological, and with the methods of translating these values into International Units.

Many reports are in agreement that vitamin A potency or carotene content of tomatoes increases during ripening (House *et al.* 1929; Jones and Nelson 1930; Morgan and Smith 1928). House *et al.* (1929) indicated that vine-ripened fruits are more potent sources than fruits detached while partially green and ripened in air or ethylene, although ripe fruits are richer than green fruits regardless of method of ripening (Morgan and Smith 1928; Smith, Undated). However, ethylene- or air-ripened fruits are equivalent to vine-ripened fruits in vitamin A potency.

The vitamin A potency also varies markedly with variety. The red varieties being much more potent than the so-called pink or white varieties. Environmental conditions, such as exposure to light, are also important. From a practical standpoint the stage of ripeness is not an important consideration here because tomatoes are usually canned or consumed only when ripe and because of this the method of ripening seems to have little effect. To improve the vitamin A potency of tomatoes, the utilization of fully ripened fruit of the proper varieties seems the most promising procedure (Hamner and Maynard 1942).

Factors Affecting Retention of Nutrients in Tomato Products

The vitamin C and vitamin A content of processed tomato products manufactured without fortification are necessarily less than or nearly equal to that of the fresh tomatoes from which they were produced.

Processing Effects on Vitamin C. The maintenance of high levels of ascorbic acid products during processing has received considerable emphasis by food technologists. In the manufacture of tomato juice, ascorbic acid is destroyed, mainly by oxidation. Ascorbic acid is oxidized to dehydroascorbic acid, which is further oxidized to degradation products with no vitamin C activity. The oxidation may be enzymatic or nonenzymatic, and is catalyzed by copper ions. The rate of oxidation is dependent on the dissolved oxygen, enzyme content, dissolved copper, and temperature of the juice. The longer the tomato juice is held at optimum conditions for oxidation the lower will be the retention of ascorbic acid after processing.

Clifcorn and Peterson (1947) reviewed the retention of ascorbic acid during tomato juice manufacture. They reported that an average of 63 to 70% retention was found during three separate plant surveys and that in some plants retention as high as 94% had been achieved. They emphasized that in plants where retention was high, total elapsed canning time was short (2 to 3 min) and those conditions increasing the oxidation rate were minimized.

The temperature to which tomato products such as tomato juice are heated in the presence of air is the most important factor in the rate of ascorbic acid destruction; it has been found that the rate of ascorbic acid destruction increases with increased temperature in the presence of air. It is therefore important that juice be brought to the desired temperature as quickly as possible and held for only a short period at high temperature. Guerrant *et al.* (1945) showed that retention of ascorbic acid was greater (92%) after 15 sec preheat before extraction at 57°C (135°F). Retention decreased to 54% after 35 min preheat at 88°C (190°F). While cold extraction at 49°C (120°F) has been suggested, later work has shown that retention is nearly equal for hot- and cold-break processes if the juice is not held at high temperatures while exposed to air for long periods of time prior to extraction (Clifcorn 1945; Clifcorn and Peterson 1947).

Enzymatic destruction of ascorbic acid in tomato juice is minimal, as the oxidative enzymes occur in the pulpy portion of the tomato and are destroyed or removed during extraction (Lamb 1946). The copper catalytic effect, however, also occurs in nonenzymatic oxidation and can only be prevented by absolute removal of all copper equipment.

Any unit operation which incorporates air into the juice will accelerate oxidation of ascorbic acid. Therefore, the process line should be as simple as possible, with few pumping stations. Extraction and homogenization equipment should be designed to eliminate dissolved air. Similarly, improper filling that allows an excessive headspace and low vacuums will allow air to dissolve in the juice increasing loss of ascorbic acid.

Concentrated products present a further problem in retaining vitamin C. Tomato paste has been reported to have ascorbic acid content of 49 mg per 100 g on a solids percentage equal to that of tomato juice; however, concentrated products usually contain less ascorbic acid than do whole canned tomatoes or tomato juice (Lamb *et al.* 1951; Hummel and Okey 1950). Bolcato (1936) reported less loss of ascorbic acid in the preparation of tomato concentrates by heating at 60° to 70°C (140° to 158°F) than at temperatures either higher or lower.

It is evident from the literature on canned tomato products that a good deal of ascorbic acid may be lost during the processing unless care is exercised, but that most of the vitamin potency may be preserved if suitable precautions are taken. Tomatoes lend themselves to the production of juice high in ascorbic acid more readily than many other fruits, but the importance of proper processing methods is obvious. It is also clear that an important factor in improving the vitamin C content of commercial products is the utilization of fruits in the canning process which are initially high in ascorbic acid (Guerrant *et al.* 1945).

Processing Effects on Vitamin A and Other Vitamins. Canned tomatoes and tomato juices have vitamin A potencies equal to the fresh material (De Caro and Perling 1936). Poe *et al.* (1939) found no marked differences in the vitamin A potency of 15 brands of commercially canned tomato juice. Prolonged heating accompanied by exposure to air resulted in some destruction of vitamin A potency and thus concentrated juice may be rather low (Fellers 1940; De Caro and Perling 1936). Most of the vitamin A value is present in the solid matter, and thus filtered tomato juices are very low in it.

Prolonged heating also decreased retention of the B vitamins. Cameron reported an average retention of 89% for thiamin, 97% for riboflavin, and 98% for niacin.

Retention of Vitamins During Storage

Since tomato packs are usually consumed over an extended period of time, the shelf life of nutrients, especially ascorbic acid, in processed tomato products has received considerable study. Length of storage, temperature, and the initial nutrient concentration have been shown to affect retention.

Food processors surveyed in 1971 reported their tomato products had a shelf life of 2 to 3 years (Anon. 1971). In terms of nutrient retention these figures may be overly optimistic. In early work by Guerrant *et al.* (1945), 75 to 85% of the initial ascorbic acid in tomato juice after processing was retained after 1 year at the high temperature range expected in warehouses in this country 27° to 29°C (80° to 85°F). After 2 years storage at 21°C (70°F) and at higher temperatures of 27° to 29°C (80° to 85°F) retention of ascorbic acid has been reported at 60 to 70% or below (Brenner *et al.* 1948; Moschette *et al.* 1947; Sheft *et al.* 1949; Lamb *et al.* 1951). When stored at normal room temperatures, loss of ascorbic acid may be expected at a rate of 1% per month. At higher temperatures of 27° to 29°C (80° to 85°F) loss will be 2 to 5% per month.

Temperature of storage is extremely important to retention of nutrients in tomato products. When the temperature was held in the range of 4° to 10°C (40° to 50°F) retentions of ascorbic acid were observed from 92 to 100% after 2 years (Cameron 1955; Feaster *et al.* 1949; Guerrant *et al.* 1945). Retentions decrease directly with increasing temperatures from 50°F up to approximately 90°F (see Fig. 26.2). Above this temperature loss of ascorbic acid increases as log function of temperature increase (Brenner *et al.* 1948). When fluctuating temperatures were investigated, Feaster *et al.* (1949) supported early work showing retentions near 100% during periods of storage at 10°C (50°F) or lower, and accelerated loss above 10°C (50°F), with very rapid loss during periods of storage above 32°C (90°F).

Since tomato products are usually held in warehouses that are not refrigerated, the vitamin concentration will be less than in the fresh product or immediately after processing. To maintain vitamin levels at or above that of fresh tomatoes, fortification of tomato products has been proposed (Bauerfeind and Pinkert 1970; Kuryloski 1971, 1972; Siemers 1971; Bunnell 1968). Bauerfeind's review of the literature suggests that a minimum of 30 mg of ascorbic acid per 4 fluid oz (26 mg per 100 ml) could easily be maintained in tomato juice (Bauerfeind and Pinkert 1970). Although the current Standard of Identity for Tomato Juice does not allow fortification with ascorbic acid, the USDA (Kuryloski 1972) has asked for special packs of tomato juice fortified to 50 mg per 100 ml for the Needy Families Program.

The shelf life of tomato juice fortified with ascorbic acid is dependent not only on temperature and length of storage, as is unfortified juice, but also upon the ascorbic acid concentration. When stored at 21°C (70°F) tomato juice fortified to 30 mg per 100 ml retained 70 to 85% ascorbic acid after 1 year, only slightly lower than expected for unfortified juice (Bauerfeind and Pinkert 1970). However, when fortified to 60 mg per 100 ml retention was down to 70 to 85% after only 6 months storage at room temperature (Rehnfeld and Pratt 1969).

The interactions of time, temperature, and fortification level can be seen in Fig. 26.2. Percent retention is highest when tomato juice is not fortified, temperature is below 10°C (50°F), and storage time is shorter however, it can be a successful carrier of added ascorbic acid if levels do not exceed 40 to 50 mg per 100 ml juice (Pope 1972).

Nutritional labeling of tomato products requires the processor to take into account the conditions under which the product is to be stored and the length of storage. If ascorbic acid (vitamin C) is to be labeled, the processor should determine the initial level after processing and cooling. If the unfortified product is not to be exposed to extreme temperatures for long periods of time (above 32°C or 90°F for over 2 weeks) 75% of the initial level can be expected to remain after 2 years. When tomato juice is fortified to 50 mg per

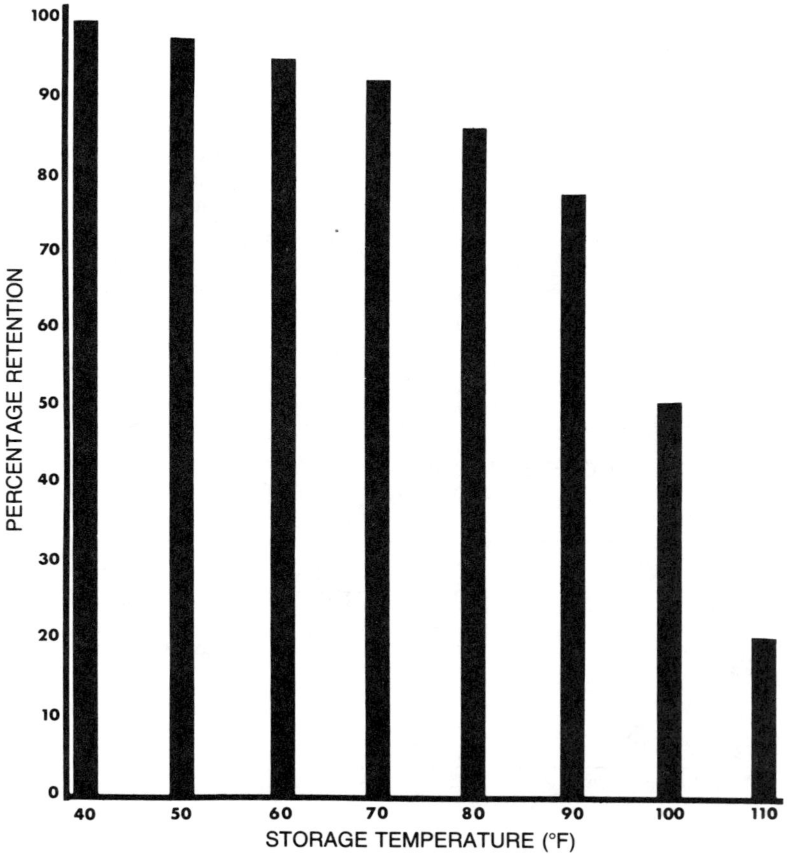

FIGURE 26.2. PERCENTAGE RETENTION OF ASCORBIC ACID IN TOMATO JUICE AFTER 1 YEAR STORAGE AT VARIOUS STORAGE TEMPERATURES.

100 g, 60% (30 mg per 100) may be expected to remain after 1 year under normal storage conditions. Fortification above 50 mg per 100 g can successfully increase ascorbic acid retention above 30 mg per 100 g only if the juice is refrigerated.

The FDA nutrient labeling regulations state that testing for compliance will be on basis of a 12-can composite sample, each can being drawn from a separate shipping case. The suggested labels, as shown in Table 26.5 and 26.6, are based on USDA Handbook 8 values and guidelines prepared by the National Food Proc. Assoc. It should be emphasized that variation due to variety, culture, areas of the country, processing methods, and storage may affect these values. Therefore, the processor should follow a sampling plan as provided by FDA and analyze for these nutrients to assure truthful and not misleading statements on his label.

TABLE 26.5. SUGGESTED NUTRITIONAL INFORMATION FOR LABELS FOR CANNED TOMATOES

		% U.S. Recommended Daily Allowance (U.S. RDA)	
Calories	50	Protein	2
Protein	2 g	Vitamin A	40
Carbohydrate	10 g	Vitamin C	40
Fat	1 g	Thiamin	6
Sodium	0	Riboflavin	2
		Niacin	4
		Calcium	4
		Iron	4

TABLE 26.6. SUGGESTED NUTRITIONAL INFORMATION FOR LABELS FOR TOMATO JUICE

(6 fluid oz Portion)

		% U.S. Recommended Daily Allowance (U.S. RDA)	
Calories	30	Protein	[a]
Protein	1 g	Vitamin A	15
Carbohydrates	7 g	Vitamin C	25
Fat	0	Thiamin	4
		Riboflavin	2
		Niacin	6
		Calcium	[1]
		Iron	4

[a]Contains less than 2% of the U.S. RDA for this nutrient.

TABLE 26.7. TOMATO VOLATILES

	Aliphatic	Aromatic and Heterocyclic	Terpenoid	Total
Aldehydes	21	4	1	26
Ketones	8	—	10	18
Alcohols (including phenols)	15	10	4	29
Acids	4	—	—	4
Esters	5	1	—	6
Acetals	3	1	—	4
Lactones	4	—	—	4
Heterocyclics	—	6	—	6
Hydrocarbons	2	5	4	11
Sulfur compounds	7	1	—	8
Others	1	1	—	2
Total	70	29	19	118

Source: Buttery *et al.* (1971).

Tomato Flavor

The volatile flavor components of the tomato have attracted the attention of many researchers. To date about 118 components have been reported (Table 26.7). Of interest is the large proportion of aldehydes, ketones, and alcohols and the small number of esters. Many components are produced as a result of cooling, notably the sulfur compounds, especially dimethyl sulfide. Compounds which are important to tomato flavor are hex-*cis*-3-enal, beta-ionone, deca-*trans,trans*-2, 4-dienal, and 2-isobutylthiazole. However, it is clear that with such a complex mixture, a good understanding of tomato flavor is not to be gained overnight (Hall 1963; Kazeniac and Hall 1970; Nelson and Hoff 1969).

REFERENCES

ANDERSON, E.E., FAGERSON, I.S., HAYES, K.M. and FELLERS, C.R. 1954. Ascorbic acid and sodium chloride content of commercial canned tomato juice. J. Am. Diet. Assoc. *30*, 1250.

ANON. 1971. Food stability survey. Rutgers. State Univ. *2*, 115.

BACHARACH, A.L. 1941. The Distribution of nicotinic acid in human and animal foods. Nutr. Abstr. Rev. *10*, 459.

BAKER, A.Z. and WRIGHT, M.C. 1935. The vitamin B_1 content of foods. Biochem. J. *29*, 1802.

BAUERFEIND, J.D. and PINKERT, D.M. 1970. Food processing with added ascorbic acid. Adv. Food Res. *18*, 219.

BOLCATO, B. 1936. The preparation of concentrated tomato extracts in relation to the keeping qualities of vitamin C. I. Experiments on the juices. Ind. Ital. Conserv. Aliment. *11*, 89.

BOOKER, L.E., HARTZLER, E.R. and MARSH, R.L. 1940. A Table of the Vitamin Content of Foods. U.S. Dep. Agric. Bur. Home Econ., Oct.

BRENNER, S., WODICKA, V.O. and DUNLOP, S.G. 1948. Effect of high temperature storage on retention of nutrients in canned foods. Food Technol. *2*, 207.

BROOKS, R.E. and MacGILLIVRAY, J.H. 1928. Studies on tomato quality II. Effect of soil moisture upon the percentage of dry matter in the fruit. J. Assoc. Off. Agric. Chem. *11* (3) 389.

BUNNELL, R.H. 1968. Enrichment of fruit products and fruit juices. J. Agric. Food Chem. *16*, 177.

BUTTERY, R.G., SEIFERT, R.M., GUADAGNI, D.G. and LING, L.C. 1971. Characterization of additional volatile components of tomato. J. Agric. Food Chem. *19*, 524.

CAMERON, E.J. 1955. Retention of Nutrients During Canning. Natl. Food Processors Assoc., Washington, DC.

CLIFCORN, L.E. 1945. Variables influencing vitamin content of processed foods. Continental Can Co. Res. Dep. Bull. *6*.

CLIFCORN, L.E. and PETERSON, G.T. 1947. The retention of vitamin C in tomato juice. Continental Can Co. Res. Dep. Bull. *12*.

CREAN, D.E. 1966. Acid components of fruit and vegetables. 1. Production of acids by thermal processing. J. Food Technol. *1*, 55.

CREAN, D.E. 1969. A study of the consistency of tomato juice as influenced by pH and cell wall components. Ph.D. Dissertation. Ohio State Univ., Columbus.

CURRENCE, T.M. 1939. A comparison of tomato varieties for vitamin C content. Am. Soc. Hortic. Sci. Proc. *37*, 901.

DANIEL, E.P. and RUTHERFORD, M.B. 1936. Effect of home canning and storage on ascorbic acid content of tomatoes. Food Res. *1*, 342.

DAVEY, B.L. *et al.* 1956. Utilization of ascorbic acid in fruits and vegetables, 1. Plan of study and ascorbic acid content of 24 foods. J. Am. Diet. Assoc. *32*, 1064.

DAVIES, J.N. 1968. Chromatographic studies of the free sugars of the tomato juice. Glasshouse Corp. Res. Inst. Rep., London.

DeCARO, L. and PERLING, G. 1936. Vitamin A, B, and C content of tomato conserves. Quad. Nutr. *3*, 64 (Nutr. Abstr. Rev. *6* 57).

FEASTER, J.F., TOMPKINS, N.D. and PEARCE, W.E. 1949. Effect of storage on vitamins and quality in canned foods. Food Res. *14*, 25.

FELLERS, C.R. 1940. Effect of processing on the vitamin A content of foods. Inst. Food Tech. Proc. *1*, 97.

FODA, Y.H. and McCOLLUM, J.P. 1970. Viscosity as affected by various constituents of tomato juice. J. Food Sci. *35*, 333.

GOLDBERGER, J. and WHEELER, G.A. 1927. A study of the pellagra-preventive action of the tomato, carrot, and rutabaga turnip. U.S. Public. Health Serv. Rep. *42*, 1299 (Reprint *1157*).

GOOSE, G.P. and BINSTEAD, R. 1964. Tomato Paste, Puree, Juice, and Powder. Food Trade Press, London.

GUERRANT, N.B. *et al.* 1945. The nutritive value of canned foods—Changes in the vitamin content of foods during storage. J. Nutr. *32*, 435.

HALL, C.B. 1963. Cellulose in tomato fruits. Nature (London) *200*, 1010.

HAMNER, K.C. and MAYNARD, L.A. 1942. Factors influencing the nutritive value of the tomato. A review of the literature. U.S. Dep. Agric. Misc. Publ. *502*.

HAMNER, K.C., BERSTEIN, L. and MAYNARD, L.A. 1945. Effects of light intensity, day length, temperature and other environmental factors on the ascorbic acid content of tomatoes. J. Nutr. *29*, 85.

HANNING, F. 1936. Comparison of the biological and chemical methods for the determination of vitamin C in canned strained vegetables and a study of its variation from year to year. J. Nutr. *12*, 405.

HANUCK, H.N. 1938. Vitamin C content of home-canned tomato juice. J. Home Econ. *30*, 183.

HESTER, J.B. 1940. The influence of potash fertilization upon the yield and quality of tomatoes. Am. Fert. *93*, 5.

HESTER, J.B. 1941. Manganese and vitamin C. Science *93*, 401.

HESTER, J.B. and KOHMAN, E.F. 1940. The influence of soil type and fertilization upon the yield and composition of tomatoes. Soil Sci. Soc. Am. Proc. *5*, 395.

HODSON, A.Z. 1940. Riboflavin content of some common vegetables and fruits. Food Res. *5*, 395.

HOUSE, M.C., NELSON, P.M. and HABER, E.S. 1929. The vitamin A, B, and C content of artificially versus naturally ripened tomato. J. Biol. Chem. *81*, 495.

HUMMEL, M. and OKEY, R. 1950. Relation of concentration of canned tomato products to storage loses of ascorbic acid. Food Res. *15*, 405.

HUSSEMANN, D.L. 1935. Vitamin C in home canned tomato juice. Wisc. Agric. Exp. Stn. Annu. Rep. 1933−34, Bull. *430*.

JONES, D.B. and NELSON, E.M. 1930. Vitamin content of ethylene-treated and untreated tomatoes. Am. J. Public. Health *20*, 387.

JUDGE, E.E. & SONS. The Almanac, of the Canning, Freezing, Preserving Industries. Edward E. Judge & Son, Westminster, MD.

KAZENIAC, S.J. and HALL, R.M. 1970. Flavor chemistry of tomato volatiles. J. Food Sci. *35*, 519.

KOHMAN, E.F., EDDY, W.H. and ZALL, E. 1930. Vitamins in canned foods. IX. Tomato products. Ind. Eng. Chem. *22*, 1015.

KURYLOSKI, D. 1971. Consumer and Marketing Service, Needy Families Program, Canned Tomato Juice. Announcement *FV-538*, Program *MMP 45a*. U.S Dep. Agric., U.S. Govt. Printing Office, Washington, DC.

KURYLOSKI, D. 1972. Consumer and Marketing Service, Needy Families Program, Canned Tomato Juice. Announcement *FV-584*. U.S. Dep. Agric., U.S. Govt. Printing Office, Washington, DC.

LAMB, F.C. 1946. Nutritive value of canned foods—Factors effecting ascorbic acid content of canned juices. Ind. Eng. Chem. *38*, 860.

LAMB, F.C., LEWIS, L.D. and WHITE, D.G. 1951. The nutritive value of canned foods—The effect of storage on ascorbic acid content of canned tomato juice and tomato paste. Food Technol. *5*, 269.

LANFORD, C.S., FINKELSTEIN, B. and SHERMAN, H.C. 1941. Riboflavin content of some typical fruits. J. Nutr. *21*, 175.

LEVEDEV, A. 1935. The presence of vitamin C in canned fruits and vegetables. Konservn. Promst. *2*, 18 (Chem. Abstr. *33*, 7915).

LO COCO, G. 1945. Composition of Northern California tomatoes. Food Res. *10*, 114.

LOPEZ, A. and WILLIAMS, H.L. 1981. Essential elements in fresh and canned tomatoes. J. Food Sci. *46* (2) 432−444.

MACLINN, W.A. and FELLERS, C.R. 1938. Ascorbic acid (vitamin C) in tomatoes and tomato products. Mass. Agric. Exp. Stn. Bull. *354*.

MILADI, S. 1970. Introduction to the Composition and Technology of the Tomato. National Institut of Nutrition, Tunsis, Tunisia. (French)

MILADI, S., GOULD, W.A. and CLEMENTS, R.L. 1969. Heat processing effect of starch, sugars, proteins, amino acids of tomato juice. Food Technol. *23*, 93.

MORGAN, A.F. and SMITH, L.L.W. 1928. Development of vitamin A during ripening of tomatoes. Soc. Exp. Biol. Med. Proc. *26*, 44.

MOSCHETTE, D.S., HINMAN, W.F. and HALLIDAY, E.G. 1947. Nutritive value of canned food. XXVII. Ind. Eng. Chem. *39*, 994–999.

MUNSELL, H.E. 1940. Vitamins and their occurrence in foods. Milbank Mem. Fund. Q. *18*, 311.

NATL. ACAD. SCI.—NATL. RES. COUNC. 1968. Recommended Dietary Allowances, 7th Edition. Food Nutr. Board Publ. *1694*. Natl. Acad. Sci., Washington, DC.

NELSON, P.E. and HOFF, J.E. 1969. Tomato volatiles: Effect of variety, processing and storage time. J. Food Sci. *34*, 53.

OZA, A.M. and RANGNEDAI, Y.G. 1972. Effect of application of gibberellie acid on the ascorbic acid content of tomato indian. J. Agric. Sci. *39*, 980.

POE, E.F., GANT, O.K. and GRIFFIN, E. 1939. Vitamin A content of some commercially canned tomato juice. Fruit Prod. J. and Am. Vinegar Ind. 19, 73.

POPE, G. 1972. Effect of time, temperature and fortification level on the shelf-life of ascorbic acid in tomato juice. Ph.D. Dissertation. Ohio State Univ., Columbus.

REHNFELD, B.M. and PRATT, D.E. 1969. Influence of organophosphate insecticides on ascorbic acid oxidation in aqueous solutions. J. Food Sci. *34*, 551.

REYNOLDS, H. 1938. Ascorbic acid content of Arkansas tomato juice. Ark. Agric. Exp. Stn. Annu. Rep. *50*, Bull. *368*.

SATLER, A., ALI, M. and MUHAMMED, A. 1970. Effect of gamma irradiation on postharvest behavior of tomatoes. Food Irradiat. *4*, 2726.

SAYWELL, L.G. and CRUESS, W.V. 1932. The Composition of Canning Tomatoes. Univ. Calif., Coll. Agric., Agric. Exp. Stn. Bull. *545*.

SCOTT, L.E. and KRAMER, A. 1949. Effect of storage upon the ascorbic acid content of tomatoes harvested at different stages of maturity. Am. Soc. Hortic. Sci. *54*, 277.

SHEFT, B.B., GRISWOLD, R.M., TARLOWSKY, T. and HALLIDAY, E.G. 1949. Nutritive value of canned foods. XXVI. Ind. Eng. Chem. *41*, 144.

SIEMERS, G.F. 1971. Methods of increasing the nutritive value of foods. A. Enrichment with vitamins and minerals. *In* Nutritional Evaluation of Food Processing. R.S. Harris and H. Von Loesecke (Editors). AVI Publishing Co., Westport, CT.

SMITH, H.R. (Undated.) The consistency of tomato catsup. Natl. Food Processors Assoc. Res. Lab., Mimeo Rep. *310415*.

SMITH, H.R., BIGELOW, W.D. and GREENLEAF, C.A. 1950. Tomato Products Pulp, Paste, Catsup and Chili Sauce. Natl. Canners Assoc. Res. Lab. Bull. *27-L* (Revised).

TRESSLER, D.K. and CUBRAN, K.M. 1938. The cause of loss of vitamin C from bottled toamto juice. J. Home. Econ. *30*, 487.

UMALE, L.L. 1969. Ascorbic acid stability as influenced by gas hydrates and freezing. Diss. Abstr. Int. B: *30* (3) 1190.

U.S. DEP. AGRIC. 1963. Composition of foods. U.S. Dep. Agric., Agric. Handb. *7*.

WILLIAMS, K.T. and BEVENUE, A. 1954. Some carbohydrate components of tomato. J. Agric. Food Chem. *2*, 472.

APPENDIX A – U.S. Standards of Identity and Grades, Fill of Container, Factors of Quality, Definitions, Inspection and Score Sheets

52.2101 through 52.2112	Tomato Catsup
52.2191 through 52.2202	Chili Sauce
52.2371 through 52.2381	Tomato Sauce
52.3421 through 52.3432	Okra and Tomato
52.3621 through 52.3630	Tomato Juice
52.5041 through 52.5051	Tomato Paste
52.5081 through 52.5091	Tomato Purée
52.5161 through 52.5171	Tomatoes
52.5201 through 52.5210	Concentrated Tomato Juice

UNITED STATES STANDARDS FOR GRADES OF TOMATO CATUSP[1]

Effective August 31, 1953
Including Proposed Amendments of June 1991

IDENTITY AND GRADES

52.2101 Identity.
52.2102 Grades of tomato catsup.

FILL OF CONTAINER

52.2103 Recommended fill of container for tomato catsup.

[1]The requirements of these standards shall not excuse failure to comply with the provisions of the Federal Food, Drug, and Cosmetic Act.

FACTORS OF QUALITY

52.2104 Ascertaining the grade.
52.2105 Ascertaining the rating for the factors which are scored.
52.2106 Color.
52.2107 Consistency.
52.2108 Absence of defects.
52.2109 Flavor.

DEFINITIONS

52.2110 Definitions of terms used in these standards.

LOT INSPECTION AND CERTIFICATION

52.2111 Ascertaining the grade of lot.

SCORE SHEET

52.2112 Score sheet for tomato catsup.

IDENTITY AND GRADES

§ 52.2101 *Identity*. Tomato catsup means the product as defined in the standard of identity for catsup, ketchup, catchup (21 CFR 155.194) issued pursuant to the Federal Food, Drug, and Cosmetic Act.

§ 52.2102 *Grades of tomato catsup*. (a) "U.S. Grade A" or "U.S. Fancy" is the quality of tomato catsup that possesses a good color; that possesses a good consistency; that is practically free from defects; that possesses a good flavor; that possesses a good finish; that has a total solids content of not less than 33 percent by weight; and that scores not less than 85 points when scored in accordance with the scoring system outlined in this subpart.

(b) "U.S. Grade B" or "U.S. Extra Standard" is the quality of tomato catsup that possesses a good color; that possesses a good consistency; that is practically free from defects; that possesses a good flavor; that possesses a good finish; that has a total solids content of not less than 29 percent, by weight; and that scores not less than 85 points when scored in accordance with the scoring system outlined in this subpart: *Provided*, That the tomato catsup may score not less than 18 points for the factor of consistency if the total score is not less than 85 points.

(c) "U.S. Grade C" or "U.S. Standard" is the quality of tomato catsup that possesses a fairly good color; that possesses a fairly good consistency; that is fairly free from defects; that possesses a good finish; that possesses a fairly good flavor; that has a total solids content of not less than 25 percent, by weight; and that scores not less than 70 points when scored in accordance with the scoring system outlined in this subpart.

(d) "Substandard" is the quality of tomato catsup that fails to meet the requirements of U.S. Grade C or U.S. Standard.

FILL OF CONTAINER

§ 52.2103 *Recommended fill of container for tomato catsup.* The recommended fill of container is not incorporated in the grades of the finished product since fill of container, as such, is not a factor of quality for the purposes of these grades. It is recommended that each container of tomato catsup be filled as full as practicable without impairment of quality and that the product occupy not less than 90 percent of the capacity of the container.

FACTORS OF QUALITY

§ 52.2104 *Ascertaining the grade.* (a) The grade of tomato catsup is ascertained by considering in conjunction with the requirements of the respective grade, the respective ratings for the factors of color, consistency, absence of defects, and flavor. The relative importance of each factor which is scored is expressed numerically on the scale of 100. The maximum number of points that may be given such factors are:

Factors:	*Points*
Color	25
Consistency	25
Absence of defects	25
Flavor	25
Total score	100

§ 52.2105 *Ascertaining the rating for the factors which are scored.* The essential variations within each factor which is scored are so described that the value may be ascertained for each factor and expressed numerically. The numerical range within each factor which is scored is inclusive for example, 17 to 20 points means 17, 18, 19 or 20 points.

§ 52.2106 *Color*—(a) *General.* The amount of red in the tomato catsup is determined by comparing the color of the product with that produced by spinning a combination of the following Munsell color discs:

Disc 1—Red (5R 2.6/13) (glossy finish).
Disc 2—Yellow (2.5YR 5/12) (glossy finish).
Disc 3—Black (N1) (glossy finish).
Disc 4—Gray (N4) (mat finish).

Any method or device approved by the USDA (including electric colormeters) which given equivalent results may be used.

(b) *(A) and (B) classification.* Tomato catsup that possesses a good color may be given a score of 21 to 25 points. "Good color" means that the color is typical of tomato catsup made from well ripened red tomatoes and which has been properly prepared and properly processed. Such color contains as much or more red than that produced by spinning the specified Munsell color discs in the following combinations: 65 percent of the area of Disc 1; 21 percent of the area of Disc 2; 14 percent of the area of either Disc 3 or Disc 4, or 7 percent of the area of Disc 3 and 7 percent of the area of Disc 4 whichever most nearly matches the reflectance of the tomato catsup. To receive a score in this classification, tomato catsup, when packed in glass, shall show no discoloration in the "neck" of the bottle.

(c) *(C) classification.* If the tomato catsup possesses a fairly good color, a score of 17 to 20 points may be given. Tomato catsup that falls into this classification shall not be graded above U.S. Grace C, regardless of the total score for the product (this is a limiting rule). "Fairly good color" means that the color is typical of tomato catsup and contains as much or more red than that produced by spinning the specified Munsell color discs in the following combinations: 53 percent of the area of Disc 1; 28 percent of the area of Disc 2; 19 percent of the area of either Disc 3 or Disc 4, or 9½ percent of the area of Disc 3 and 9½ percent of the area of Disc 4 whichever most nearly matches the reflectance of the tomato catsup.

(d) *(SStd) classification.* Tomato catsup that fails to meet the requirements of paragraph (c) of this section may be given a score of 0 to 16 points and shall not be graded above Substandard, regardless of the total score for the product (this is a limiting rule).

§ 52.2107 *Consistency*—(a) *General.* The factor of consistency refers to the viscosity of the product and the tendency to hold its liquid portion in suspension.

(b) *(A) and (B) classification.* Tomato catsup that possesses a good consistency may be given a score of 22 to 25 points. "Good consistency" means that the tomato catsup shows not more than a slight separation of free liquid when poured on a flat grading tray; is not excessively stiff; and flows not more than 3 centimeters not more than 7 centimeters in 30 seconds at 20° Centigrade in the Bostwick consistometer. Within this range, the higher consistency catsup will receive the higher score points.

(c) *(C) classification.* If the tomato catsup possesses a fairly good consistency, a score of 18 to 21 points may be given. Tomato catsup that falls into this classification shall not be graded above U.S. Grade C, regardless of the total score for the product (this is a limiting rule). "Fairly good consistency" meand that the tomato catsup may show a noticeable, but not excessive, separation of free liquid when poured on a flat grading tray; is not excessively stiff; and flows not less than 2 centimeters nor more than 10 centimeters in 30 seconds at 20° Centigrade in the Bostwick consistometer. Within this range, catsup approaching "good conditions" would receive the highest score points.

(d) *(SStd) classification.* Tomato catsup that fails to meet the requirements of paragraph (c) of this section may be given a score of 0 to 17 points and shall not be graded above Substandard, regardless of the total score for the product.

§ 52.2108 *Absence of defects*—(a) *General.* The factor of absence of defects refers to the degree of freedom from defects such as: dark specks or scale-like particles, seeds, particles of seed, tomato peel, core material, or other similar substances. This factor is evaluated by observing a layer of the product on a smooth white, flat surface. Such layer is prepared by drawing a scraper with a clearance of 3/32 inch high by 7 inches long rapidly through the product in two horizontal planes so as to form an approximate square.

(b) *(A) and (B) classification.* Tomato catsup that is practically free from defects may be given a score of 21 to 25 points. "Practically free from defects" means that any defects present do not more than slightly affect the appearance or eating quality of the product.

(c) *(C) classification.* If the tomato catsup is fairly free from defects, a score of 18 to 20 points may be given. Tomato catsup that falls into this classification shall not be scored above U.S. Grade C or U.S. Standard, regardless of the total score for the product (this is a limiting rule). "Fairly free from defects" means that any defects present may be noticeable but are not so large, so numerous, or of such contrasting color as to seriously affect the appearance or eating quality of the product.

(d) *(SStd) classification.* Tomato catsup that fails to meet the requirements of paragraph (c) of this section may be given a score of 0 to 17 points and shall not be graded above Substandard, regardless of the total score for the product (this is a limiting rule).

§ 52.2109 *Flavor*—(a) *(A) and (B) classification.* Tomato catsup that possesses a good flavor may be given a score of 21 to 25 points. "Good flavor" means a good, distinct flavor characteristic of good quality ingredients. Such flavor is free from scorching or any objectionable flavor of any kind.

(b) *(C) classification.* If the tomato catsup possesses only a fairly good flavor, a score of 17 to 20 points may be given. Tomato catsup that falls into this classification shall not be graded above U.S. Grade C or U.S. Standard, regardless of the total score for the product (this is a limiting rule). "Fairly good flavor" means a flavor characteristic of the ingredients in which there may be slight traces of undesirable flavor such as scorched, bitter, or astringent, but is free from objectionable or off-flavor of any kind.

(c) *(SStd) classification.* Tomato catsup that fails to meet the requirements of paragraph (b) of this section may be given a score of 0 to 16 points and shall not be graded above Substandard, regardless of the total score for the product (this is a limiting rule).

DEFINITIONS

§ 52.2110 *Definitions of terms used in these standards.* (a) "Total solids" in tomato catsup for the purposes of these standards is the refractometric sucrose value of the filtrate determined in accordance with the International Scale of Refractive Indices of Sucrose Solutions to which value is added 1 percent.

(b) "Good finish" means that the product has a uniform, smooth texture.

LOT INSPECTION AND CERTIFICATION

§ 52.2111 *Ascertaining the grade of the lot.* The grade of a lot of tomato catsup covered by these standards is determined by the procedures set forth in the Regulations Governing Inspection and Certification of Processed Fruits and Vegetables, Processed Products Thereof, and Certain Other Processed Food Products (§ § 52.1 to 52.87; 22 F. R. 3535).

SCORE SHEET

§ 52.2112 *Score sheet for tomato catsup.*

Type of container		....
Container size		....
Container mark		....
Label		....
....		....
Net weight or volume		....
Total solids		....
Vacuum readings		....
Factors	**Score point**	
Color	25 (A–B) 21–25 (C) [1]17–20 (SStd.) [1]0–16	
Consistency	25 (A–B) 22–25 (C) [1]18–21 (SStd.) [1]0–17	
Absence of defects	25 (A–B) 21–25 (C) [1]18–20 (SStd.) 0–17	
Flavor	25 (A–B) 21–25 (C) 17–20 (SStd.) [1]0–16	
Total score	100	
Normal flavor and odor		....
Grade		....

[1]Indicates limiting rule.

Recodified in the Federal Register of December 9, 1953 (18 F.R. 8069)
Section 52.2111 amended May 22, 1957 (22 F.R. 3535)

UNITED STATES STANDARDS FOR GRADES OF CHILI SAUCE[2]

Effective October 20, 1953

PRODUCT DESCRIPTION AND GRADES

Sec.
52.2191 Product description.
52.2192 Grades of chili sauce.

FILL OF CONTAINER

52.2198 Recommended fill of container for chili sauce.

FACTORS OF QUALITY

52.2194 Ascertaining the grade.
52.2195 Ascertaining the rating for the factors which are scored.
52.2196 Color.
52.2197 Consistency.
52.2198 Character.
52.2199 Absence of defects.
52.2200 Flavor.

LOT INSPECTION AND CERTIFICATION

52.2201 Ascertaining the grade of a lot.

SCORE SHEET

52.2202 Score sheet for chili sauce.

PRODUCT DESCRIPTION AND GRADES

§ 52.2191 *Product description.* Chili sauce is the product prepared from mature, clean, sound, tomatoes of the red or reddish varieties which are peeled and chopped or crushed, or all (or a portion) of the tomatoes may be chopped, crushed, or macerated and the peelings screened out in a manner so that at least a substantial portion of the seed remains in the product, to which are added salt, spices, vinegar, nutritive sweetening ingredients, and to which may be added vegetable flavoring ingredients such as chopped onion, chopped green or red pepper, chopped green tomatoes, chopped celery, and sweet pickle relish in such quantities as will not materially alter the appearance of the product with respect to the predominance of the tomato ingredient, and any other ingredients permissible under the provisions of the Federal Food, Drug, and Cosmetic Act. The chili sauce is processed in accordance with good commercial practice; is packed in hermetically sealed containers; and is sufficiently processed by heat, before or after sealing, to assure preservation of the product. The refractive index of the filtrate of the chili sauce at 20 degrees C is not less than 1.3784.

[2]The requirements of these standards shall not excuse failure to comply with the provisions of the Federal Food, Drug, and Cosmetic Act.

§ 52.2192 *Grade of chili sauce.* (a) "U.S. Grade A" or "U.S. Fancy" is the quality of chili sauce that possesses a good color; that possesses a good consistency; that possesses a good character; that is practically free from defects; that possesses a good flavor; and that scores not less than 85 points when scored in accordance with the scoring system outlined in this subpart: *Provided*, That the chili sauce may score not less than 14 points on the factor of character if the total score is not less than 85 points.

(b) "U.S. Grade C" or "U.S. Standard" is the quality of chili sauce that possesses a fairly good color; that possesses a fairly good consistency; that possesses a fairly good character; that is fairly free from defects; that possesses a fairly good flavor; and that scores not less than 70 points when scored in accordance with the scoring system outlined in this subpart.

(c) "Substandard" is the quality of chili sauce that fails to meet the requirements of "U.S. Grade C" or "U.S. Standard."

FILL OF CONTAINER

§ 52.2193 *Recommended fill of container for chili sauce.* The recommended fill of container is not incorporated in the grades of the finished product since fill of container, as such, is not a factor of quality for the purposes of these grades. It is recommended that each container of chili sauce be filled as full as practicable without impairment of quality and that the product occupy not less than 90 percent of the capacity of the container.

FACTORS OF QUALITY

§ 52.2194 *Ascertaining the grade.* (a) The grade of the chili sauce is ascertained by considering, in conjunction with the requirements of the respective grade, the respective ratings for the factors of color, consistency, character, absence of defects, and flavor. The relative importance of each factor which is scored is expressed numerically on the scale of 100. The maximum number of points that may be given for each such factor is:

Factors:	*Points*
Color	20
Consistency	20
Character	20
Absence of defects	20
Flavor	20
Total score	100

§ 52.2195 *Ascertaining the rating for the factors which are scored.* The essential variations within each factor which is scored are so described that the value may be ascertained for each factor and expressed numerically. The numerical range within each factor which is scored is inclusive (for example, "17 to 20 points" means 17, 18, 19, and 20 points).

§ 52.2196 *Color*—(a) *(A) classification.* Chili sauce that possesses a good color may be given a score of 17 to 20 points. "Good color" means that the color of the chili sauce is bright; the color of the tomato ingredient is predominant and characteristic of properly prepared, well ripened, properly processed tomatoes; and that the added seasoning ingredients do not materially detract from the appearance of the product.

(b) *(C) classification.* If the chili sauce possesses a fairly good color, a score of 14 to 16 points may be given. Chili sauce that scores in this classification shall not be graded above U.S. Grade C or U.S. Standard, regardless of the total score for the product (this is a limiting rule). "Fairly good color" means that the color of the chili sauce may be slightly dull but not off color; the color of the tomato ingredient is characteristic of properly prepared, fairly well ripened, properly processed tomatoes; and that the added seasoning ingredients do not seriously detract from the appearance of the product.

(c) *(SStd) classification.* Chili sauce that fails to meet the requirements of paragraph (b) of this section may be given a score of 0 to 13 points and shall not be graded above Substandard, regardless of the total score for the product (this is a limiting rule).

§ 52.2197 *Consistency*—(a) *(A) classification.* Chili sauce that possesses a good consistency may be given a score of 17 to 20 points. "Good consistency" means that the chili sauce is heavy bodied and when emptied from the container to a flat surface forms a moderately mounded mass and shows not more than a slight separation of free liquid at the edges of the mass.

(b) *(C) classification.* If the chili sauce possesses only a fairly good consistency a score of 14 to 16 points may be given. Chili sauce that falls into this classification shall not be graded above U.S. Grade C or U.S. Standard, regardless of the total score for the product (this is a limiting rule). "Fairly good consistency" means that the chili sauce, when emptied from the container to a flat surface, may tend to level itself, or may show a moderate separation of free liquid at the edges of the mass, but is not excessively stiff or excessively liquid.

(c) *(SStd) classification.* Chili sauce that fails to meet the requirements of paragraph (b) of this section may be given a score of 0 to 13 points and shall not be graded above Substandard, regardless of the total score for the product (this is a limiting rule).

§ 52.2198 *Character*—(a) *General.* The factor of character refers to the degree of disintegration of the tomatoes and the tenderness and texture of the onion, celery, pickle relish, or other similar ingredients.

(b) *(A) classification.* Chili sauce that possesses a good character may be given a score of 17 to 20 points. "Good character" means that the product does not have a finely comminuted appearance and that the onion, celery, pickle relish, and other similar ingredients are tender, reasonably firm, or crisp in texture.

(c) *(C) classification.* If the chili sauce possesses a fairly good character a score of 14 to 16 points may be given. "Fairly good character" means that the product may be finely comminuted and that the other vegetable ingredients may be only fairly tender.

(d) *(SStd) classification.* Chili sauce that fails to meet the requirements of paragraph (c) of this section may be given a score of 0 to 13 points and shall not be graded above Substandard, regardless of the total score for the product (this is a limiting rule).

§ 52.2199 *Absence of defects*—(a) *General.* The factor of absence of defects refers to the degree of freedom from dark specks or scale-like particles, discolored seeds or pieces of abnormally discolored ingredients, tomato peel, and other defects.

(b) *(A) classification.* Chili sauce that is practically free from defects may be given a score of 17 to 20 points. "Practically free from defects" means that any defects present do not more than slightly affect the appearance or eating quality of the product.

(c) *(C) classification.* If the chili sauce is fairly free from defects, a score of 14 to 16 points may be given. Chili sauce that falls into this classification shall not be graded above U.S. Grade C or U.S. Standard, regardless of the total score for the product (this is a limiting rule). "Fairly free from defects" means that any defects present may be noticeable but are not so large, so numerous, or so prominent as to seriously affect the appearance or eating quality of the product.

(d) *(SStd) classification.* Chili sauce that fails to meet the requirements of paragraph (c) of this section may be given a score of 0 to 13 points and shall not be graded above Substandard, regardless of the total score for the product (this is a limiting rule).

§ 52.2200 *Flavor*—(a) *(A) classification.* Chili sauce that possesses a good flavor may be given a score of 17 to 20 points. "Good flavor" means a good distinct flavor characteristic of chili sauce properly prepared from good quality ingredients. Such flavor is free from scorching or any objectionable flavor or any kind.

(b) *(C) classification.* If the chili sauce possesses a fairly good flavor, a score of 14 to 16 points may be given. Chili sauce that falls into this classification shall not be graded above U.S. Grade C or U.S. Standard, regardless of the total score for the product (this is a limiting rule). "Fairly good flavor" means that the product may be lacking in good characteristic flavor, but is free from objectionable or off-flavors of any kind.

(c) *(SStd) classification.* Chili sauce that fails to meet the requirements of paragraph (b) of this section may be given a score of 0 to 13 points and shall not be graded above Substandard, regardless of the total score for the product (this is a limiting rule).

LOT INSPECTION AND CERTIFICATION

§ 52.2201 *Ascertaining the grade of a lot.* The grade of a lot of chili sauce covered by these standards is determined by the procedures set forth in the Regulations Governing Inspection and Certification of Processed Fruits and Vegetables, Processed Products Thereof, and Certain Other Processed Food Products (§ § 52.1 through 52.87; 22 F. R. 3535).

SCORE SHEET

§ 52.2202 *Score sheet for chili sauce.*

Type of container			
Container size			
Label			
Net weight or volume			
Refractive index			
Vacuum readings			
Factors	Score points		
Color	20	(A) 17–20 (C) [1]14–16 (SStd) [1]0–13	
Consistency	20	(A) 17–20 (C) [1]14–16 (SStd) [1]0–13	
Character	20	(A) 17–20 (C) 14–16 (SStd) [1]0–13	
Absence of defects	20	(A) 17–20 (C) [1]14–16 (SStd) [1]0–13	
Flavor	20	(A) 17–20 (C) [1]14–16 (SStd) [1]0–13	
Total score	100		
Normal flavor and odor			
Grade			

[1]Indicates limiting rule.

UNITED STATES STANDARDS FOR GRADES OF TOMATO SAUCE[3]

Effective May 10, 1954 as Amended October 21, 1960

PRODUCT DESCRIPTION AND GRADES

Sec.

52.2371 Product description.
52.2372 Grades of Tomato Sauce.

[3]The requirements of these standards shall not excuse failure to comply with the provisions of the Federal Food, Drug, and Cosmetic Act.

FILL OF CONTAINERS

52.2373 Recommended fill of container for tomato sauce.

FACTORS OF QUALITY

52.2374 Ascertaining the grade.
52.2375 Ascertaining the rating for the factors which are scored.
52.2376 Color.
52.2377 Consistency.
52.2378 Defects.
52.2379 Flavor.

LOT INSPECTION AND CERTIFICATION

52.2380 Ascertaining the grade of a lot.

SCORE SHEET

52.2381 Score sheet for tomato sauce.

AUTHORITY: § §2.2371 to 52.2381 issued under sec. 205, 60 Stat. 290, Pub Law 156, 83d Cong.; 7 U. S. C. 1624.

PRODUCT DESCRIPTION AND GRADES

§ 52.2371 *Product description.* Tomato sauce is the concentrated product prepared from the liquid extracted from mature, sound, whole tomatoes, the sound residue from preparing such tomatoes for canning, or the residue from partial extraction of juice, or any combination of these ingredients, to which is added salt and spices and to which may be added one or more nutritive sweetening ingredients, a vinegar or vinegars, and onion, garlic, or other vegetable flavoring ingredients. The refractive index of the tomato sauce at 20 degrees Centigrade is not less than 1.3461.

§ 52.2372 *Grades of tomato sauce.* (a) "U.S. Grade A" or "U.S. Fancy" is the quality of tomato sauce that possesses a good color; that possesses a good consistency; that is practically free from defects; that possesses a good flavor; that possesses a good finish; and that scores not less than 85 points when scored in accordance with the scoring system outlined in this subpart.

(b) "U.S. Grade C" or "U.S. Standard" is the quality of tomato sauce that possesses a fairly good color; that possesses a fairly good consistancy; that is fairly free from defects; that possesses a fairly good flavor; that possesses a good finish; and that scores not less than 70 points when scored in accordance with the scoring system outlined in this subpart.

(c) "Substandard" is the quality of tomato sauce that fails to meet the requirements of "U.S. Grade C" or "U.S. Standard."

FILL OF CONTAINER

§ 52.2373 Recommended fill of container for tomato sauce. The recommended fill of container is not incorporated in the grades of the finished product since fill of container, as such, is not a factor of quality for the purposes of these grades. It is recommended that each container of tomato sauce be filled as full as practicable without impairment of quality and that the product occupy not less than 90 percent of the capacity of the container.

FACTORS OF QUALITY

§ 52.2374 *Ascertaining the grade.* (a) The grade of tomato sauce is ascertained by considering the factor of finish, which is not scored, and the ratings for the factors of color, consistency, defects, and flavor, which are scored. The relative importance of each factor which is scored is expressed numerically on the scale of 100. The maximum number of points that may be given for each such factor is:

Factors:	*Points*
Color	25
Consistency	25
Defects	25
Flavor	25
Total score	100

(b) "Good finish" means that the product is evenly comminuted and has a uniform, smooth texture.

§ 52.2375 *Ascertaining the rating for the factors which are scored.* The essential variations within each factor which is scored are so described that the value may be ascertained for each factor and expressed numerically. The numerical range within each factor which is scored is inclusive (for example, "21 to 25 points" means 21, 22, 23, 24, and 25 points).

§ 52.2376 *Color*—(a) *General.* The amount of red in the tomato sauce is determined by comparing the color of the product with that produced by spinning a combination of the following Munsell color discs:

Disc 1—Red (5R 26/13) (glossy finish).
Disc 2—Yellow (2.5YR 5/12) (glossy finish).
Disc 3—Black (N1) (glossy finish).
Disc 4—Gray (N4) (mat finish).

(b) *(A) classification.* Tomato sauce that possesses a good color may be given a score of 21 to 25 points. "Good color" means that the color is typical of tomato sauce made from well ripened red tomatoes and which has been properly prepared and properly processed. Such color contains as much or more red than that produced by spinning the specified Munsell color discs in the following combinations: 65 percent of the area of Disc 1; 21 percent of the area of Disc 2; 14 percent of the area of Disc 3 or of Disc 4, or 7 percent of the area of Disc 3, and 7 percent of the area of Disc 4, whichever most nearly matches the reflectance of the tomato sauce.

(c) *(C) classification.* If the tomato sauce possesses a fairly good color, a score of 17 to 20 points may be given. Tomato sauce that falls into this classification shall not be graded above U.S. Grade C or U.S. Standard, regardless of the total score for the product (this is a limiting rule). "Fairly good color" means that the color is typical of tomato sauce and contains as much or more red than that produced by spinning the specified Munsell color discs in the following combinations: 53 percent of the area of Disc 1; 28 percent of the area of Disc 2; 19 percent of the area of either Disc 3 or 4, or 9½ percent of the area of Disc 3, and 9½ percent of the area of Disc 4, whichever most nearly matches the reflectance of the tomato sauce.

(d) *(SStd) classification.* Tomato sauce that fails to meet the requirements of paragraph (c) of this section may be given a score of 0 to 16 points and shall not be graded above Substandard, regardless of the total score for the product (this is a limiting rule).

§ 52.2377 *Consistency*—(a) *General.* The factor of consistency refers to the viscosity of the product and the tendency to hold its liquid portion in suspension.

(b) *(A) classification.* Tomato sauce that possesses a good consistency may be given a score of 22 to 25 points. "Good consistency" means that the tomato sauce shows not more than a slight separation of free liquid when poured on a flat grading tray; is not excessively stiff; and flows not more than 14 centimeters in 30 seconds at 20 degrees Centigrade in the Bostwick consistometer.

(c) *(C) classification.* If the tomato sauce possesses a fairly good consistency, a score of 18 to 21 points may be given. Tomato sauce that falls into this classification shall not be graded above U.S. Grade C or U.S. Standard, regardless of the total score for the product (this is a limiting rule). "Fairly good consistency" means that the tomato sauce may show a noticeable, but not excessive, separation of free liquid when poured on a flat grading tray; is not excessively stiff; and flows not more than 18 centimeters in 30 seconds at 20 degrees Centigrade in the Bostwick consistometer.

(d) *(SStd) classification.* Tomato sauce that fails to meet the requirements of paragraph (c) of this section may be given a score of 0 to 17 points and shall not be graded above Substandard, regardless of the total score for the product (this is a limiting rule).

§ 52.2378 *Defects*—(a) *General.* The factor of defects refers to the degree of freedom from defects, such as dark specks or scale-like particles, seeds, particles of seed, tomato peel, core material, or other similar substances. This factor is evaluated by observing a layer of the product on a smooth white, flat surface. Such layer is prepared by drawing a scraper with a clearance of 3/32 inch high by 7 inches long rapidly through the product in two horizontal planes so as to form an approximate square.

(b) *(A) classification.* Tomato sauce that is practically free from defects may be given a score of 21 to 25 points. "Practically free from defects" means that any defects present do not more than slightly affect the appearance or eating quality of the product.

(c) *(C) classification.* If the tomato sauce is fairly free from defects, a score of 18 to 20 points may be given. Tomato sauce that falls into this classification shall not be scored above U.S. Grade C or U.S. Standard, regardless of the total score for the product (this is a limiting rule). "Fairly free from defects" means that any defects present may be noticeable but are not so large, so numerous, or of such contrasting color as to seriously affect the appearance or eating quality of the product.

(d) *(SStd) classification.* Tomato sauce that fails to meet the requirements of paragraph (c) of this section may be given a score of 0 to 17 points and shall not be graded above Substandard, regardless of the total score for the product (this is a limiting rule).

§ 52.2379 *Flavor*—(a) *(A) classification.* Tomato sauce that possesses a good flavor may be given a score of 21 to 25 points. "Good flavor" means a good, distinct flavor characteristic of good quality ingredients. Such flavor is free from scorching or any objectionable flavor of any kind.

(b) *(C) classification.* If the tomato sauce possesses only a fairly good flavor, a score of 17 to 20 points may be given. Tomato sauce that falls into this classification shall not be graded above U.S. Grade C or U.S. Standard, regardless of the total score for the product (this is a limiting rule). "Fairly good flavor" means a flavor characteristic of the ingredients in which there may be slight traces of undesirable flavor, such as scorched, bitter, or astringent, but is free from objectionable or off-flavors of any kind.

(c) *(SStd) classification.* Tomato sauce that fails to meet the requirements of paragraph (b) of this section may be given a score of 0 to 16 points and shall not be graded above Substandard, regardless of the total score for the product (this is a limiting rule).

LOT INSPECTION AND CERTIFICATION

§ 52.2380 *Ascertaining the grade of a lot.* The grade of a lot of tomato sauce covered by these standards is determined by the procedures set forth in the Regulations Governing Inspection and Certification of Processed Fruits and Vegetables, Processed Products Thereof, and Certain Other Processed Food Products (§ § 52.1 to 52.87; 22 F. R. 3535).

SCORE SHEET

§ 52.2381 *Score sheet for tomato sauce.*

Type of container		....
Container size		....
Label		....
Net weight or volume		....
Refractive index		....
Vacuum readings		....
Factors	**Score points**	
Color	25 (A) 21–25; (C) [1]17–20; (SStd) [1]0–16	
Consistency	25 (A) 22–25; (C) [1]18–21; (SStd) [1]0–17	
Defects	25 (A) 21–25; (C) [1]18–20; (SStd) [1]0–17	
Flavor	25 (A) 21–25; (C) [1]17–20; (SStd) [1]0–16	
Total score	100	
Good finish		....
Grade		....

[1]Indicates limiting rule.

Effective time. The United States Standards for Grades of Tomato Sauce (which is the first issue) contained in this subpart will become effective thirty (30) days after the date of publication of this subpart in the FEDERAL REGISTER.

Dated: April 5, 1954.

[SEAL] ROY W. LENNARTSON
Deputy Administrator, Marketing Services.

[F.R. Doc. 54-2649; Filed, Apr. 8, 1954; 8:47 a.m.]

Published in the Federal Register April 9, 1954 (19 F.R. 2061)
Section 52.2380 amended May 22, 1957 (22 F.R. 3535)

UNITED STATES STANDARDS FOR GRADES OF CANNED TOMATOES AND OKRA AND CANNED OKRA AND TOMATOES[4]

Effective December 24, 1957

TITLE 7—AGRICULTURE

Chapter 1—Agricultural Marketing Service (Standards, Inspections, Marketing Practices), Department of Agriculture

[4]Compliance with the provisions of these standards shall not excuse failure to comply with the provisions of the Federal Food, Drug, and Cosmetic Act.

PART 52—PROCESSED FRUITS AND VEGETABLES, PROCESSED PRODUCTS THEREOF, AND CERTAIN OTHER PROCESSED FOOD PRODUCTS

SUBPART—UNITED STATES STANDARDS FOR GRADES OF CANNED TOMATOES AND OKRA AND CANNED OKRA AND TOMATOES[4]

On July 13, 1957, a notice of proposed rule making was published in the FEDERAL REGISTER (22 F. R. 5541) regarding a proposed issuance of the United States Standards for Grades of Canned Tomatoes and Okra and Canned Okra and Tomatoes.

After consideration of all relevant matters presented, including the proposal set forth in the aforesaid notice, the following United States Standards for Grades of Canned Tomatoes and Okra and Canned Okra and Tomatoes are hereby promulgated pursuant to the authority contained in the Agricultural Marketing Act of 1946 (60 Stat. 1087 et seq., as amended; 7 U. S. C. 1621 et seq.).

PRODUCT DESCRIPTION, PROPORTION OF INGREDIENTS, AND GRADES

Sec.
52.3421 Product description.
52.3422 Proportion of ingredients.
52.3423 Grades.

FILL OF CONTAINER

52.3424 Fill of container.

FACTORS OF QUALITY

52.3425 Ascertaining the grade.
52.3426 Ascertaining the rating for the factors which are scored.
52.3427 Color.
52.3428 Flavor.
52.3429 Defects.
52.3430 Character.

LOT INSPECTION AND CERTIFICATION

52.3431 Ascertaining the grade of a lot.
52.3432 Score sheet for canned tomatoes and okra and canned okra and tomatoes.

AUTHORITY: § § 52.3421 to 52.3432 issued under sec. 205, 60 Stat. 1090, as amended; 7 U. S. C. 1624.

PRODUCT DESCRIPTION, PROPORTION OF INGREDIENTS, AND GRADES

§ 52.3421 *Product description.* Canned tomatoes and okra and canned okra and tomatoes, hereinafter called canned tomatoes and okra, is the product prepared from clean, sound, properly matured tomatoes of red or reddish varieties which may be canned, and the clean, sound, succulent, immature pods of either the green or white varieties of the okra plant which may be processed. The tomatoes have been prepared by trimming, washing, sorting, peeling, and coring. The okra is prepared by trimming, washing, and sorting, and may be whole or cut into pieces. The tomatoes and okra may be packed with or without a small amount of tomato juice or tomato puree and with or without such other ingredients permissible under the provisions of the Federal Food, Drug, and Cosmetic Act in accordance with good commercial practice. The product is sufficiently processed by heat to assure preservation in hermetically sealed containers.

§ 52.3422 *Proportion of ingredients.* (a) Canned tomatoes and okra shall consist of 50 percent or more, by weight, of the tomato ingredient and not less than 12.5 percent nor more than 50 percent, by weight, of whole or cut okra.

(b) Canned okra and tomatoes shall consist of 50 percent or more, by weight, of whole or cut okra and not less than 12.5 percent nor more than 50 percent, by weight, of the tomato ingredient.

(c) Compliance with the requirement for proportions of ingredients will be determined by averaging the percent, by weight, of each ingredient in all of the containers in the sample: *Provided,* That any deviation from the prescribed proportion of ingredients in any one container is within the limits of good commercial practice.

§ 52.3423 *Grades.* (a) "U.S. Grade A" or "U.S. Fancy" is the quality of canned tomatoes and okra that possess a good flavor; that possess a good color; that are practically free from defects; that possess a good character; and that score not less than 85 points when scored in accordance with the scoring system outlined in this subpart.

(b) "U.S. Grade C" or "U.S. Standard" is the quality of canned tomatoes and okra that possess a fairly good flavor; that possess a fairly good color; that are fairly free from defects; that possess a fairly good character; and that score not less than 70 points when scored in accordance with the scoring system outlined in this subpart.

(c) "Substandard" is the quality of canned tomatoes and okra that fail to meet the requirements of U.S. Grade C or U.S. Standard.

FILL OF CONTAINER

§ 52.3424 *Fill of container.* The recommended fill of container for canned tomatoes and okra is not incorporated in the grades of the finished product since fill of container, as such, is not a factor of quality for the purpose of these grades. It is recommended that each container be filled as full of tomatoes and okra as practicable without impairment of quality and that the product shall fill the container to not less than 90 percent of its total capacity.

FACTORS OF QUALITY

§ 52.3425 *Ascertaining the grade*—(a) *General.* In addition to considering other requirements outlined in the standards the following quality factors are evaluated in ascertaining the grade of the product:

(1) *Factor not rated by score points.* (i) Proportion of ingredients.

(2) *Factors rated by score points.* The relative importance of each factor which is rated is expressed numerically on the scale of 100. The maximum number of points that may be given each such factor is:

Factors:	*Points*
Color	20
Flavor	20
Defects	30
Character	30
Total score	100

§ 52.3426 *Ascertaining the rating for the factors which are scored.* The essential variations within each factor which is scored are so described that the value may be ascertained for each factor and expressed numerically. The numerical range within each factor which is scored is inclusive (for example, "26 to 30 points" means 26, 27, 28, 29, or 30 points).

§ 52.3427 *Color*—(a) *(A) classification.* Canned tomatoes and okra that possess a good color may be given a score of 17 to 20 points. "Good color" means that the tomato ingredient possesses a reasonably good red color that is typical of reasonably well ripened tomatoes and that the okra possesses the color of young tender okra, which ingredients have been properly prepared and properly processed and meet the following additional requirements:

(1) *Tomatoes.* Not less than 50 percent, by weight, of the drained tomato ingredient possesses as much or more red than USDA Tomato Red Color Standard: *Provided,* That the color of not more than 10 percent, by weight, of the remaining portion of the drained tomato ingredient may be yellow and none is green in color.

(2) *Okra.* The outer surface of the okra pods or pieces of pods possesses a color that is typical of young tender okra.

(b) *(C) classification.* Canned tomatoes and okra that possess a fairly good color may be given a score of 14 to 16 points. Canned tomatoes and okra that fall into this classification shall not be graded above U.S. Grade C or U.S. Standard, regardless of the total score for the product (this is a limiting rule). "Fairly good color" means that the tomato ingredient possesses a fairly good red color typical of fairly well ripened tomatoes and that the okra possesses the color of reasonably tender okra, which ingredients have been properly prepared and properly processed.

(c) *(SStd) classification.* Canned tomatoes and okra that fail to meet the requirements of paragraph (b) of this section may be given a score of 0 to 13 points and shall not be graded above Substandard, regardless of the total score for the product (this is a limiting rule).

§ 52.3428 *Flavor*—(a) *(A) classification.* Canned tomatoes and okra that possess a good flavor may be given a score of 17 to 20 points. "Good flavor" means that the product has a characteristic flavor and odor typical of well ripened tomatoes and young tender okra.

(b) *(C) classification.* Canned tomatoes and okra that possess a fairly good flavor may be given a score of 14 to 16 points. Canned tomatoes and okra that fall into this classification shall not be graded above U.S. Grade C or U.S. Standard, regardless of the total score for the product (this is a limiting rule). "Fairly good flavor" means that the product may be lacking in good flavor and odor but is free from objectionable flavors and objectionable odors of any kind.

(c) *(SStd) classification.* Canned tomatoes and okra that fail to meet the requirements of paragraph (b) of this section may be given a score of 0 to 13 points and shall not be graded above Substandard, regardless of the total score for the product (this is a limiting rule).

§ 52.3429 *Defects*—(a) *General.* The factor of defects refers to the degree of freedom from sand, grit, or silt; harmless extraneous material; tomato peel; tomato core material; small pieces of okra; and from damaged or seriously damaged tomato or okra ingredient.

(1) "Sand, grit, or silt" means any particle of earthy material.

(2) "Harmless extraneous material" means any extraneous vegetable substance (including, but not limited to, a leaf, unattached stem, or portions thereof) that is harmless.

(3) "Damaged" means that the appearance and eating quality of the product is materially affected by the presence of scarred, discolored, or blemished tomato ingredient and/or scarred, poorly trimmed okra ingredient, and small pieces of pod less than ⅜ inch in length, or other damage.

(4) "Seriously damaged" means that the appearance and eating quality of the product is seriously affected by the presence of damaged ingredients.

(b) *(A) classification.* Canned tomatoes and okra that are practically free from defects may be given a score of 25 to 30 points. "Practically free from defects" means that the presence of sand, grit, or silt; harmless extraneous material; tomato peel; core material; and damaged and seriously damaged ingredients, individually or collectively, does not materially affect the appearance and eating quality of the product.

(c) *(C) classification.* Canned tomatoes and okra that are fairly free from defects may be given a score of 21 to 24 points. Canned tomatoes and okra that fall into this classification shall not be graded above U.S. Grade C or U.S. Standard, regardless of the total score for the product (this is a limiting rule). "Fairly free from defects" means that the presence of sand, grit, or silt; harmless extraneous material; tomato peel; core material; and damaged and seriously damaged ingredients, individually or collectively, does not seriously affect the appearance and eating quality of the product.

(d) *(SStd) classification.* Canned tomatoes and okra that fail to meet the requirements of paragraph (c) of this section may be given a score of 0 to 20 points and shall not be graded above Substandard, regardless of the total score for the product (this is a limiting rule).

§ 52.3430 *Character*—(a) *General.* The factor of character refers to the texture and wholeness of the tomato ingredient and the tenderness and texture of the okra pods or pieces of pods.

(1) "Large piece of tomato ingredient" means a piece of tomato flesh that weighs not less than 1 ounce.

(2) "Small piece of tomato ingredient" means a piece of tomato flesh that weighs less than 1 ounce and that does not pass through a United States Standard ½-inch sieve.

(b) *(A) classification.* Canned tomatoes and okra that possess a good character may be given a score of 26 to 30 points. "Good character" means that the tomato ingredient is composed of whole, almost whole, or large pieces of tomato flesh and that the okra pods or pieces of pods are fleshy and tender and practically intact, that the okra seed is in the early stages of development, and that not more than 2 percent, by count, of the pods or pieces of pods may possess fibrous material which is materially objectionable upon eating.

(c) *(C) classification.* Canned tomatoes and okra that possess a fairly good character may be given a score of 21 to 25 points. Canned tomatoes and okra that fall into this classification on account of fibrous material shall not be graded above U.S. Grade C or U.S. Standard, regardless of the total score for the product (this is a partial limiting rule). "Fairly good character" means that the tomato ingredient may be predominantly in small pieces and that the okra pods and pieces of pods may have lost to a considerable extent their fleshy texture, that the units may be fairly tender but not excessively mushy, that the seed may be in the later stages of maturity but are not hard, and that not more than 5 percent, by count, of the pods or pieces of pods may possess fibrous material which is materially objectionable upon eating.

(d) *(SStd) classification.* Canned tomatoes and okra that fail to meet the requirements of paragraph (c) of this section with respect to fibrous material may be given a score of 0 to 20 points and shall not be graded above Substandard, regardless of the total score for the product (this is a partial limiting rule).

LOT INSPECTION AND CERTIFICATION

§ 52.3431 *Ascertaining the grade of a lot.* The grade of a lot of canned tomatoes and okra covered by these standards is determined by the procedures set forth in the Regulations Governing Inspection and Certification of Processed Fruits and Vegetables, Processed Products Thereof, and Certain Other Processed Food Products (§ § 52.1 to 52.87).

SCORE SHEET

§ 52.3432 *Score sheet for canned tomatoes and okra and canned okra and tomatoes.*

Size and kind of container			....
Container mark or identification			....
Label			....
Net Weight (ounces)			....
Vacuum readings (inches)			....
Proportion of ingredients,—% tomatoes,—% okra			....
Whole or cut okra			....
Factors		Score points	
Color	20	(A) 17–20 (C) [1]14–16 (SStd.) [1]0–13	
Flavor	20	(A) 17–20 (C) [1]14–16 (SStd.) [1]0–13	
Defects	30	(A) 25–30 (C) [1]21–24 (SStd.) [1]0–20	
Character	30	(A) 26–30 (C) [2]21–25 (SStd.) [2]0–20	
Total score	100		
Grade			....

[1]Indicates limiting rule.
[2]Indicates partial limiting rule

The United States Standards for Grades of Canned Tomatoes and Okra and Canned Okra and Tomatoes (which is the first issue) contained in this subpart shall become effective 30 days after publication hereof in the FEDERAL REGISTER.

Dated: November 20, 1957.

[SEAL] ROY W. LENNARTSON,
Deputy Administrator, Marketing Services.

Published in the Federal Register, November 23, 1957 (22 F.R. 9379).

UNITED STATES STANDARDS FOR GRADES OF CANNED TOMATO JUICE[5]

Effective July 22, 1985

PRODUCT DESCRIPTION AND GRADES

Sec.
52.3621 Product description.
52.3622 Grades of canned tomato juice.

FILL OF CONTAINERS

52.3623 Recommended fill of container.

FACTORS OF QUALITY

52.3624 Ascertaining the grade of a sample unit.
52.3625 Color.
52.3626 Consistency.
52.3627 Defects.
52.3628 Flavor.

LOT INSPECTION AND CERTIFICATION

52.3629 Ascertaining the grade of a lot.

SCORE SHEET

52.3630 Score sheet for canned tomato juice.

AUTHORITY: § § 52.3621 to 52.3630 issued under sec. 205, 60 Stat. 1090, as amended; 7 U. S. C. 1624.

PRODUCT DESCRIPTION

§ 52.3621

(a) Tomato juice is the product as defined in the Standards of Identity for Tomato Juice (21 CFR 156.145), issued under the Federal Food, Drug, and Cosmetic Act.

(b) Tomato juice from concentrate is the product as defined in the Standards of Identity for Tomato Juice (21 CFR 156.145), issued under the Federal Food, Drug, and Cosmetic Act.

[5]Compliance with the provisions of these standards shall not excuse failure to comply with the provisions of the Federal Food, Drug, and Cosmetic Act.

DEFINITIONS OF TERMS.

§ 52.3622 In these U.S. standards, unless otherwise required by the context, the following terms shall be construed, respectively, to mean:

(a) Color.

(1) General. The amount of red in tomato juice and tomato juice from concentrate is determined by comparing the color of the product with the Munsell color discs with specified percentages. The availability and procedure for using the Munsell color discs may be obtained from the supplier licensed by the U.S. Department of Agriculture:

Munsell Color Co.
P.O. Box 230
Newburgh, NY 12551-0230

Any method or device approved by the USDA (including electronic color meters) which gives equivalent results may be used.

(2) Good color means a color that is bright and characteristic of tomato juice, made from mature red tomatoes, which is not affected by caramelization, oxidation or other similar causes. Such color contains as much red as, or more red than, that produced by spinning the specified Munsell color discs in the following combinations or an equivalent of such composite color:

65 percent of the area of Red (5R 2.6/13) (glossy finish);

21 percent of the area of Yellow (2.5 YR 5/12) (glossy finish); and

14 percent of the area of either Black (N1) (glossy finish) or Grey (N4) (mat finish); or

7 percent of the area of Black (N1) (glossy finish) and

7 percent of the area of Grey (N4) (mat finish), whichever most nearly matches the reflectance of the product.

(3) Reasonably good color means a color that is typical of red tomato juice which may be slightly affected by caramelization, oxidation or other similar causes.

(i) To score 25 points for color, the juice shall contain as much red as, or more red than, that produced by spinning the specified Munsell color discs in the following combinations or an equivalent of such composite color:

59 percent of the area of Red (5R 2.6/13) (glossy finish);

24-1/2 percent of the area of Yellow (2.5 YR 5/12) (glossy finish); and

16-1/2 percent of the area of either Black (N1) (glossy finish) or Grey (N4) (mat finish); or

8-1/4 percent of the area of Black (N1) (glossy finish) and

8-1/4 percent of the area of Grey (N4) (mat finish), whichever most nearly matches the reflectance of the product.

(ii) To score 23 or 24 points for color, the juice shall contain as much red as, or more red than, that produced by spinning the specified Munsell color discs in the following combinations or an equivalent of such composite color:

53 percent of the area of Red (5R 2.6/13) (glossy finish);

28 percent of the area of Yellow (2.5 YR 5/12) (glossy finish); and

19 percent of the area of either Black (N1) (glossy finish) or Grey (N4) (mat finish); or

9-1/2 percent of the area of Black (N1) (glossy finish) and

9-1/2 percent of the area of Grey (N4) (mat finish), whichever most nearly matches the reflectance of the product.

(b) Consistency means the viscosity of tomato juice or tomato juice from concentrate including the degree of separation of the insoluble solids.

(1) Good consistency means that the juice flows readily; has a normal amount of insoluble tomato solids in suspension; and that there is little solids separation.

(2) Reasonably good consistency means that the juice flows readily; has a normal amount of insoluble tomato solids in suspension; and that there is not a marked degree of solids separation.

(c) Defects means dark specks or scale-like particles, seeds, particles of seed, tomato peel, core material, or other similar substances.

(1) Practically free from defects means that any defects present do not more than slightly affect the appearance or drinking quality of the product.

(2) Reasonably free from defects means that any defects present may be noticeable, but are not so large, so numerous, or of such contrasting color as to seriously affect the appearance or drinking quality of the product.

(d) Flavor is evaluated based upon characteristics normal to tomato juice or tomato juice from concentrate.

(1) Good flavor means a distinct tomato juice flavor and odor characteristic of good quality tomatoes which is not materially affected by stems, leaves, crushed seeds, cores, immature tomatoes, or the effects of improper trimming or processing.

(2) Reasonably good flavor means a characteristic tomato juice flavor which may be materially, but not seriously, affected by stems, leaves, crushed seeds, cores, immature tomatoes, or the effects of improper trimming or processing.

(e) Tomato soluble solids (TSS) means the sucrose value as determined by the method prescribed in 21 CFR 156.3.

RECOMMENDED SAMPLE UNIT SIZES

§52.3623 The requirements for all factors of quality are based on the following:

(a) The entire contents of a container;

(b) A representative portion of the contents of a container;

(c) A combination of the contents of two or more containers; or

(d) A representative portion of processed product stored or held in bulk containers.

GRADES

§52.3624

(a) U.S. Grade A is the quality of tomato juice or tomato juice from concentrate that meets the applicable requirements of Tables I or II.

(b) U.S. Grade B is the quality of tomato juice or tomato juice from concentrate that meets the applicable requirements of Tables I or II.

(c) Substandard is the quality of tomato juice or tomato juice from concentrate that fails to meet the requirements for "U.S. Grade B."

RECOMMENDED FILL OF CONTAINER

§52.3625 Fill of container is not incorporated in the grades of the finished product, since fill of container, as such, is not a factor of quality for the purposes of these grades. It is recommended that each container of tomato juice be filled as full as practicable without impairment of quality, and that the product occupies not less than 90 percent of the capacity of the container.

FACTORS OF QUALITY AND ANALYSIS

§52.3626

(a) The grade of a lot of tomato juice or tomato juice from concentrate is based on evaluation and analysis of the juice for the following quality and analytical factors:

(1) Color;
(2) Consistency;
(3) Defects;
(4) Flavor; and
(5) Tomato soluble solids (tomato juice from concentrate).

(b) The relative importance of each quality factor is expressed numerically on the scale of 100. The maximum number of points that may be given each factor is:

Factors	Points
Color	30
Consistency	15
Defects	15
Flavor	40
Total Score	100

(c) The essential variations within each factor are so described that the value may be determined for each factor and expressed numercially. The numerical range for the rating of each factor is inclusive (for example, 13 to 15 points means 13, 14, or 15 points) and the score points shall be prorated relative to the degree of excellence for each factor.

REQUIREMENTS FOR GRADES

§52.3627 Requirements for grades.

TABLE I
Tomato Juice

QUALITY FACTORS	FACTOR DESCRIPTION	GRADE	SCORE POINT RANGE
Color	Good	"A"	26 - 30
	Reasonably Good	"B"	25 1/ 23 - 24 2/
Consistency	Good	"A"	13 - 15
	Reasonably Good	"B"	10 - 12 3/
Defects	Practically Free	"A"	13 - 15
	Reasonably Free	"B"	10 - 12 4/
Flavor	Good	"A"	33 - 40
	Reasonably Good	"B"	27 - 32 4/
MINIMUM TOTAL SCORE		"A"	85 points
		"B"	70 points

1. Tomato juice assigned 15 points for the factor of color can be graded U.S. Grade A, provided all other requirements for U.S. Grade A are met and that the total score for the product is not less than 85 points.

2. Tomato juice assigned 23 or 24 points for the factor of color cannot be graded above U.S. Grade B regardless of the total score for the product.

3. Tomato juice possessing only "reasonably good consistency" (10-12 points) can be graded U.S. Grade A, provided all other requirements for U.S. Grade A are met and that the total score for the product is not less than 85 points.

4. Cannot be graded U. S. Grade B regardless of the total score for the product.

TABLE II
Tomato Juice From Concentrate

QUALITY FACTORS	FACTOR DESCRIPTION	GRADE	SCORE POINT RANGE
Color	Good	"A"	26 - 30
	Reasonably Good	"B"	25 1/ 23 - 24 2/
Consistency	Good	"A"	13 - 15
	Reasonably Good	"B"	10 - 12 3/
Defects	Practically Free	"A"	13 - 15
	Reasonably Free	"B"	10 - 12 4/
Flavor	Good	"A"	33 - 40
	Reasonably Good	"B"	27 - 32 4/
MINIMUM TOTAL SCORE		"A"	85 points
		"B"	70 points
Tomato Soluble Solids (Minimum)		5.0 percent by weight	

1. Tomato juice from concentrate assigned 25 points for the factor of color can be graded U.S. Grade A, provided all other requirements for U.S. Grade A are met and that the total score for the product is not less than 85 points.

2. Tomato juice from concentrate assigned 23 or 24 points for the factor of color cannot be graded above U.S. Grade B regardless of the total score for the product.

3. Tomato juice from concentrate possessing only "reasonably good consistency" (10-12 points) can be graded U.S. Grade A, provided all other requirements for U.S. GradeA are met and that the total score for the product is not less than 85 points.

4. Cannot be graded above U.S. Grade B regardless of the total score for the product.

SAMPLE SIZE

§52.3028 The sample size used to determine whether the requirements of these standards are met shall be as specified in the sampling plans and procedures in the "Regulations Governing Inspection and Certification of Processed Fruits and Vegtables, Processed Products Thereof, and Certain Other Processed Food Products" (7 CFR 52.1 through 52.83) for lot grading and on-line grading, as applicable.

LOT QUALITY REQUIREMENTS

§52.3629 A lot of tomato juice or tomato juice from concentrate is considered as meeting the requirements for quality if:

(a) The requirements specified in Table I or II, as applicable, are met; and

(b) The sampling plans and procedures in 7 CFR 52.1 - 52.83 are met.

Washington, D.C., on: July 16, 1985

William T. Manly
Deputy Administrator
Marketing Programs

UNITED STATES STANDARDS FOR GRADES OF TOMATO PASTE[6]

Effective September 19, 1977

PRODUCT DESCRIPTION, CONCENTRATION, TEXTURE, GRADES

Sec.

52.5041 Product description.
52.5042 Concentration.
52.5043 Texture.
52.4044 Grades.

[6] Compliance with the provisions of these standards shall not excuse failure to comply with the provisions of the Federal Food, Drug, and Cosmetic Act or with applicable state laws and regulations.

FILL OF CONTAINER

52.5045 Fill of container.

FACTORS OF QUALITY

52.5046 Ascertaining the grade of a sample unit.
52.5047 Ascertaining the rating for each factor.
52.5048 Color.
52.5049 Defects.

LOT COMPLIANCE

52.5050 Ascertaining the grade of a lot.

SCORE SHEET

52.5051 Score sheet for tomato paste.

AUTHORITY: The provisions of this subpart issued under secs. 202–208, 60 Stat. 1087, as amended; 7 U. S. C. 1621–1627.

PRODUCT DESCRIPTION, CONCENTRATION, TEXTURE, GRADES

§ 52.5041 Product description.

(a) "Tomato Paste" is the clean, sound wholesome product as defined in the Standard of Identity for Tomato Paste (21 CFR 155.191) issued pursuant to the Federal Food, Drug and Cosmetic Act and contains not less than 24 percent of natural tomato soluble solids.

(b) The standards in this subpart cover canned tomato paste processed by heat to assure preservation in hermetically sealed containers as well as tomato paste which may be preserved by other means.

§ 52.5042 Concentration.

The degree of concentration is not considered a factor of quality for the purposes of these standards but the following designations of concentration may be used in connection with these standards for the applicable natural tomato soluble solids group:

NATURAL TOMATO SOLUBLE SOLIDS

Extra heavy concentration.	39.3 percent or more.
Heavy concentration.	32 percent or more, but less than 39.3 percent.
Medium concentration.	28 percent or more, but less than 32 percent.
Light concentration.	24 percent or more, but less than 28 percent.

§ 52.5043 Texture.

(a) *General.* Texture is the degree of fineness or coarseness of the product. Texture is classified when the product is diluted with water to between 8 percent and 9 percent, inclusive, of natural tomato soluble solids.

(b) *Kinds of texture.* (1) "Fine" texture means a smooth, uniform finish (2) "Coarse" texture means a coarse, slightly granular finish.

§ 52.5044 Grades.

(a) "U.S. Grade A" (or "U.S. Fancy") is the quality of tomato paste that (1) has a good flavor and odor, (2) has a good color, (3) is practically free from defects, and (4) scores not less than 90 points when scored in accordance with the scoring system outlined in this subpart.

(b) "U.S. Grade C" (or "U.S. Standard") is the quality of tomato paste that (1) has at least a fairly good flavor and odor, (2) has at least a fairly good color, (3) is at least fairly free from defects, and (4) scores not less than 80 points when scored in accordance with the scoring system outlined in this subpart.

(c) "Substandard" is the grade of tomato paste that fails to meet the requirements of "U.S. Grade C."

FILL OF CONTAINER

§ 52.5045 Recommended fill of container.

It is recommended that containers of tomato paste be filled as full as practicable without impairment of quality.

FACTORS OF QUALITY

§ 52.5046 Ascertaining the grade of a sample unit.

(a) *General.* The grade of a sample unit of tomato paste is ascertained by considering the factor of flavor and odor which is not scored, the ratings for the factors of color and defects which are scored, the total score, and the limiting rules which apply.

(b) *Factors not rated by score points*—(1) *Flavor and odor.* (i) The flavor and odor of the product is determined on undiluted samples and also after dilution with water to between 8 percent and 9 percent, inclusive, of natural tomato soluble solids.

(ii) "Good flavor and odor" means a distinct tomato paste flavor characteristic of ripe, good quality tomatoes and inclusive of any optional ingredient which may have been added. Such flavor and odor may be no more than slightly affected, by any one or combination of the following: stems, sepals, leaves, crushed seeds, cores; by immature, soured, or overripe tomatoes; from the effects of unsatisfactory preparation, processing, or storage; or from any other factor not specifically mentioned.

(iii) "Fairly good flavor and odor" means a characteristic tomato paste flavor and inclusive of any optional ingredient which may have been added. Such flavor and odor may be affected, but not to a serious degree, by any one or combination of the following: stems, sepals, leaves, crushed seeds, cores; by immature, soured, or overripe tomatoes; from the effects of unsatisfactory preparation, processing, or storage; or from any other factor not specifically mentioned.

(iv) "Off flavor and odor" means tomato paste flavor that fails to meet the requirements of "fairly good flavor and odor" or which possesses a flavor and/or odor which is seriously objectionable.

(c) *Factors rated by score points.* The relative importance of each scoring factor is expressed numerically on the scale of 100. The maximum number of points that may be given each factor is:

Factors:	*Points*
Color	50
Defects	50
Total score	100

§ 52.5047 Ascertaining the rating for the factors which are scored.

The essential variations within each factor which is scored are so described that the value may be ascertained numerically. The numerical range within each factor is inclusive. (For example, "45 to 50 points" means, 45, 46, 47, 48, 49 or 50 points.)

§ 52.5048 Color.

(a) *General.* The amount of red in tomato paste is determined by comparing the color or the product, diluted with water to between 8 percent and 9 percent, inclusive, of natural tomato soluble solids, with that produced by spinning a combination of the following Munsell color discs:

Disc 1—Red (5 R 2.6/13) (glossy finish).
Disc 2—Yellow (2.5 YR 5/12) (glossy finish).
Disc 3—Black (N1) (glossy finish).
Disc 4—Grey (N4) (mat finish).

Such comparison is made under a diffused light source of approximately 250 foot-candle (candela) intensity and having a spectral quality of approximating that of daylight under a moderately overcast sky, and a color temperature of 7,500° Kelvin ±200°. With the light source directly over the disc and diluted product, observation is made at an angle of 45° and at a distance of 12 or more inches from the product.

(b) *Availability of color references.* The colors referred to in this section are available from the approved supplier under a license from the U.S. Department of Agriculture:

Munsell Color Co., Newburgh, New York.

(c) *Grade A Classification.* Tomato paste that has a good color may be given a score of 45 to 50 points. "Good color" means a bright, typical, red tomato paste color. Such color, when the product is diluted and observed as specified in this section, is as red as, or more red than, that produced by spinning the specified Munsell color discs in the following combinations or an equivalent of such composite color:

65 percent of the area of Disc 1;
21 percent of the area of Disc 2; and
14 percent of the area of either Disc 3 or Disc 4; or
7 percent of the area of Disc 3 and
7 percent of the area of Disc 4, whichever most nearly matches the appearance of the diluted sample.

(d) *Grade C Classification.* Tomato paste that has at least a fairly good color may be given a score of 40 to 44 points. Tomato paste that falls into this classification shall not be graded above U.S. Grade C, regardless of the total score for the product (this is a limiting rule). "Fairly good color" means a typical red tomato paste color which may be slightly dull or may have a slightly brownish cast. Such color, when the product is diluted and observed as specified in this section, is as red as, or more red than, that produced by spinning the specified Munsell color discs in the following combinations or an equivalent of such composite color:

53 percent of the area of Disc 1;
28 percent of the area of Disc 2; and
19 percent of the area of either Disc 3 or Disc 4; or

9½ percent of the area of Disc 3 and
9½ percent of the area of Disc 4, whichever most nearly matches the appearance of the diluted sample.

(e) *Substandard Classification.* Tomato paste that fails to meet the requirements of paragraph (d) of this section may be given a score of 0 to 39 points and shall not be graded above Substandard, regardless of the total score for the product (this is a limiting rule).

§ 52.5049 Defects.

(a) *General.* This factor is concerned with the degree of freedom from defects such as dark specks or scale-like particles, seeds or objectionable particles of seeds, objectionable tomato peel, harmless extraneous material, or any other similar substances.

(b) *Grade A Classification.* Tomato paste that is practically free from defects may be given a score of 45 to 50 points. "Practically free from defects" means that any defects present do not more than slightly affect the appearance or usability of the product.

(c) *Grade C Classification.* Tomato paste that is fairly free from defects may be given a score of 40 to 44 points. Tomato paste that falls into this classification shall not be graded above U.S. Grade C, regardless of the total score of the product (this is a limiting rule). "Fairly free from defects" means that any defects present may be noticeable but are not so large, so numerous, nor of such contrasting color or nature as to seriously affect the appearance or usability of the product.

(d) *Substandard Classification.* Tomato paste that fails to meet the requirements of paragraph (c) of this section may be given a score of 0 to 39 points and shall not be graded above Substandard, regardless of the total score for the product (this is a limiting rule).

LOT COMPLIANCE

§ 52.5050 Ascertaining the grade of a lot.

The grade of a lot of tomato paste covered by these standards is determined by the procedures set forth in the Regulations Governing Inspection and Certification of Processed Fruits and Vegetables, Processed Products Thereof, and Certain Other Processed Food Products (sections 52.1 to 52.87).

§ 52.5051 *Score sheet for tomato paste.*

Size and kind of container ....
Label ....
Container mark or identification ....
Net weight ....
Vacuum (inches) ....
Texture (fine or coarse) ....
Natural tomato soluble solids (percent) ....

Factors		Score points	
Color	50	(A)	45–50
		(C)	[1]40–44
		(SStd)	[1]0–39
Defects	50	(A)	45–50
		(C)	[1]40–44
		(SStd)	[1]0–39
Total score	100		

Flavor and ordor .... Good / Fairly Good / Off

Grade ....

[1]Indicates limiting rule.

The U.S. Standards for grades as hereby amended shall become effective on the date of publication in the FEDERAL REGISTER and thereupon will supersede:

U.S. Standards for Grades of Canned Tomato Paste which have been in effect since August 22, 1964; and

(Sec. 205, 60 Stat. 1090, as amended, 7 U.S.C. 1624)

Dated: February 17, 1970.

G.R. GRANGE,
Deputy Administrator, Marketing Services

Published in the Federal Register of July 22, 1964 (29 F.R. 9836)
Amended, Federal Register February 25, 1970 (35 F.R. 3651)

UNITED STATES STANDARDS FOR GRADES OF CANNED TOMATO PUREE[7]

Effective May 1, 1978

PRODUCT DESCRIPTION, CONCENTRATION, TEXTURE, GRADES

Sec.
52.5081 Product description.
52.5082 Concentration.
52.5083 Texture.
52.5084 Grades.

[7] Compliance with the provisions of these standards shall not excuse failure to comply with the provisions of the Federal Food, Drug, and Cosmetic Act or with applicable state laws and regulations.

FILL OF CONTAINER

52.5085 Fill of container.

FACTORS OF QUALITY

52.5086 Ascertaining the grade of a sample unit.
52.5087 Ascertaining the rating for each factor.
52.5088 Color.
52.5089 Defects.

LOT COMPLIANCE

52.5090 Ascertaining the grade of a lot.

SCORE SHEET

52.5091 Score sheet for tomato puree.

AUTHORITY: The provisions of this subpart issued under secs. 202– 208, 60 Stat. 1087, as amended; 7 U. S. C. 1621–1627.

PRODUCT DESCRIPTION, CONCENTRATION, TEXTURE, GRADES

§ 52.5081 Product description.

"Tomato Puree" (or "Tomato pulp"), hereinafter referred to as "Tomato Puree", is the clean, sound, wholesome product as defined in the standard of Identity for Tomato Puree (Tomato Pulp) (21 CFR 155.192) issued pursuant to the Federal Food, Drug, and Cosmetic Act.

§ 52.5082 Concentration.

(a) The degree of concentration is not considered a factor of quality for the purposes of these standards, but tomato puree contains not less than 8 percent, but less than 24 percent of natural tomato soluble solids.

(b) The following designations of concentration may be used in connection with these standards for the applicable natural tomato soluble solids group:

NATURAL TOMATO SOLUBLE SOLIDS

Extra heavy concentration.	15.0 percent or more, but less than 24.0 percent.
Heavy concentration.	11.3 percent or more, but less than 15.0 percent.
Medium concentration.	10.2 percent or more, but less than 11.3 percent.
Light concentration.	8 percent or more but less than 10.2 percent.

§ 52.5083 Texture.

(a) *General.* Texture is the degree of fineness or coarseness of the product. Texture is classified when the product is diluted with water, when necessary, to between 8 percent and 9 percent, inclusive, of natural tomato soluble solids.

(b) *Kinds of texture.* (1) "Fine" texture means a smooth, uniform finish. (2) "Coarse" texture means a coarse, slightly granular finish.

§ 52.5084 Grades.

(a) "U.S. Grade A" (or "U.S. Fancy") is the quality of tomato puree that (1) has a good flavor and odor, (2) has a good color, (3) is practically free from defects, and (4) scores not less than 90 points when scored in accordance with the scoring system outlined in this subpart.

(b) "U.S. Grade C" (or "U.S. Standard") is the quality of tomato puree that (1) has at least a fairly good flavor and odor, (2) has at least a fairly good color, (3) is at least fairly free from defects, and (4) scores not less than 80 points when scored in accordance with the scoring system outlined in this subpart.

(c) "Substandard" is the grade of tomato puree that fails to meet the requirements of "U.S. Grade C".

FILL OF CONTAINER

§ 52.5085 Recommended fill of container.

It is recommended that containers of tomato puree be filled as full as practicable without impairment of quality.

FACTORS OF QUALITY

§ 52.5086 Ascertaining the grade of a sample unit.

(a) *General.* The grade of a sample unit of tomato puree is ascertained by considering the factor of flavor and odor which is not scored, the ratings for the factors of color and defects which are scored, the total score, and the limiting rules which apply.

(b) *Factors not rated by score points*—(1) *Flavor and odor.* (i) The flavor and odor of the product is determined in undiluted samples and also after dilution with water to between 8 percent and 9 percent, inclusive, of natural tomato soluble solids.

(ii) "Good flavor and odor" means a distinct tomato puree flavor characteristic of ripe, good quality tomatoes, and inclusive of salt which may be added. Such flavor and odor may be no more than slightly affected by any one or combination of the following: stems, sepals, leaves, crushed seeds, cores; by immature, soured or overripe tomatoes; from the effects of unsatisfactory preparation, processing, or storage; or from any other factor not specifically mentioned.

(iii) "Fairly good flavor and odor" means a characteristic tomato puree flavor, inclusive of salt which may be added. Such flavor and odor may be affected, but not to a serious degree, by any one or combination of the following: stems, sepals, leaves, crushed seeds, cores; by immature, soured, or overripe tomatoes; or from the effects of unsatisfactory preparation, processing, or storage; or from any other factor not specifically mentioned.

(iv) "Off flavor and odor" means tomato puree flavor that fails to meet the requirements of "fairly good flavor and odor" or which possesses a flavor and/or odor which is seriously objectionable.

(c) *Factors rated by score points.* The relative importance of each scoring factor is expressed numerically on the scale of 100. The maximum number of points that may be given each factor is:

Factors:	*Points*
Color	50
Defects	50
Total	100

§ 52.5087 Ascertaining the rating for the factors which are scored.

The essential variations within each factor which is scored are so described that the value may be ascertained numerically. The numerical range within each factor is inclusive. (For example, "45 to 50 points" means 45, 46, 47, 48, 49, or 50 points.)

§ 52.5088 Color.

(a) *General.* The amount of red in tomato puree is determined by comparing the color of the product, diluted with water, when necessary, to between 8 percent and 9 percent, inclusive, of natural tomato soluble solids, with that produced by spinning a combination of the following Munsell color discs:

Disc 1—Red (5R 2.6/13) (glossy finish).
Disc 2—Yellow (2.5 YR 5/12) (glossy finish).
Disc 3—Black (N1) (glossy finish).
Disc 4—Grey (N4) (mat finish).

Such comparison is made under a diffused light source of approximately 250 foot-candle (candela) intensity and having a spectral quality approximating that of daylight under a moderately overcast sky, and a color temperature of 7,500° Kelvin ±200°. With the light source directly over the disc and diluted product, observation is made at an angle of 45° and at a distance of 12 or more inches from the product.

(b) *Availability of color references.* The colors referred to in this section are available from the approved supplier under license from the U.S. Department of Agriculture:

Munsell Color Co., Newburgh, New York.

(c) *Grade A Classification.* Tomato puree that has a good color may be given a score of 45 to 50 points. "Good color" means a bright, typical, red tomato puree color. Such color, when the product is of the proper concentration and observed as specified in this section, is as red as, or more red than, that produced by spinning the specified Munsell color discs in the following combinations or an equivalent of such composite color:

65 percent of the area of Disc 1;
21 percent of the area of Disc 2; and
14 percent of the area of either Disc 3 or Disc 4; or
7 percent of the area of Disc 3 and
7 percent of the area of Disc 4, whichever most nearly matches the appearance of the sample

(d) *Grade C Classification.* Tomato puree that has at least a fairly good color may be given a score of 40 to 44 points. Tomato puree that falls into this classification shall not be graded above U.S. Grade C, regardless of the total score for the product (this is a limiting rule). "Fairly good color" means a typical red tomato puree which may be slightly dull or have a slight yellow, yellow-orange, or brownish cast. Such color, when the product is of the proper concentration and observed as specified in this section, is as red as, or more red than, that produced by spinning the specified Munsell color discs in the following combinations or an equivalent of such composite color:

53 percent of the area of Disc 1;
28 percent of the area of Disc 2; and
19 percent of the area of either Disc 3 or Disc 4; or
9½ percent of the area of Disc 3 and
9½ percent of the area of Disc 4, whichever most nearly matches the appearance of the sample.

(e) *Substandard Classification.* Tomato puree that fails to meet the requirements of paragraph (d) of this section may be given a score of 0 to 39 points and shall not be graded above Substandard, regardless of the total score for the product (this is a limiting rule).

§ 52.5089 Defects.

(a) *General.* This factor is concerned with the degree of freedom from defects such as dark specks or scale-like particles, seeds or objectionable particles of seeds, objectionable tomato peel, harmless extraneous material, or any other similar substances.

(b) *Grade A Classification.* Tomato puree that is practically free from defects may be given a score of 45 to 50 points. Tomato puree that falls into this classification shall not be graded above U.S. Grade C, regardless of the total score of the product (this is a limiting rule). "Fairly free from defects" means that any defects present do not more than slightly affect the appearance or usability of the product.

(c) *Grade C Classification.* Tomato puree that is at least fairly free from defects may be given a score of 40 to 44 points. Tomato puree that falls into this classification shall not be graded above U.S. Grade C, regardless of the total score of the product (this is a limiting rule). "Fairly free from defects" means that any defects present may be noticeable but are not so large, so numerous, nor of such contrasting color or nature as to seriously affect the appearance or usability of the product.

(d) *Substandard Classification.* Tomato puree that fails to meet the requirements of paragraph (c) of this section may be given a score of 0 to 39 points and shall not be graded above Substandard, regardless of the total score for the product (this is a limiting rule).

LOT COMPLIANCE

§ 52.5090 Ascertaining the grade of a lot.

The grade of a lot of tomato puree covered by these standards is determined by the procedures set forth in the Regulations Governing Inspection and Certification of Processed Fruits and Vegetables, Processed Products Thereof, and Certain Other Processed Food Products (§ § 52.1 to 52.87).

SCORE SHEET

§ 52.5091 *Score sheet for tomato puree.*

Size and kind of container ....
Label ....
Container mark or identification ....
Net weight ....
Vacuum (inches) ....
Texture (fine or coarse) ....
Natural tomato soluble solids (percent) ....

Factors		Score points	
Color	50	(A)	45–50
		(C)	[1]40–44
		(SStd)	[1]0–39
Defects	50	(A)	45–50
		(C)	[1]40–44
		(SStd)	[1]0–39
Total score	100		

Flavor and odor .... Good / Fairly Good / Off

Grade ....

[1]Indicates limiting rule.

The U.S. Standards for grades as hereby amended shall become effective on the date of publication in the FEDERAL REGISTER and thereupon will supersede:

U.S. Standards for Grades of Canned Tomato Puree which have been in effect since August 22, 1964;

(Sec. 205, 60 Stat. 1090, as amended, 7 U.S.C. 1624)

Dated: February 17, 1970.

G.R. GRANGE,
Deputy Administrator, Marketing Services

Published in the Federal Register of July 22, 1964 (29 F.R. 9838)
Amended, Federal Register February 25, 1970 (35 F.R. 3651)

UNITED STATES STANDARDS FOR GRADES OF CANNED TOMATOES[8]

Effective April 13, 1990

PRODUCT DESCRIPTION AND GRADES

Sec.
52.5161 Product description.
52.5162 Grades of canned tomatoes.

FILL OF CONTAINER

52.5163 Fill of container.

FACTORS OF QUALITY

52.5164 Ascertaining the grade of a sample unit.
52.5165 Ascertaining the rating for each factor which is scored.
52.5166 Drained weight.
52.5167 Wholeness.
52.5168 Color.
52.5169 Defects.

LOT COMPLIANCE

52.5170 Ascertaining the grade of a lot.

SCORE SHEET

52.5171 Score sheet for canned tomatoes.

AUTHORITY: The provisions of this subpart issued under Secs. 202– 208, 60 Stat. 1087, as amended; 7 U. S. C. 1621–1627.

PRODUCT DESCRIPTION

§52.5161 The products to which these standards apply are defined in the Standards of Identity issued under the Federal Food, Drug, and Cosmetic Act as:

(a) Canned tomatoes (21 CFR 155.190); and
(b) Canned Stewed tomatoes (21 CFR 155.190).

[8]Compliance with the provisions of these standards shall not excuse failure to comply with the provisions of the Federal Food, Drug, and Cosmetic Act or with applicable state laws and regulations.

STYLES

§52.5162

(a) Whole means tomatoes, peeled or unpeeled, of any size that are substantially whole or almost whole.

(b) Sliced means tomatoes, peeled or unpeeled, that have been cut into units of approximately uniform thickness.

(c) Halves means tomatoes, peeled or unpeeled, that have been cut into two approximately equal halves.

(d) Wedges means tomatoes, peeled or unpeeled, that have been cut into approximate quarters or wedge-shaped sectors.

(e) Diced means tomatoes, peeled or unpeeled, that have been cut into approximate cube-shaped units.

TYPE OF PACK

§52.5163 Regular pack means tomatoes packed in a medium consisting of tomato juice, tomato puree, or tomato paste.

DEFINITIONS OF TERMS

§52.5164

(a) Character means the degree of firmness normally found when tomatoes have been processed using good manufacturing practices as defined in 21 CFR part 110. Canned and stewed tomatoes that, when fully cooked, are excessively soft or mushy are considered to lack character. Excessively soft is further defined for all styles and means that the unit may disintegrate upon handling, has evidence of sloughing (erosion of the tomato tissue) or has ragged edges, and has lost ability to retain shape. For the purposes of these grade standards, character is classified as good, reasonably good and fairly good. There is no substandard classification for the quality factor of character.

(1) Good character is whole, halves, wedges, or sliced tomatoes is defined as tomatoes in which not more than 15 percent by count are excessively soft or mushy. Good character in diced tomatoes is defined as diced tomatoes in which not more than 15 percent by weight are excessively soft or mushy.

(2) Reasonably good character in whole, halves, wedges, or sliced tomatoes is defined as tomatoes in which not more than 25 percent by count are excessively soft or mushy. Reasonably good character in diced tomatoes is defined as diced tomatoes in which not more than 25 percent by weight are excessively soft or mushy.

(3) Fairly good character in whole, halves, wedges, or sliced tomatoes is defined as tomatoes in which more than 25 percent by count are excessively soft or mushy. Fairly good character in diced tomatoes is defined as diced tomatoes in which more than 25 percent by weight are excessively soft or mushy.

(b) Color.

(1) USDA Tomato Red means the color of an USDA approved plastic color comparator.

(2) Minimum Red for Canned Tomatoes means the equivalent of any of the colors produced by blending the combinations of the following Munsell Color discs of equal diamter when placed as indicated.

Red-Munsell 5 R 2.6/13 (glossy finish)
Yellow-Munsell 2.5 YR 5/12 (glossy finish)
Black-Munsell N 1/ (glossy finish)
Grey-Munsell N 4 (mat finish)

(i) The discs are placed so that one-third of the area of the Red disc, and not more than one third of the area of the Yellow disc, are exposed. The exposed areas of the Black and Gray discs make up the remainder of the area.

(ii) To determine U.S. Grade C Color for canned and stewed tomatoes, apply the test for strength and redness of color outlined in 21 CFR 155.190(b).

(c) Defects refer to objectionable core material, tomato peel, extraneous vegetable material (EVM), blemished areas, and discolored portions that affect the appearance and eating quality of the product as further described below.

(1) Blemished areas are abnormal areas on the tomatoes which contrast strongly in color and/or texture with normal tomato tissue. Blemished areas may show objectionable discoloration ranging from light to dark discoloration. Blemished areas include but are not limited to scarred, raised, scabby tissue; darkened, tough areas which remain around the core area; internal or external dark tissue around blossom ends of the tomatoes; or any other areas that are objectionable or unsightly.

(2) Discolored portions are imperfections of the tomato tissue which may not be of strong contrasting color with respect to the tomato tissue but which detract slightly from the appearance of the product. Examples include sunburned areas, internal browning or tobacco mosaic, cloudy spots or ghost spots.

(3) Extraneous vegtable material (EVM) is vegetable material that is not harmful and includes, but is not limited to, whole or parts of stems, calyx bracts, tomato leaves or portions thereof, sprouted seeds, and other similar vegetable material which may not be part of the tomato plant.

(4) Objectionable core material is tomato material associated with the core of the tomato which detracts from the appearance of edibility of the product and includes tough fibers and tough or slightly discolored tomato tissue.

(5) Peel means the loose or attached skin of the tomato.

(d) Drained tomatoes means all of the tomato material that remains on the sieve after draining as prescribed in 21 CFR 155.190.

(e) Flavor and odor.

(1) Normal flavor and odor means a typical characteristic flavor and odor of mature tomatoes.

(2) Off-flavor and off-odor means any flavor or odor other than typical characteristic flavor and odor of mature tomatoes that is objectionable.

(f) Whole or almost whole, for whole style only, regardless of size, means that:

(1) The contour of the tomato is not materially affected by coring, trimming, or other means;

(2) The tomato may be cracked or split but not to the extent that there is material loss of seeds or placenta (gelatinous mass filling the seed cavity); and

(3) The tomato units may be restored to their approximate original shape during handling.

FILL OF CONTAINER

§52.5165 The standard of fill of container for canned tomatoes [21 CFR 155.190 (c)]requires that the product occupy not less than 90 percent of the water capacity of the container.

MINIMUM DRAINED WEIGHT AVERAGES

§52.5166

(a) The average minimum drained weight requirements for U.S. Grades are listed in Table I in §52.5170.

(b) The drained weight must not be less than 50 percent of the weight of water required to fill the container.

(c) The minimum drained weight averages are based on equalization of the product 15 days or more after the product has been canned.

(d) The method for ascertaining drained weight for all styles of canned and stewed tomatoes is found in 21 CFR 155.190 except that the U.S. Standard Number 8 circular sieve is used in lieu of a U.S. Standard Number 2 circular sieve for "sliced" and "diced" styles of canned and stewed tomatoes.

(e) Meeting drainged weight averages. A lot of canned tomatoes is considered as meeting the minimum drained weight averages if the following criteria are met:

(1) The average of the drained weights from all the containers in the sample meets the average drained weight in Table I in § 52.5170.

(2) The drained weights from the individual containers which do not meet the minimum are:

(i) For No. 2 1/2 can size and smaller, all styles except diced: not more than 0.7 ounce lower than the minimum average.

(ii) For no. 2 1/2 can size and smaller, diced style: not more than 0.5 ounce lower than the minimum average.

(iii) For No. 10 cans, all styles: not more than 2.0 ounces below the minimum average; and

(3) The number of containers in the sample which do not meet the requirements of paragraph (e) (2) of this section, does not exceed the acceptance number prescribed for the sample size in 7 CFR 52.38, Table I.

SAMPLE UNIT SIZES

§ 52.5167

(a) For "whole", "halves", and "wedges" styles, the sample unit size is the amount of product specified to be used for inspection in 7 CFR 52.38, Table I.

(b) For "sliced" and "diced" styles, the sample unit size is the same as indicated in paragraph (a) of this section EXCEPT in containers Number 10 (No. 10) can size and larger, an optional sample unit size of 32 ounces per container (drained weight) is permitted.

GRADES

§ 52.5168

(a) U.S. Grade A is the quality of canned tomatoes that meets the applicable requirements of Tables I, II, III, IV, and V of –52.5170, and scores not less than 90 points.

(b) U.S. Grade B is the quality of canned tomatoes that meets the applicable requirements of Tables I, II, III, IV, and V of –52.5170, and scores not less than 80 points.

(c) U.S. Grade C is the quality of canned tomatoes that meets the applicable requirements of Tables I, II, III, IV, and V of –52.5170, and scores not less than 70 points.
(d) Substandard is the quality of canned tomatoes that fails the requirements for U.S. Grade C.

FACTORS OF QUALITY

§ 52.5169 The grade of canned tomatoes is based on the following quality factors:

(a) Drained weight;
(b) Character;
(c) Color;
(d) Wholeness ("whole style" only);
(e) Flavor and odor; and
(f) Defects.

REQUIREMENTS FOR GRADES

§ 52.5170

TABLE I

MINIMUM DRAINED WEIGHT AVERAGES
FOR CANNED TOMATOES AND STEWED TOMATOES
ALL GRADES, ALL STYLES
(English (avoirdupois) weights)

Container	Container Dimensions 1/	Average drained weight (ounces) U.S. Grade A and B	U.S. Grade C
8 oz Tall	211 X 304	5.0	4.3
No. 300	300 X 407	8.8	7.6
No. 303	303 X 406	9.8	8.5
No. 2	307 X 409	11.9	10.3
No. 2-1/2	401 X 411	17.3	14.9
No. 10	603 X 700	63.5	54.7

1. The first figure in this column represents the diameter of the container and the second figure represents the height. The first digit in each number represents inches and the second two digits represent sixteenths of an inch. Thus 307 is three and seven-sixteenths inches.

TABLE II
MAXIMUM DEFECTS PERMITTED IN EACH GRADE
ALL STYLES

		In a single container		
	Defect (aggregate area)	In cans of less than 2 pounds total contents (in any container)	In cans of 2 or more pounds total contents (equivalent amount per pound of contents of any container).	In total sample representing a lot; in cans of any size [per pound of total contents of all containers (average)].
Grade A	Peel	2 square inches.	1 square inch.	1/2 square inch.
	Blemished areas	1/8 square inch.	1/16 square inch.	1/16 square inch.
	Discolored portions	1/2 square inch.	1/4 square inch.	1/4 square inch.
Grade B	Peel	3 square inches.	2 square inches.	1.06 square inches.
	Blemished areas	1/4 square inch.	1/8 square inch.	1/8 square inch.
	Discolored portions	1 square inch.	1/2 square inch.	1/2 square inch.
Grade C	Peel	No limit.	No limit.	1.06 square inches.
	Blemished areas	1/2 square inch.	1/4 square inch.	1/4 square inch.
	Discolored portions	1-1/2 square inches.	3/4 square inch.	3/4 square inch.

TABLE III - QUALITY REQUIREMENTS
FOR CANNED TOMATOES AND STEWED TOMATOES 1/
WHOLE STYLE

FACTOR	Grade A	Grade B	Grade C
Average Drained weight:	See Table I	See Table I	See Table I
Character: (by count)	Good character	Reasonably good character	Fairly good character
Score	18 - 20 points	16 - 17 points 2/	14 - 15 points 3/
Color: (by surface area)	At least 90% USDA Tomato Red, but not more than 5% less red than Minimum Red for Canned Tomatoes or yellow, and none may be vivid green.	At least 50% USDA Tomato Red, but not more than 10% less red than Minimum Red for Canned Tomatoes or yellow, and none may be vivid green.	Meets FDA requirements for strength and redness of color [21 CFR 155.190 (b)].
Score	27 - 30 points	24 - 26 points 2/	21 - 23 points 3/
Wholeness:	80% by weight whole or almost whole.	70% by weight whole or almost whole.	Less than 70% whole or almost whole.
Score	18 - 20 points	16 - 17 points 4/	14 - 15 points 3/
Flavor:	Normal	Normal	Normal

1/ Stewed tomatoes are graded excluding other vegetable ingredients.
2/ Cannot be graded above U.S. Grade B, regardless of the total score.
3/ Cannot be graded above U.S. Grade C, regardless of the total score.
4/ Can be graded U.S. Grade A, if the total score requirement is met.

TABLE III - QUALITY REQUIREMENTS
FOR CANNED TOMATOES AND STEWED TOMATOES 1/
WHOLE STYLE

(Continued)

FACTOR	Grade A	Grade B	Grade C
Defects:			
Core material	Trace	Slight	Moderate
EVM	Trace	Slight	Moderate
Peel	See Table II	See Table II	See Table II
Blemished areas	See Table II	See Table II	See Table II
Discolored portions	See Table II	See Table II	See Table II
Score	27 - 30 points	24 - 26 points 2/	21 - 23 points 3/
Total Score	90 - 100 points	80 - 89 points	70 - 79 points

1/ Stewed tomatoes are graded excluding other vegetable ingredients.
2/ Cannot be graded above U.S. Grade B, regardless of the total score.
3/ Cannot be graded above U.S. Grade C, regardless of the total score.

TABLE IV - QUALITY REQUIREMENTS
FOR CANNED TOMATOES AND STEWED TOMATOES 1/
HALVES AND WEDGES STYLES

FACTOR	Grade A	Grade B	Grade C
Average Drained weight:	See Table I	See Table I	See Table I
Character: (by count)	Good character	Reasonably good character	Fairly good character
Score	18 - 20 points	16 - 17 points 2/	14 - 15 points 3/
Color: (by surface area)	At least 90% USDA Tomato Red, but not more than 5% less red than Minimum Red for Canned Tomatoes or yellow, and none may be vivid green.	At least 50% USDA Tomato Red, but not more than 10% less red than Minimum Red for Canned Tomatoes or yellow, and none may be vivid green.	Meets FDA requirements for strength and redness of color [21 CFR 155.190 (b)].
Score	27 - 30 points	24 - 26 points 2/	21 - 23 points 3/
Flavor:	Normal	Normal	Normal
Defects:			
Core material	Trace	Slight	Moderate
EVM	Trace	Slight	Moderate
Peel	See Table II	See Table II	See Table II
Blemished areas	See Table II	See Table II	See Table II
Discolored portions	See Table II	See Table II	See Table II
Score	27 - 30 points	24 - 26 points 2/	21 - 23 points 3/
Total Score 4/	90 - 100 points	80 - 89 points	70 - 79 points

1/ Stewed tomatoes are graded excluding other vegetable ingredients.
2/ Cannot be graded above U.S. Grade B, regardless of the total score.
3/ Cannot be graded above U.S. Grade C, regardless of the total score.
4/ For Halves and Wedges styles, the final total score is adjusted by dividing the total score by 80 then multiplying by 100 to allow for the absence of the factor of wholeness.

TABLE V - QUALITY REQUIREMENTS
FOR CANNED TOMATOES AND STEWED TOMATOES 1/
SLICED AND DICED STYLES

FACTOR	Grade A	Grade B	Grade C
Average Drained weight:	See Table I	See Table I	See Table I
Character: Sliced: by count Diced: by weight	Good character	Reasonably good character	Fairly good character
Score	18 - 20 points	16 - 17 points 2/	14 - 15 points 3/
Color: (by surface area)	At least 90% USDA Tomato Red, but not more than 5% less red than Minimum Red for Canned Tomatoes or yellow, and none may be vivid green.	At least 50% USDA Tomato Red, but not more than 10% less red than Minimum Red for Canned Tomatoes or yellow, and none may be vivid green.	Meets FDA requirements for strength and redness of color [21 CFR 155.190 (b)].
Score	27 - 30 points	24 - 26 points 2/	21 - 23 points 3/
Flavor:	Normal	Normal	Normal
Defects:			
Core material	Trace	Slight	Moderate
EVM	Trace	Slight	Moderate
Peel	See Table II	See Table II	See Table II
Blemished areas	See Table II	See Table II	See Table II
Discolored portions	See Table II	See Table II	See Table II
Score	27 - 30 points	24 - 26 points 2/	21 - 23 points 3/
Total Score 4/	90 - 100 points	80 - 89 points	70 - 79 points

1/ Stewed tomatoes are graded excluding other vegetable ingredients.
2/ Cannot be graded above U.S. Grade B, regardless of the total score.
3/ Cannot be graded above U.S. Grade C, regardless of the total score.
4/ For Sliced and Diced styles, the final score is adjusted by dividing the total score by 80 then multiplying by 100 to allow for the absence of the factor of wholeness.

DETERMINING THE GRADE OF A LOT

§52.5171 The grade of a lot of canned tomatoes covered by these standards is determined by the procedures found in the "Regulations Governing Inspection and Certification of Processed Fruits and Vegetables, Processed Products Thereof, and Certain other Processed Food Products" (7 CFR 52.1 through 52.83).

Dated: March 9, 1990.

/s/

Kenneth C. Clayton
Acting Administrator

APPENDIX B – Food & Drug Administration Part 53 – Tomato Products

156.145 Tomato juice; identity.
156.147 Yellow tomato juice; identity.
155.194 Catsup, ketchup, catchup, identity; label statement of optional ingredients.
155.192 Tomato puree, tomato pulp; identity; label statement of optional ingredients.
155.191 Tomato paste; identity; label statement of optional ingredient.
155.190 Canned tomatoes; identity; label statement of optional ingredients; quality; fill of container.

§ 156.145 Tomato juice; identity.

Tomato juice is the unconcentrated liquid extracted from mature tomatoes of red or reddish varieties, with or without scalding followed by draining. In the extraction of such liquid, heat may be applied by any method which does not add water thereto. Such liquid is strained free from skins, seeds, and other coarse or hard substances, but carries finely divided insoluble solids from the flesh of the tomato. Such liquid may be homogenized, and may be seasoned with salt. When sealed in a container it is so processed by heat, before or after sealing, as to prevent spoilage.

§ 156.147 Yellow tomato juice.

Yellow tomato juice is the unconcentrated liquid extracted from mature tomatoes of yellow varieties. It conforms, in all other respects, to the definition and standard of identity for tomato juice prescribed in § 156.145.

§ 155.194 Catsup.

(a) Catsup, ketchup, catchup, is the food prepared from one or any combination of two or all of the following optional ingredients.

(1) The liquid obtained from mature tomatoes of red or reddish varieties.

(2) The liquid obtained from the residue from preparing such tomatoes for canning, consisting of peelings and cores with or without such tomatoes or pieces thereof.

(3) The liquid obtained from the residue from partial extraction of juice from such tomatoes.

Such liquid is obtained by so straining such tomatoes or residue, with or without heating, as to exclude skins, seeds, and other coarse or hard substances. Prior to straining, food-grade hydrochloric acid may be added to the tomato material at a rate to obtain a pH no lower than 2.0 ± 0.2. Such acid is then neutralized with food-grade sodium hydroxide so that the treated tomato material is restored to a pH of 4.2 ± 0.2 prior to straining. It is concentrated and seasoned with salt (sodium chloride formed during acid neutralization shall be considered added salt), a vinegar or vinegars, spices or flavorings or both, and onions or garlic or both, and is sweetened with sugar, or dextrose, or corn sirup (including dried corn sirup), or glucose sirup (including dried glucose sirup), or any mixture of these; provided that when the solids of corn sirup, or dried corn sirup, or glucose sirup, or dried glucose sirup (or any combination of these) used contains less than 58 percent by weight of reducing sugars calculated as anhydrous dextrose, then such corn sirup or glucose sirup shall be mixed with sugar or dextrose or both, in such quantity that the weight of the solids of such corn sirup or dried corn sirup or both, or glucose sirup, or dried glucose sirup or both, is not more than one-third of the weight of the solids of such mixture. When sealed in a container, it is so processed by heat, before or after sealing, as to prevent spoilage.

(b)(1) For the purposes of this section, the term "corn sirup" means refined corn sirup (including dried corn sirup) the solids of which contain not less than 40 percent by weight of reducing sugars calculated as anhydrous dextrose.

(2) The term "glucose sirup" means a clarified, concentrated, aqueous solution of the products obtained by incomplete hydrolysis of any edible starch. The solids of glucose sirup contain not less than 40 percent by weight of reducing sugars calculated as anhydrous dextrose. "Dried glucose sirup" means the product obtained by drying "glucose sirup."

(c) When optional ingredient specified in paragraph (a)(2) of this section, is present, in whole or in part, the label shall bear the statement "Made from ——" (or "Made in part from ———", as the case may be), "Residual tomato material from canning." When optional ingredient specified in paragraph (a)(3) of this section is present, in whole or in part, the label shall bear the statement "Made from ———" (or "Made in part from ———", as the case may be) "Residual tomato material from partial extraction of juice." If both such ingredients are present, such statements may be combined in the statement "Made from ———" (or "Made in part from ———", as the case may be) "Residual tomato material from canning and from partial extraction of juice." Wherever the name "Catsup," "Ketchup" or "Catchup" appears on the label so conspicuously as to be easily seen under customary conditions of purchase, the statement or statements specified in this paragraph showing the optional ingredients present shall immediately and conspicuously precede or follow such name, without intervening written, printed, or graphic matter.

§ 155.192 Tomato puree

(a) Tomato puree, tomato pulp, is the food prepared from one or any combination of two or all of the following optional ingredients:

(1) The liquid obtained from mature tomatoes of red or reddish varieties.

(2) The liquid obtained from the residue from preparing such tomatoes for canning, consisting of peelings and cores with or without such tomatoes or pieces thereof.

(3) The liquid obtained from the residue from partial extraction of juice from such tomatoes.

(4) Salt.

Such liquid is obtained by so straining such tomatoes or residue, with or without heating, as to exclude skins, seeds, and other coarse or hard substances. Prior to straining, food-grade hydrochloric acid may be added to the tomato material at a rate to obtain a pH no lower than 2.0 ± 0.2. Such acid is then neutralized with food-grade sodium hydroxide so that the treated tomato material is restored to a pH of 4.2 ± 0.2, prior to straining. It is concentrated and may be seasoned with salt (sodium chloride formed during acid neutralization shall be considered added salt). When sealed in a container, it is so processed by heat, before or after sealing, as to prevent spoilage. It contains not less than 8.0 percent, but less than 24.0 percent, of natural tomato soluble solids, as determined by the following method: Determine the refractive index of the clear serum obtained from the product, corrected for temperature, converting the resultant index to "% Sucrose" in accordance with the "International Scale of Refractive Indices of Sucrose at 20°C.," pages 931–33, 935, Reference Tables 47.012, 47.013, and 47.015 of the book "Official Methods of Analysis of the Association of Official Analytical Chemists," 11th edition 1971. If no salt has been added, this percent sucrose from the reference table shall be considered the percent of natural tomato soluble solids. If salt has been added, determine the percent of sodium chloride by the method prescribed on page 561, section 32.017, under "Sodium Chloride—Official," of said book. Subtract the percent of sodium chloride from the percent of total soluble solids found and multiply the difference by 1.016. The product shall be considered the percent of natural tomato soluble solids.

(b)(1) When the optional ingredient specified in paragraph (a)(2) of this section is present, in whole or in part, the label shall bear the statement "Made from residual tomato material from canning" or "Made in part from residual tomato material from canning," as the case may be. When the optional ingredient specified in paragraph (a)(3) of this section is present, in whole or in part, the label shall bear the statement "Made from residual tomato material from partial extraction of juice" or "Made in part from residual tomato material from partial extraction of juice," as the case may be. If both such ingredients are present, such statements may be combined in the statement "Made from residual tomato material from canning and from partial extraction of juice" or "Made in part from residual tomato material from canning and from partial extraction of juice," as the case may be. When the optional ingredient specified in paragraph (a)(4) of this section is present, the label shall bear the statement "Salt added" or "With added salt."

(2) The name specified for the food covered by this section is "tomato puree" or alternatively "tomato pulp"; however, if the only optional tomato ingredient used is the ingredient specified in paragraph (a)(1) of this section and the food contains not less than 20.0 percent of natural tomato soluble solids, the name "concentrated tomato juice" may be used in lieu of the names "tomato puree" or "tomato pulp."

(3) Wherever the name "Tomato puree" or "Tomato pulp" appears on the label so conspicuously as to be easily seen under customary conditions of purchase, the statement or statements specified in this section, showing the optional ingredients present, shall immediately and conspicuously precede or follow such name, without intervening written, printed, or graphic matter.

§ 155.191 Tomato paste.

(a) Tomato paste is the food prepared from one or any combination of two or all of the following optional ingredients:

(1) The liquid obtained from mature tomatoes of red or reddish varieties.

(2) The liquid obtained from the residue from preparing such tomatoes for canning, consisting of peelings and cores with or without such tomatoes or pieces thereof.

(3) The liquid obtained from the residue from partial extraction of juice from such tomatoes.

Such liquid is obtained by so straining such tomatoes or residue, with or without heating, as to exclude skins, seeds, and other coarse or hard substances. Prior to straining, food-grade hydrochloric acid may be added to the tomato material at a rate to obtain a pH no lower than 2.0 ± 0.2. Such acid is then neutralized with food-grade sodium hydroxide so that the treated tomato material is restored to a pH of 4.2 ± 0.2, prior to straining. It is concentrated and may be seasoned with one or more of the optional ingredients:

(4) Salt (sodium chloride formed during acid neutralization shall be considered added salt).

(5) Spice.

(6) Flavoring.

It may contain, in such quantity as neutralizes a part of the tomato acids, the optional ingredient:

(7) Baking soda.

When sealed in a container it is so processed by heat, before or after sealing, as to prevent spoilage. It contains not less than 24.0 percent of natural tomato soluble solids as determined by the following method. Determine the refractive index of the clear serum obtained from the product, corrected for temperature, converting the resultant index to "% Sucrose" in accordance with the "International Scale of Refractive Indices of Sucrose at 20°C.," pages 828–30, Reference Tables 43.008 and 43.009 of the book "Official Methods of Analysis of the Association of Official Agricultural Chemists," 10th edition, 1965. If no salt has been added, this percent sucrose from the reference table shall be considered the percent of natural tomato soluble solids. If salt has been added, determine the percent of sodium chloride by the method prescribed on page 519, section 30.009, under "Sodium Chloride-Official," of said book. Subtract the percent of sodium chloride found from the percent of total soluble solids found and multiply the difference by 1.016. The product shall be considered the percent of natural tomato soluble solids.

(b) When the optional ingredient specified in paragraph (a)(2) of this section is present, in whole or in part, the label shall bear the statement "Made from" (or "Made in part from," as the case may be) "residual tomato material from canning." When the optional ingredient specified in paragraph (a)(3) of this section is present in whole or in part, the label shall bear the statement "Made from" (or "Made in part from", as the case may be) "residual tomato material from partial extraction of juice". If both such ingredients are present, such statements may be combined in the statement "Made from" (or "Made in part from", as the case may be) "residual tomato material from canning and from partial extraction of juice". When the optional ingredient specified in paragraph (a)(4), (5), or (6) of this section is present, the label shall bear the statement or statements "Salt added" or "With added salt", "Spice added" or "With added spice", "Flavoring added" or "With added flavoring", as the case may be. When the optional ingredient specified in paragraph (a)(7) of this section is present, the label shall bear the statement "Baking soda added". If two or all of the optional ingredients specified in paragraphs (a) (4), (5), (6), and (7) of this section are present, such statements may be combined; for example, "Salt, spice, flavoring, and baking soda added". In lieu of the word "salt", "spice", or "flavoring" in such statement or statements, the common or usual name of such salt, spice, or flavoring may be used.

(c) Wherever the name "Tomato paste" appears on the label so conspicuously as to be easily seen under customary conditions of purchase, the statement or statements specified in this section, showing the optional ingredients present shall immediately and conspicuously precede or follow such name, without intervening written, printed, or graphic matter.

§ 155.190 Canned tomatoes.

(a) *Identity*—(1) *Description.* (i) Canned tomatoes is the food prepared from mature tomatoes conforming to the characteristics of the fruit *Lycopersicum esculentum* P. Mill, of red or reddish varieties. The tomatoes may or may not be peeled, but shall have had the stems and calicies removed and shall have been cored, except where the internal core is insignificant to texture and appearance.

(ii) Canned tomatoes may contain one or more of the safe and suitable optional ingredients specified in paragraph (a)(2) of this section, be packed without any added liquid or in one of the optional packing media specified in paragraph (a)(3) of this section and be prepared in one of the styles specified in paragraph (a)(4) of this section. Such food is sealed in a container and before or after sealing is so processed by heat as to prevent spoilage.

(2) *Optional ingredients.* One or more of the following safe and suitable ingredients may be used:

(i) Calcium salts in a quantity reasonably necessary to firm the tomatoes, but the amount of calcium in the finished canned tomatoes is not more than 0.045 percent of the weight, except that when the tomatoes are prepared in one of the styles specified in paragraph (a)(4)(iv) to (vi) of this section the amount of calcium is not more than 0.08 percent of the weight of the food.

(ii) Organic acids for the purpose of acidification.

(iii) Dry nutritive carbohydrate sweeteners whenever any organic acid provided for in paragraph (a)(2)(ii) of this section is used, in a quantity reasonably necessary to compensate for the tartness resulting from such added acid.

(iv) Salt.

(v) Spices, spice oils.

(vi) Flavoring and seasoning.

(vii) Vegetable ingredients such as onion, peppers, and celery, that may be fresh or preserved by physical means, in a quantity not more than 10 percent by weight of the finished food.

(3) *Packing media.* (i) The liquid draining from the tomatoes during and after peeling or coring.

(ii) The liquid strained from the residue from preparing tomatoes for canning consisting of peels and cores with or without tomatoes or pieces thereof.

(iii) The liquid strained from mature tomatoes (tomato juice).

(iv) Tomato paste, or tomato puree, or tomato pulp complying with the compositional requirements of §§ 155.191 and 155.192.

(4) *Styles.* (i) Whole.

(ii) Diced.

(iii) Sliced.

(iv) Wedges.

(5) *Name of the food*, (i) The name of the food is "tomatoes", except that when the tomatoes are not peeled that name is "unpeeled tomatoes".

(ii) The following shall be included as part of the name or in close proximity to the name of the food.

(*a*) A declaration of any flavoring that characterizes the product as specified in § 101.22 of this chapter.

(*b*) A declaration of any added spice, seasoning, or vegetable ingredient that characterizes the product, (e.g., "with added ——————" or, "with ——————" the blank to be filled in with the word(s) "spice(s)", "seasoning(s)", or the name(s) of the vegetable(s) used or in lieu of the word(s) "spice(s)" or "seasoning(s)" the common or usual name(s) of the spice(s) or seasoning(s) used) except that no declaration of the presence of onion, peppers, and celery is required for stewed tomatoes.

(*c*) The word "stewed" if the tomatoes contain characterizing amounts of at least the three optional vegetables listed in paragraph (a)(2)(vii) of this section.

(*d*) The styles: "Diced", "sliced", or "wedges" as appropriate.

(*e*) The name of the packing medium: "tomato paste", "tomato puree", or "tomato pulp" as provided in paragraph (a)(3)(iv) of this section, or "strained residual tomato material from preparation for canning" as provided for in paragraph (a)(3)(ii) of this section, as appropriate. The name of the packing medium shall be preceded by the work "with."

(iii) The following may be included as part of the name or in close proximity to the name:

(*a*) The word "whole" if the tomato ingredient is whole or almost whole, and the weight of such ingredient is not less than 80 percent of the drained weight of the finished food as determined in accordance with the method prescribed in paragraph (b)(2) of this section.

(*b*) The words "solid pack" when none of the optional packing media specified in paragraph (a)(3) of this section are used.

(*c*) The words "in tomato juice" if the packing medium specified in paragraph (a)(3)(iii) of this section is used.

(6) *Label declaration of optional ingredients*. The name of each optional ingredient used shall be declared on the label as required by the applicable sections of Part 101 of this chapter.

(b) *Quality*. (1) The standard of quality for canned tomatoes is as follows:

(i) The drained weight, as determined by the method prescribed in paragraph (b)(2)(i) of this section, is not less than 50 percent of the weight of water required to fill the container, as determined by the general method for water capacity of the containers prescribed in §130.12(a) of this chapter.

(ii) The strength and redness of color as determined by the method prescribed in paragraph (b)(2) of this section, are not less than that of the blended color of any combination of the color discs described in such method in which one-third the area of disc 1, and not more than one-third the area of disc 2, is exposed;

(iii) Peel per kilogram (2.2 pounds) of the finished food covers an area of not more than 15 cm^2 (2.3 square inches) which is equivalent to 6.8 cm^2 (1.06 square inches) per pound based on an average of all containers examined provided, however, that the area of peel is not a factor of quality for canned unpeeled tomatoes labeled in accordance with paragraph (a)(5)(i) of this section; and

(iv) Blemishes per kilogram (2.2 pounds) of the finished food cover an area of not more than 3.5 cm^2 (0.54 square inch) which is equivalent to 1.6 cm^2 (0.25 square inch) per pound based on an average of all containers examined.

(2) Canned tomatoes shall be tested by the following method to determine whether or not they meet the requirements of paragraph (b)(1) (i) and (ii) of this section:

(i) Remove lid from container, but in the case of a container with lid attached by double seam, do not remove or alter the height of the double seam. Tilt the opened container so as to distribute the contents over the meshes of a circular sieve which has previously been weighed. The diameter of the sieve used is 20.3 centimeters (8 inches) if the quantity of the contents of the container is less than 1.4 kilograms (3 pounds) or 30.5 centimeters (12 inches) if such quantity is 1.4 kilograms (3 pounds) or more. The meshes of such sieve are made by so weaving wire of 1.4 mm (0.054 inch) diameter as to form square openings 11.3 mm by 11.3 mm (0.446 inch by 0.446 inch). Without shifting the tomatoes, so incline the sieve as to facilitate drainage of the liquid. Two minutes from the time drainage begins, weigh the sieve and drained tomatoes. The weight so found, less the weight of the sieve, shall be considered to be the drained weight.

(ii) Remove from the sieve the drained tomatoes, cut out and segregate successively those portions of least redness until 50 percent of the drained weight has been so segregated. Comminute the segregated portions to a uniform mixture without removing or breaking the seeds. Fill the mixture into a black container to a depth of at least 25.4 mm (1 inch). Free the mixture from air bubbles, and skim off or press below the surface all visible seeds. Compare the color of the mixture, in full diffused daylight or its equivalent, with the blended color of combinations of the following concentric Munsell color discs of equal diameter, or the color equivalent of such discs:

(*a*) Red—Munsell 5 R 2.6/13 (glossy finish).

(*b*) Yellow—Munsell 2.5 YR 5/12 (glossy finish).

(*c*) Black—Munsell N 1/ (glossy finish).

(*d*) Grey—Munsell N 4 (mat finish).

(3) *Sampling and acceptance procedure.* A lot is to be considered acceptable when the number of "defectives" does not exceed the acceptance number in the sampling plans given in paragraph (b)(3)(ii) of this section.

(i) Definitions of terms to be used in the sampling plans in paragraph (b)(3)(ii) of this section are as follows:

(*a*) Lot. A collection of primary containers or units of the same size, type, and style manufactured or packed under similar conditions and handled as a single unit of trade.

(*b*) Lot size. The number of primary containers or units in the lot.

(*c*) Sample size (n). The total number of sample units drawn for examination from a lot.

(*d*) Sample unit. A container, the entire contents of a container, a portion of the contents of a container, or a composite mixture of product from small containers that is sufficient for examination or testing as a single unit.

(*e*) Defective. Any sample unit shall be regarded as defective when the sample unit does not meet the criteria set forth in the standards.

(*f*) Acceptance number (*c*). The maximum number of defective sample units permitted in the sample in order to consider the lot as meeting the specified requirements.

(*g*) Acceptable quality level (*AQL*). The maximum percent of defective sample units permitted in a lot that will be accepted approximately 95 percent of the time.

(ii) Sampling plans and acceptance procedure:

Acceptable quality level 6.5

Lot size (primary containers)	Size of container	
	Net weight equal to or less than 1 kg (2.2 lb)	
	n	*c*
4,800 or less	13	2
4,801 to 24,000	21	3
24,001 to 48,000	29	4
48,001 to 84,000	48	6
84,001 to 144,000	84	9
144,001 to 240,000	126	13
Over 240,000	200	19
	Net weight greater than 1 kg (2.2 lb.) but not more than 4.5 kg (10 lb)	
	n	*c*
2,400 or less	13	2
2,401 to 15,000	21	3
15,001 to 24,000	29	4
24,001 to 42,000	48	6
42,001 to 72,000	84	9
72,001 to 120,000	126	13
Over 120,000	200	19
	Net weight greater than 4.5 kg (10 lb)	
	n	*c*
600 or less	13	2
601 to 2,000	21	3
2,001 to 7,200	29	4
7,201 to 15,000	48	6
15,001 to 24,000	84	9
24,001 to 42,000	126	13
Over 42,000	200	19

n = number of primary containers in sample.
c = acceptance number.

(4) If the quality of canned tomatoes falls below the standard prescribed in paragraph (b)(1) of this section, the label shall bear the general statement of substandard quality specified in § 130.14(a) of this chapter in the manner and form therein specified; if, however, the quality of canned tomatoes falls below standard with respect to only one of the factors of quality specified by paragraph (b)(1)(i) to (iii) of this section, there may be substituted for the second line of such general statement of substandard quality ("Good Food—Not High Grade") a new line, appropriate for the corresponding subparagraph designation of paragraph (b)(1) of this section which the canned tomatoes fail to meet, to read as follows: (i) "Poor color" or (ii) "Excessive peel" or (iii) "Excessive blemishes."

(c) *Fill of container.* (1) The standard of fill of container for canned tomatoes is a fill of not less than 90 percent of the total capacity of the container, as determined by the general method for fill of containers prescribed in § 130.12(b) of this chapter.

(2) If canned tomatoes fall below the standard of fill of container prescribed in paragraph (c)(1) of this section, the label shall bear the general statement of substandard fill specified in § 130.14(b) of this chapter, in the manner and form therein specified.

[42 FR 14449, Mar. 15, 1977, as amended at 43 FR 12858, Mar. 28, 1978; 43 FR 30274, July 14, 1978]

APPENDIX C – Quality Control and Evaluation Forms

C.1 Raw Product Receiving Report
C.2 Daily Preparation Quality Control Report
C.3 Daily Canned Tomato Quality Control Report
C.4 Daily Tomato Juice Quality Control Report
C.5 Daily Double Seam Quality Control Report
C.6 Daily Process Record Quality Control Report
C.7 Canned Tomato Score Sheet
C.8 Canned Tomato Juice Score Sheet
C.9 Tomato Catsup Score Sheet
C.10 Tomato Puree Score Sheet
C.11 Daily Sanitation Report
C.12 Tomato Variety Evaluation Report

Date ______________________

Line ______________________

Plant ______________________

Load	Grower	Amount (lbs)	U.S. Grades				Color Index	Defects		Holding		Remarks
			% 1's	% 2's C	% 2's D	% Culls		Kind	% Mold	Time	Conditions	

C.1. RAW PRODUCT RECEIVING REPORT

Plant ______________________ Date ______________________

Location ______________________ Product _____TOMATOES_____

DAILY QUALITY CONTROL REPORT FOR TOMATOES AND TOMATO PRODUCTS

Time									
Raw Product Quality									
Color Score									
Defects - %									
Defects - Type									
Condition									
Flume									
Detergent - ppm									
Agitation - psi									
Washer - Efficiency									
Water Pressure									
Temperature									
Lye Concentration									
Peel - Time									
Temperature									
Peel Efficiency									
% Peel									
% Core									
% Defects									
% Wholeness									
% Color									
Juice Stock Condition									
Trim Efficiency									
Color Score									
% Defects									
Kinds of Defects									

C.2. DAILY PREPARATION QUALITY CONTROL REPORT

Plant ______________________ Date ______________________

Location ______________________ Product CANNED TOMATOES

DAILY QUALITY CONTROL REPORT FOR CANNED TOMATOES

Time											
Packing Efficiency											
Net Weight											
Color Score											
Salt Tablet Size											
Fill Juice Temperature											
Exhausting Time											
Headspace											
Center Can Temp.											
Code											
Process Time											
Process Temperature											
Cooling Time											
Cooling Water	Ingoing										
Temperature:	Outgoing										
Chlorine ppm											
Condition of Cans											
Remarks											

Quality Control Technologist ______________________

C.3. DAILY CANNED TOMATO QUALITY CONTROL REPORT

Plant ____________________ Date ____________________

Location ____________________ Product TOMATO JUICE

DAILY QUALITY CONTROL REPORT FOR TOMATO JUICE

Time									
Preheat Temperature									
Preheat Time									
Vibrating Screen – % Waste									
Extractor – % Waste									
Pasteurizer Time									
Temperature									
Hold Time									
Hold Temperature									
Discharge Temperature									
Filler Temperature									
Headspace									
Code									
Cooling of Cans									
Temperature									
Time									
Chlorine ppm									
Condition of Cans									
Remarks									

C.4. DAILY TOMATO JUICE QUALITY CONTROL REPORT

Date ______ Line No. ______ Can Size ______ Can Code ______

Time	No. Head	Width			Body Hook			Overlap			Coverhook			Tightness Rating	Countersink			Remarks
		Min	Max	Ave	Min	Max	Ave	Min	Max	Ave	Min	Max	Ave		Min	Max	Ave	

Operator ______

C.5. DAILY DOUBLE SEAM QUALITY CONTROL REPORT

Product ______ Type of Container ______ Company Name ______ Location ______

Product (and/or) Style of Pack	Code	Retort No.	Container		Min. I.T. °F	Time Steam On	Vent Closed	Time Temp. Up	Time Steam Off	Process		Thermometers		Recorder Chart No.
			Size	Approx. No. per Code			°F Time &/or Temp.			Time	°F Temp.	Mercury in Glass	Recorder Chart	

Operator ______

C.6. DAILY PROCESS RECORD QUALITY CONTROL REPORT

Date ______________________

No., Size, and Kind of Container

Label											
Container Mark or Identification		Cans/Glass Case									
Net Weight (ounces)											
Vacuum (inches)											
Drained Weight (ounces)											
Factors				Score Points							
Drained Weight	20	(A)	(A Whole)	18-20							
		(B)		16-17							
		(C)		14-15**							
		(SStd)		0-13*							
Wholeness	20	(A Whole)		20							
		(A)		18-19							
		(B)		16-17							
		(C)		14-15*							
Color	30	(A)	(A Whole)	27-30							
		(B)		24-26*							
		(C)		21-23*							
		(SStd)		0-20*							
Defects	30	(A)	(A Whole)	27-30							
		(B)		24-26*							
		(C)		21-23*							
		(SStd)		6-20							
Total Score	100										
Flavor and Odor Good, Normal, or Off											

*Limiting Rule

**Special Limiting Rule

Technologist ______________________

U.S. Department of Agriculture, Agricultural Marketing Service, Score Sheet for Canned Tomatoes

C.7. CANNED TOMATO SCORE SHEET

Date ____________

Time								
Can Size								
Code								
Net Weight								
Vacuum								
Head Space								
% Solids								
Viscosity								
pH								
Total Acid								
Vitamin C								
Agtron F Color								
Mold Count								
Fly Egg								
Factors: Color 30 (A) 26–30 (C) 23–25** (D) 0–22*								
Consistency 15 (A) 13–15 (C) 10–12 (D) 0– 9*								
Absence of Defects 15 (A) 13–15 (C) 10–12* (D) 0– 9*								
Flavor 40 (A) 33–40 (C) 27–32 (D) 0–26*								
Total Score								

*Limiting Rule
**Partial Limiting Rule

Technologist ____________

C.8. CANNED TOMATO JUICE SCORE SHEET

Date ____________

Time								
Can Size								
Code								
Net Weight								
Vacuum								
Head Space								
% Solids								
Viscosity								
pH								
Total Acid								
Agtron F Color								
Mold Count								
Fly Egg								
Factors: Color 25 (A) 21–25 (C) 17–20* (D) 0–16*								
Consistency 25 (A) 22–25 (C) 18–21** (D) 0–17*								
Absence of Defects 25 (A) 21–25 (C) 18–20 (D) 0–17*								
Flavor 25 (A) 21–25 (C) 17–20 (D) 0–16*								
Total Score								

*Limiting Rule
**Partial Limiting Rule

Technologist ____________

Grade catsup scores not less than 85 points, but has total solids content of not less than 29%, by weight.

C.9. TOMATO CATSUP SCORE SHEET

Date ______________________

Time									
Can Size									
Code									
Net Weight									
Vacuum									
Head Space									
% Solids									
Viscosity									
pH									
Total Acid									
Agtron F Color									
Mold Count									
Fly Egg									
Factors: Color 50 (A) 45–50 (C) 40–44* (D) 0–39*									
Absence of Defects 50 (A) 45–50 (C) 40–44* (D) 0–39*									
Total Score 100									

*Limiting Rule

Technologist ______________________

C.10. TOMATO PUREE SCORE SHEET

Plant __________ Line __________ Product __________ Date __________

Time				
Receiving Areas:				
Flumes				
Dumpers				
Conveyors				
Containers				
Storage				
Processing Areas:				
Soak Tank				
Washer				
Scalder – Peeler				
Inspection Belt				
Waste Belt				
Buckets – Pans – Trays – Knives				
Packing – Filling				
Exhaust Box				
Chopper				
Preheater				
Vibrating Screen				
Extractor				
Pulper				
Finisher				
Pasteurizer				
Deaerator				
Closing Machines				
Salter				
Spinner – Cooler				
Pumps				
Piping				
Cooling Canal				
Can Tracks				
Retorts – Cookers				
General Plant Areas:				
Floors				
Walls				
Ceiling				
Rest Rooms				
Supplies				
Wash Basins				
Toilets and Urinals				
Floors and Walls				
Lunch Area				
Personnel				
Cleanliness				
Head Covering				
Containers				
Water Disposal				
Premises				
Remarks				

Quality Control Technologist __________

C.11. DAILY SANITATION REPORT (S - SATISFACTORY; U - UNSATISFACTORY; NI - NEEDS IMPROVEMENT)

VARIET / OR CULTIVAR________ SEED SOURCE________

DIRECT SEED DATE______TRANSPLANT DATE______HARVEST DATE______

PLANT HABIT: DETERMINATE____SEMI DETERMINATE____INDETERMINATE____DWARF____

LEAF TYPE: NORMAL______POTATO______OTHER______

FOLIAGE COVER: DENSE_____MEDIUM_____FAIR_____POOR_____

UNIFORMITY OF FRUIT SET IN PERCENT______COMMENTS______

UNIFORMITY OF RIPENING IN PERCENT______COMMENTS______

FRUIT HOLDING ABILITY ON VINES: EXCELLENT____GOOD____FAIR____POOR____

EASE OF FRUIT REMOVAL: EXCELLENT____GOOD____FAIR____POOR____

MATURITY IN NUMBER OF DAYS:______

YIELD PER PLANT______YIELD PER ACRE______

FRUIT SHAPE: GLOBE____PEAR____OBLONG____FLAT____OTHER______

FRUIT SURFACE: SMOOTH_____SEMI SMOOTH_____IRREGULAR_____OTHER_____

FRUIT SIZE: DIAMETER IN INCHES_____LENGTH IN INCHES_____

FRUIT SIZE (COUNT PER 25 POUNDS)_____WEIGHT PER FRUIT IN OUNCES_____

FRUIT SIZE CLASSIFICATION: SMALL_____MEDIUM_____LARGE_____

STEM SCAR: LESS THAN 1/4 INCH_____1/4 to 1/2 INCH_____OVER 1/2 INCH_____

STYLAR SCAR: NONE____1/8 INCH____1/4 INCH____3/8 INCH____OVER 3/8 INCH____

STEM SEPARATION: JOINTLESS______NON JOINTLESS______

LENGTH OF STEM IN INCHES IF PRESENT______

PERCENT RIPE FRUITS BY HARVESTS: ONCE OVER____1st PICK____2nd PICK____3rd PICK____

PERCENT CRACK RESISTANT AT HARVEST______

PERCENT BROKEN FRUIT AT FACTORY______

RAW PRODUCT QUALITY:

EXTERNAL COLOR IN %: FULL RED_____GREEN SHOULDERS_____OTHER_____

NUMBER OF LOCULES: ______

WALL THICKNESS IN MM: ______

SIZE OF LOCULES: SMALL_____MEDIUM_____LARGE_____

SIZE OF CORE: NONE_____SMALL_____MEDIUM_____LARGE_____

INTERNAL COLOR SCORE______

pH______TITRATABLE ACIDITY______

REFRACTIVE INDICE______

SOLUBLE SOLIDS/ACID RATIO______

FIRMNESS______

% DEFECTS______TYPES OF DEFECTS______

OTHER ATTRIBUTES______

CANNED PRODUCTS QUALITY:

U S GRADE WHOLE PACK______TOTAL SCORE______

U S GRADE TOMATO JUICE______TOTAL SCORE______

U S GRADE TOMATO PULP______TOTAL SCORE______

U S GRADE TOMATO PASTE______TOTAL SCORE______

COMMENTS:

AREA ADAPTABILITY______

MACHINE HARVEST ADAPTABILITY______

OTHER______

RECOMMENDATIONS:______

C.12. TOMATO VARIETY EVALUATION REPORT

INDEX

Anatomy, frontispiece
Acidity, acidification, 192, 206, 208
  acids, 195
  acidity, 345
  application, 195-196
  total, 345-346
Adams Consistometer, See Consistency
Agtron, 140-147
Alcohol insoluble solids, 317
Amino acids, 434-435
Aphids, 77

Bacillus coagulans, 420-423
Bacillus thermoacidurans, See Bacillus coagulans
Beetle, flea, 77
Blotter test, 318
Bostwick consistometer, see consistency,
Break, cold, 202
  hot, 203, 204
Breeding, challenges, 94-95
  general, 88
  improvements, 88-89
  methods, 85
  phenotypes, 86
  regulations, 96
  Tomato Genetics Cooperative, 85
Brix, 206
Bulk storage, 227-228

Calcium salt, 183-184
Carotene, 439-441
Catchup, see catsup,
Catsup constitutents,
  defects, 356-357
  defined, 233-234, 454
  formulation 236-237   manufacture, 237
  mold counting, 416
  See, also, standards
Caustic, 165-169
Chlorination, 121, 373-375
Chopping, 202-204
Citric acid, 192-195
Clostridium botulinum, 108
Cold break, see break
Color, See also, light and lighting, 297
  Agtron, see Agtron, 311
  CIE system, 303
  factors affecting, 298
  Hunter Color and Color Difference Meter, 148-151, 307-310
  Light and Lighting, 300-301
  Macbeth-Munsell Colorimeter, 305
  Macbeth disk colorimetry, 303
    noitations, 303
    chroma, 304-305
    hue, 304-305
    value, 304-305
  perception, 299
  systems of measurement, 302
  USDA Hunterlab D6 Tomato Colorimeter, 148-150
Composition, 433, 437
  acids, 434-435
  amino acid, 434
  carbohydrates, 433
  labeling, 446
  minerals, 436
  nutrients, 439-443
  pectin, 436, 438
  proteins, 434
  solids, 433
Concentration, paste, 220
  pulp, 220
Consistency, Adams, 334-335
  blotter, 334
  Bostwick, 329-330
  Brookfield, 331-333
  capillary, 333
  catsup, 332
  chili sauce, 331-332
  classification, 323-324
  efflux-tube viscometer, 327
  factors affecting, 343
  Fisher Electroviscometer, 341

FMC consistometer, 341
Gardner Mobilometer, 342
GOSUC, 327
Stormer, 327
tomato juice, 325
tomato paste, 336
tomato pulp, 339
tomato sauce, 340
tomato soup, 340
Viscometer, capillary, 326
potentiometric viscometer, 338
USDA, 326
Viscosity, 323
Consumption, 12-13
Containers, 209
Continuous cooker, 190
cooler, 192
Cooking, catsup, 237-238
tomato juice, 211-214
tomatoes, 192
Coring, 161, 163-164
hand, 161
machine, 162-163
Critical control points, 197
Crushing, see chopping
Cultivar, 32-34
Cultivation, 39
Cutworms, 77

Davis Sampler, See Sampling
Deaeration, 161
Defects, 353
Material other than tomatoes MOT, 353
sand, organic matter, 354
specs, 354
Direct seeding, 36-37
Disease chart, 74-75
Diseases, 41-68, 387
Alternaria, 404
Anthracnose, 63-64
Aspergillus, 404
Bacterial canker, 57-59
Bacterial speck, 51-53
Bacterial spot, 54-56
Bacterial wilt, 60-62
Black mold, 63
Botrytis (gray mold), 67
Buckeye rot, 66
Early blight, 42-44
Late blight, 45-47
Septoria, 48-50
Southern blight (sclerotium rot), 62
Rhizoctonia, 65-66
Pythium rot, 65, 67
Drained weight, 184
Drosophila, activity, 369
AOAC method, 378-380
eggs, 370
control, 372-374
detection, 375
FDA action level, 376
GOSUL method, 376-378
life cycle, 369
removal, 373-375

Efflux tube viscometer, See Consistency
Enzymes, 202-204
Eriopersicon, 84
Ethephone (Ethrel), 79
Eulycopersicon, 84
Exhausting, 189
Extraction, 205

Fertilizer, 28-30
Field selection, 19-21
Fill, catsup, 239
juice, 208-209
nomograph, 185
paste, 227
tomatoes, 184
Filling, 181
Finishing, 222
Firmness, 183
firming agents, 186
Flavor, 206, 353, 447
interpretation, 361-367
methods, 359-361
volatiles, 446
Flea beetle, 77
Flocron, 187
Flow chart, 182
Formulation, catsup, 236-237
soup, 245-247

Genetics, See Breeding
Grade Standards, See Standards
Grading, belt, 129
equipment, see Laboratory equipment,
platforms, 131
Standards, 132-139
tables, 134, 139
Grasshoppers, 77

Handling practices, bulk, 110, 118-122
hampers, 117
harvester, 104, 107
lug or field boxes, 117-118
mechanical, 103-104
plastic boxes, 118
preparation, 78
water, 119-121
water tanks, 119-120
History, processing, 7-14
tomato, 3-7
Homogenization, 210
Hornworms, 77
Hot break, See Break
Howard Mold Count Method, See Mold
Hunter Color and Color Difference Meter, See Color
Hydrout, See Coring

Insect, chart, 76
control, 76-78
determination, 380-383,
fragment, 380-384
Also See *Drosophila*
Inspection, California, 139
raw product, 128-129
peeled, 147
USDA, 134-138
Irrigation, 41

Ketchup, See Catsup

Larvae, See *Drosophila,* life cycle
Lycopersicon, 3, 83
Lye, 165-169, See, also, Peeling

Macbeth-Munsell Colorimeter, See Color
Mechanical harvesting, 107-114
Microbiology, 107-108
Microscope, 387-392
Mold characteristics, 387, 401-404
counting AOAC, 408-416
FDA action levels, 416
genera, 403-405
histology, 392-395
Howard Mold Count Method, 396-401
inclined cover glass technique, 412
parallel cover glass technique, 412
sampling plans, 413-416

Nutrient composition, 433
Ascorbic Acid, 439
B Complex, 440
Carotene, Vitamin A, 439
factors affecting, 440
Nutrition, 29

Organic acids, 435
Organization chart, 4

Pectin, See Composition, Pectin
Peeling, 164
cryogenic scalding, 173-175
infrared, 169-173
lye, 165-169
steam, 164-165
pH, 347-349
classification, 350
determination, 347
indicators, 348
maturity effect, 351
variety effect, 351
Plant populations, 24, 36, 89
Planting, 34
Precipitate weight method, 319
Process time and temperature
catsup, 237-239
juice, 212-214
tomatoes, 190-192
Pulping, 220-221
Pupae, See *Drosophila,* life cycle

Quality, assurance, 253
control, 285
problem solving, 287
Brainstorming, 287
Cause and effect, 289-291
Pareto, 288
definition, 254
evaluation, 293
laboratory, 264-265
measurement, 264-272
methods, 256
organization, 258
personnel, 258
purposes, 257
samples, 259-261
standards, 254, 261-263

Raw product, color, 140 to 149
firmness, 151
inspections, 128-131

Record forms, AC, 518-529

Refractive index, method, 314, 321

Salting, juice, 208-209
  tablets, 184, 186-189
  tomatoes, 183-184
Sampling, bulk handling systems, 125
  Davis, 126
  flume system, 126
  rates, 127
  restricted, 125
  Yuba City, 127
Seeding direct, 34-38
Serum separation, 319
Shelf-life, 443-445
Size, 153
Soaking, 154-156
Soils, liming, 27-28
  nutrients, 24-27
  pH, 27
  preparation, 22-24
  selection, 21
  starter solutions, 30-32
  testing, 27
  types, 22
Solids, alcohol insoluble solids, 317-318
  soluble, 314-316
  total, 313-314
  water insoluble solids, 317
Sorting, 106
  central sort, 107
  dry, 153
  final, 158-161
Soup, 243-248
Species, 84
Specific gravity, 226, 319-320
Spoilage, 285, 427-429
  catsup, 426
  flat sour, 420
  types, 427-429
  tomatoes, 425
Standards, fill, 294
  grade, catsup, AA, 453-458
    chili sauce, AA 459-462
    okra and tomatoes, 468-474
    tomato canned, 494-505
      concentrated juice
      juice, 475-482
      paste, 482-488
      pulp, 488-493
      sauce, 463-468
  identity, catsup, AB, 508-509
    juice, 507
    paste, 510-512
    pulp and paste, 509-510
    tomato, canned, 512-516
  raw, 110, 215

Starter solution, 30-32
Statistical quality control, 271-284
  process control, 271
Storage, aseptic, 227-228

Tomato anatomy, frontispiece
  acreage, 10-15
  breeding, See Breeding
  catsup, catchup, ketchup, See catsup
  chili sauce, 241-242
  color index, See Color
  control, 240
  composition, 313
  cultivars, 32-34
  cultivation, 39-40
  culture, 19
  definition, 233
  diced, 196
  histology
  history, 3-14
  juice, containers, 209
    composition, 313
    crushing, 202-205
    definition, 201
    homogenization, 210
    manufacture, 234
    milling, 238
    reconstituted, 214
    sterilizing, 239
    thermal process, 210-214
  paste, definition, 224
    manufacture, 225-227
    storage, bulk, 227-228
  peeled (canned tomatoes)
    acidification, 192-196
    cooling, 192
    exhausting, 189-190
    filling, 181
    processing, 190-191
    salting, 183-189
    styles of pack, 198
  physiological disorders, 75
  planting, 34-39
  process, flow diagram, 194

pulp-puree, definition, 219
manufacture, 220-223
soups, 243
formulations, 244-248
species, 84
statistics, 9-15
total solids, 313
value, 15
varieties, 7
waste, 249
yield, 15
Transplanting, 34-35
Total solids, 223-313

Vacuum, gauge, 267
kettle, 221
pan, 221
Varieties (cultivars), 7
Vinegar, 236
Vinegar gnat, 369
Viscosity, See Consistency
Vitamin A, See Carotene
Vitamin C, See Ascorbic acid

Washing, 154, 156-157, 161
Waste, See Tomato wastes
Water activity, 424
Water capacity of container, 294-295
Water insoluble solids, 317
Water tanks and systems, 120-121
Wetting agents, 163
Wildman trap flask method, See Insect determination
Wireworm, 77

Yuba City Sampler. See Sampling